P9-CBM-481

The Pollution Crisis

Official Documents

Edited and Arranged by

Edward H. Rabin
Professor of Law
University of California, Davis

Mortimer D. Schwartz
Professor of Law and Law Librarian
University of California, Davis

1972
Oceana Publications, Inc.
Dobbs Ferry, New York

Library of Congress Cataloging in Publication Data

Rabin, Edward H. 1937- comp.
The pollution crisis.
1. Pollution--U.S. I. Schwartz, Mortimer D., joint comp. II. Title.
TD180.R3 628'.5'0973 73-37009
ISBN 0-379-00163-2

Manufactured in the United States of America

PREFACE

Ecology is a new religion and many a false prophet is abroad in the land. Despite the plethora of news releases, books, and magazine articles on our environment it is still difficult to find authoritative unbiased information prepared by competent experts. This is unfortunate since the known facts tell a story which needs no sensational embellishment to be disturbing.

From the vast quantity of available literature we have selected and edited those materials which were interesting, readable, and scientifically accurate. The materials which follow consist of studies made by or for the federal government. With deliberate scientific caution they tell of a world already grossly polluted and becoming more so every day.

Although we believe that this collection of official United States government reports will be of interest to the general public, we believe that it will be particularly useful to the citizen who is actively engaged in the struggle for a cleaner environment. Too many well meaning "activists" are armed only with facts supplied by self-proclaimed experts whose credentials are suspect and whose objectivity is nonexistent. Arguments in support of environmental protection will be most effective when they are based upon the official reports of the United States Government. If our work makes such reports more readily available for use in the struggle for a cleaner environment we will be content.

Edward H. Rabin

Mortimer D. Schwartz

Davis, California

I

IN GENERAL

Environmental Quality, The First Annual Report of the Council on Environmental Quality, Transmitted to the Congress, August 1970. Pages 5-18, 231-41.

President Nixon's first official act of the decade was to sign into law the National Environmental Policy Act of 1969 on January 1, 1970. This can be found in 83 United States Statutes at Large 852 and in 42 United States Code Annotated section 4321. Section 202 of the Act created in the Executive Office of the President a Council on Environmental Quality. The Council is composed of three members who are appointed by the President with the advice and consent of the Senate. Each member of the Council must be:

> "a person who, as a result of his training, experience, and attainments, is exceptionally well qualified to analyze and interpret environmental trends and information of all kinds; to appraise programs and activities of the Federal Government in the light of the policy set forth in title I of this Act; to be conscious of and responsive to the scientific, economic, social, esthetic, and cultural needs and interests of the Nation; and to formulate and recommend national policies to promote the improvement of the quality of the environment."

The nominees of the President were confirmed by the Senate on February 6, 1970. They seem to fit the job description prescribed by the statute just quoted.

The Chairman, Russell E. Train, is a former Under Secretary of the Interior, and a former president of The Conservation Foundation. He has had ample governmental experience, and in 1968 served as chairman of a special task force on environmental problems.

The other members of the Council, Robert Cahn and Gordon J.F. MacDonald, are also well qualified. Mr. Cahn, on leave from the Washington Bureau of The Christian Science Monitor, was awarded a Pulitzer Prize for National Reporting for his series of articles on national parks. Dr. MacDonald has been a member of the President's Science Advisory Committee, the Committee on Atmospheric Sciences of the National Academy of Sciences, chairman of the Environmental Studies Board, and chairman of the Executive Committee of the Earth Sciences Division of the National Research Council.

Under section 201 and 204 of the Act it is the duty of the Council to report to the President at least once each year on the state of the environment, and it is the duty of the President to transmit this report as part of his own Environmental Quality Report to the Congress.

Environmental Quality, The First Annual Report of the Council on Environmental Quality, consists of 336 pages, including appendices. In addition, there are twenty five pages of introductory material, including President Nixon's eleven page Message to the Congress. The Report is an excellent summary of the major environmental issues facing the nation, and is especially useful in providing an introduction to environmental issues for someone who is beginning his studies in this field. The excerpts reproduced below establish a conceptual framework or background against which one can judge specific environmental proposals.

I

Understanding Environmental Problems

HISTORIANS MAY one day call 1970 the year of the environment. They may not be able to say that 1970 actually marked a significant change for the better in the quality of life; in the polluting and the fouling of the land, the water, and the air; or in health, working conditions, and recreational opportunity. Indeed, they are almost certain to see evidence of worsening environmental conditions in many parts of the country.

Yet 1970 marks the beginning of a new emphasis on the environment—a turning point, a year when the quality of life has become more than a phrase; environment and pollution have become everyday words; and ecology has become almost a religion to some of the young. Environmental problems, standing for many years on the threshold of national prominence, are now at the center of nationwide concern. Action to improve the environment has been launched by government at all levels. And private groups, industry, and individuals have joined the attack.

No one can say for sure just how or why the environment burst into national prominence in 1970. Certainly national concern had been mounting for a long time, and the tempo has increased greatly in the last decade.

Early environmentalists—Henry David Thoreau, George Perkins Marsh, John Muir, Gifford Pinchot, Theodore Roosevelt, Aldo Leopold—and a legion of dedicated citizens contributed to the rise in awareness. In its early days, the conservation movement aimed primarily at stemming the exploitation of natural resources and preserving wildlife and important natural areas. By the 1950's, Federal air and water pollution laws had been enacted, and the pace of environmental legislation quickened dramatically in the decade of the 1960's. Now the conservation movement has broadened to embrace concern for the totality of man's environment, focusing on pollution, population, ecology, and the urban environment.

The public has begun to realize the interrelationship of all living things—including man—with the environment. The Santa Barbara oil spill in early 1969 showed an entire nation how one accident could temporarily blight a large area. Since then, each environmental issue—the jetport project near Everglades National Park, the proposed pipeline across the Alaskan wilderness, the worsening blight of Lake Erie, the polluted beaches off New York and other cities, smog in mile-high Denver, lead in gasoline, phosphates in detergents, and DDT—flashed the sign to Americans that the problems are everywhere and affect everyone. Millions of citizens have come to realize that the interdependent web of life—man, animals, plants, earth, air, water, and sunlight—touches everyone.

A deteriorating environment has awakened a lively curiosity in Americans about exactly what is meant by an ecosystem, a biome, or the biosphere. Citizens who are now aware of environmental problems want to know the full extent of the environmental crisis and the nature of the factors that have contributed to it. They are anxious to learn what can be done to correct the mistakes that have led to the current condition of the environment. This report attempts to answer some of these questions.

ECOLOGY AND CHANGE

Ecology is the science of the intricate web of relationships between living organisms and their living and nonliving surroundings. These interdependent living and nonliving parts make up *ecosystems*. Forests, lakes, and estuaries are examples. Larger ecosystems or combinations of ecosystems, which occur in similar climates and share a similar character and arrangement of vegetation, are *biomes*. The

Arctic tundra, prairie grasslands, and the desert are examples. The earth, its surrounding envelope of life-giving water and air, and all its living things comprise the *biosphere*. Finally, man's total *environmental system* includes not only the biosphere but also his interactions with his natural and manmade surroundings.

Changes in ecosystems occur continuously. Myriad interactions take place at every moment of the day as plants and animals respond to variations in their surroundings and to each other. Evolution has produced for each species, including man, a genetic composition that limits how far that species can go in adjusting to sudden changes in its surroundings. But within these limits the several thousand species in an ecosystem, or for that matter, the millions in the biosphere, continuously adjust to outside stimuli. Since interactions are so numerous, they form long chains of reactions. Thus small changes in one part of an ecosystem are likely to be felt and compensated for eventually throughout the system.

Dramatic examples of change can be seen where man has altered the course of nature. It is vividly evident in his well-intentioned but poorly thought out tampering with river and lake ecosystems. The Aswan Dam was primarily built to generate electric power. It produced power, but it also reduced the fish population in the Mediterranean, increased the numbers of disease-bearing aquatic snails, and markedly lowered the fertility of the Nile Valley.

In the United States, the St. Lawrence Seaway has contributed significantly to the economic growth of the Great Lakes region. Yet it has done so at a high and largely unforeseen cost to the environment. The completion of the Welland Canal let the predatory sea lamprey into the Great Lakes. Trout, which had been the backbone of the lakes' fishing industry, suffered greatly from the lamprey invasion. By the mid-1950's the trout and some other large, commercial predatory fish were nearly extinct. And with their near extinction, smaller fish, especially the alewife, normally kept under control by these predators, proliferated. The aggressive alewife dominated the food supply and greatly reduced the numbers of small remaining native fish, such as the lake herring. The alewife became so numerous, in fact, that on occasion great numbers died and the dead fish along the shore caused a major public nuisance.

Man attempted to restore the ecological balance by instituting sea lamprey control in the 1950's and 1960's and by stocking the lakes with coho salmon beginning in 1965—to replace the lost native predatory fish. Feeding on the abundant alewife, the salmon multiplied rapidly and by 1969 had become important both as a commercial and sport resource. Some of the salmon, however, were contaminated by excessive concentrations of DDT and were taken off the commercial market.

The lesson is not that such activities as the St. Lawrence Seaway must be halted, but that the consequences of construction must be carefully studied in advance of construction. Planners and managers must begin to appreciate the enormous interrelated complexity of environmental systems, weigh the tradeoffs of potential environmental harm against the benefits of construction, look at alternatives, and incorporate environmental safeguards into the basic design of new developments.

The stability of a particular ecosystem depends on its diversity The more interdependencies in an ecosystem, the greater the chances that it will be able to compensate for changes imposed upon it. A complex tropical forest with a rich mosaic of interdependencies possesses much more stability than the limited plant and animal life found on the Arctic tundra, where instability triggers frequent, violent fluctuations in some animal populations, such as lemmings and foxes. The least stable systems are the single crops—called monocultures—created by man. A cornfield or lawn has little natural stability. If they are not constantly and carefully cultivated, they will not remain cornfields or lawns but will soon be overgrown with a wide variety of hardier plants constituting a more stable ecosystem.

The chemical elements that make up living systems also depend on complex, diverse sources to prevent cyclic shortages or oversupply. The oxygen cycle, which is crucial to survival, depends upon a vast variety of green plants, notably plankton in the ocean. Similar diversity is essential for the continued functioning of the cycle by which atmospheric nitrogen is made available to allow life to exist. This cycle depends on a wide variety of organisms, including soil bacteria and fungi, which are often destroyed by persistent pesticides in the soil.

TYPES OF PROBLEMS

Pollution

Although pollution may be the most prominent and immediately pressing environmental concern, it is only one facet of the many-sided environmental problem. It is a highly visible, sometimes dangerous sign of environmental deterioration. Pollution occurs when materials accumulate where they are not wanted. Overburdened natural processes cannot quickly adjust to the heavy load of materials which man, or sometimes nature, adds to them. Pollution threatens natural systems, human health, and esthetic sensibilities; it often represents valuable resources out of place. DDT, for instance, is a valuable weapon in combating malaria. But DDT, when out of place—for example in lakes and streams—concentrates in fish, other wildlife, and the smaller living things on which they depend.

Historically, man has assumed that the land, water, and air around him would absorb his waste products. The ocean, the atmosphere, and even the earth were viewed as receptacles of infinite capacity. It is clear now that man may be exceeding nature's capacity to assimilate his wastes.

Most pollutants eventually decompose and diffuse throughout the environment. When organic substances are discarded, they are attacked by bacteria and decompose through oxidation. They simply rot. However, some synthetic products of our advanced technology resist natural decomposition. Plastics, some cans and bottles, and various persistent pesticides fall into this category. Many of these materials are toxic, posing a serious health danger.

Some pollutants, which may be thinly spread throughout the environment, tend to reconcentrate in natural food chains. Pesticides tend to diffuse in ocean water. The physical effects of 1 pound of a well-mixed pesticide in 10 billion pounds of water may seem negligible. But many sea animals filter out particular kinds of chemical compounds, including pesticides, and collect them in certain parts of their bodies at concentrations far higher than in the water in which they live. Algae may concentrate some component of a pesticide which is then concentrated further in the fish that eat the algae. In turn, still further concentrations may occur in the birds that eat the fish. When the accumulation of the toxic substances reaches a high enough level, it may kill the organism directly or interfere with its reproduction. A well-known example of such accumulation occurred in 1957 at Clear Lake, Calif. DDD (similar to DDT) was diffused through the water in a concentration of only 0.02 parts per million. The lake's plant and animal organisms, however, had stored residues of DDD at 5 parts per million—250 times greater than the concentration in the water itself. Fish, which consumed large quantities of these small organisms, accumulated DDD concentrations in their body tissues of over 2,000 parts per million. And there was heavy mortality among grebes which fed on the fish.

Radioactive fallout from the air also concentrates through food chains. Arctic lichens do not take in food through their roots but instead absorb their mineral nutrition from dust in the air. Radioactive fallout tends therefore to collect in the lichens and is further concentrated by grazing caribou, which eat huge quantities of lichen. Caribou meat is a major part of the Eskimo's diet. Although reconcentration of radioactive fallout at low levels has not been proved damaging to health, the effects of long-term, low-level exposure to radioactive pollutants are still not well known.

Water pollution is a problem throughout the country, but is most acute in densely settled or industrial sections. Organic wastes from municipalities and industries enter rivers, where they are attacked and broken down by organisms in the water. But in the process, oxygen in the river is used up. Nutrients from cities, industries, and farms nourish algae, which also use up oxygen when they die and decompose. And when oxygen is taken from the water, the river "dies." The oxygen is gone, the game fish disappear, plant growth rots, and the stench of decay reaches for miles.

Air pollution is now a problem in all parts of the United States and in all industrialized nations. It has been well known for some time to Los Angeles residents and visitors who have long felt the effect of highly visible and irritating smog from automobile exhaust. Now Los Angeles's local problem is becoming a regional problem, because noxious air pollution generated in the Los Angeles Basin has spread beyond the metropolitan area. This same problem, which seemed unique to Los Angeles in the 1950's is today common to major cities in the United States and abroad. Smog is but one of the many types of air pollution that plague the United States, especially its cities.

Land Use

Urban land misuse is one of today's most severe environmental problems. The character of our urban areas changes rapidly. Old buildings and neighborhoods are razed and replaced by structures designed with little or no eye for their fitness to the community's needs. A jumble of suburban developments sprawls over the landscape. Furthermore, lives and property are endangered when real estate developments are built on flood plains or carved out along unstable slopes.

Unlimited access to wilderness areas may transform such areas into simply another extension of our urban, industrialized civilization. The unending summer flow of automobiles into Yosemite National Park has changed one of nature's great wilderness areas into a crowded gathering place of lessened value to its visitors. The worldwide boom in tourism, teamed with rapid and cheap transportation, threatens the very values upon which tourist attraction is based.

The proposed jetport west of Miami and north of the Everglades National Park raised a dramatic land use problem. The jetport, together with associated transportation corridors, imperiled a unique ecological preserve. Planners for the jetport had considered density of population, regional transportation needs, and a host of other related variables. But they gave slight consideration to the wildlife and recreational resources of the Everglades. The jetport could have spawned a booming residential, commercial, and industrial complex which would have diminished water quality and without question drastically altered the natural water cycle of Southern Florida. This in turn would have endangered all aquatic species and wildlife within the park and beyond.

Natural Resources

Natural resource depletion is a particular environmental concern to a highly technological society which depends upon resources for energy, building materials, and recreation. And the methods of exploiting resources often create problems that are greater than the value of the resources themselves.

A classic case was the Federal Government's decision to permit oil drilling in California's Santa Barbara Channel. There, primary value was placed on development of the oil resources. The commercial, recreational, esthetic, and ecological values, which also are important to the residents of Santa Barbara and to the Nation, were largely ignored. The President recently proposed to the Congress that the Federal Government cancel the 20 Federal leases seaward from the State sanctuary extending 16 miles along the Santa Barbara Channel. This is where the blowout erupted in January 1969, spreading a coat of oil across hundreds of square miles including the sanctuary. This action illustrates a commitment to use offshore lands in a balanced and responsible way.

CAUSES OF PROBLEMS

Environmental problems seldom stem from simple causes. Rather they usually rise out of the interplay of many contributing circumstances.

Misplaced Economic Incentives

Many individuals cite selfish profit seekers for environmental degradation, rather than laying much of the blame—where it belongs—to misplaced incentives in the economic system. Progress in environmental problems is impossible without a clearer understanding of how the economic system works in the environment and what alternatives are available to take away the many roadblocks to environmental quality.

Our price system fails to take into account the environmental damage that the polluter inflicts on others. Economists call these damages—which are very real—"external social costs." They reflect the ability of one entity, e.g., a company, to use water or air as a free resource for waste disposal, while others pay the cost in contaminated air or water. If there were a way to make the price structure shoulder these external costs—taxing the firm for the amount of discharge, for instance—then the price for the goods and services produced would reflect these costs. Failing this, goods whose production spawns pollution are greatly underpriced because the purchaser does not pay for pollution abatement that would prevent environmental damage. Not only does this failure encourage pollution but it warps the price structure. A price structure that took environmental degradation into account would cause a shift in prices, hence a shift in consumer preferences and, to some extent, would discourage buying pollution-producing products.

Another type of misplaced incentive lies imbedded in the tax structure. The property tax, for example, encourages architectural design that leans more to rapid amortization than to quality. It may also encourage poor land use because of the need for communities to favor industrial development and discourage property uses, such as high-density housing, which cost more in public services than they produce in property taxes. Other taxes encourage land speculation and the leapfrog development that has become the trademark of the urban-rural fringe.

Values

American have placed a high priority on convenience and consumer goods. In recent times they have learned to value the convenience and comfort of modern housing, transportation, communication, and recreation above clean earth, sky, and water. A majority, like a prodigal son, have been willing to consume vast amounts of resources and energy, failing to understand how their way of life may choke off open space, forests, clean air, and clear water. It is only recently that the public has become conscious of some of the conflicts between convenience and a deteriorating environment.

In the early days of westward expansion, a period in which many national values were shaped, choices did not seem necessary. The forests, minerals, rivers, lakes, fish, and wildlife of the continent seemed inexhaustible. Today choices based on values must be made at every turn. Values can be gauged to some degree by the costs that the Nation is collectively willing to incur to protect them. Some of the costs of environmental improvements can be paid with local, State, and Federal tax money. But paying taxes and falling back on government programs is not enough. People may ultimately have to forgo some conveniences and pay higher prices for some goods and services.

Population

Americans are just beginning to measure the magnitude of the impact of population and its distribution on their environment. The concept that population pressures are a threat to the Nation's well-being and to its environment is difficult to grasp in a country which, during its formative decades, had an ever receding western frontier. That frontier ended at the Pacific many years ago. And it is at the western end of the frontier that some of the most serious problems of population growth emerge most clearly.

California continues to lure large numbers of Americans from all over the country, in large part because of its climate and its beauty. But as the people come, the pressures of population mount. Smog, sprawl, erosion, loss of beaches, the scarring of beautiful areas, and the congestion of endless miles of freeways have caused thoughtful Californians to consider stemming the continued uncontrolled development of their State. When the Governor's Conference on California's Changing Environment met last fall, it agreed that there was now a need "to deemphasize growth as a social goal and, rather, to encourage development within an ideal and quality environment."

The magnitude of the press of population, although significant, must be put in perspective. This is a vast country, and its potential for assimilating population is impressive, although there is disagreement over what level of population would be optimum. Some authorities believe that the optimum level has already been passed, others that it has not yet been reached. More troublesome, population control strikes at deeply held religious values and at the preference of some Americans for large families.

Population density outside metropolitan areas is not high. There is a desire—indeed an almost inevitable compulsion—to concentrate population in urban areas—primarily in the coastal and Great Lakes regions. If the trend continues, 70 to 80 percent of all Americans will be concentrated in five large urban complexes by the year 2000. The pressures that cause environmental problems that the Nation now confronts—water and air pollution and inefficient land use—will only increase.

Population growth threatens the Nation's store of natural resources. Currently the United States, with about 6 percent of the world's population, uses more than 40 percent of the world's scarce or nonreplaceable resources and a like ratio of its energy output. Assuming a fixed or nearly fixed resource base, continued population growth embodies profound implications for the United States and for the world.

Technology

The major environmental problems of today began with the Industrial Revolution. Belching smoke from factory stacks and the dumping of raw industrial wastes into rivers became the readily identified, but generally ignored hallmarks of "progress" and production. They are no longer ignored, but the extraordinary growth of the American economy continues to outpace the efforts to deal with its unwanted byproducts.

The growth of the economy has been marked not just by greater production but also by an accelerating pace of technological innovation. This innovation, although it has provided new solutions to environmental problems, has also created a vast range of new problems. New chemicals, new uses for metals, new means of transportation, novel consumer goods, new medical techniques, and new industrial processes all represent potential hazards to man and his surroundings. The pace of technological innovation has exceeded our scientific and regulatory ability to control its injurious side effects. The environmental problems of the future will increasingly spring from the wonders of 20th-century technology. In the future, technology assessment must be used to understand the direct and secondary impacts of technological innovation.

Mobility

The extraordinary, growing mobility of the American people constitutes another profound threat to the environment—in at least three major ways. The physical movement of people crowds in on metropolitan centers and into recreation areas, parks, and wild areas. Mobility permits people to live long distances from their places of employment, stimulating ever greater urban and suburban sprawl. The machines of this mobility—particularly automobiles and aircraft—themselves generate noise, air pollution, highways, and airports—all in their way affecting the environment.

The automobile freed Americans from the central city and launched the flight to the suburbs. As a consequence, thousands of acres of undeveloped land fall prey each year to the bulldozer. More single-family, detached homes shoulder out the open spaces. Many of these developments are drab in design and wasteful of land. They denude the metropolitan area of trees and thus affect climate; they cause erosion, muddy rivers, and increase the cost of public services.

Limitations of Government Units

Most government agencies charged with solving environmental problems were not originally designed to deal with the severe tasks they now face. And their focus is often too narrow to cope with the broad environmental problems that cut across many jurisdictions. Agencies dealing with water pollution, for example, typically do not have jurisdiction over the geographic problem area—the watersheds. Control is split instead among sewerage districts, municipalities, and a multitude of other local institutions. To attack water pollution effectively may require establishing new river basin authorities or statewide basin agencies with the power to construct, operate, and assess for treatment facilities.

Public decisions, like private decisions, suffer from the inadequate balancing of short-run economic choices against long-term environmental protection. There is a nearly irresistible pressure on local governments to develop land in order to increase jobs and extend the tax base—even if the land is valuable open space or an irreplaceable marsh. The problem is amplified by the proliferation of agencies, all competing narrowly, without consideration of broader and often common goals. The development that generates economic benefits in a town upstream may create pollution and loss of recreation in a town downstream.

Information Gap

Sometimes people persist in actions which cause environmental damage because they do not know that they are causing it. Construction of dams, extensive paving of land surfaces, and filling of estuaries for industrial development have in many cases been carried out with incomplete or wrong information about the extent of the impact on the environment. Furthermore, change in the environment has often been slow and exceedingly difficult to detect, even though piecemeal changes may eventually cause irreversible harm. Widespread use of certain types of pesticides, mercury pollution, and the use of dangerous substances such as asbestos occurred without advance recognition of their potential for harm.

EFFECTS OF PROBLEMS

Health

The impact of environmental deterioration on health is subtle, often becoming apparent only after the lapse of many years. The speed of change in a rapidly altering technological society and the complex causes of many environmental health problems produce major uncertainty about what environmental changes do to human well-being. Nevertheless, it is clear that today's environment has a large and adverse impact on the physical and emotional health of an increasing number of Americans.

Air pollution has been studied closely over the past 10 years, and its tie to emphysema and chronic bronchitis is becoming more evident.

These two diseases are major causes of chronic disability, lost work-days, and mortality in industrial nations. Estimates of deaths attributable to bronchitis and emphysema are beset with doubts about cause; nevertheless, physicians have traced 18,000 more deaths in the United States to these two causes in 1966 than 10 years earlier—an increase of two and one-half times. The increase of sulfur oxides, photochemical oxidants, and carbon monoxide in the air is related to hospital admission rates and length of stay for respiratory and circulatory cases.

Whether the accumulation of radioactive fallout in body tissues will eventually produce casualties cannot be predicted now, but close surveillance is needed. Nor has a direct correlation between factors in the urban environment and major malignancies of the digestive, respiratory, and urinary tracts been established. But the frequency of these diseases is much higher in cities than in nonurban environments.

Esthetics

The impact of the destruction of the environment on man's perceptions and aspirations cannot be measured. Yet today citizens are seeking better environments, not only to escape pollution and deterioration but to find their place in the larger community of life. It is clear that few prefer crowding, noise, fumes, and foul water to esthetically pleasing surroundings. Objections today to offensive sights, odors, and sounds are more widespread than ever. And these mounting objections are an important indicator of what Americans are unwilling to let happen to the world about them.

Economic Costs

The economic costs of pollution are massive—billions of dollars annually. Paint deteriorates faster, cleaning bills are higher, and air filtering systems become necessary. Direct costs to city dwellers can be measured in additional household maintenance, cleaning, and medical bills. Air pollution causes the housewife to do her laundry more often. The farmer's crop yield is reduced or destroyed. Water pollution prevents swimming, boating, fishing, and other recreational and commercial activities highly valued in today's world.

Natural Systems

Vast natural systems may be severely damaged by the improvident intervention of man. The great Dust Bowl of the 1930's was born in the overuse of land resources. Many estuarine areas have been altered and their ecology permanently changed. On a global scale, air pollution could trigger large-scale climatic changes. Man may also be changing the forces in the atmosphere through deforestation, urban construction, and the spilling of oil on ocean waters.

SOLVING PROBLEMS

In the short run, much can be done to reverse the deadly downward spiral in environmental quality. Citizens, industries, and all levels of government have already begun to act in ways which will improve environmental quality. The President's February 10 Message on the Environment spelled out some specific steps which can be taken now.

It is clear, however, that long-range environmental improvement must take into account the complex interactions of environmental processes. In the future, the effects of man's actions on complete ecosystems must be considered if environmental problems are to be solved.

Efforts to solve the problems in the past have merely tried—not very successfully—to hold the line against pollution and exploitation. Each environmental problem was treated in an ad hoc fashion, while the strong, lasting interactions between various parts of the problem were neglected. Even today most environmental problems are dealt with temporarily, incompletely, and often only after they have become critical.

The isolated response is symptomatic of the environmental crisis. Americans in the past have not adequately used existing institutions to organize knowledge about the environment and to translate it into policy and action. The environment cuts across established institutions and disciplines. Men are beginning to recognize this and to contemplate new institutions. And that is a hopeful sign.

XIII

Present and Future Environmental Needs

THIS REPORT has looked closely at particular environmental problems. It has looked at what is being done now to combat them. And it has looked at what might be done in the days, months, and years ahead. The agenda for urgent action is long. Much has already been done, but much more must still be done with current management tools wielded by existing institutions. Moreover, the pace of change in programs underway promises, over the next few years, to brake even further what has seemed a headlong careening toward environmental decay.

The pressing need for tomorrow is to know much more than we do today. We lack scientific data about how natural forces work on our environment and how pollutants alter our natural world. We lack experience in innovating solutions. We lack tools to tell us whether our environment is improving or deteriorating. And most of all, we lack an agreed upon basic concept from which to look at environmental problems and then to solve them.

NEEDED—A CONCEPTUAL FRAMEWORK

A problem is said to exist when our view of what conditions are does not square with our view of what they should be. Problems, in short, are products of our values. People agree—for example, that a river should not be polluted. And when they see that it is, water pollution becomes a problem. But some of the values dealt with in this report are not unanimously agreed upon. The chapter on land use is critical of urban sprawl; yet many Americans choose to live in dwellings which abet such sprawl. This uncertainty about what values are relevant to environmental questions and how widely or strongly they are held throws up a major obstacle to conceiving environmental problems. How much value do Americans place on the natural environment as against the man-made environment of cities? How much do people value esthetics? Do they agree about what is esthetically desirable? These and a host of similar questions must be raised when trying to align priorities for coping with environmental decline.

Our ignorance of the interrelationship of separate pollution problems is a handicap in devising control strategies. Is pollution directly related to population or to land use or to resources? If so, how? Indeed, does it do any good to talk about pollution in general, or must we deal with a series of particular pollution problems—radiation, pesticides, solid waste? A systems approach is needed, but what kind of system? The pollution system, the materials and resources use system, the land use system, the water resources or atmospheric system? In this report the Council has suggested tentative answers to some of these questions. But much more thought is necessary before we can be confident that we have the intellectual tools necessary to delineate accurately the problems and long-range strategies for action.

Experience will help resolve some of the conceptual problems. We already know what problems are most pressing. Clearly we need stronger institutions and financing. We need to examine alternative approaches to pollution control. We need better monitoring and research. And we need to establish priorities and comprehensive policies.

NEEDED—STRONGER INSTITUTIONS

Most of the burden for dealing with environmental problems falls to governments at all levels. And the Nation's ability to strengthen these institutions is central to the struggle for environmental quality. To make them stronger, fundamental changes are necessary at Federal, State, and local levels of government. Chapter II of this report treats in detail the President's proposed improvements in the Federal Government for better environmental policy development and management. Although these changes will not be the final answer, they do lay the base for a comprehensive and coordinated Federal attack on environmental problems.

States play a key role in environmental management because of their geographic scope and broad legal powers. Many have reorganized to focus comprehensively on environmental problems. Many are helping municipalities build sewage treatment plants; some are planning statewide treatment authorities to construct and operate plants. And California has led the Nation in trying to curb automobile air pollution since the 1950's. In land use control many States are carving out larger responsibilities for land use decisions of regional scope.

In many respects local government, of all the levels, most needs institutional improvement. It has suffered from fragmentation, from skyrocketing demands and costs for public services, and from generally inelastic tax sources. The financial burden of environmental improvement staggers local governments. Most of the costs of water pollution control, both capital and operating expenses, come from their budgets. In some cities, efforts to deal with combined sewer overflows raise almost insuperable financial and technical hardships. Solid waste disposal is a major expense for most local governments, and the costs grow as disposal techniques are upgraded, as land grows scarcer, and as wages spiral. On top of its financial headaches, local government is caught in a tangled web of overlapping and conflicting jurisdictions that hamstring solutions to land use and air pollution.

Existing institutions must be made better and, in some cases, new institutions created to deal with the environment. Occasionally more funds, personnel, and public support are all that is necessary. Other cases call for a more fundamental restructuring. This may mean extending geographic coverage and operational capabilities. Air and water pollution, for example, do not respect political boundaries, so institutions covering entire watersheds or airsheds may be necessary to cope with them. Important aspects of land use planning, review, and control may need to be shifted to regional or State levels as the only way to tie land use needs together over wide areas. And new forms of land use criteria may be necessary to reverse the current myopia of local government zoning.

Many environmental problems cross not only local, state, and regional boundaries, but international boundaries as well. Control of pollution of the seas and the atmosphere requires new forms of international cooperation—for monitoring, research, and regulation.

NEEDED—FINANCIAL REFORM

Financing for environmental quality is in need of dramatic overhaul. Liquid and solid waste collection and disposal by local governments represent an indispensible service—not unlike electricity and water. Yet rarely do the users of these services, industry and homeowners, bear the full costs of operation and amortization. Rather, financially beleaguered local governments subsidize these services. The current method of financing, therefore, is not only inequitable; it encourages a greater accumulation of waste by industries because they do not bear the full costs of disposal. It deprives local governments of needed funds to operate and maintain waste disposal facilities properly. In short, it contributes to the sorry performance of facilities in the United States in treating sewage and disposing of solid wastes. If future demands for environmental improvement are to be funded adequately, better methods of financing must be developed.

NEEDED—POLLUTION CONTROL CURBS

This report discusses many tools for curbing pollution. Most have been regulatory. For centuries authority to regulate has been wielded to a limited extent—more broadly by the middle of the 20th century.

But there is considerable debate whether regulation represents the best course of action. Economic incentives have won increasing support as a pollution control weapon. Charges or taxes on the volume of pollutants—say, 10 cents a pound on oxygen-demanding material—are another lever that might spur industry to reduce wastes. The charge system, some say, would not only be more economic but also more effective compared to the traditionally cumbersome enforcement process.

In this report the Council urges stricter and more systematic enforcement of air and water standards. That cannot be done, however, without better monitoring and data—as well as clear-cut, enforcement policies that will leave no doubt of responsibilities on the part of the private sector.

The Council believes that economic incentives offer promise, especially if backed up by regulatory power. It believes that they should be selectively demonstrated. And it believes that effluent or emission charges should be evaluated as a supplementary method of stimulating abatement measures.

NEEDED—MONITORING AND RESEARCH

Effective strategy for national environmental quality requires a foundation of information on the current status of the environment, on changes and trends in its condition, and on what these changes mean to man. Without such information, we can only react to environmental problems after they become serious enough for us to see. But we cannot develop a long-term strategy to prevent them, to anticipate them, and to deal with them before they become serious. For example, we became aware of the mercury problem only after it had become critical in some areas and had probably done environmental damage. Yet we still do not know the extent or significance of that damage. Our attack on the problem can now be but a cure or a cleanup. It has already happened. However, if we had possessed an adequate environmental early warning system, we would have been able to anticipate mercury pollution and take early action to stop it at its sources.

We do not know what low-level exposure to most pollutants does to man's health over the long term. Nor do we know how people react to changes in their environment. The challenge to the social sciences is to develop entirely new gauges to measure environmental stress.

What do crowding, urban noise, and automation do to man? These are critical questions. We do not understand enough about the interactions of different environmental forces such as urbanization, land use, and pollution. We do not even understand many of the natural processes that play critical roles in environmental well being—such as changes in world climate.

To obtain such information, a comprehensive program is required. It involves nationwide environmental monitoring, collection, analysis, and—finally—effective use of the information. In the case of some pollution, such monitoring should be international.

The first step is to identify the environmental parameters—things in the environment which are or should be measured. These range from substances such as DDT, sulfur dioxide, and lead to percentages of open space in the cities, visitor use of parks, and survival of species. Once identified, the parameters must be monitored—measured on a regular, repeated, continuing basis. In this way, baselines of the present status can be determined and changes from that base detected.

Environmental indices can be developed from these data. Indices are data aggregated to provide a picture of some aspect of environmental quality—for example, the quality of air as it affects human health. They are not unlike the cost of living index by which economists measure the status of the economy and by which housewives measure their budgets. Some environmental indices—and the parameters on which they are based—are easily identified and measured. For example, conditions that clearly affect human health in air or drinking water can be easily detected. Other indices and parameters are based on value judgments and are much more difficult to deal with. The quality of National Parks and scenic beauty are examples. To develop indices, the information from monitoring must be collected, translated into a usable form, and analyzed. Good indices do two important things. They inform the general public of the quality of the environment, and they inform the government and other decision-makers who can take action. Good indices show the current environmental quality on a national or local scale and whether this condition is improving or degrading.

At present no nationwide environmental monitoring and information system exists. Federal, State, and local agencies now collect a variety of data. Many of these data, however, are obtained for limited program purposes or for scientific understanding. They are fragmentary and not comparable on a nationwide basis. Although it may be possible to use some of these in the comprehensive system which is needed, at present they do not provide the type of information or coverage necessary to evaluate the condition of the Nation's environment or to chart changes in its quality and trace their causes.

Therefore, a major national objective must be to develop a comprehensive nationwide system of environmental monitoring, information, and analysis. The Council has initiated a study of the nature and requirements for the early development of such a system. However, even after we have developed a system, we must then have additional knowledge to enable us to understand and interpret the data we get. We are not yet in a position to understand the significance of the monitoring results to man and to natural systems. More research is needed on how the environmental systems operate and on the impact of man on the environment and its impact on him. Consequently, augmenting such research must take a high national priority.

NEEDED—A SYSTEM FOR PRIORITIES

It is difficult, given the current state of environmental knowledge, to set long-term priorities for the future. Relevant measures of environmental quality are often not available or, if available, are inadequate. These difficulties are compounded by great regional differences. For the present we can use our limited current data to identify pressing problems for immediate attention. In the future, the difficult task of deciding the Nation's environmental priorities, however, must be faced. Resources for combating environmental blight and decay are limited. Choices will have to be made on which problems have first claim on these resources. Four main criteria should determine this priority:

- The intrinsic importance of the problems—the harm caused by failing to solve them.
- The rate at which the problems are going to increase in magnitude and intensity over the next few years
- The irreversibility of the damage if immediate action is not taken
- The measure of the benefits to society compared to the cost of taking action.

The process of setting priorities is difficult. There is deep conflict over which problems are most important. And the inertia of on-going activities is a major obstacle. There are conflicts between the needs of industry and the needs of the environment. And the public yearning for more conveniences clashes often with the best interests of the ecology. Nor will the priorities of the Federal Government always coincide with those of State and local governments. The Federal priorities will be broad and national. States and localities, however, will often give higher priority to other aspects of environmental quality. As long as these other levels of government at least meet national standards, the imposition of higher standards in some areas is welcome. Whatever the divergences, diligent application of priorities will be necessary to make any real progress toward a high quality environment.

NEEDED—COMPREHENSIVE POLICIES

As priorities are developed, policies must be devised to translate them into action. These policies may consist of a mix of activities aimed at a particular goal. Dealing with many environmental problems will require a battery of economic incentives, regulations, research, and assistance programs. In some areas, policies cannot be developed until more information is available. In other areas, they can and should be developed now.

For example, the need for a national energy policy is clear. As the demand for power increases rapidly, new power facilities have to be built. Power plants will pollute the air with oxides of sulfur and nitrogen, the water with heat, and the landscape with mammoth towers and obtrusive power lines.

This environmental harm cannot be wholly averted now, but it can be limited. For the short term, the design and siting of power generating facilities and transmission lines must be better planned and controlled. But for the longer run, a national energy policy should be developed. It would require a comprehensive analysis of energy resources and actual needs. It would provide for wise use of fuels, both conserving them for the future and lessening environmental damage. For example, wider use of nuclear fuel, natural gas, or low sulfur coal and oil would lower sulfur oxide levels in critical areas.

As national transportation policies are shaped, air pollution is one among several critical environmental factors that must be considered. Although air pollution can be abated by enforcing emission standards, control devices for individual vehicles and other technological solutions may not be enough in the long run to keep air pollution from worsening as population and the number of automobiles continue to increase in the cities. One part of a transportation policy should be the continued examination of alternative means of curbing auto emissions, such as the development and use of systems combining the flexibility of the individual automobile with the speed of modern mass transportation.

Control over land use, a critical need of the seventies, is lodged for the most part in local governments. And often local solutions are piecemeal and haphazard. The local property tax favors the single-family residence on a large lot over types of housing less wasteful of land. Planning often fails to take into account the impact of development on the natural surroundings and often is not heeded by local governments. All these factors together lead to a series of local zoning decisions and regulatory action that perpetuate urban sprawl.

The State role in land use control has traditionally been small because most of the authority has been delegated to local governments. And direct Federal control over local land use is smaller still. However, the Federal Government can influence how land is used through planning and capital grants. Under existing programs the Federal Government, by its actions, could spur more modern land use methods. It could encourage cluster zoning and timed development. It could identify natural areas for preservation and encourage channeling of future growth in more rational patterns.

The problems of land use are complicated and diffuse. And the challenge is to center all the capabilities of all levels of government in a coordinated attack on them. The problems and the challenge together argue for a national land use policy.

Population growth and economic growth are potential wellsprings of environmental decay. They increase the demands upon limited natural resources. The U.S. population will continue to grow for the next few decades. But environmental quality is difficult to achieve if population growth continues. The President has appointed a commission on Population Growth and the American Future, headed by John D. Rockefeller III, which will explore the policy implications of future population growth.

The development of knowledge will doubtless indicate many new areas in which national policies are appropriate. And as these policies are developed, specific programs for implementation must then be formulated.

CONCLUSIONS

The year 1970 represents a pivotal year in our battle for a clean environment. The Nation is committing resources at all levels of government and in the private sector. Public support is at an all-time high. And the President's proposal for consolidation of anti-pollution programs, coupled with the Council's policy advisory and coordinating role, provide an opportunity to look at environmental quality in new ways.

This report emphasizes the need to move aggressively now to deal with problems that can be dealt with within existing knowledge and by existing institutions. For the long term, we need much more knowledge of values; the scope and nature of environmental problems; status and trends in the environment; the workings of natural processes; and the effects of pollutants on man, animals, vegetation, and materials. As we gain this knowledge, we will need to develop the institutions and financing mechanisms, the priorities, the policies, and finally, the programs for implementation. Without such a systematic approach, the current piecemeal, unrelated efforts will achieve only partial and unsatisfactory progress in meeting environmental problems of tomorrow.

This report emphasizes that changes in one part of the environment inevitably trigger changes in other parts. These complex interactions of environmental processes must be looked at as a whole. While keeping in mind the indivisibility of the environment and its intricate interrelationships, it is also necessary that some segments be treated separately when attacking environmental decay. Water pollution caused by a specific source may affect an entire ecosystem. But enforcement action must be taken against the particular source, not against the ecosystem. The major portion of this report has dealt separately with interrelated environmental problems, but only because of the inadequacy of our current framework for considering the environment and the need to focus attention on particular problem areas.

The National Environmental Policy Act of 1969 clearly stresses the necessity of approaching environmental problems as a totality. The act requires that Federal decision making incorporate environmental values along with technical and economic values; that both short- and long-term effects be given careful consideration; and that irreversible actions and commitments be carefully weighed.

National environmental goals must be developed and pursued in the realization that the human environment is global in nature, and that international cooperation must be a principal ingredient to effective environmental management.

All levels of government should function in two distinct ways: Within their geographic scope and needs, they must consider and plan for the environment as an interrelated system. But at the same time they must make specific decisions and take specific actions to remedy environmental problems. These two levels apply to action by individual citizens and private institutions as well. Our view of the environment and its value is changing and will continue to change. But these changes have effect only as they relate to specific choices by local communities, by particular industries, and by individuals. People in the end shape the environment. If a better environment is passed down to future generations, it will be because of the values and actions of people—all of us—today.

II

Water Pollution

A. Inland Waters

<u>Clean Water for the 1970's, A Status Report, June 1970</u>. U.S. Department of the Interior, Federal Water Quality Administration. Pages 1-78.

Water pollution was the first type of pollution of serious concern to the federal government. Section 13 of the Rivers and Harbors Act of 1899 (33 United States Code Annotated sec. 407), prohibited the discharge of any kind of solid refuse matter but did not prohibit liquid sewage from being discharged into the navigable waters of the United States. Since that early statute the regulation of water polluting activities has become more comprehensive and sophisticated.

The first comprehensive statute aimed specifically at the water pollution control problem was enacted in 1948. Additional laws were adopted in 1952, 1956, 1961, 1965, 1966, and 1970. All of this legislation can be found in 33 United States Code Annotated sections 1151-1175. Undoubtedly, much new legislation will be passed in the years ahead.

Originally the Federal Water Quality Administration was known as the Federal Water Pollution Control Administration and was part of the Department of Health, Education and Welfare. In 1956 it was transferred to the Department of the Interior under Reorganization Plan No. 2 of 1966.

In 1970 it became known as the Federal Water Quality Administration and was made a part of the Environmental Protection Agency pursuant to Reorganization Plan No. 3 of 1970.

The sources and manner of water pollution are myriad. Road building and construction may produce sediment; agriculture may produce animal waste, pesticides, and chemical pollution; nuclear power plants may produce thermal or radioactive pollution; ships and drilling operations may produce oil pollution; residential areas produce municipal or domestic waste; and manufacturing plants produce various kinds of chemical pollutants. The question is so vast that it is difficult to find a study which touches upon all of these problems and yet avoids the perils of superficiality. It is believed that the following study by the Federal Water Quality Administration most nearly satisfies these requirements.

WATER POLLUTION AND THE ENVIRONMENT

Almost any day, in the waters near any large population center in the United States and, increasingly, in the countryside, we can see the signs of water pollution. It comes from many sources and exists in many forms to assail the eyes and the nose and the taste buds. Standing by the banks of an urban river—if one can actually get past the warehouses and wharfs and weeds to see the river—pollution may appear as surface oil slicks, in which old tires and debris and someone's picnic remnants are trapped and float sluggishly by, or as the public health notices warning the citizen not to swim or wade in the water at his feet. Pollution may be manifested in less obvious ways by masses of aquatic weeds and bad taste in the drinking water supplies. Even more subtle will be the—often unseen—changes in the aquatic life of the river, the loss of sport fish and the ascendence of sludge worms and other "tolerant" life-forms such as carp.

This urban example is repeated throughout the Nation. As our society and economy have grown, the wastes generated by our population and our technology have caused staggering amounts of pollution. Use of our waters to receive and carry away wastes has seriously damaged our ability to enjoy other water uses, such as swimming and boating, sport and commercial fishing. Other water uses, such as domestic, agricultural and industrial water supply, are possible, but often only after considerable advance treatment. Growing public awareness and concern with mounting pollution of the Nation's streams, lakes and coastal waters have stimulated a vast and vigorous national effort to control and abate water pollution.

Water quality problems caused by pollution are prevalent in every region of the country. The two areas where water quality and uses have been most seriously damaged are in the Northeastern States and the Great Lakes. In the Northeast, tremendous urban and industrial growth occurred during the 19th and early 20th centuries when little or no provision was made to control municipal or industrial waste flows to surface waters; the water was expected to "purify itself" and the wastes would float on downstream to become someone else's problem. The result was a legacy of pollution. The Northeastern States have the largest amount of untreated municipal and industrial waste discharges and the largest backlog of waste treatment facility needs.

In the Great Lakes, the discharge of large volumes of wastes, principally from municipal and industrial sources, has greatly accelerated the natural aging process of lakes. The most seriously affected of the lakes, Lake Erie, is now in a state of advanced eutrophication or aging—choked with plants, algae and other organic material. Although Lake Erie is not, as some experts have asserted, "dead," it is certain that very great expenditures for water pollution abatement are necessary to restore the fishery of the lake and reopen beaches closed because of pollution.

There are a number of other pollution problems caused by certain industries and sectors of the economy which have led to serious water quality damage in other parts of the country. Animal wastes from feedlots or runoff from irrigated and fertilized fields and areas where pesticides are used are an increasing cause of pollution, particularly in the Midwest and Southwest. The Colorado River becomes more saline every year as a result of irrigation return flows full of salts leached from the fields. The Annual Federal Water Quality Administration (FWQA) report on *Pollution Caused Fish Kills* chronicles the tremendous aquatic life mortality from agricultural pollution in Kansas and Missouri.

Acid drainage from abandoned mines has destroyed life in many streams in Appalachia and the Ohio Basin generally. Domestic and vessel wastes have polluted many coastal waters where sensitive shellfish were harvested; each year more areas are closed to private and commercial harvesting. Oil spills from vessels and leaks from offshore oil drilling facilities have resulted in several spectacular oil pollution incidents in the last few years, among them the TORREY CANYON and OCEAN EAGLE spills, the Santa Barbara offshore well leaks and the recent fire and oil leaks from drilling in the Gulf of Mexico. Less spectacular oil spills are occurring almost daily in navigable waters across the Nation.

How Has All This Pollution Happened?

Population growth is one major factor. In 1967, the Nation's population passed the 200 million mark. This number of people is expected to double in the next 50 to 60 years. Staggering demands will be placed on our natural resources to support this population. Waters are needed for consumptive purposes, such as public water supply, food production and processing, and some industrial uses, as well as for non-consumptive uses, such as reaction, industrial cooling, and sport and commercial fishing. At the same time that demands for water will increase, so will production of wastes that threaten the environment.

Not only the rate but the pattern of population growth concentrates and magnifies pollution. Urban and suburban sprawl covers green spaces and reduces clean environment in the very areas where people most need it. Intensive development has occurred particularly along the Nation's coastline, in the very estuarine areas that are most sensitive to environmental degradation.

Higher individual incomes and expectations have led to increasing demands for food and consumer goods, for better housing and highways, for a whole range of conveniences. In most cases, production of wastes is "built in" to our technology; as industrial production increases, with attendant damands for water, so does the per capita production of wastes. The public's demand for "throw-away" containers and other convenience items, as well as the tendency toward planned obsolescence, further accelerate this trend.

Consumer use and production of goods have greatly increased the demand for electric power—power production has doubled every ten years since World War II and this rate is expected to increase. Great amounts of water are used in producing electricity, and waste heat from both fossil fueled and nuclear generating plants constitutes a serious, and increasing, threat to the Nation's waters. For example, the famous salmon runs of the Pacific Northwest are threatened by thermal pollution.

Not only is the volume of industrial production increasing, but the very complexity of the products and wastes creates severe challenges for waste treatment technology. New chemical products are coming on the market every day, most often without sufficient research into the environmental consequences of using them. Widespread use of detergents has led to great increases in the release of phosphate nutrients to the waters, stimulating tremendous and noxious growths of aquatic weeds which cause severe problems in many areas. Radioactive and physiologically-active chemicals, which pose vexing problems, can only increase. Effects which cannot be predicted may be profound and irreversible.

Mining and transporting natural resources also pose increasing dangers for the environment. Greater use of supertankers and pipelines to transport oil and other materials, as well as increasing use of offshore and underwater mining, will greatly increase the dangers of accidental oil pollution and other hazards.

The growing popularity of deep well disposal of wastes presents yet another serious threat to our water resources. Although in some cases carefully controlled deep well injection may contribute to groundwater management, improperly carried out, this method of disposal may result in the contamination of groundwater or interconnected surface water supplies. The greatest problem in dealing with subsurface disposal is that the effects of underground pollution and the fate of the injected materials are uncertain with the limited knowledge available today.

Production of greater quantities of better quality food for American citizens has caused increasing pollution problems. Higher agricultural productivity has been based on irrigation and use of chemical fertilizers and pesticides. Runoff carries salts and chemicals, many of which are highly toxic and have long-lasting environmental effects, into streams. These diffuse waste sources are most difficult to control or treat. The possibility of irreparable and disastrous ecological consequences, particularly from persistent pesticides, has led to increasing demands for controlling or eliminating their use; no one can predict with certainty the impact of such a move on agricultural productivity.

Population growth and greater prosperity have brought a rising demand for beef and other meats. To increase productivity and profits, the trend has been toward raising heavier livestock and concentrating animals in large feedlots, thereby increasing and concentrating the agricultural waste problems.

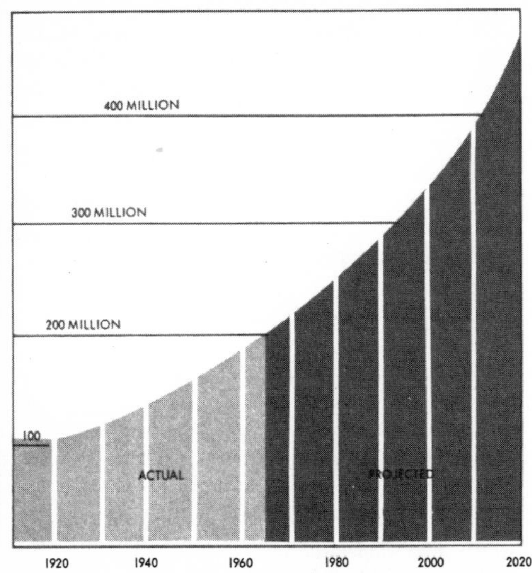

U.S. POPULATION GROWTH 1910-2020

The New York Skyline, viewed across the polluted Hudson River, exemplifies the growth and concentration of population and waste in the Nation's cities. Huge outpourings of wastes must be treated to improve water quality.

In summary, neither the institutions nor the technology of our society has been effectively utilized to prevent widespread pollution from occurring. To provide a better understanding of the specific challenges that control of pollution involves, the sources of pollution are discussed in greater detail in the following sections. These discussions will provide some indication of the magnitude of these sources of pollution and the estimated dimension and costs of clean-up.

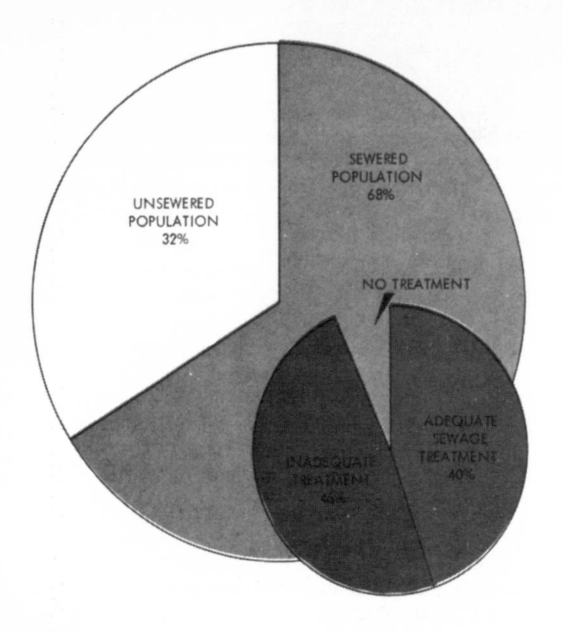

Aquatic growths, forming dried cakes along a lake shore, are caused by excessive nutrients in wastes and constitute a major aesthetic nuisance.

Municipal Wastes

The two largest sources of waste discharges to the Nation's waters are sewered municipal wastes and industrial wastes. Besides being a large source of organic material, which lowers the dissolved oxygen content of water and increases the concentration of bacteria, municipal waste also contains nutrients that fertilize algae and thus accelerate eutrophication of lakes.

Today, the number of sewered communities in the United States is just under 13,000; 68% of the Nation's population lives in such communities. Raw or inadequately treated sewage from millions of people still flows into our streams. Fortunately, we have the technological knowledge to deal effectively with municipal wastes. However, this technology has not been applied to the extent needed to prevent pollution. Although many communities have been installing and improving their waste treatment facilities, over 1000 communities outgrow their treatment systems every year.

The economic analyses contained in the FWQA's annual report to the Congress on the costs of clean water indicate that only about 40% of the Nation's treatment systems are adequate. An estimated 46% of the sewered population is now served by treatment plants that are overloaded or in need of major upgrading. Seven percent of the sewered population lives in communities which provide no treatment.

Generally speaking, the greatest municipal waste problems exist in the areas with the heaviest concentrations of population. Past neglect, however, has led to a greater backlog of waste treatment facility needs in the Northeast than in other parts of the Nation. The six New England States, New York and Pennsylvania contain just over 20% of the Nation's population but 52% of the sewered population that is not provided with waste handling facilities.

The cost studies indicate that a major investment, totalling about $10 billion, will be necessary over the next five years to overcome this legacy of neglect and achieve adequate levels of treatment for the Nation's municipal wastes. After that, significant annual investments will still be necessary to expand and replace plants as population growth continues. Treatment of domestic-type wastes from Federal facilities will also require significant expenditures by Government agencies; the waste treatment needs for sanitary and other wastes generated by Federal sources have been estimated at $246.5 million.

The waste loads from municipal systems are expected to increase nearly four times over the next 50 years. Even if municipal and industrial waste loads are substantially reduced through treatment, pollution problems may continue to exist in densely populated and highly industrialized areas where the assimilative capacity of receiving waters is exceeded. In these areas, higher and higher levels of treatment, approaching 100%, will probably be necessary, and water supply demands will lead to ever increasing use of renovated wastewaters.

Other municipal waste problems that will become more ... treatment ... are those ... and by ... conven- ... bined ... with ... wer ... ds. ... to ... he ... st ... e, ... e ..., ... d ...

I ... E ... a ... th ... E ... re ... va ...

ma ... the ... prin ... pou ... smal ... imals ... and a ... ing al ... lakes ...

Alt ... from agricul ... municipal wastes contribute the major load. Already nutrient pollution has led to the imposition of very high treatment requirements for waste discharges to the Great Lakes and several other areas; the cost of meeting these requirements is included in the investment totals noted above. In future years, the need for nutrient removal at other cities will greatly increase the costs of waste treatment.

Industrial Wastes

Industries discharge the largest volume and most toxic of pollutants. Industrial waste discharges are the source of an enormous variety of materials found in our water. Our 1969 report, *The Cost of Clean Water and Its Economic Impact,* listed a total of fifty-one agents being introduced into our Nation's waters as a result of industrial processes—and the list is known to be partial rather than comprehensive. For purposes of quantification, the common substances can be reduced to two general classes of materials, settleable and suspended solids and oxygen demanding organic materials. Major water-using industries are believed to discharge, on the average, about three times the amount of each class of waste as is discharged by all of the sewered persons in the United States.

There are over 300,000 water-using factories in the United States. Although there is as yet no detailed inventory of industrial wastes, general indications are that over half the volume of the wastes discharged to water comes from four major groups of industries—paper manufacturing, petroleum refining, organic chemicals manufacturing and blast furnaces and basic steel production.

The areas where the greatest quantities of industrial wastes are discharged to water are the Northeastern States, the Ohio River Basin, the Great Lakes States and the Gulf States. Lesser, but significant, volumes of industrial wastes are discharged in some areas of the Southeast and in the Pacific Coast States. Like municipal wastes, industrial waste sources are concentrated in certain areas, for factories, like people, tend to be found in clusters.

The volume of industrial wastes is growing several times as fast as that of sanitary sewage as a result of the growing per capita output of goods, declining raw materials concentrations and increasing degrees of processing per unit of product. Given the necessary expenditures, a large percentage of this volume can be treated efficiently, much of it, after pre-treatment in some cases, in the municipal treatment system. Whereas factories which used large volumes of water traditionally discharged wastes directly back to the stream, more stringent pollution control requirements and cost factors have led to increasing use of public treatment systems by a variety of industries. Most wastes from food-processing industries can be treated in public plants, and wastes from paper and pulp mills, chemical, pharmaceutical, plastics, textile and rubber plants have successfully been treated in municipal plants. Some combinations of municipal and industrial wastes actually improve the treatment process by, for instance, reducing the nutrients in waste discharges.

Increased use of joint municipal-industrial treatment systems will facilitate abatement of industrial pollution, and feasible treatment processes have been developed for many types of industrial wastes. Although the lack of an industrial waste inventory makes estimates difficult, the increasing level of investment in industrial treatment facilities appears indicative of progress towards meeting water quality standards. FWQA's economic studies have estimated the annual investment need for manufacturing industries at $650 million for each of the next five years.

Although, overall, this continued level of investment for treatment of present industrial pollution is encouraging, certain types of industrial pollution present much more complex abatement problems. The trends towards increasing production and use of complex chemical products and radioactive materials have greatly increased the possibility of releasing exceedingly dangerous wastes to the environment. Many of the new chemicals are a challenge to detect, much less control. There is fear that too little caution and study precede the processing or marketing of these materials.

This municipal discharge carried domestic sewage and industrial wastes into the Missouri River.

Thermal Pollution

The growing demands for electric power will require a tremendous expansion of power generating facilities. Water is used in the production of almost all electric power now generated —whether by hydroelectric, fossil fueled or nuclear power plants. Two of these generating methods, fossil and nuclear fueled steam electric plants, produce large amounts of waste heat.

As the amount of waste heat from steam electric power plants discharged to water bodies has increased, concern over thermal pollution and its effects has increased. As usually defined, thermal pollution means the addition of heat to natural waters to such an extent that it creates adverse conditions for aquatic life; accelerates biological processes in the streams, reducing the dissolved oxygen content of the water; increases the growth of aquatic plants, contributing to taste and odor problems; or otherwise makes the water less suitable for domestic, industrial, and recreational uses. Not the least important of the effects of heated wastewater is the reduced utility of the water for further cooling. An increasing number of authorities are beginning to believe that this waste heat may be the most serious contemporary source of water pollution.

The electric power industry is one of the most dynamic industries in the United States, and it has had a growth rate which has exceeded that of the gross national product for a number of years. The technology of electric power generation and distribution is changing rapidly. Larger-sized units have become economically feasible because of load growth and the increasing inter-connection and coordination of power systems via extra high voltage transmission facilities. In recent years, a large number of nuclear fueled plants have been planned and put under construction.

The principal use of water in steam electric generating plants is for condenser cooling purposes. The amount of water required for condenser flows depends upon the type of plant, its efficiency, and the designed temperature rise within the condensers. The temperature rise of cooling water condensers is usually in the range of 10° to 20° F, and the average rise is about 13°F. Currently, large nuclear steam electric plants require about 50% more condenser water for a given temperature rise than fossil fueled steam electric plants of equal size. It is estimated that by 1980, the electric power industry will use the equivalent of one-fifth of the total fresh water runoff of the United States for cooling.

Both fresh and saline water are used for

TOTAL GENERATION

NUCLEAR-FUELED STEAM

OTHER

HYDRO

1980 2000 2020

PROJECTED ELECTRIC GENERATION 1965-2020

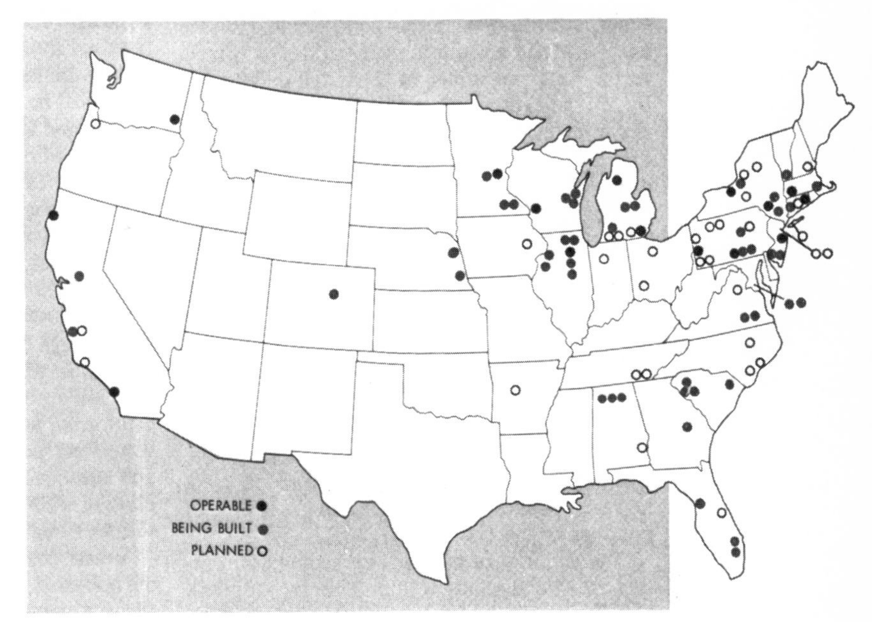

LOCATION OF NUCLEAR POWER PLANTS

cooling; in some cases, sewage effluents are used. Water for condenser use may be withdrawn from rivers, lakes, reservoirs, canals, tidewater, or groundwater. When adequate water supplies are available and allowable discharge temperatures permit, the water is usually passed through the condensers once and returned to the source body of water. The economic desirability of once through cooling has traditionally been a factor in locating power plants. Sites have usually been selected where large quantities of water were available for cooling at all times. Such sites in inland areas, however, are limited in number, and the increasing density of power plants on rivers and estuaries will require utilities to find effective means of controlling thermal discharges. Two factors can limit the adverse environmental effects of new power plants: better selection of sites and improved design of plants and equipment to reduce the discharge of heated wastewaters.

With the tremendous pollution potential of projected power production, it is exceedingly fortunate that waste heat from power generation is amenable to treatment or control at a reasonable cost. The amount of waste heat discharged to waterways can be reduced by improving the efficiency of the thermal plants, by making productive use of heat, or by using cooling towers, cooling ponds or spray ponds. The impact of thermal pollution control on the consumer cost of electricity is relatively minor.

The selection of appropriate sites for locating power plants so as to minimize environmental damage poses a significant challenge to both the industry and government. Environmental concerns will necessitate the consideration of many more factors in the planning of power production facilities than has been the practice in the past. In addition to thermal pollution control, a number of other critical selection factors make siting very complicated—aesthetic impact, availability of water supply, safety (for example, potential of earthquakes), air pollution control, access to transportation and others. These factors compete in some ways, and the tendency in the past was to give primary attention to producing power at low cost to the consumer rather than to environmental considerations. Installation of facilities, such as long discharge lines or cooling towers to control thermal pollution will affect cost factors and require more space for the plant and may make it more difficult to meet aesthetic goals. The increasing use of nuclear power adds another potential hazard to the environment—radiation. Siting is likely to become an increasingly difficult and controversial factor in the continued growth of power production.

Oil and Hazardous Substances

Dumping and accidental spilling of oil and other hazardous materials continue to increase each year and constitute major pollution threats to the water resources of the Nation. Pollution by oil and other hazardous substances may occur in any of our waterways and coastal areas, or on the high seas as a result of deliberate dumping, accidental spills, leaks in pipelines, drilling rigs and storage facilities, or the breakup of transportation equipment.

Damages caused by oil pollution are both significant and diverse. Such pollution can destroy or limit marine life, ruin wildlife habitat, kill birds, limit or destroy the recreational value of beach areas, contaminate water supplies, and create fire hazards. Damages caused by other hazardous substances can be just as significant and diverse as those caused by oil pollution. The sheer volume of oil transported or used, however, makes oil the largest single source of pollution of this type.

The majority of oil spills exceeding 100 barrels involve discharge from vessels. Approximately one-third of the incidents involve pipelines, oil terminals, bulk storage facilities, etc.

Reported Oil Spills in U.S. Waters (Over 100 barrels)

	1968	1969
Vessels	347	532
Shore facilities	295	331
Unidentified	72	144
Total	714	1007

Oil pollution may come from several different sources. Gasoline service stations dispose annually of 350 million gallons of used oil. Two hundred thousand miles of pipelines carry more than a billion tons of oil and hazardous substances. The pipelines cross waterways and reservoirs and are subject to cracks, punctures, corrosion, and other causes of leakage. Offshore oil and gas exploration and production occur mainly in the Gulf of Mexico, Southern California coastal waters, Cook Inlet in Alaska, the Great Lakes, and the East Coast. The blowout of wells, the dumping of drilling muds and oil-soaked wastes, and the demolition of offshore drilling rigs by storms and vessel collisions are significant potential pollution sources. In 1969 a massive oil spill occurred off Santa Barbara, California, with severe damage to the coastline, waterfowl and beaches. More recently a fire and subsequent oil blowout on an offshore production well in the Louisiana Gulf presented a serious threat to our marine environment.

Vessel casualties, too, are a prime source of oil pollution, and the damage can be extensive when several million gallons of oil enter the water at one time. The largest spill to date was over 30 million gallons in 1967 from the TORREY CANYON. England, alone, spent $8 million on clean-up following this casualty. In Tampa Bay on February 13, 1970, the tanker DELIAN APOLLON ran aground and spilled over ten thousand gallons of fuel oil into the bay, and some 100 square miles of area were contaminated as a result. Discharge of either oily ballast water or "slop oil" recently occurred offshore of Alaska, causing extensive

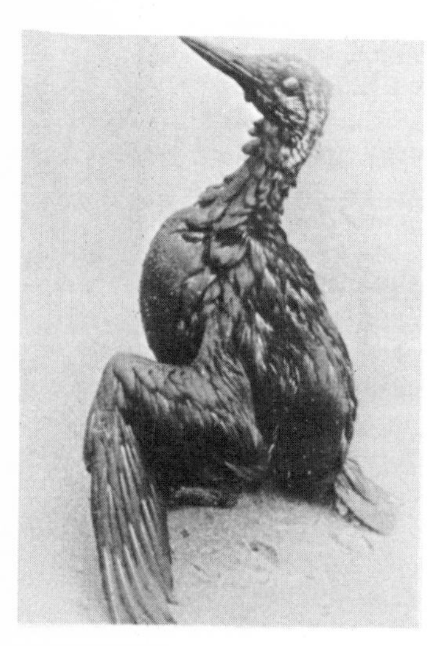

Birds suffered most from the more than 200,000 gallons of crude oil that escaped from this ruptured Santa Barbara well.

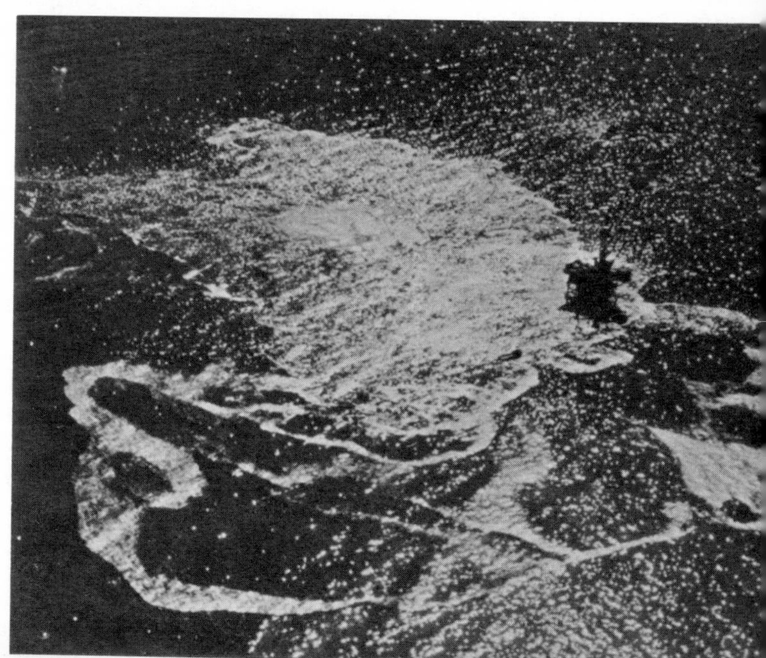

waterfowl mortalities and contamination of fur seals and sea lions.

Hazardous substances can enter our waters in many of the same ways as oil. Spills caused by accidents or ruptures of containers are important sources. For example, a train wreck on January 2, 1968, at Dunreith, Indiana, spilled a cyanide compound into Bucks Creek, a tributary of the Big Blue River. The cyanide moved with the flow of the stream and an estimated 1,600 pounds passed the town of Carthage on the Big Blue River, downstream from the site of the accident. The cyanide caused fish kills in the affected streams; more than 25 cattle were reported killed; at least one industrial plant temporarily ceased operations; and groundwater supplies were contaminated.

Incidents similar to the cyanide spill are not uncommon and can cause serious consequences in the affected areas. Presently, an estimated 10,000 spills of oil and hazardous materials occur annually in the navigable waters of the Nation. With the increasing volumes of these materials being transported, the number of spills may grow. Some increase in the number of spills reported can be expected since discovery and notification systems are improved continually and spills, that heretofore have gone unreported, will now be recorded. Unfortunately, the potential magnitude of each individual spill will increase as the size of the carrier increases. For instance, the UNIVERSE IRELAND, a ship launched in August, 1968, has a cargo capacity of over 90 million gallons of oil. The construction of even larger ships is under consideration. The potential pollution from a ship of that capacity is about three times greater than that resulting from the TORREY CANYON spill.

Mine Drainage

Mine drainage, one of the most significant causes of water quality degradation and destruction of water uses in Appalachia and the Ohio Basin States, as well as in some other mining areas of the United States, degrades water primarily by chemical pollution and sedimentation. Acid formation occurs when water and air react with the sulfur-bearing minerals in the mines or refuse piles to form sulfuric acid and iron compounds. The acid and iron compounds then drain into ponds and streams. About 60 percent of the mine drainage pollution problem is caused by mines which have been worked and then abandoned. Coal mines idle for 30 to 50 years may still discharge large quantities of acid waters.

Although acid pollution is usually limited to coal field areas, suspended solids and sedimentation damage can extend much further downstream. Mine drainage pollution may degrade municipal and industrial water supplies; reduce recreational uses of waters; lower the aesthetic quality of waterbodies and corrode boats, piers and other structures. During 1967, over a million fish were reported killed by mine discharges, ranking mine drainage as one of the primary causes of fish kills in the United States.

Total unneutralized acid drainage from both active and unused coal mines in the United States is estimated to amount to over 4 million tons of sulfuric acid equivalent annually. Although about twice this amount of acid is actually produced, roughly one-half is neutralized by natural alkalinity in mines and streams. In Appalachia alone, where an estimated 75 percent of the coal mine drainage problem occurs, approximately 10,500 miles of streams are reduced below desirable levels of quality by acid mine drainage. About 6,700 miles of these streams are continuously degraded; the remainder are degraded some of the time. Acid mine drainage problems also occur from other types of mining throughout the Nation, such as phosphate, sand and gravel, clay, iron, gold, copper and aluminum mines.

It is estimated that 3.2 million acres of land in the United States had been disturbed by surface (strip and auger) mine operations prior to January 1, 1965. Of these 3.2 million acres, approximately 2 million acres are either unreclaimed or only partially reclaimed. An additional 153,000 acres have since been disturbed each year, only part of which are reclaimed annually. In addition to contributing to the acid pollution problem, surface mines also contribute large quantities of sediment to the Nation's streams.

Sediment yields from strip-mined areas average nearly 30,000 tons per square mile annually—10 to 60 times the amount of sedimentation from agricultural lands. At this rate, the 2 million acres of strip-mined land in need of reclamation could be the source of 94 million tons of sediment a year.

In addition to mine drainage, refuse piles, tailings ponds and washery preparation residues are also important indirect sources of pollution from mining. For many minerals, such as phosphate, the pollution from processing operations exceeds that resulting directly from the mining operation. The pollution from coal mines in Indiana and Illinois, for example, stems primarily from refuse piles, tailings ponds and preparation plants. No national estimates are available, however, which show the volume or relative importance of pollution from these sources.

Prevention of acid and sediment drainage from surface mines can be accomplished through renovation of the mined area. Regrading and revegetation can be very effective means of mine drainage control, and reclaimed mining areas can be used for recreation and other beneficial uses. Other methods of control may involve sealing mines, diversion and/or control of underground drainage and use of chemicals or biological inhibitors to reduce the formation of acid. Neutralization is the most common method of treating acid drainage.

Although many methods have been applied and others are being tested, the problems of mine drainage have been very difficult to deal with, largely because of the costs involved in achieving significant levels of control. Recent cost estimates for pollution control and land reclamation in the mining States total as much as $7 billion. Moreover, the distribution of the mine pollution problem is such that a large percentage of this investment would have to be made in some of the most economically depressed areas in the Nation, involving mines that are no longer operating or producing any revenues.

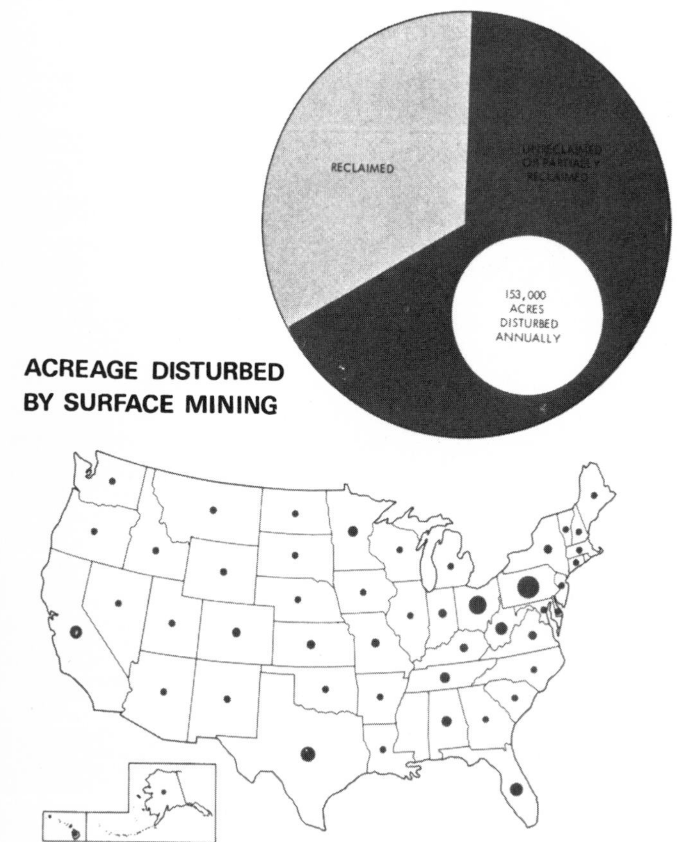

ACREAGE DISTURBED BY SURFACE MINING

Sedimentation and Erosion

Sediments produced by erosion are the most extensive pollutants of surface waters. It is estimated that suspended solids loadings reaching our waters are at least 700 times the loadings from sewage discharge. The dirty brown or gray appearance of a river or reservoir after a rainstorm is due to sediments washed in from croplands, unprotected forest soils, overgrazed pastures or the bulldozed "developments" of urban areas. The presence of sediment generally increases the cost of water purification and reduces the value of water recreation, and nutrients adsorbed on sediment particles contribute to undersirable conditions in lakes.

Sediments adversely affect commercial and game fish habitats, power turbines, pumping equipment and irrigation distribution systems. Deposited during floods, sediments damage crops and, if coarse-textured, may reduce the productivity of the soil. Channels and drainage facilities may be impaired, and the clean-up and removal of sediments from residential and other developed areas is costly. Sediments are also depleting the capacity of artifical reservoirs in this country, and potential storage sites to replace these depleted reservoirs are limited.

Erosion rates of lands are increased 4 to 9 times by agricultural development, and may be increased as many as 100 times by construction activities. Paving and drainage facilitate flushing of urban areas. The 470,000 miles of rural and secondary roads in the United States also contribute significantly to sediment pollution. Erosion is a serious problem on at least 300,000 miles of the Nation's stream banks and along many of the 470,000 miles of rural and secondary roads. As has been discussed, sedimentation from stripped mining lands is also considerable.

Construction is a large contributor to the sedimentation problem if erosion control is not provided. According to the 1969 report, *The Cost of Clean Water and Its Economic Impact,* the average sediment yield during a rainstorm at highway construction sites is about 10 times greater than that for cultivated land, 200 times greater than for grass areas, and 2000 times greater than for forest areas, depending upon the rainfall, land slope and the exposure of the bank. Similar rates of sediment production occur from commercial and industrial construction in urban areas. The Potomac River Basin discharges about 2.5 million tons of sediment a year into the Potomac estuary, a large share due to disturbance of land surfaces by construction in urban areas.

Sources of sediment are diffuse and therefore often difficult or costly to control. Where feasible, erosion prevention provides the most effective method for sediment control. In certain remote arid areas of United States, however, such measures would be extremely expensive, and on certain construction sites, completely impractical.

With regard to agricultural land, erosion control by such means as contour cultivation or crop rotation may achieve many benefits—reduction of sediment pollution of streams and damage to water uses, and conservation of productive soil and vegetation resources. Gully erosion may require costly measures of filling, seeding or damming.

Excessive sediment runoff from highway construction can be controlled by reducing the amount of time ground is exposed and/or using measures such as grassing or channeling to prevent sediment from reaching streams. Similar control measures can be used to prevent erosion at other types of construction sites.

Erosion control practices may add about $1000 to the cost of each mile of new highway and $1000 per highway construction project for overhead. For the 470,000 miles of secondary and rural roads which need erosion control measures, costs may range from $275 up to $15,000 per mile, with an additional $50 per mile per year required for maintenance. In total, the initial costs to control erosion from roads may range from $130 million to $7 billion, with annual maintenance thereafter costing $23 million. Much of the construction costs and all the maintenance costs would be non-Federal.

Control of erosion at urban construction projects could cost from $100 to $1000 per project depending on size and location. Thus preventing water pollution from construction activities may add somewhat to the cost of buying a house.

LESS THAN 270 PPM
270 TO 1900 PPM
1900 OR MORE PPM

CONCENTRATION OF SEDIMENT IN STREAMS

Control of streambank and streambed erosion may require construction of special stabilization structures, riprap of streambanks and sloping and vegetating eroded banks. These measures, however, may not be compatible with other water uses. Estimates of the cost of renovating the eroded streambanks in the United States range from $200 million to $3 billion.

In summary, the sources of water pollution from sedimentation are exceedingly diverse and diffuse. Much can be done to reduce this cause of pollution, but control and prevention will be very costly.

Erosion control reduces sediment pollution. Here an eroded gully has been transformed into a productive farm pond.

Feedlot Pollution

Both the increasing number of animals raised and the modern methods of raising these animals contribute to the increased pollution of waters from animal wastes. Beef cattle, poultry and swine feeding operations, along with dairy farms, are the major sources of actual or potential water pollution from animal wastes.

In the past two decades production of animal products has been increasing rapidly. The technology of this increasing production requires that animals be confined in a minimum space and fed a concentrated ration, both of which increase the pollution potential of animal wastes. The heavy concentration of wastes precludes their natural decomposition and assimilation on pastures as is the case where animals are more dispersed. The heavy concentration also makes it difficult to find nearby farmland that can use manure as an economical source of fertilizer. In addition to being heavily concentrated in small areas, wastes from concentrated feeding operations have a high oxygen demand when they are being degraded, and they may contain a high proportion of roughages.

When animal wastes find their way into water, they can contribute to pollution in several ways. Heavy concentrations of animal wastes in water may: add excessive nutrients that unbalance natural ecological systems, causing excessive aquatic plant growth and fish kills; load water filtration systems with solids, complicating water treatment; cause undersirable tastes and odors in waters; add chemicals that are detrimental to both man and animals; increase consumption of dissolved oxygen, producing stress on aquatic populations and occasionally resulting in septic conditions; and add microorganisms that are pathogenic to animals and to man.

The magnitude of the livestock pollution problem is primarily dependent upon the number of animals that are needed to meet the demand for their products. The average population increase in the United States is about 2.5 million people per year. At 1966 consumption rates, each additional million people will require another 172,000 beef cattle, 24,500 dairy cattle and 433,000 hogs. Thus, it can be seen that if these consumption rates continue, the amount of animal wastes will continue to increase significantly. In addition, the trend toward increased use of confined feeding and concentrated rations will continue to add to the pollution potential of the animal wastes.

Agricultural waste sources are scattered across the country, with large amounts of cattle being produced in the Midwest, West and

CONCENTRATION OF WASTE PRODUCING FARM ANIMALS

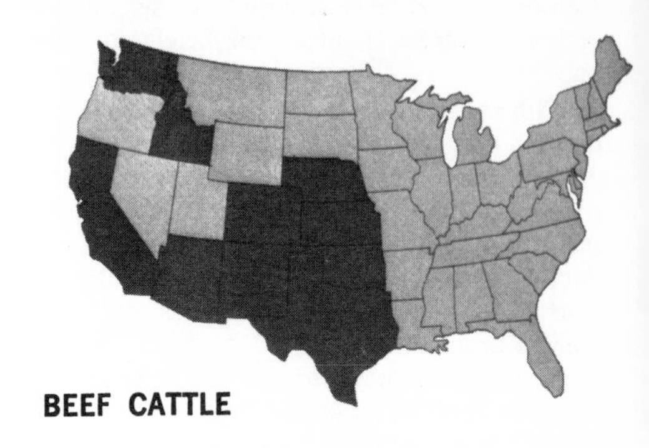

BEEF CATTLE

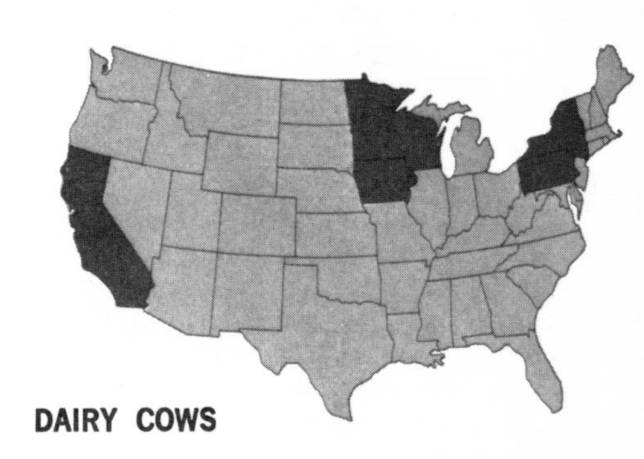

DAIRY COWS

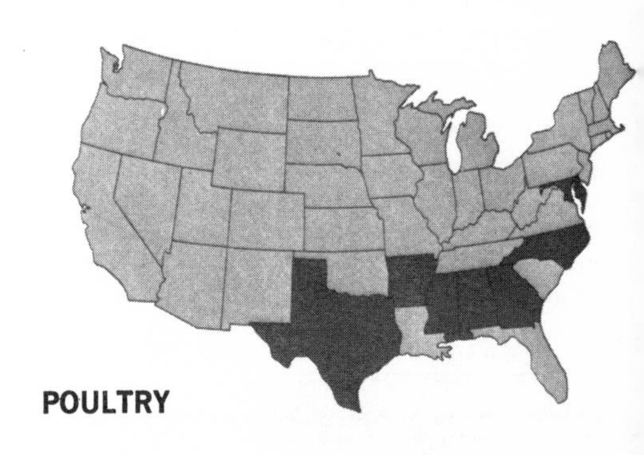

POULTRY

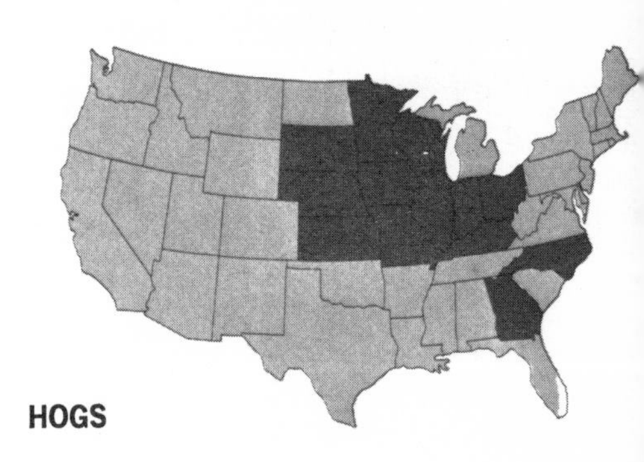

HOGS

Southeast; poultry in the South and some of the Middle Atlantic States; and hogs in the Midwest and South. Although there have been few detailed assessments of the distribution of agricultural waste problems, feedlot pollution appears to be a particularly severe water pollution problem in certain parts of the States where large cattle feedlots are located.

A number of waste handling and control methods are available, which vary widely in complexity and cost. Many States are just beginning to survey feedlot operations and other agricultural operations to determine the pollution potential and necessary measures to deal with the problem.

Other Agricultural Wastes

Other pollution problems are caused by farming operations, in addition to those related to erosion and animal wastes which have been discussed. There is increasing concern about the short-and long-term environmental effects of runoff from farmlands which contains a variety of chemicals including pesticides, herbicides, insecticides and fertilizers. The soil conservation methods discussed earlier in relation to sediment control also help to control runoff. A number of Federal agencies are cooperating on research devoted to the search for chemicals, or biological control methods, which will sustain agricultural productivity while reducing the possibility of environmental damage and destruction of aquatic life and wildlife.

In some areas, serious water quality degradation has occurred as a result of runoff from irrigated lands. Water returned from irrigated areas usually has a much higher concentration of dissolved solids than does streamflow, because the diverted water leaches additional solids from the canals and fields, and because evaporation from the soil and transpiration by the crops concentrates these dissolved solids into a smaller flow of water. Thus, as the concentration of dissolved solids in surface water increases with each irrigation diversion and drainage return, the quality of the water deteriorates and its suitability for further irrigation diversion or other beneficial uses is impaired. This degradation of water quality is evident in many of the river basins where irrigation is practiced and must be taken into account in consideration of any further development.

Particular problems have been encountered in the Colorado River Basin. While agricultural productivity in parts of the Basin has been impressive as a result of irrigation, the Colorado River is becoming more saline every year. Its agricultural usefulness in parts of the lower basin has been seriously impaired.

Some methods to control leaching by irrigation, such as lining canals, are available, and in some areas the possibility of using desalination plants is being studied. Overall, however, the water quality problems caused by irrigation return flows are difficult and expensive to control. Degradation by agricultural practices of the water resource on which that agricultural development depends may place previously unconsidered limitations on the extent to which further massive irrigation schemes are practicable.

Wastes from Watercraft

The problems of water pollution incidents, often spectacular, caused by vessel accidents which release oil or other hazardous materials has been discussed. But vessels (and marinas) also contribute to pollution of the Nation's waters in a number of other ways. It has been determined that approximately 46,000 Federally registered commercial vessels, 65,000 unregistered commercial fishing vessels, 1600 Federally owned vessels and 8 million recreational watercraft use the navigable waters of the United States. The potential pollution from sewage from these vessels is estimated to be equivalent to just over 500,000 persons, comparable to a city the size of San Diego. In major harbors such as the Hamptom Roads, Virginia area, sewage discharges from vessels contribute significantly to water pollution, damaging shellfish harvesting and recreation.

At the present time, a very small percentage of watercraft are equipped with sewage treatment devices. Sewage equipment for use aboard watercraft is available in the form of holding tanks which collect sewage for disposal onshore, incinerators and biological treatment facilities. Estimates of the costs to install control devices on vessels to prevent sewage pollution come to about $660 million.

Other significant pollution from vessels is often evident where ships discharge bilge and ballast water containing oils and a variety of other substances. Poor "housekeeping" practices may cause a good deal of environmental degradation. Even if vessels go beyond the territorial waters to discharge bilge and ballast and solid wastes in the open ocean, aesthetic and other damages often result, as witnessed by Thor Heyerdahl and his crew aboard the RA.

A WATER POLLUTION CONTROL PROGRAM FOR THE 1970'S

Water pollution control has traditionally been a multi-agency, multi-program effort with the localities and industries having the principal responsibility for installing and operating pollution control facilities; the State water pollution control agencies having the basic regulatory programs; and the Federal government backing up the localities with treatment facility grants and backing up the States with additional enforcement authority, technical, financial and planning assistance, training and research and development. These basic arrangements have provided for a valuable division of effort and responsibility to build on and strengthen for the future. In looking to the future, it is necessary to keep in mind this wide basic underpinning of pollution programs and intergovernmental relations that has been established over the years. These ongoing activities, which will be fully described in this report, provide the basis and the background for the areas of acceleration—those major program thrusts—which are now necessary to meet the challenge of the 1970's.

These major program thrusts are aimed at immediate implementation of the technology available today to substantially reduce municipal and industrial pollution over the next few years. While research and technical studies must continue on methods of dealing with other complex pollution problems, immediate emphasis must be given to the regulatory and financial assistance programs needed to abate urgent municipal and industrial problems without further delay. Thus, far-reaching proposals to strengthen both of these basic programs within the context of the existing Federal-State-local partnership represent the keystone of the Nixon Administration's water pollution control program.

Better Financing of Municipal Treatment

The proposed legislative program for the 1970's calls for strengthening the present construction grants program with a major new investment in municipal waste treatment facilities, providing a strong and guaranteed program of Federal waste treatment works construction grants. Economic estimates by the Federal Water Quality Administration (FWQA), have pointed to a need for at least $10 billion worth of investment in municipal facilities to achieve the treatment goals contained in the water quality standards all across the Nation. The proposed Federal share would be $4 billion—$1 billion over each of the next four years. The States would be encouraged to share the total cost of projects with the Federal government and the localities in the present grant program, through continuation of the incentives that allow projects to receive a larger Federal share if the States contribute funds, and through new provisions in the proposed formula for allocating funds.

In addition to providing for more Federal funds for waste treatment works construction, the proposals would also strengthen the capacity of the construction grants program to assure that facilities are built according to the best designs and in accordance with basin and regional planning requirements. The formula for allocating grant funds would be revised to permit more funds to be spent for plant construction in areas where the need is greatest and where the greatest improvements in water quality will be realized.

The Secretary of the Interior has recently published proposed regulations in the Federal Register, which will help assure that the treatment plants constructed with Federal assistance will be well-built and well-maintained through more stringent requirements for design, operation and maintenance. Moreover, the regulations will require comprehensive river basin programs that would relate construction of treatment facilities to the magnitude and types of other pollution problems. In other words, the aim would be to assure that municipal treatment plants are built in areas where there is a positive program to clean up other kinds of water pollution. In line with this kind of comprehensive approach, the grants program would encourage development of regional treatment facilities that handle municipal and other wastes on an area-wide basis and which provide for treatment of many kinds of industrial wastes, as well as municipal sewage.

Better Standards and Enforcement Authority

One of the chief mechanisms for achieving an accelerated pollution abatement program is effective use of regulatory powers. The Federal government has had an enforcement program since 1956; its accomplishments will be discussed in this report. The present authority, however, is limited, and the procedures under present law are time-consuming. Although Federal-State water quality standards have been set which contain abatement requirements for all municipal and industrial waste sources on interstate waters, the Federal government does not have jurisdiction to enforce standards without the permission of the Governor if pollution occurs in only one State.

Legislation has been proposed to apply the regulatory provisions of the Federal Water Pollution Control Act expressly to boundary waters, as well as to interstate and navigable waters, the tributaries of these waters, ground-waters, the waters of the Contiguous Zone and, under certain circumstances, the high seas. Water quality standards, which now consist of water quality criteria and a plan for their implementation and enforcement, would include a third element: water quality requirements controlling discharges, or effluent requirements. The abatement authority would be made directly applicable to discharges which violate water quality standards in any or all of their three elements. A Governor's consent would no longer be required in cases of intrastate standards violations, nor in cases of enforcement conferences and postconference court action involving intrastate pollution. The court could impose a penalty on violators in both types of actions of up to $10,000 a day, and the second stage in the present three-stage enforcement process, the public hearing, would be eliminated. In addition, the Secretary of the Interior could seek an immediate injunction in an emergency situation in which there is an imminent and substantial danger to the health or welfare of persons or possible irreparable damage to water quality or the environment. The Administration's proposal would also provide other new enforcement tools.

The proposed legislation is not meant to override the responsibility of the State agencies to enforce pollution control regulations; rather, it is intended to provide a backstop to the States' authorities. The Federal government will continue to encourage the States to carry out their responsibilities by providing better financial and technical assistance to the States, in addition to the promise of Federal involvement when the States fail to act.

Better Assistance To The States

The challenge of carrying out an accelerated pollution control program and implementing water quality standards has placed increased responsibilities for monitoring, enforcement and technical activities on the States, as well as on the Federal government. The responsibilities of the States will further be increased by the recently enacted legislation which requires State certification of Federally-licensed activities; and acceleration of waste treatment works construction will place yet another heavy burden on State pollution control agencies.

For some years, the Federal government has assisted in supporting the administrative expenses of the State and interstate water pollution control programs through program grants, which are now at a $10 million level. To aid the States in expanding their programs, the proposed legislation would increase the authorization for State program grants each year on a sliding scale from $12.5 million in FY 1971 up to $30 million in FY 1975. Emphasis for using the augmented grant funds would be placed on certain program improvements, such as establishing effective waste discharge permit systems, improving sewage treatment facilities programs, and setting up programs for training and developing water pollution control personnel.

Besides providing financial assistance, the Federal government will continue to help the States through joint water quality monitoring activities, technical support and training programs for State personnel.

Better Programs For Prevention and Abatement Of Pollution From Federal Facilities

One of the primary tasks of the Federal government in pollution control is to assure that the facilities owned by the government and activities carried out or licensed by the government do not contribute to water or air pollution. In a move to strengthen the Federal commitment to pollution control, President Nixon issued a new Executive Order on pollution control from Federal facilities on February 4, 1970. This Order requires that all projects or installations owned or leased by the Federal government be designed, operated and maintained in conformance with present and future water quality standards. The Executive Order provides for strict compliance and establishes a deadline by which existing facilities must comply with environmental standards. This comprehensive plan for pollution abatement includes control, not only of water pollution, but also of air pollution by Federal facilities.

In a subsequent Executive Order issued on March 7, implementing the landmark National Environmental Policy Act of 1969, the President set forth additional procedures to assure that Federal programs will meet national environmental goals. He directed that attention be given to Federal policies, including administration of loans, grants, contracts, and licenses, to minimize their pollution impact.

Enactment by the Congress of the Water Quality Improvement Act of 1970 adds further force to this effort by requiring that applicants for Federal permits, for activities such as construction of nuclear facilities or reservoirs, meet applicable water quality standards.

Programs To Deal With Emerging Problems

At the same time that a massive effort to employ present technology to clean up municipal and industrial water pollution is being initiated, the water pollution control program for the 1970's looks to expanding its capacity to deal with other complex pollution problems. One of the most significant emerging programs is in oil pollution control, where substantial expansion of Federal prevention, control and enforcement activities is called for under the 1970 Act. In conjunction with development of plans to prevent and control oil spills, planning has been undertaken to handle accidents of other hazardous substances.

Increased attention has been given to methods of preventing and controlling pollution caused by vessels. The Water Quality Improvement Act provides for Federal performance standards for water pollution control equipment on commercial and private vessels.

With the greatly increased growth of electric power producing facilities, thermal pollution control has emerged as a major pollution problem. The water pollution program for 1970's anticipates much more stringent controls on the discharge of heated effluents, a greater research effort to improve thermal standards and abatement technology, and an active participation in planning studies to locate power facilities in areas where environmental damage would be minimized.

Another problem which is becoming increasingly significant is that of pollution caused by persistent pesticides. Under the 1970 Act, the FWQA will be developing, within the next two years, the scientific knowledge necessary for the development of water quality criteria for pesticides. This will require increased research on the effects of pesticides and the search for less harmful pesticides, expanded monitoring and investigation to identify critical areas and closer interagency coordination with the Departments of Agriculture and Health, Education and Welfare to assure full utilization of regulatory authorities to achieve environmental protection.

The expanded use of deep-well and other subsurface waste disposal practices poses a new challenge, particularly for protecting the purity of groundwater supplies. Meeting this challenge will require increased research on groundwater quality and movement and on the effects of wastes, investigations of present disposal sites and tighter regulation of subsurface waste disposal practices.

The activities and problems just described will receive increasing emphasis in the coming months. How these areas fit into the full water pollution control program will be described in greater detail below. As noted at the beginning of this section, the financial assistance and regulatory programs must rest upon a broad base of planning and research, technical studies, manpower development and other programs. It must also be clear that the Federal program is but one aspect of a nationwide network of State, local and, increasingly, regional activities. The greatest challenge of the 1970's may well be intergrating these programs to form a comprehensive nationwide attack on pollution of our environment.

PROGRAMS FOR WATER POLLUTION CONTROL

REGULATORY PROGRAMS

Strong, effective, and equitable regulatory activity is the most essential element in the nationwide pollution control effort. President Nixon in his environmental message has declared that "strict standards and strict enforcement are necessary—not only to assure compliance, but also in fairness to those who have voluntarily assumed the often costly burden while their competitors have not." Such effective nationwide enforcement requires a complementary State-Federal regulatory effort.

From the initiation of the Federal water pollution control program, the Congress has recognized the basic role of the States in implementing and enforcing water pollution control regulations. The Federal Act, however, asserts broad jurisdiction for the application of Federal regulatory authority to back up the States and to assure effective pollution control. Over the years, this Federal regulatory role has been expanded and strengthened to include: water pollution enforcement authority on interstate and, under certain circumstances, navigable waters; authority to establish and enforce water quality standards on interstate waters; and administration of the Oil Pollution Act of 1924. In addition, there has been a growing emphasis on control of pollution from Federal facilities.

Through its role in administering or participating in these programs, the Federal Water Quality Administration (FWQA) has emerged as the principal water pollution regulatory agency in the Federal government. Recently enacted and proposed legislative changes will further strengthen FWQA's regulatory authority. Passage of the Water Quality Improvement Act of 1970 adds significantly to Federal authority to control vessel and oil pollution and to requirements for control of water pollution from Federally licensed activities. Equally significant, the Administration's legislative proposal would result in far-reaching improvements designed to provide a comprehensive, swift and equitable regulatory authority. These measures will vastly strengthen the Federal government's capacity to control water pollution.

Water Quality Standards and Enforcement

Federal enforcement authority on interstate and navigable waters has been strengthened over the years since initial enactment of the Federal Water Pollution Control Act in 1956. The most significant increase in these authorities stemmed from the Water Quality Act of

1965, authorizing the establishment and enforcement of water quality standards for interstate waters, including coastal waters.

Today, action to abate pollution of interstate or navigable waters which endangers the health or welfare of persons may be taken at State request or on Federal initiative. The Governor's request is required in cases of intrastate pollution of such waters. However, action may be taken on Federal initiative to abate pollution, whether inter- or intrastate, of such waters which impairs the marketing in interstate commerce of shellfish or shellfish products. Action to abate international pollution may be taken under certain circumstances.

Two abatement procedures are provided in the Act. A three-stage enforcement procedure is set out in the law—conference, public hearing, court action—the succeeding stage to be reached only if adequate progress is not made at the previous stage. In a case of violation of water quality standards, direct court action may be sought 180 days from the date of notification of violation: the 180-day period is to be used for obtaining voluntary compliance if at all possible.

The water quality standards authorized by the 1965 legislation are the keystone of America's clean water program. The Act called upon the States to establish standards for their interstate waters. These State standards could then be accepted as Federal standards by the Secretary of the Interior. To set standards, the States had to make crucial decisions involving the desired uses of their water resources, the quality of water to support these uses and specific plans for achieving such levels of quality. The standards are, in effect, blueprints for the national program.

Water quality standards are composed of two parts: the criteria designed to protect present and future water uses of interstate waters through establishment of quality levels which must be maintained, and a plan of implementation which outlines the pollution abatement measures which will be required to meet those criteria. First responsibility for implementing and enforcing water quality standards rests with the States. But, once accepted by the Secretary of the Interior, the standards become Federal standards and are subject, if necessary, to Federal enforcement. In the absence of timely and acceptable action by a State to adopt water quality standards on interstate streams, the Secretary of the Interior can initiate action to establish Federal standards.

The standards of all of the States have now been approved by the Secretary of the Interior. With the establishment of these standards, there is for the first time a specified set of conditions for the enhancement and protection of the water quality of interstate waters throughout the country to which waste dischargers must adhere. The goal of providing nationwide, systematic and comprehensive water quality standards, however, which are tailored to the particular use and quality of the specific waters, is far from being accomplished.

The Secretary excepted from initial approval portions of the standards of over half the States, where certain aspects of the standards were not stringent enough to assure adequate water quality protection. For example, the temperature criteria of a number of States have been excepted, because they did not provide adequate safeguards against thermal pollution. In other cases, implementation plans have not received approval because the abatement measures required or schedules established were deemed inadequate.

During the past year, heavy emphasis has been placed on resolving these exceptions so that State standards can be fully approved. Negotiations have been underway with the States concerned and a number of States have agreed to improve their standards. In two instances, where such agreement could not be reached, the Secretary has taken initial action toward direct establishment of Federal standards, under procedures specified by the Act. A conference to consider the establishment of water quality standards for certain interstate waters of Iowa convened at Davenport on April 8 and at Council Bluffs on April 15, 1969. Regulations setting forth the Federal standards have been published in the Federal Register and will be adopted if the State does not adopt acceptable standards within the specified time period. A conference to consider the establishment of water quality standards for Virginia's interstate waters was called for December 9–11, 1969, and subsequently postponed when the State Water Control Board indicated it would act on the Secretary's recommendations. During the year ahead, a principal objective will be elimination of the exceptions from the standards of all the States, by agreement or direct Federal action.

Even where standards have been approved, there is a need to refine and improve certain of the water quality criteria to assure that the criteria applied will adequately protect the intended water uses. Continued emphasis must be given to improving our knowledge of water quality characteristics and requirements and incorporating this information in approved criteria.

Towards this end, FWQA, the Atomic Energy Commission and the Department of Health, Education and Welfare are working together to develop standard radiological criteria for natural waters. The radiological criteria currently established in water quality standards possess certain shortcomings insofar as providing complete coverage of all radioactive pollutants and maximum protection for all water uses. These established criteria do provide reasonably adequate protection from the sources of radiological wastes currently in place, but with the expected growth of the nuclear power industry, the nuclear fuel reprocessing industry and other peaceful uses of nuclear materials, such as those being developed through Operation Plowshare, much more precise and restrictive criteria for water will be required. The radiological criteria being developed are aimed at this objective. Also, they will complement the radiological effluent and emission standards presently set by the Atomic Energy Commission for nuclear power plants and other users of nuclear materials.

The increasing impact of pesticides on the environment has pointed to the need for both stricter regulation of pesticide uses and the es-

Stricter control of pesticides will be needed to protect wildlife.

tablishment of specific, quantified pesticide criteria for natural waters. Under the Federal Insecticide, Fungicide and Rodenticide Act, the authority to regulate the uses and labeling of pesticides resides with the Secretary of Agriculture. An interdepartmental agreement has recently been established among the Departments of Agriculture, Interior and Health, Education and Welfare through which environmental, fish and wildlife, and public health interests in pesticide uses are factored into the Department of Agriculture's registrations. With respect to pesticide criteria for interstate waters, this responsibility and authority rests with the Secretary of the Interior under the Federal Water Pollution Control Act.

General criteria on all toxic materials have been incorporated in all of the water quality standards adopted and approved pursuant to the Act; however, specific quantified criteria for the various pesticides in current use have not been made a part of these standards. Under a provision of the Water Quality Improvement Act of 1970, FWQA will be developing specific and quantified information on pesticides to be subsequently incorporated into water quality standards.

Most important, a vigorous State and Federal enforcement program is needed to obtain compliance with water quality standards and to assure that treatment schedules are being met. Development of strengthened and accelerated enforcement efforts has been a major objective during the past year. Where the States are prepared to exercise their authorities, FWQA stands ready to provide any assistance they may require. A number of States are moving aggressively against polluters. Illinois has not hesitated to initiate proceedings against the very giants of industry. Pennsylvania successfully carried through on the first test of its Clean Stream Law. And, with the passage of the Porter-Cologne Water Quality Act in 1969, California has vastly strengthened and stepped up its regulatory activity.

At the Federal level, the record of enforcement activity compiled under the new Administration reflects a commitment to a vigorous enforcement program equally and fairly applied.

In this same year, FWQA initiated the first enforcement actions to abate violations of water quality standards under procedures provided by the Water Quality Act of 1965. As mentioned before, the procedure provided in the law is direct court action, preceded by a 180–day notice to the alleged violator. On August 30, 1969, the Secretary issued such 180–day notices to six alleged violators. The first involved the Eagle-Picher Industries, Inc., whose mining operations resulted in discharges violating water quality standards established for Spring River in Kansas and Oklahoma. The other five actions were taken to abate violations of Lake Erie water quality standards and involved the City of Toledo and Interlake Steel on the Maumee River and Republic Steel Co., U.S. Steel, and Jones and Laughlin on the Cuyahoga River. Hearings were held with all six of the alleged violators. All six sources have indicated that they will comply.

FWQA's enforcement conference activity under previously established procedures has also been stepped up. The initiation of the Biscayne Bay conference in February, 1970, brought to 50 the total of such actions taken since 1956. Five of these—Lake Superior, Escambia River Basin, Perdido Bay, Mobile Bay, and Biscayne Bay—have been held since January 1, 1969. In addition, eight conferences were reconvened and three progress meetings held to put renewed emphasis on progress in obtaining compliance.

The enforcement conference has been an effective mechanism for the solution of complex and long-standing pollution situations. At the recently reconvened Potomac River conference, for example, agreement was reached on cooperative programs of remedial action which include the most stringent waste treatment requirements yet fixed for a metropolitan area. The Lake Michigan conference, reconvened in 1969 and again in March, 1970, has dealt with control of the more diffuse wastes, such as nutrients, thermal pollution, and agricultural wastes.

More recently, in February, 1970, a Federal-State enforcement conference was held at Biscayne Bay, Florida, regarding local damages to aquatic plant and animal populations of lower Biscayne Bay attributed to the heated effluent from the Turkey Point plant of the Florida Power and Light Company. Because of the selection of the site of the plant at Turkey Point, considerable technical difficulties are being encountered in the disposal of the heated cooling water. Present and proposed treatment measures were found to be inadequate and the conferees have recommended that the excessive waste heat load being discharged from the Turkey Point power plant be reduced to specified levels so that the quality of the waters, including the biological balance of Biscayne Bay, will not be impaired to the detriment of the full enjoyment and use of the Bay.

Subsequently, Secretary Hickel requested the Attorney General to bring suit against the Florida Power and Light Company on the basis of Section 13 of the River and Harbor Act of 1899, known as "The Refuse Act," and other authorities for injunctions against discharges contrary to the heat criteria of the applicable water quality standards, and to restrain construction and operation of power plants which would cause such discharges.

The character of the pollution situation governs the application of the Federal Water Pollution Control Act's authorities and procedures. The Mobile Bay conference of December, 1969, was called under the "shellfish" authority of the Act. Shellfishing areas at Mobile have been closed by the State of Alabama for eight of the past sixteen years. Through this conference, a specific regulatory program for control of municipal and industrial wastes polluting the Bay is being developed.

The Refuse Act, administered by the Secretary of the Army through the Corps of Engineers, extends Federal authority to intermittent discharges of waste into navigable waters and provides a valuable additional enforcement tool. FWQA and the Corps of Engineers coordinate the enforcement of the Refuse Act with the enforcement of the Federal Water Pollution Control Act. Through this coordination and the use of the Refuse Act, regulatory authority can be extended to intrastate waters where no Federal water quality standards apply, as well as to interstate standards violations. The Refuse Act has also been used effectively against "one-time" dumpings of pollutants.

There are limitations in existing enforcement authority which prevent the Federal government from playing a fully effective role. The Federal government may act on its own or at State request to enforce the abatement of pollution which is interstate. In the case of pollution of interstate or navigable waters which occurs only in one State and has its effects only in that State, however, Federal enforcement assistance must be requested by that State. This important distinction results in real complications. Enforcement action on Lake Superior was initiated by the Secretary on his own authority on the basis of interstate pollution which was occurring in tributary border streams. The principal pollution source to Lake Superior, however, was the Reserve Mining Company taconite operations at Silver Bay, Minnesota. To establish enforcement conference jurisdiction over this source, it was necessary to show interstate effects of the pollution from Reserve's operations. If the interstate effect had not been established through FWQA studies, the enforcement conference would have had no jurisdiction over the taconite discharges.

The procedures for enforcement actions also present several limitations on Federal authority. At the conference stage, no direct Federal relation is established with individual polluters. Such parties may not even be compelled to be present at the conference, as no subpoena authority is provided. The Federal authority deals directly with the polluter at the public hearing stage, but, again, there is no subpoena authority to compel the presence of witnesses.

During the post-conference and post-public hearing periods, the States are directed to obtain compliance under their own laws and authorities. The Act directs that a reasonable time, which cannot be less than six months, must be provided to the States for obtaining such compliance. This means that in bringing a recalcitrant polluter to terms, the Federal government's hands initially are tied for at least a whole year. This year stretches to a minimum of 18 months when the time needed to prepare the filing of court action is taken into account.

Despite the acceleration in Federal enforcement activity, deficiencies in the existing legislation have become increasingly apparent. To further strengthen the Federal regulatory role, the Secretary of the Interior has proposed legislative changes in the Act which would provide substantial new authority for FWQA enforcement activities.

Specificially, water quality standards would be strengthened by the addition of effluent requirements and by extending the applicability of these standards to all navigable as well as interstate and certain other waters. These discharge requirements would be established by the States as were the original water quality standards. If the Secretary of the Interior determined that these requirements met the requirements of the Federal Act, they would be enforceable as an element of the Federal, as well as the State standards. The extension of the water quality standards program in terms of more specific requirements and in terms of waters included is a logical progression, building upon the water quality criteria and plans of implementation already in force in all fifty States.

Another significant change would be the extension of geographic coverage of enforcement authority to include all navigable and certain other waters. As has been pointed out, under existing law an enforcement action may not be taken in the absence of an interstate pollution effect without the request of the Governor of the State. Under these circumstances, the availability of Federal enforcement authority depends on the geographic accidents of pollution crossing interstate boundaries. The Administration's proposal would remove the distinction between interstate and intrastate waters and pollutional effect. Federal enforcement authority would be available in any case where the Secretary of the Interior believes water quality standards are being violated or the health or welfare of persons is being endangered.

In addition, the new proposal would extend the coverage of the Act to include the authority to set and enforce standards for groundwaters and for ocean waters beyond the Territorial Sea, two important components of the water environment that need increasing protection.

Furthermore, at the conclusion of an enforcement conference, remedial measures could be required directly of individual polluters. The hearing board phase of enforcement would be eliminated and the government could proceed directly to court enforcement. Fines of up to $10,000 a day for violation of water quality standards or enforcement conference requirements would be authorized. Substantial investigatory authorities would be provided to permit the Secretary to subpoena records and witnesses, to enter and inspect plants and installations and to require testimony. Further, the Secretary would be authorized to request the Attorney General to bring suit under a new injunctive authority to stop waste discharges immediately in cases of serious damages, real or threatened.

Even though the proposed legislation would increase FWQA's regulatory authority, it is intended to back up the enforcement activity of the States, which continue to have primary responsibility. Though at a much accelerated pace and with a much larger scope of enforcement activity, FWQA and the States would continue to work as partners to obtain cleaner waters.

Control of Oil Pollution

With the grounding of the TORREY CANYON in 1967, the breakup of the OCEAN EAGLE in Puerto Rican waters in 1968, and the Santa Barbara offshore oil well leak in 1969, oil pollution has become recognized as a serious national and worldwide problem. These incidents were spectacular in terms of the damages they caused, the control and clean-up efforts and expenditures they necessitated, and the public concern they generated. Of even greater significance, however, is the fact that these major disasters are matched by the aggregate of large and small incidents that occur every day throughout the Nation's coastal and inland waters.

It is estimated that there are annually over 10,000 spills of polluting materials into our Nation's waters. About three-fourths of these spills are oil; the remainder are other hazardous materials, such as chlorine and anhydrous ammonia. The sources of these incidents are vessels, pipelines, rail and highway carriers, land- and water-based storage tanks, refining and other manufacturing operations, the jettisoning of fuel tanks by aircraft, on and offshore petroleum loading and unloading terminals, on and offshore petroleum drilling and production operations, and various other facilities and activities. The problem of accidental spills of oil is further compounded by discharges of oily ballast waters from tankers and other vessels. Pollution from oil and hazardous materials is an everyday occurrence and affects all our waters.

Of particular significance are the potentially large and damaging oil spill accidents that might easily result from the increase in shipping and pipeline transport of oil. The emergence of supertankers as the prime oceanic movers of crude oil imports, the construction of a large pipeline, such as the Trans-Alaska Pipeline System from the new Alaska North Slope oil fields, and the greater development of offshore oil are all contributing factors to the oil spill problem. This rapid increase of oil traffic and the expansion of the offshore production of oil only intensifies the possibility of more frequent and larger accidents and of significantly greater damage to the environment.

Presently, the technology for coping with oil and hazardous materials spills is woefully inadequate. Prevention of accidents is the only sure way of protecting the environment. The Santa Barbara incident and subsequent similar spill situations have shown conclusively that no completely effective techniques are available to control oil spills in the open ocean or lake waters. Wind and wave actions neutralize the effectiveness of oil spill containment devices, such as floating booms. Vacuum or scoop equipment to remove floating oils from the water does not accomplish the job, being effective only in rarely occurring calm seas. Chemical dispersants, sinking agents, and other materials are often ineffective and frequently very toxic to marine and wildlife. Common straw, which soaks up oil so that it can be removed, is still the standard material for fighting and cleaning up oil spills.

Compounding these technological shortcomings, the legal and institutional devices available for handling oil and hazardous material spills have been less than adequate. The Oil Pollution Act of 1924, as amended—the principal Federal legislation in this area of pollution control—prohibited and provided penalties for only the "grossly negligent and willful" spilling or discharging of oils and oily materials. This restrictive legal language essentially precluded enforcement of the Act. This has been rectified by passsage of the Water Quality Improvement Act of 1970, which repeals the 1924 Act and greatly increases the regulatory controls for oil pollution incidents. Many State and local governments, however, are still lacking in oil pollution control authority.

In addition to lack of adequate legal tools, well-organized and well-equipped governmental forces have not always been available to respond in a timely manner to oil pollution incidents. Many of the smaller incidents go undiscovered or ignored by local, State, and Federal agencies; only the larger incidents generally receive the type of response necessary to assure adequate control and clean-up. The usual procedure is to encourage or require the party responsible for the spill to procure the equipment, materials and personnel and to bear the expense of control and clean-up. In some cases, these resources may not be available in the local area, adding yet another problem.

Since the TORREY CANYON incident, and particularly during the aftermath of the Santa Barbara incident, FWQA has played a principal role in organizing and coordinating the Federal, State, and local effort in the control of oil and hazardous materials pollution. This has included development of contingency plans and reporting and response capabilities, pursuit of research and development of new and improved technology, study of potential oil pollution threats—as in the case of proposed exploration and production of oil in Lake Erie—and participating in strengthening of the Federal regulations covering the drilling for and production of oil and gas on the Outer Continental Shelf.

During 1969 and early 1970, the National Multi-Agency Oil and Hazardous Materials Contingency Plan was re-assessed and revisions to strengthen it were undertaken. The first Plan was prepared in 1968, at the request of the President by the Departments of the Interior, Transportation, Defense and Health, Education and Welfare and by the Office of Emergency Preparedness. Together with the supplementary regional contingency plans, the National Plan provides the organizational and communications mechanisms for welding Federal, State and local efforts into a coordinated response to oil and hazardous materials incidents. The Secretary of the Interior has been responsible for the preparation and administration of the National Plan, and FWQA has acted as the lead agency in carrying out this responsibility. The National and regional plans provide for on-scene commanders, operating teams, communication centers, lines of responsibility and other organizational features necessary to bring about an immediate and effective response to major pollution disasters and lesser incidents. The National and regional plans were put into effect during the Santa Barbara incident and proved to be decidedly important in the control and clean-up of that disaster. FWQA is continuing to provide guidance in extending the coverage of contingency plans, particularly in local areas, such as harbor and oil on-loading/off-loading areas, where the threat of oil pollution is greatest. The contingency plans have and are continuing to overcome the institutional shortcomings for coping with spills; and they are becoming increasingly more effective in ensuring that the supply of equipment, materials and other resources, including communications and technical advice, needed to combat oil and hazardous materials accidents becomes immediately available.

In the implementation of the contingency plans in coastal waters, the Great Lakes and the major inland navigable waters, the Coast Guard has provided the on-scene commanders and the principal operating resources, including personnel, ships, equipment and communications systems. FWQA participates by providing advice on containment and clean-up techniques, including the use of dispersants and other chemicals. In other waters of the Nation, FWQA has the lead operating role.

Another important accomplishment during 1969 was the strengthening of the regulation covering the exploration and production of oil and gas on the Outer Continental Shelf. The Secretary of the Interior is authorized to lease

The OCEAN EAGLE breaks up on the rocks, spilling its oily cargo into the sea near Puerto Rico.

lands on the Shelf for oil, gas and mineral extraction and is responsible for regulating these operations, which are in the coastal waters outside of State jurisdiction. The Santa Barbara incident clearly indicated that adequate consideration had not been given to the environmental impact of offshore oil operations. In recognition of this, Secretary Hickel ordered the suspension of pending lease offerings and revisions of the Federal regulations applicable to offshore leasing.

The revisions made call for, among other things, the evaluation of potential environmental effects of offshore oil operations prior to lease offerings. Under this feature, FWQA and other Federal agencies concerned with the protection of marine resources are given the opportunity to assess the impact of offshore oil and gas activities. The Secretary of the Interior is authorized to make appropriate decisions on leasing and lease requirements based upon these recommendations. Other revisions of the regulations pertain to the inclusion of the National Contingency Plan and to lessee's responsibilities for pollution prevention, control and clean-up, for the reporting of spills and for the provision of equipment, materials and resources to cope with pollution incidents. The aim of the Department of the Interior is to assure adequate water and environmental quality protection in its management of the Outer Continental Shelf lands and waters, and the strengthened regulations promulgated by Secretary Hickel are directed toward this objective.

With regard to offshore oil and gas regulations, it is also important to note that Secretary Hickel recently recommended to the Justice Department that a grand jury be convened to investigate the violations of Federal regulations by a lessee off the Louisiana coast. The reported failure to provide storm chokes and other protective features required by the regulations is believed to have led to the oil well fires and the large oil discharge from several wells operated by the Chevron Oil Company.

In the area of research and development, the Federal agencies have divided among themselves the work necessary to find new and improved technology to deal with oil and hazardous materials pollution. FWQA has taken on the primary tasks pertaining to prevention, containment and clean-up in sheltered and inland waters, the fate and ecological effects in these waters, and the technology for cleaning oil contaminated beaches. The Departments of Transportation, Defense and Health, Education and Welfare, as well as other agencies of the Department of the Interior, are assuming primary responsibility for other pertinent areas of research, including the combating of oil pollution in open waters.

FWQA's research activities are being carried out under grants and contracts, as well as through in-house work centered in its laboratory at Edison, New Jersey. One project consists of investigating the use of gelling agents. These could be released into the oil cargo of a tanker to form a semi-solid material when an accident causes a rupture in the vessel. This material either would not leak out of the ruptured tanks or, if released, could more easily be contained and picked up. Other efforts are aimed at developing and demonstrating oil containment and recovery equipment, barrier devices to protect marinas and other water areas from incoming oil slicks, and techniques for cleaning oil from beaches and disposing of the material removed.

In its day-to-day operations, FWQA operates a teletype communications systems covering the Headquarters and Regional Offices to handle reports and information on oil and hazardous materials spills, as well as other emergency situations, such as fish kills. Under the contingency plan, to the extent possible, personnel in Regional and field offices respond to pollution incidents by inspecting and collecting samples and information on the situation, by providing technical advice on control techniques, and by participating in the direction of control activities. In these activities, particularly in coastal waters, FWQA and the Coast Guard and/or the Corps of Engineers work together—each agency performing those tasks which it is best organized and equipped to handle.

Although FWQA has not had the resources to respond to most spill incidents, it has responded to all major episodes. Substantial on-scene effort was put into the Santa Barbara disaster. This was followed by responses to the many serious pollution problems resulting from Hurricane Camille; to the large release of oil from a ruptured storage tank at Seawarren, New Jersey; to a number of oil spill incidents in Alaska, including the recent oil disaster affecting 1,000 miles of shoreline along the coast of Kodiak Island; and to some 130 other incidents, about 40 of which were hazardous materials situations.

Although a considerable amount of attention is devoted to reporting and response activities, a

Oil pollution on the Jackson River in Virginia.

significant effort has been and is directed to other program activities. These include contingency planning; evaluation of potential pollution situations and impacts, including those associated with offshore oil drilling and production; testing of hazardous materials and the neutralizing or combating agents needed to deal with them when a spill occurs; participation in international meetings on oil spill prevention; and technical assistance to State and local agencies and other groups.

Along these lines, several significant actions were undertaken in 1969. The bunker oil from the grounded motorship, NORDMEER, which threatened to rupture and spill its contents into Lake Huron, was removed to prevent a serious incident. This was the first effort of its kind by FWQA.

In the case of the Kodiak Island incident, Secretary Hickel has appealed to ten major oil companies to enter into a voluntary "no discharge" agreement to halt the oil pollution caused by vessels pumping their oily ballast waters into the high seas outside of the 50 mile limit. These areas are not addressed under international controls. Investigations by FWQA have shown that the oil-contaminated ballast waters released by commercial tankers enroute to terminal facilities in Cook Inlet were the most probable cause of the Kodiak Island disaster, which involved the destruction of an estimated 10,000 waterfowl. The discharge of oily ballast waters on the high seas is a frequent source of pollution. Many stretches of shoreline along both coasts are affected by oil believed to have drifted in from offshore ballast water pumping operations and it is the goal of the Department of the Interior to prevent these incidents by proper handling of ballast waters.

Proposed drilling for oil and gas in Lake Erie was studied, and, as a result, recommendations were made to the State of New York and the International Joint Commission opposing oil production and encouraging the strictest regulation of gas production in order to protect the valuable water supply, fishery and other uses of the Lake. Considerable attention has also been devoted to a study of the Alaska North Slope oil development and the Trans-Alaska Pipeline System to assure that adequate consideration for maximum protection of the unspoiled environment is taken in the design, construction and operation of these facilities. Along similar lines, technical assistance was given to the State of Maine in its preparation of comprehensive regulations for the prevention and control of potential pollution in all types of oil operations.

These activities and others were essentially wholly aimed at pollution prevention, a goal which FWQA believes must be ultimately achieved through fail-safe systems and practices if real control of oil pollution is to be attained.

Experts attempt to control one of eight blazing oil wells off the Louisiana coast with dynamite. Until these wells were controlled, about 1,000 barrels of oil a day threatened the Louisiana Coast.

The recent passage of the Water Quality Improvement Act of 1970 substantially strengthens the Federal law and authority to prevent and control oil pollution. Most importantly, this new legislation removes the restrictive definition of illegal spills and discharges and provides notification requirements and substantial penalites and liabilities for oil spills. These features, including the requirement for the showing of financial responsibility—or liability insurance—will promote greater care and effort on the part of the oil and oil transportation industries in the prevention of spills. Other provisions authorize greater effort by the Federal agencies in developing strengthened contingency plans, directing or fully undertaking the containment and clean-up of oil spills and providing a revolving fund to cover the costs of the latter. FWQA recently created the Office of Oil and Hazardous Materials and is expanding its staff to handle the increased work load resulting from the new legislation.

Control of Vessel Wastes

The discharge of wastes from ships, barges, houseboats, pleasure craft and other types of watercraft has been receiving increased attention in the nationwide effort to clean up polluted waterways and preserve clean streams, lakes and coastal waters. Until recently, the effect of vessel wastes has been obscured by the pollution resulting from municipal and industrial waste discharges and other causes. With the progress anticipated in abating municipal and industrial waste discharges, the significant increase in the number of toilet and galley equipped vessels—particularly pleasure craft—plying the Nation's waterways and lakes, and the greater demands for high quality recreation and sport fishery waters in those areas most used by both commercial and non-commercial watercraft, vessel wastes have emerged as significant source of water quality impairment. Accordingly, vessel waste discharges are currently a concern in the navigable waters of this country, including even mountain lakes where the intensity of vessel use is relatively low but the need for the protection of the high quality water is great.

In June 1969, FWQA completed a report of its San Diego Bay Vessel Pollution Study Project following intensive field and laboratory activity. The purpose of this project was to determine the magnitude, extent and kinds of pollutional effects to be expected from the discharges of shipboard sanitary wastes and the pollution abatement measures required to reduce or eliminate these discharges. The findings were illustrative of this problem: vessel waste discharges were found to cause serious bacterial pollution, to be responsible for bottom sludge deposits and floating waste material and to cause violations of the water quality standards established for San Diego Bay. The pollution was directly attributable to the high numbers of military, commercial and pleasure vessels using the Bay.

Investigations by State agencies and FWQA have discovered similar conditions in other bodies of water across the United States. Bacterial pollution and the attendant impairment of recreational water uses are the principal adverse effects of untreated vessel waste discharges, but the occurrence of aesthetically displeasing floating material follows close behind in pollutional importance.

It will not be an easy task to remedy vessel waste pollution. The weight and volume of waste treatment devices or waste handling tanks cause considerable installation problems, particularly on existing vessels, especially if they are military. The expense of control devices, particularly to pleasure craft owners, is also a factor. A considerable amount of research and development is underway by Federal agencies including FWQA, the Navy and the Coast Guard to find adequate and adaptable waste control systems. Consideration is being given to incineration devices, modified versions of conventional waste treatment methods, recirculation systems, chemical-toilets—such as are used on commercial aircraft—and other devices. Good progress is being made, and there appears to be little doubt that American ingenuity can and will develop the technology required to adequately handle vessel waste pollution problems.

Within recent years, many of the States have enacted or strengthened their legislation or regulations pertaining to the control of vessel

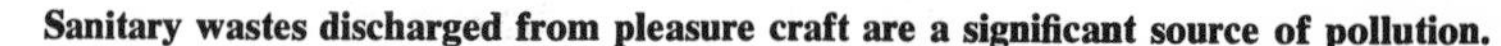
Sanitary wastes discharged from pleasure craft are a significant source of pollution.

wastes. Unfortunately, the non-uniformity of the waste treatment and control requirements imposed by these States has presented some significant compliance problems for vessels which travel between States. Also, in many cases the State regulations do not apply to or are ineffective in their coverage of interstate and international carriers and Federal vessels. In response to these basic problems, the Congress recently enacted comprehensive Federal legislation—the first legislation of this type—covering the control of vessel wastes.

The Water Quality Improvement Act of 1970 provides for the establishment of performance or effluent standards for the sanitary waste discharges from all classes of watercraft. These standards are to be set by the Secretary of the Interior. The amendment further provides for the establishment and enforcement of regulations to implement these standards by the Secretary of Transportation, under whose administration the Coast Guard comes. This Federal statute applies to new and existing vessels and provides for penalties for the failure of vessel owners and manufacturers to provide adequate shipboard treatment or control of sanitary wastes. Importantly, this new legislation provides for uniform, nationwide regulation of watercraft waste discharges. This will promote a comprehensive attack on vessel pollution problems by FWQA and the Coast Guard, who will join in carrying out this task.

During the past year, FWQA has been preparing for its role under the new legislation. Research, development and demonstration of vessel waste treatment devices have been pursued and considerable assistance has been given to other Federal agencies, including the Navy, the Corps of Engineers and the Coast Guard, in the development, testing and installation of treatment and control equipment on Federal vessels. With the eactment of the new legislation, FWQA's activities in the vessel wastes area will be expanded. FWQA is planning to consult with the boating industry, the manufacturers, and others concerned with treatment devices and will hold public hearings prior to the establishment of standards. In addition, assistance will be given to the Coast Guard in establishing both the regulations necessary to implement the performance standards and an adequate certification program. Finally, assistance will be provided Federal agencies in equipping Federal vessels with adequate control equipment. The new legislation provides the means to fully abate the pollution arising from watercraft sanitary wastes, and FWQA plans to move rapidly forward to meet this objective.

Control of Pollution from Federal Activities

The Federal government is involved in many activities which have an impact on the quality of our Nation's waters. These operations include the maintenance of Federal facilities, such as military bases, lighthouses and post offices; management of Federal lands; and diverse activities, such as dredging, nuclear energy development, and pest control. Today, in the United States, there are approximately 20,000 Federal real properties, many of which have an impact on the environment. In addition, Federal lands comprise one-third of the United States, and the use of these lands has a bearing on progress in achieving national goals of clean water and a quality environment.

Abatement and prevention of pollution from these sources is a major Administration goal. On February 4, 1970, the President issued Executive Order 11507, establishing a new and aggressive approach to the problem of keeping the Federal house clean. The Order superceded earlier Executive Orders on water and air pollution control.

In issuing this Order, the President gave more specific direction to Federal agencies in the conduct of their activities with regard to environmental protection than had any previous Order. To establish the Federal government as a true leader in the battle to save the environment, he required that all projects or installations owned by or leased to the Federal government would have to be designed, operated and maintained so as to conform with water and air quality standards. For the first time, a conformance date for Federal compliance, December 31, 1972, was established and written into the Order. The Presidential statement accompanying the Order set forth a $359 million program for obtaining this objective. To insure that these funds, once appropriated, were utilized for the purposes intended, the Order contained a section which, in effect, prevented use of the appropriated funds for purposes other than pollution control.

FWQA has an important role to play in working with the other Federal agencies concerned to assure that the objectives of the Executive Order are met. FWQA has primary responsibility for reviewing and approving permissible limits of waste discharges from such installations and for coordinating the water pollution control activities of Federal, State, and local programs. The new order contains important provisions to insure this role will be an effective one and to correct some of the administrative problems brought about by earlier Orders. Rather than have professional staff at all levels of government review plans and specifications for improved abatement facilities, the Order requires that specific performance requirements for each facility be set by the agency and approved by the Secretary of the Interior. In evaluating the adequacy of the performance requirements, the Secretary is to take into consideration water quality standards where such standards exist. The Secretary is also given, for the first time, the authority to issue regulations establishing water quality standards for the purposes of the Order where such do not exist. More importantly, the Secretary is also authorized to establish more stringent requirements for Federal facilities than contained in existing standards. Both of these actions are to be taken after consultation with appropriate Federal, State, interstate and local agencies.

FWQA has taken a number of steps to meet these and related responsibilities. The staff assigned to work with the other Federal agencies has been restructured and enlarged. Increased emphasis has been placed on better channels of communication and cooperative relationships with the other Federal agencies. Fruitful meetings and seminars have been held at which Federal programs have been reviewed, information exchanged, and advice both sought and given.

FWQA conducts on-site inspections of waste-water treatment and disposal practices at Federal installations to advise the agencies concerned as to the adequacy and effectiveness of such measures. This information is used by agency planners to develop and update plans for corrective actions. Whenever possible, these inspections are conducted jointly with State officials to promote better Federal-State relationships.

From the information collected on such inspections, a system of recording and reporting information on Federal installations and their waste treatment needs and accomplishments was developed in 1969. This system will be the basis for a comprehensive inventory of Federal installations, which will streamline the review process and provide better information on which to recommend nationwide priorities to the Bureau of the Budget and Congress.

To facilitate budgeting for corrective measures, Federal agencies are required to present to the Bureau of the Budget a plan for installing improvements needed to meet the target date. FWQA reviews the agencies' plans and recommends priorities for funding to the Bureau of the Budget. Each project is ranked in the order of its priority to ensure that the most significant problems will receive first attention.

Emphasis has been placed on conferences to ensure that information on improvements in waste treatment technology would be available to Federal agencies. In this regard, a seminar was held for representatives of other agencies on new advances in waste treatment technology and was geared to problems routinely faced at Federal installations. Attendance of agency personnel at seminars conducted by FWQA's Research and Development program has been encouraged. A field trip was arranged for officials of the Department of Defense in order to familiarize them with the new treatment technology being developed at the Blue Plains sewage treatment plant in Washington. Reports of completed FWQA research projects are being made available to the appropriate Federal agencies for their consideration in the development of new facilities, and incorporation of these newly developed techniques in remedial work is being highly encouraged.

Correction of conventional municipal and industrial waste problems from Federal facilities is only a part of the job in ensuring that the wide-ranging activities of the Federal establishment have a minimum impact on the environment. New opportunities for pollution abatement are continually being brought to the attention of other agencies. As the wastes from conventional point sources are brought under control or eliminated, the wastes from nonpoint sources come to the forefront as significant problems.

One such area receiving recent attention was related to management practices on Federal lands. In the past year FWQA chaired a Department of the Interior task force established to assess the effect of Federal land management practices on water quality. A pilot review study conducted in Oregon showed a major need and opportunity to reduce water pollution associated with Federal land management practices and conservation measures. The report, *Federal Land Management Practices and Water Quality Control*, found serious damage to the environment stemming from long-established practices, as well as from more recent practices involving pesticides, fertilizers, and other chemical applications. The report specifically identified 12 kinds of land management practices and 22 conservation measures having an impact on water quality. These would be reviewed by agencies and altered whenever necessary to conform with national environmental goals.

Operation Plowshare, the Atomic Energy

Commission's program to develop peaceful uses of atomic energy, represents another activity which must be carefully monitored and controlled to avoid unwanted effects on the environment. This program has and will involve nuclear explosions designed to stimulate gas production in oil and gas bearing formations, to fracture mineral formations to enable extraction by leaching, to develop storage for water or other materials. To assure that the program, as planned, provides adequate safeguards for water quality, FWQA provides review and advice to the Commission concerning these experiments. Careful planning of the program, as well as pre- and post-detonation surveillance, is essential because of the potentially great hazards involved.

The Corps of Engineers' dredging activities in the Great Lakes and elsewhere are yet another cause for concern. For more than 100 years the Corps of Engineers has been dredging material from the harbors of the Great Lakes and depositing most of the dredged material in designated dumping areas in the open waters. Growing concern over the resulting effect on the Lakes led to completion last year of a Corps of Engineers' pilot program related to dredging and water quality problems in the Great Lakes. Among the conclusions of the Corps' study were that heavily polluted sediments when transported to the open waters must be considered presumptively undesirable because of their possible long-term effects on the ecology of the Great Lakes, as evidenced by bio-assays of the effects on bottom organisms and plankton, and that disposal in diked areas would be the least costly effective method of withholding pollutants associated with dredgings from the Lakes.

On April 15, the President sent a message to the Congress, proposing legislation to discontinue open water disposal of polluted dredge spoil in the Great Lakes. The legislation would authorize the Corps to construct and maintain contained disposal facilities, in cooperation with States and other non-Federal interests. Dredge spoils from Federal and non-Federal operations would be disposed of in these enclosed areas under appropriate cost-sharing arrangements.

We also must be increasingly alert to the environmental impact of such diverse activities as Forest Service timber sales in Alaska, use of persistent pesticides for quarantine control at Federal airports, and proposed development of oil shale lands in Colorado, Wyoming, and Utah. FWQA will place increasing emphasis on working with the agencies concerned to correct deficiencies and to prevent environmental problems from arising in the future.

Dredging, often necessary to keep navigation channels open, is a source of pollution when spoils are dumped in open waters.

Control of Pollution from Federally Licensed and Supported Activities

Closely related to pollution resulting from direct Federal activities, is the environmental impact of the various functions conducted under loans, grants, contracts, leases and permits from the Federal government. These diverse activities range from the nuclear power plants receiving licenses from the Atomic Energy Commission to urban renewal projects financed by the Department of Housing and Urban Development. Combined, these Federally supported and licensed activites constitute a real and potential threat to the environment, which cuts across the full spectrum of the Nation's economic life. They also reflect an unusual opportunity for the Federal government to extend the exercise of its responsibilities for pollution control.

Two landmark pieces of legislation and an implementing Executive Order promise effective action. The National Environmental Policy Act of 1969 called for all agencies of the Federal government to give full attention to environmental protection in their planning activities and decision making. In furtherance of this legislation, the President issued an Executive Order on March 5, 1970. This Order directed the heads of all Federal agencies to review their statutory authority, administrative regulations, policies and procedures, including those relating to loans, grants, contracts, leases, licenses or permits, in order that they might identify deficiencies and inconsistencies which keep each agency from full compliance with the national

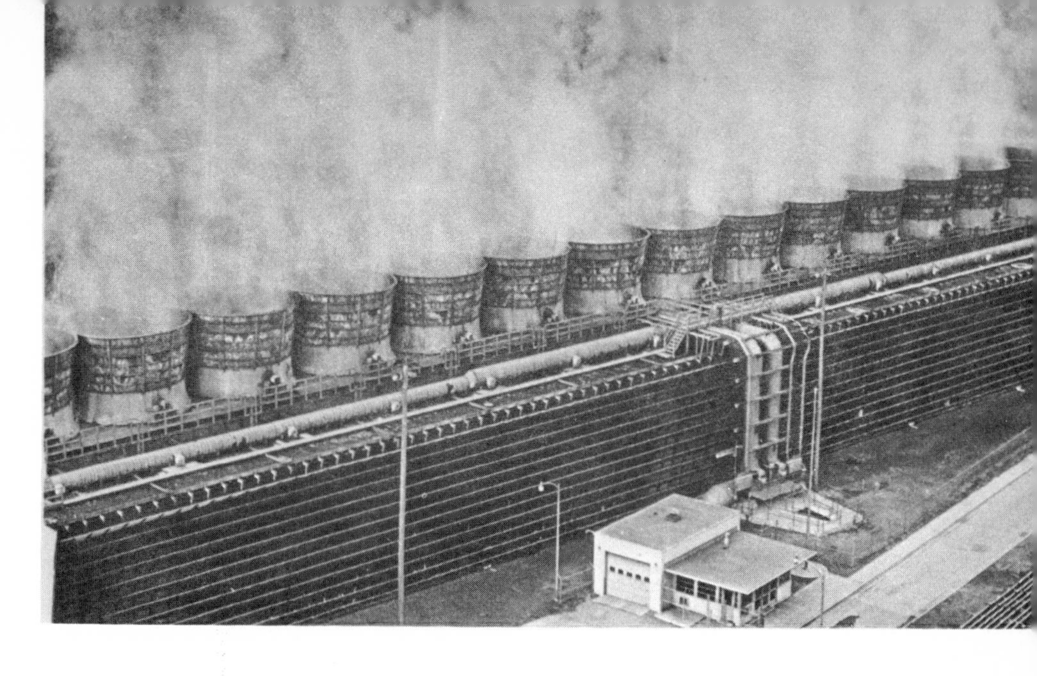

Cooling towers are used to dissipate huge quantities of heat produced by steam electric power plants.

environmental goals established by the Act. The Order requires a report to the Council on Environmental Quality on the results of this review along with corrective actions taken and planned.

Recent enactment of the Water Quality Improvement Act of 1970 gave further impetus to this trend. The Act provides that any applicant for a Federal permit or license to construct or operate any facility which may result in any discharge into the navigable waters of the United States shall provide certification from the State in which the discharge originates that such facilities or related activities can be expected to comply with applicable water quality standards. The Act further provides that no license or permit shall be granted without such certification and such conditions as the State may reasonably require, including but not limited to provision for suspension or termination of any issued license or permit for failure to be in compliance with applicable water quality standards. It also provides special conditions under which the views of an adjacent State will be obtained; or an interstate agency or the Secretary of the Interior, if appropriate, may provide the certification.

The legislation is clear in its intent that the States are to exercise primary responsibility for the administration of the water quality standards for their waters and for the assurance that State-Federal water quality standards are met by anyone who uses these waters, and that FWQA is to cooperate with other Federal agencies, with State and interstate agencies, and with water users in assuring that appropriate control measures are applied to meet the water quality standards. The legislation provides that the Secretary of the Interior shall provide, upon the request of any Federal department or agency, or State or interstate agency or applicant, any relevant information on applicable water quality standards and comment on any methods of complying with such standards.

The major and most significant activities to receive immediate attention under this legislation are those of the Atomic Energy Commission, which issues construction permits and operating licenses for nuclear power plants; those of the Federal Power Commission which licenses hydroelectric power plants and whose approval must be secured before changes can be made in those projects, including use of project waters and construction across project lands; and those of the Corps of Engineers which issue permits for dredging and construction in the navigable waters of the United States (except where hydroelectric power production is contemplated and licensed by the Federal Power Commission).

Prior to the enactment of recent legislation and the issuance of the Executive Order, cooperative arrangements had been made with the Atomic Energy Commission, the Federal Power Commission and the Corps of Engineers to review materials submitted in request of Federal permits or licenses for activities which could result in water pollution. These reviews have been conducted in coordination with other Department of the Interior agencies concerned with environmental protection. FWQA has reviewed these applications to determine the possible effects of the activity, as proposed, upon water quality. Recommendations have been made as to the need for additional control facilities and any provisions which should be included in the permit or license to ensure that water pollution would be controlled. These activities have led to an increasingly well-coordinated and cooperative effort to ensure that water pollution control measures are considered in connection with the issuance of a Federal license.

These arrangements have been satisfactory, however, only in part. For example, there has been a serious inadequacy in procedures for review of environmental factors in design and site selection for new fossil fueled or nuclear power plants.

With respect to nuclear plants, the Atomic Energy Commission receives comments on environmental factors from the Department of the Interior in accordance with established administrative procedures. These comments are forwarded to the applicants for consideration. The Atomic Energy Commission, however, has held that it lacks regulatory authority to incorporate in its licenses for nuclear plants requirements for measures to protect the environment beyond radiation safety hazards. This position has been supported by the Department of Justice and also affirmed in a court decision.

Fossil fueled plants are licensed by State regulatory authorities and require no Federal license whatever. With public concern about the environmental impact of power developments running high, a number of utilities have entered into voluntary discussions of projects under consideration with concerned State and Federal agencies. There has been, however, little or no opportunity for the Department of the Interior to require environmental protection measures in the plans for power plants, both nuclear and fossil fueled, unless they used water from the reservoir of a licensed hydroelectric project.

By contrast, there have been adequate procedures for environmental review in the category of hydroelectric power plants. Over the years, the Federal Power Commission prior to issuing a license for the construction of hydro plants has increasingly incorporated environmental protection requirements. These have included, for example, minimum flows for fisheries and water quality below licensed dams, fish screens and spawning channels, and the making available of project lands for public recreation.

Many proposals for incorporating these measures come from Department of the Interior agencies. FWQA has the opportunity to review license applications made to the Federal Power Commission and to propose changes in construction and operation plans on behalf of water quality improvement. It has received excellent cooperation from the Federal Power Commission in incorporating recommended measures in its licensing procedures.

A prime example of the application of this policy is the Blue Ridge case on West Virginia's Kanawha River. Although this case is still pending before the Federal Power Commission, the preliminary finding provided for the development of a project which would require the power company to provide flows for maintenance of water quality in the downstream reaches of the Kanawha River.

The activities involving dredging and construction in navigable waters of the United States and requiring permits from the Corps of Engineers constitute another category of pollution. The discharge of dredged materials into the Great Lakes by private dredgers is directly comparable in effect to the discharge of dredgings from Corps operations. This illustrates the importance of applying the same stringent environmental controls to Federally licensed activities as to the Federal agencies themselves.

FWQA and the other Interior agencies concerned review thousands of applications for such permits annually. Comments to the Corps of Engineers have resulted in inclusion of provisions to protect water quality in some permits and in the withholding of other permits. However, major difficulties have remained. The inclusion of specific provisions relative to control of pollution in Corps of Engineers' permits has been contested in the courts. A lower court decision that the Corps of Engineers is not authorized to include such restrictions in its permits is being contested by the Corps of Engineers.

Enactment of the Water Quality Improvement Act of 1970 represents a major improvement in procedures and methods. The Act's emphasis on compliance with water quality standards as the basic mechanism for ensuring water quality protection is of great significance. Nevertheless to adequately ensure the effectiveness of these new requirements, FWQA must place continued emphasis on development of adequate standards. At present, there are no standards that adequately ensure protection of water quality from the impacts of dredging, and the temperature standards of many States remain unimproved. In order to provide effective implementation of the Water Quality Improvement Act, within the concepts outlined by the Congress, FWQA must and will accelerate its efforts to obtain adequate water quality standards.

ASSISTANCE PROGRAMS

From the very start of the water pollution control program, the Congress has made it clear that the responsibility for preventing and controlling water pollution begins at the State and local levels. And, although the Federal government has been given an increasingly greater hand in dealing with the problem, the States and communities continue to bear a major share of the responsibility.

The job of controlling pollution, as indicated earlier, is an enormous one both in terms of costs and in terms of manpower requirements. Few, if any, State and local governments have revenues large enough to meet the many and increasing demands, including water pollution control, confronting them. The largest share of the Federal program's resources are therefore spent for direct assistance to States and communities—grants for treatment plant construction and program development, technical assistance, and manpower development—to help meet the national goal of clean water.

President Nixon has proposed in his program of "New Federalism" that State and local governments play an increasingly important role in meeting national needs. At the same time, he has recognized the need for vigorous Federal leadership, through solid backup of State and local actions, in restoring the environment.

To ensure more effective working relationships, the President has directed nine Federal departments and agencies to work together to modernize the management of their presently complex systems of providing financial and technical assistance to State and local governments. The Department of the Interior is working to implement the objectives and goals of the Federal Assistance Review (FAR) program. One of the primary objectives sought is the simplification of the Department's grant programs —streamlining of the application process and organizational structure of assistance programs for efficiency, economy, and responsiveness to State and local needs. The Federal Water Quality Administration (FWQA) has responded to the challenge. A detailed analysis of the administrative requirements of the Construction Grants and the State and Interstate Pollution Control Grants programs is currently underway.

Secretary Hickel has also stressed the need to improve Federal working relationships with the States. In order to improve communications with States, FWQA representatives are attending public meetings of the State water pollution control boards and other appropriate meetings, such as those of legislative committees. Technical assistance is also being increased to make more of an effort to meet State needs within available resources. Increasing emphasis will be placed on coordinating State and Federal program planning to ensure the most effective pooling of resources.

Assistance to Municipalities

Rapid growth of population and its continuous trend toward urban centers has resulted in a tremendous increase in the volume of municipal wastes and in the need for an enormous investment in waste treatment facilities. National attention was focused on this problem in 1956, when the Congress, in the first permanent Federal Water Pollution Control Act, initiated the program to provide Federal grant assistance to communities to improve or build sewage treatment facilities. Amendments since that time have helped step up construction activity by making more money available and on a more liberal basis.

Under today's legislation, a community can get financial help in the construction of a municipal waste treatment plant with a Federal grant of at least 30 percent of the construction cost. Under certain conditions, such as matching State financial aid, approved water quality standards, and a comprehensive plan for approaching the problem, the Federal share may be much higher.

Since 1957, the Federal government has provided nearly $1.5 billion for construction and expansion of over 10,000 municipally owned and operated sewage treatment facilities. These funds have assisted the States and communities in the construction of $6.4 billion of treatment works.

In the thirteen years in which such grants have been available, the population served by some degree of waste treatment has increased by more than 51 million persons. More than 92 percent of the population served by sewers is connected to a waste treatment plant, as contrasted with 57 percent in 1956. These represent significant accomplishments.

Despite this progress, the Nation still lags far behind in providing modern waste treatment for its cities. Many of the works constructed were designed to provide levels of treatment which subsequently have proved inadequate to protect receiving waters. Other works have become overloaded and need major expansion. Improper operation and maintenance of many of these plants has resulted in discharge of wastes little reduced in polluting content and in breakdown and early obsolescence of facilities. Other plants have been poorly located and have resulted in fragmented, rather than systematic, regional solutions. Population growth has added additional needs; during the same years that the construction grants program was underway, the population connected to sewers for which treatment must be provided increased by 37.5 million persons. Increasing standards of living and the rising use of household chemicals and appliances, such as garbage grinders, have added an additional dimension. In many river basins, progress in treating the wastes from some of the communities has been offset by failure to deal with other waste sources.

Construction needs have far outpaced Federal, State and local funds and there have been recent efforts to increase available funds. A number of States have enacted measures to financially assist their communities. At the Federal level, the Congress this year appropriated a record $800 million to finance the Federal share of doing the job. It will not be enough, however, to merely provide additional funds under existing formulas and methods. A number of basic improvements are needed.

The FWQA has become increasing aware that major revisions in this key program—its legislative structure, funding, regulations, and administration—are necessary if the nationwide goals of providing adequate waste treatment and meeting water quality standards are to be accomplished efficiently and in the near future. A major objective over the past year has been to review the program in depth to determine what changes were needed. The General Accounting Office has also had the program under review and has made a number of recommendations for improvement.

Our review contributed to the formulation of the proposed new legislation and regulations to administer the program on a more systematic basis. These are an essential element of the Administration's environmental program. This review clearly indicated that there were three basic objectives which should be met to achieve an equitable and fully effective Federal financing program. First, the level of financing should be adequate to enable the Nation's communities to get abreast of their pollution problems. Second, the method of financing should be an assured one, in order to enable State and local governments and the construction industry to plan and gear up for the necessary effort. Third, the program must be designed to ensure that the funds will be spent efficiently to achieve the best results in cleaning up our waters.

Treatment facilities such as this one are built with the help of Federal construction grants.

The legislation proposed to the Congress by Secretary Hickel is designed to provide funds adequate to do the job. The legislation calls for a four-year Federal contribution of $4 billion in a construction program of $10 billion, the Federal share to be matched by $6 billion in State and local funds.

This is based on the determination, through FWQA's recently completed cost studies, that a $10 billion investment in waste treatment facilities is needed to meet the country's municipal waste treatment needs in the years immediately ahead. Although these cost studies, the most comprehensive ever completed, indicate that $10 billion will be enough, President Nixon has said more money would be available if necessary. The proposed legislation provides for a reassessment in 1974 to evaluate needs for the following five years. The legislative proposal would also revise the present method of allocating grants to permit a higher degree of flexibility in directing funds to areas where the need is greatest and where they can be most effectively used.

The proposed legislation also stresses measures to provide assurance to States and communities that Federal funding will be forthcoming as planned. The lag between Federal authorization and appropriations in the present legislation created a condition of confusion and uncertainty that has hampered the engineering and construction industry from gearing up for a sustained level of effort.

Ensured funding is a key component of the

Aeration tanks and trickling filters are components of municipal waste treatment processes which reduce organic loadings to streams.

proposed legislation; it would enable the Federal government to enter "grant agreements" with municipalities at the rate of $1 billion a year for four years. Pursuant to these agreements, the Federal government would be obliged to appropriate funds to satisfy obligations under these grant agreements, just as the Federal government must satisfy any other of its debt obligations. This change would assure communities of full Federal support and allow planning and construction to proceed without the traditional gap between funds authorized and funds appropriated.

The Administration has further emphasized its intent to provide assurance of funding and to alleviate State and local uncertainty by resolving the reimbursement issue. To permit States and communities to move ahead with construction of waste treatment works before full Federal funding became available, the 1966 amendments to the Act provided that the allotments of a State could be used for reimbursement of projects which went ahead with less than the full Federal share and on which construction was initiated after June 30, 1966, provided that such projects met all other Federal requirements. As a result of this provision, a number of the States went ahead with bond issues or other provisions for prefinancing the Federal share on those projects which proceeded with either no Federal funds or less than the full Federal share. As of December 31, 1969, a total of 880 such projects had proceeded. The amounts earned for Federal reimbursement were $322 million. When all these projects are completed, eligible reimbursements will be about $814 million.

Federal intentions with respect to repayment of these funds has been one of the vexing problems facing States and communities which had moved ahead on their own. In addressing the State Governors' Conference in Washington this February, President Nixon expressed his position regarding repayment of these funds. He stated that, "any State that went forward after the Clean Water Restoration Act of 1966 relying on what the Federal government had indicated, went forward in its own program, should not be penalized because it took that initiative. As a matter of fact, it should be rewarded."

Under the proposed legislation, reimbursement would be accomplished through the larger appropriations; through improvements in the reallotment procedure which would more quickly funnel funds to areas of greatest need; and through use of discretionary authority, which would permit the Secretary to assign a portion of each year's available funds to such areas.

An additional element of the Administration's program will help assure that State and local bodies will be able to borrow the necessary funds to do their share. The Department of the Interior's proposed legislation would be supplemented by a Treasury Department proposal to establish an Environmental Financing Authority (EFA). EFA would have authority to buy the waste treatment bonds of those municipalities who are unable to sell their bonds on the open market. EFA would ensure the availability of local financing for construction of waste treatment plants, so all communities would be able to participate in the construction grants program.

Higher appropriations and revised legislation are only part of the answer in accelerating the systematic construction of municipal waste treatment plants to achieve effective results in cleaning up pollution. Development of measures to ensure efficient use of funds to achieve that result has been a key element of FWQA's new approach to the administration of the program.

Secretary Hickel has said, "The job ahead will be costly. We want to ensure that the Federal funds invested in the clean-up will be spent effectively and fairly." Towards this end he has published proposed regulations in the Federal Register. The proposed new rules are that:

—Comprehensive river basinwide programs for pollution abatement must be developed, and new treatment works must fit in with such programs, as well as with metropolitan and regional plans, to be eligible for Federal aid.

—In evaluating new applications, the FWQA may demand detailed data on all sources of pollution in the entire river basin, including the volume of discharge from each source, character of effluent, present treatment, water quality effect and other items.

—If some industrial wastes are to be treated as part of a municipal system's operations, industry must pretreat those wastes if they would interfere with efficient operation of the community system. Further, a system of "cost recovery" must be required if some industrial wastes are to be treated in a new plant built with Federal aid. Such cost recovery by the municipality would assess the industries a share of the operating costs and costs of amortizing the debt, in proportion to their contributions to the cost of waste treatment.

—State water pollution control agencies must inspect new Federally-aided facilities for efficiency and economy at least once each year for the first three years of operation and periodically thereafter under standards set by FWQA.

—Design of any new Federally-aided treatment plant would have to be approved in advance as being economical, efficient, and effective under FWQA requirements.

Adequately treated sewage effluent pours into a river without harmful effects.

In addition to these changes in the substantive elements of the construction grants program, FWQA has established a study project to review grant procedures and to determine any changes necessary to streamline those procedures to assure efficient and effective grant administration. A task force including management consultants and FWQA personnel is in the process of preparing a report concerning needed improvements.

All together, the proposed legislation, the amended regulations and the continued efforts to streamline administrative procedures will result in an overall improvement of the construction grants program and will provide financial assistance to the Nation's communities, which will be fully adequate to meet the needs of the years ahead.

Assistance to Industry

With the acceleration of the Nation's cleanup program, industries are faced with major pollution control expenditures. Although there are no specific Federal assistance programs directly geared to provide funds for industrial waste treatment equivalent to the Federal assistance for construction of municipal treatment plants, there are several Federal incentive programs which provide encouragement and support for industries to meet their treatment requirements.

FWQA is encouraging and supporting the treatment of industrial wastes in municipal treatment plants; municipal systems designed to receive industrial wastes are eligible for support under the construction grants program.

The practice of treating industrial wastes in municipal treatment plants has a number of advantages. First and foremost, it provides for more effective pollution control by encouraging regionalization of the waste treatment system. A community that maintains effective treatment of its sanitary wastes can still be a polluter if industrial waste discharges from its borders are uncontrolled. Joint treatment is effective too because it locates responsibility for operation and maintenance within a single authority. In addition, complementary characteristics of sewage and industrial wastes, if properly controlled, can often permit more effective waste reduction within the plant.

Meat packing wastes discharged untreated into a river.

Joint treatment facilities offer significant advantages to both communities and industries in terms of lower treatment costs through economies of scale. The inclusion of industrial wastes in municipal plants also offers special incentives to industry, as these joint facilities can be built with the help of a Federal construction grant. Industry thus can pay for its waste treatment through operating costs, rather than having to make the extensive capital investment involved in the construction of treatment facilities.

Joint treatment of municipal and industrial wastes is increasing, as is the development of technology to handle a variety of complex wastes. For example, metropolitan Seattle has adopted an ambitious program to provide treatment for all liquid wastes that occur within its extended area of jurisdiction. More and more communities are designing their facilities to accommodate a larger portion of the total waste load that is produced by factories, with the cost of construction shared by the community, industry and the Federal, and sometimes State, government.

At the same time, as part of the overall reform of construction grants requirements, FWQA is moving to eliminate certain abuses of joint treatment and to ensure that municipal and industrial systems will operate effectively. First, through pretreatment requirements in the new regulations, the discharge of wastes which would make municipal systems nonoperative or reduce their effectiveness will be controlled. Second, industries are required to reimburse the municipality concerned for the added cost which treatment of their wastes imposes; this will ensure that the municipality will have sufficient revenues to provide adequate waste treatment on a continuing basis.

FWQA also provides assistance to industry through its research and development program. Since 1966, the Federal Water Pollution Control Act, as amended, has authorized grant support of industrial demonstration projects aimed at improving waste management. Although not intended as a direct form of assistance in defraying the costs of constructing waste treatment plants, projects supported by these grants have demonstrated methods of treating industrial wastes more economically and of recovering certain portions of wastes for reuse. Other grants have been used to show the feasibility of joint treatment of municipal and industrial wastes.

Tax write-offs provide further assistance to industry. Although a number of States have enacted tax measures designed to encourage industrial waste treatment facilities, until recently there was no comparable measure in effect at the Federal level. In enacting the Tax Reform Act of 1969, however, the Congress included provisions for accelerated amortization of air and water pollution control facilities for Federal income tax purposes.

Under this law, a taxpayer is entitled to a deduction with respect to the amortization of certified air and water pollution control facilities. A certified pollution control facility is defined as a new, identifiable treatment facility which is used, in connection with a plant or other property in operation before January 1, 1969, to abate or control water or atmospheric pollution or contamination by removing, altering, disposing, or storing of pollutants, contaminants, wastes, or heat . . . and which, in the case of water pollution control facilities, is certified by the State water pollution control agency as meeting State water pollution control requirements and by the Secretary of the Interior as meeting Federal water pollution control requirements. The Secretary may not certify facilities to the extent that the cost of such a facility will be recovered over its useful life.

Regulations are being prepared by the Department of the Interior and the Department of Health, Education, and Welfare in consultation with the Treasury Department to implement the Federal certifying responsibilities.

Assistance to State and Interstate Programs

State agencies are the first line of defense in the national water pollution control effort. Many States have been able to strengthen their pollution control programs to meet the growing problems thrust on them in the past several years. Others, however, have not had adequate laws and resources to do the job. Federal program grants are available to State and interstate agencies to help them bear the costs of needed preventive and control measures. These grants are intended as realistic incentives for the State and interstate agencies to expand and improve their programs.

The program started in 1957 with an annual authorization of $2 million. The annual figure has grown to $10 million today, and the State and interstate agency expenditures have increased more than six times during that same timeframe. Many of the States have substantially strengthened their programs. Funds have been used for employing needed technical personnel, for purchasing special laboratory and field equipment, for waste treatment plant inspection programs, for more aggressive enforcement of State laws, for expanded monitoring and surveillance programs, and for training.

Many States improved their programs in the last year by passing new laws or strengthening existing authorities to provide for a more vigorous clean-up effort. For example, in Oregon, water pollution control became part of a newly created Department of Environmental Quality. A feature of this new Department is its ability to conduct an extremely successful enforcement program. With this new authority, Oregon is carrying on an aggressive abatement program for both industrial and municipal waste sources.

Also in the Pacific Northwest, the State of Washington's 1969 Legislature inserted a requirement that after July, 1974, no applicant can receive a Federal construction grant unless the project conforms to a comprehensive drainage basin plan for water pollution control. This requirement places a burden of urgency upon the State to give planning a very high priority. This change is consistent in purpose with the Secretary's recent proposals for a more systematic and comprehensive administration of the construction grants program.

In Connecticut, legislation enacted during FY 1969 furthered the Connecticut Water Resources Commission's leadership role in several ways:

1. Bonding authorization for pollution control facilities, including pre-financing of Federal grants, was raised from $100 million to $250 million.

2. To promote regionalization, the Commission is authorized to issue orders to polluters jointly after a determination that such pollution can best be abated by the action of two or more adjacent municipalities.

3. New statutes were enacted, covering all phases of oil pollution removal and prevention and containing a provision for strict liability on spillage.

Two other highlights of State accomplishments in recent years are found in New York and Pennsylvania. New York provides reimbursement to municipalities for 1/3 of the cost of operation and maintenance of sewage treatment plants when they are operated according to established standards. Every municipal sewage treatment plant is comprehensively inspected at least once each year by a sanitary engineer and a chemist to evaluate operation and maintenance and laboratory work and to determine quality parameters for raw waste, treated effluent and receiving waters. In Pennsylvania, the Department of Health regulates and administers annual payments to municipalities of 2% of construction costs toward operation of sewage treatment facilities. About 685 applications are now being processed for payment of approximately $8.1 million in 1969. Payment is toward the operation, maintenance, repairs, replacements, and other expenses relating to sewage treatment plants.

New York and Pennsylvania are two of the States that have been fortunate enough to have the resources to support their water pollution control programs. In many States this is not the case, and it is perhaps here that the impact of Federal program grants is most significant.

As a basis for receiving a Federal program grant, each State and eligible interstate agency must prepare a plan describing how the grant will be used to strengthen its pollution control program. To assure the most effective utilization of these funds, the Federal Water Quality Administration (FWQA) has developed guidelines which set forth the essential elements of an effective State and interstate program plan.

In addition to constituting a request for grant assistance, the Program Plan serves several other important purposes.

1. It provides the State's annual report on progress in implementing water quality standards.

2. It provides information essential to FWQA in developing assistance and coordinating other grants to the State or interstate agency under other provisions of the Federal Act.

3. It identifies and discusses problems and issues in extending or improving the State or interstate agency's water pollution control program and helps in evaluation and program planning.

The FWQA has worked closely with the States in this planning process. Several Regional Offices have, in response to State requests, initiated joint review and evaluation studies of individual State programs. Last year such studies were completed in South Carolina and Idaho; additional studies are planned this year. Through these studies the Federal and State agencies work together in identifying problems and needs and in proposing action programs. As a result of the South Carolina study, the program was presented to the Governor and the legislature for consideration by the General Assembly. It is anticipated that these recommendations will result in additional staff and resources for South Carolina's program and in general provide an improved program for the State.

In order to provide maximum assistance to the States, a new approach for analyzing and evaluating the effectiveness of State program performance is being tested in cooperation with State water pollution control agencies. The proposed new system will be oriented to accommodate inclusion of such detailed information as necessary to permit an objective evaluation of program performance. The State program appraisal will form the basis for evaluating basic State program resources, such as State policy, legislative authority, rules and regulations, organization, staffing, and budget; performance in terms of resource utilization; and accomplishments, such as stream miles or estuarine acres brought into compliance with water quality standards.

In addition, the system will identify State program needs and translate those needs into priorities and objectives in pollution control. The apppraisal procedure will define and identify the minimum criteria governing Federal financial assistance to State programs. It will provide for a continuous review of State programs in order to enhance coordination of State and Federal activities and will permit relating accomplishments to established goals. Finally, the appraisal system will provide for a meaningful comparison of State program performance among States.

The FWQA is also supporting a number of special activities which demonstrate the utilization of advanced techniques by State and interstate agencies. For example, the agency is giving funds to Pennsylvania to help develop a Statewide pollution information system designed to handle all water quality data. This system will provide a modern management tool to help the State systematically administer its program. A modern, automatic monitoring system of water quality parameters and the telecommunicating of information to a central processing location have been expanded by the Ohio River Valley Water Sanitation Commission using FWQA support.

The accelerated drive for clean water stemming from strengthened Federal regulatory and financing programs will also demand an increased capability on the part of the water pollution control agencies of many States.For this reason, proposed legislation to provide additional grants to State and interstate agencies is an important element of the Administration's program. The new legislation would increase the authorization each year on a sliding scale from $12.5 million for fiscal year 1971 up to $30 million for fiscal year 1975. Emphasis would be placed on development, performance, and substantial improvements to State programs. The basic grant program contained in the present Act would remain, but three new categories of grants would be authorized: program development grants, program improvement grants, and special project grants.

The new amendments are essential to increase support to States and interstate agencies to enable them to carry out and accelerate programs of water quality standards enforcement and implementation, the implementation of the Department's proposed construction grant requirements, and the accelerated construction of needed treatment facilities.

From both a long and short-range viewpoint, the State program grants are a good investment. National pollution control efforts can move ahead only as fast as the State and interstate agencies respond with imaginative and thorough programs to meet their responsibilities.

Technical Assistance

Technical assistance is another key program available to help States solve pollution problems. A great many of the water pollution problems facing the Nation call for technical study to determine the sources or causes of the pollution and to find the most appropriate abatement measures to remedy the situation. Often, the problem is complex and requires extensive field and laboratory study. Acid mine drainage, coastal and estuarine pollution, groundwater contamination, and pesticide and toxic chemical pollution are examples of pollution problems for which the most effective and appropriate corrective action is frequently unknown and for which specific technical study is necessary before abatement action can be pursued.

Federal Water Quality Administration (FWQA) assists the State and interstate water pollution control agencies in developing their technical capabilities and provides financial assistance for this purpose through the program grants previously described. These agencies conduct a great many of the technical investigations required to carry out an effective water pollution control program, but frequently they find it necessary to call for outside assistance to handle problems which exceed their capabilities. To meet these needs FWQA provides technical assistance of various kinds ranging from technical advice and consultation to extensive, long-term field and laboratory studies. Within the limits of available resources, this assistance is provided on request, primarily to the State and interstate water pollution control agencies, but also to other public agencies, including other Federal agencies.

During 1969, FWQA responded to over 300 major requests for technical assistance and numerous requests for advice, information, reviews and comments on technical problems. The following examples will serve to illustrate the program:

During the last quarter of 1969, FWQA conducted intensive water quality and waste source surveys on Perdido and Escambia Bays and tributary river basins. The study on Perdido Bay was requested by the State of Alabama to determine the cause of the declining fishery and of the occurrence of unsightly brown foam in many parts of the Bay. The work on Escambia Bay was made in response to a request by the State of Florida to determine the cause of the dozen or more fish kills that occurred in the Bay during the summer of 1969. Both studies identified offending waste sources and the remedial measures required. Upon completion of the two studies, the respective State governors re-

FWQA provides technical assistance to States to help solve difficult pollution problems, such as fish kills.

quested Federal enforcement action, and, in response, enforcement conferences were held early this year. These resulted in specific abatement recommendations and time schedules. The State of Florida has begun implementing these recommendations through the issuance of clean-up orders, and FWQA is continuing to provide technical assistance and support for these efforts.

In other cases, States have applied the findings of FWQA technical assistance directly under their own authorities. The water quality study of Hillsborough Bay, Florida, which was completed in 1969, is a good example. This study was addressed to a long-standing obnoxious odor problem resulting from the death and decay of marine algae. Discharges of municipal wastes from the City of Tampa and industrial wastes from several chemical and fertilizer plants were shown to be adding sufficient nutrients to the Bay to cause the enhanced growth of marine algae. The cause of the massive amounts of dead and decaying algae giving rise to the odor problem was thus traced back to the waste discharges. Other aspects of water quality degradation were also identified together with their causes. The findings of the study were presented in a public hearing held by Florida Air and Water Pollution Control Board in February of this year. The abatement recommendations of the study report were adopted, and the State agency has begun issuing implementation directives. The affected companies, the City of Tampa and the other local agencies have initiated the planning of remedial facilities and practices, thus demonstrating positive follow-through action based on FWQA's technical assistance study.

Another recent example of technical assistance is a field investigation of the James River below Springfield, Missouri, completed last year. This study was conducted in response to a request from the State of Missouri and was aimed at determining the causes of and corrective measures for the severe water pollution and frequent fish kills occurring in the River below Springfield. Several waste sources were identified as the cause of the problem, and abatement measures to be taken by these sources were recommended. On the basis of these findings, the State and the polluters have initiated remedial actions, some of which have been completed, and significant improvements in water quality have already been achieved.

To aid the State of Wyoming in assessing the pollutional impact of a uranium mining operation on Little Medicine Bow River, FWQA developed a radiological monitoring network and schedule for the State and performed the radio-analysis on the samples collected by the State during the summer of 1969. Radio-analysis assistance will be given again in 1970. The State intends to use the data collected to determine the need for waste treatment or control by the mining operation. This exemplifies a type of assistance widely provided by FWQA—the performance of complex analytical and bioassay tests, such as those for pesticides, organic chemicals, heavy metals, various toxicants and radionuclides.

As these examples illustrate, most of FWQA's technical assistance is devoted to investigating specific problems and finding the appropriate available corrective measures. These studies are not directed to research or development of new and improved technology; however, in some cases these studies indicate that corrective measures are not apparent. Thus fruitful topics for research are often identified and these are referred to FWQA's research and development program for follow-up.

To carry out its technical assistance activities, FWQA relies on the basic staffs of engineering, scientific and technical personnel and support laboratory facilities and equipment in each of its nine Regions. Each Region is equipped to conduct intensive field studies involving chemical, biological, microbiological, hydrologic and other disciplines. The coastal Regions are equipped to undertake oceanographic investigations in addition to other types, and the Great Lakes Region has special capabilities for performing lake investigations.

Requests for technical assistance can be expected to continue to rise in the future, despite the expanding capability of many State pollution agencies. The problem areas of subsurface and ocean waste disposal; pesticide and radiological pollution; animal feedlot, sediments, salinity and other aspects of agricultural pollution; and the complex interacting problems of environmental quality protection will require much greater attention than they have had to date. For many individual States these problems do not occur frequently enough to justify maintenance of special skills and equipment on a permanent basis. However, they arise often enough across the country to warrant attention at the national level, and the experience gained in solving pollution problems in one part of the country can be useful in dealing with similar problems elsewhere. FWQA is prepared to give continuing help to State, interstate, and local agencies so they can carry out their remedial programs.

PLANNING AND BASIC STUDIES

Flowing past the Nation's capital, the Potomac River typifies the need for regional and basinwide planning.

Just as any structure needs a good foundation, the nationwide effort to control pollution requires a variety of supporting programs to provide a sound basis for action. These basic support programs help to ensure that our action programs are soundly conceived and will yield clean water results adequate to meet present and future needs. Planning, data systems, and economic studies, all play a supporting role in the battle for clean water.

Need for such basic underpinning is clearly illustrated in the problem of assuring adequate environmental protection in connection with location of major electric power generating facilities. Strengthened procedures authorized by the Water Quality Improvement Act of 1970 to ensure that Federally-licensed power plants will meet water quality standards have been discussed. Although these procedures only apply to one aspect of the environmental impact of power production, they represent major progress.

So long as such review takes place relatively late in the process of designing a plant, however, it cannot be fully effective. There is still a major need—recognized both by the Federal government and by enlightened sectors of the electric utility industry—to provide for consideration of environmental factors in the early stages of site selection. More effective means of planning must be found, which will provide the public with full assurance of environmental protection, and which will enable the utility industry to meet growing power needs without confusion and serious last-minute delay.

The Federal Water Quality Administration (FWQA) and other Federal agencies with an interest in the development or regulation of electric power have been working with the Office of Science and Technology on a comprehensive study and evaluation of power plant siting. In December, 1968, a report, *Considerations Affecting Steam Power Plant Site Selection,* was published. Since that time, the above agencies have been giving attention to the appropriate roles of the Federal, State and other public agencies in the regulation of power plant site planning.

At the Regional level, FWQA is participating with the New England River Basins Commission (NERBC) in developing criteria for siting power plants for New England. The NERBC power/environment program got off to a positive start in late 1969 with an in-depth look at the environmental impact of the proposed Seabrook nuclear power plant site in New Hampshire. FWQA assisted in preparing the water quality impact section of the study. The Agency is participating in a similar activity in the Columbia River Basin.

Environmental Planning

Although a major part, water pollution control is but one facet of the overall program for preserving and enhancing our environment. One of the most significant occurrences during the past year has been the greatly increased awareness on the part of public officials and citizens of the interrelationships among programs to clean up air and water pollution, to manage solid wastes and conserve natural resources, and to provide parks and increased recreational opportunities.

agencies, the FWQA is establishing an Office of Environmental and Program Planning.

Environmental planning concepts, with the emphasis on long-range consideration of the effects of certain waste disposal practices, and the realization that site location practices are as vital as pollution control facilities, are increasingly incorporated in the policies and activities of a number of FWQA programs. Through effective participation in environmental planning, FWQA can best come to grips with such difficult pollution issues as thermal pollution

Taking a clean environment for granted has led to wasteful spoilage. Planning helps both to preserve and to recover our natural resources.

To focus on major environmental issues that may involve actions of a number of interrelated Federal, State and local agencies, President Nixon established on May 29, 1969, a Council on Environmental Quality comprised of Federal Cabinet officers and Citizens' Advisory Committee. Subsequently, the Congress enacted legislation giving comprehensive expression to these concerns—the National Environmental Policy Act of 1969. This Act authorized a new Council on Environmental Quality, whose members have been recently appointed, and the former Council has been redesignated the Cabinet Committee on the Environment. The Environmental Quality Improvement Act of 1970, just recently enacted, further provides for the establishment of an Office of Environmental Quality to serve as staff to the Council.

To strengthen its capability in environmental planning and to provide a focus for coordination with the new Council, as well as with other control, including the previously discussed need for better selection of sites for power generation facilities to protect environmental values; protection of groundwaters and control of underground disposal methods; reducing the impact on waters of salinity resulting from irrigation practices and water development projects; location of oil refineries and future offshore loading facilities relating to the prevention and control of oil pollution; and decreased use of phosphates in detergents.

Other needs that have been identified include development of criteria for evaluating potential airport and highway sites; studying ways in which FWQA could help improve Federal, State and local mechanisms for land-use planning, particularly in critical estuarine areas; and ways in which marshlands could be protected from indiscriminate filling and development.

Development of policies on waste handling and treatment to avoid water pollution must be carried out with the realization that, ultimately, effective waste disposal must involve integrated consideration of air and water pollution control and solid waste management. Water pollution control policies must avoid creating air pollution or solid waste problems and seek, instead, ways of combining methods for maximum reduction of waste loads. Further emphasis must be placed on effective waste management through recycling, recovery and reuse of the by-products of our technology.

Continued thought and effort must be placed on developing means of making so-called "technology assessment"—identifying the possible environmental consequences of new technology before they become widespread problems to be cured after the fact. A major challenge is finding the means whereby we will not have to wait until products, such as the phosphate-based detergents or hard pesticides, become a cause for major concern before we turn our attention to safeguards or substitute methods.

To improve the system for identifying potential or existing environmental problems, Secretary Hickel established an "Environmental Early Warning System" in the Department, clarifying the channels through which any member of the Department can highlight situations that need attention from government. FWQA has established coordinating mechanisms to work with this System and thus far has participated in studies of a number of issues.

FWQA is actively cooperating with the Forest Service and Bureau of Outdoor Recreation in evaluating a number of rivers for inclusion in the National Wild and Scenic Rivers System. This System affords a mechanism for protecting waters of unusually high quality or scenic value from degradation. Some rivers have already been designated for inclusion in the System, and measures for protecting the quality of these rivers will involve both FWQA and the State water pollution control agencies, as well as the Federal agencies which have been designated to administer these areas (Bureau of Land Management, National Park Service, Forest Service).

During the past year, FWQA has participated actively in several interagency planning efforts aimed at studying the impact of development on several areas and seeking measures to mitigate the effects of that development on the environment. One of the most significant involved plans for large-scale development of petroleum resources on the North Slope of Alaska. FWQA made significant input into establishment of guidelines on practices which the oil companies would have to use in construction of facilities, in use of pipelines and other means for transporting the oil, and in carrying out production, so that the resource could be developed without severely damaging the environment, particularly the sensitive and complex tundra areas.

Other issues have involved industrial and housing development in areas along the Eastern Coast. There is increasing realization that the harmful effects of poorly located developments on the quality of coastal waters; on sensitive aquatic resources, such as shellfish; and on marshlands and beaches are too high a price to pay for short-term economic gains in coastal areas, and that many of these effects could be avoided by better consideration of alternative locations and methods of waste handling. In one case, serious shortcomings with the location of an oil refinery near the Chesapeake Bay were brought to light in FWQA investigations, and the company subsequently changed its development plans.

A case currently being studied involves the location of a chemical complex on the South Carolina coast in an area of extremely high natural and recreational value. A German chemical company, BASF, purchased land near Hilton Head, South Carolina, to construct a large petrochemical plant; this project has received nationwide attention and has caused considerable concern to environmental agencies. After reviewing the company's proposal and the conditions of the area, Secretary Hickel wrote to BASF on March 24, 1970, to express his concern that waste discharges from such a plant or transportation of materials might damage the high quality waters and the shellfishery which are now protected by Federal-State water quality standards, and that dredging of any navigation channels would destroy very valuable aquatic habitat. He stated that the Department would oppose any action which would result in degradation of that water quality and would oppose any proposal for channel dredging which would cause environmental damage. Subsequently, on April 7, 1970, BASF announced suspension of its plans pending further consideration of necessary measures to avoid these damages.

These and other issues point strongly to the need for better ways of assessing public values and of planning development in consonance with protecting the environment.

The Big Cypress Swamp is another significant issue where FWQA resources are being used in conjunction with those of other agencies to protect an area faced with development. This swamp is a vital source of water for the Ever-

Economic development and preservation of natural resources, such as Big Cypress Swamp, must be reconciled through environmental planning.

glades National Park, and both the Everglades and Big Cypress form a unique and very valuable natural resource. Proposed construction of a jetport in the swamp has been halted; however, the larger challenge of controlling development in South Florida and providing needed facilities for a rapidly growing population and economy while still protecting the Florida environment is just beginning to be faced by a variety of Federal, State and local agencies. In a sense, South Florida is an early and compelling example of conflicts on the use of resources which we may face in many parts of the Nation before long. The beautiful Florida environment has attracted the very forces that endanger the survival of that environment, and that survival must depend, it appears, on effective long-range planning and control of development.

The above examples have concerned areas where development is threatening a high quality environment. Some of the greatest challenges and potential rewards for water pollution control are also in areas which have been degraded and where pollution clean-up may bring great recreational and other opportunities. This is particularly the case in increasing the opportunities for inner-city residents to swim, fish, picnic and enjoy the outdoors in urban areas. For example, FWQA assisted the National Park Service and Bureau of Outdoor Recreation in planning for the proposed Gateway National Recreational Area near New York City. Full use and enjoyment of this area will depend on effective pollution control.

Other proposals of this kind have been or will be increasingly explored. One FWQA research plan is to clean up pollution from combined sewers discharging to the Anacostia River within Washington, D.C., and develop a large inner-city swimming and boating area. Yet another approach—Project Cure, developed jointly by FWQA and the Bureau of Outdoor Recreation—is being considered for application in some areas, based on experience in Santee, California, with total wastewater renovation and use of the treated wastewater for recreation. One of the features of the Santee project is a series of five lakes which have been created below the treatment plant and filled with essentially pure effluent. Because of the high quality of treated wastewaters, these lakes are used for a host of recreational activities such as boating, fishing, and picnicking.

Basin and Regional Planning

Basin and regional planning is an essential element in pollution control. As President Nixon has pointed out: "A river cannot be polluted on its left bank and clean on its right. In a given waterway, abating *some* of the pollution is often little better than doing nothing at all, and money spent on such partial efforts is

The Santee Project in California renovates wastewaters and provides reuse through recreational lakes.

largely wasted." Clean water results will only be achieved by systematically controlling pollution in entire river basins. Further, we must be sure that these results are lasting ones, that our actions today are adequate to meet the needs of the future, and that we make provision for future growth of waste loads, population, industry, and water use. Otherwise, these future developments may more than offset any gains that our action programs in the years immediately ahead will achieve.

For these reasons, planning is an important element of the FWQA program. FWQA is participating in basin and regional planning in cooperation with State and other Federal agencies and is financially supporting regional planning activities at the State and local level.

The long-term impact of river basin development will be a major factor in keeping the Nation's rivers clean and useful. Changes in stream flows cause temperature increases and other water quality effects. Sustained stream flows are essential for maintaining water quality even where a high degree of treatment is practiced. Irrigation diversion and other developments often deplete these needed flows and return them in lesser quantity and quality.

A major part of the planning responsibility is to ensure that water pollution control and water quality are adequately considered in all Federal water resource development activities, such as planning or construction of reservoirs or irrigation projects. FWQA is participating in broad-scale water resource planning in association with other Federal and State water resources agencies in basin planning studies coordinated by the Water Resources Council. These interagency studies result in comprehensive water and land related resource plans, laying out a future framework for river basin development. These plans are presented by the Water Resources Council to the President and the Congress to be considered in authorizing Federal water resource development projects. FWQA participates in these framework studies to ensure that water pollution control is an integral part of the development and management of the Nation's waters.

Last year, for example, a study of the White River in Arkansas and Missouri completed under this program provided for an intensive program of development and management of water and land resources while emphasizing the continued protection and enhancement of the environment. The plan provides for the cleanup of polluted sections of the River and the maintenance of other sections at their present high quality. In addition to specific treatment facilities at present and anticipated waste sources, the plan provides for the inclusion of storage in specific Federal reservoirs to regulate stream flows to assist in the maintenance of water quality.

In addition to participation in these framework studies, FWQA is involved in a number of more specific water resource planning activities. In the Central Valley of California, where agricultural development threatens water quality in San Francisco Bay and the San Joaquin Delta, FWQA is participating with the Corps of Engineers and the Bureau of Reclamation in long-range planning studies to determine the overall regional impact of continued water development on the environment and the necessary measures to ensure protection of future quality over the long run. In the Delaware

River Basin, FWQA has participated with the Delaware River Basin Commission in the development of a plan and program for the use and upgrading of the highly polluted Delaware estuary. This program has involved the development and utilization of pioneering systems analysis techniques to model the Delaware and show the most effective and systematic approach to achieving improved water quality.

In addition to its basinwide resources planning activities, FWQA reviews proposed water resource projects on an individual basis to ensure that these projects do not have an adverse effect on water quality and that, when it can contribute to the economical control of pollution, storage for water quality is included in Federal reservoirs. As an example, plans for a Federally-assisted project on the Alcovy River, Georgia, were changed considerably after it was shown that the removal of vegetation along the stream channel would adversely affect water quality and be detrimental to fish and wildlife. Revisions made will result in maintaining much of the stream channel and its present cover, greatly reducing the amount of dredging and providing additional safeguards to minimize the removal of vegetation along the river bank.

FWQA also makes recommendations to Federal construction agencies for inclusion of storage for water quality management in proposed projects. Higher sustained streamflows are sometimes needed, in addition to adequate treatment of wastes and other controls, to meet water quality standards.

River basin planning can yield important results in developing solutions to complex pollution problems that must be dealt with along the lines of entire basins and that cannot be solved without a coordinated effort by all parties involved. The information collected and the plans developed under this program have served as a springboard for a number of State clean-up programs. For example, a mine drainage study conducted as part of a comprehensive water pollution planning effort in the Pennsylvania portion of the Susquehanna River Basin has resulted in a substantial State program to abate mine drainage pollution. The study resulted in the locating of over 1,000 mine drainage discharges causing gross water quality degradation in 1,200 miles of stream. It was found that restoration of streams polluted with mine drainage could be accomplished through a program which included mine sealing, neutralization, land treatment, and water regulation and diversion. Selective implementation of action called for by the water quality management study is underway with the aid of a conservation bond issue adopted by the Pennsylvania legislature which provides $150,000,000 for the reclamation of areas disturbed by mining and the abatement of mine drainage pollution.

In addition to direct planning activities, FWQA is supporting regional planning through grants to planning agencies at the State and local level. These grants are designed to stimulate the kind of State and local planning which is important to the implementation and improvement of water quality standards along river basin lines. This program was initiated in 1967, and 12 studies are underway with total Federal costs of over $2.5 million. The Federal share is limited to 50 percent of the costs of developing the plan. These grants afford agencies at the State and local level a unique opportunity to participate in solving their pollution problems on a coordinated, long-range basis.

Under this program, the Santa Anna Watershed Planning Agency in California is developing a pollution control plan which will provide for eventual reuse of the reclaimed water. In this watershed, available surface water flows are almost completely developed and large quantities of Colorado River water are being imported. In areas near the coast, because of heavy pumping, groundwaters are threatened by salt water intrusion. The plan being developed will consider both surface and groundwaters and provides for pollution control and wastewater reclamation and reuse as an integral part of the water supply program in the watershed.

The Miami Conservancy District in Ohio is also conducting a planning study partially funded by an FWQA planning grant. The study is utilizing an extremely sophisticated systems analysis technique to relate water quality, flood control, and other factors involved in water quality management decisions in the basin. The plan will consider the whole range of effects on water quality of such alternatives as in-stream aeration, use of abundant groundwater supplies to augment streamflow, and the regionalization of waste treatment facilities.

FWQA is also assisting a planning effort in Puerto Rico. The Commonwealth planning is aimed at developing programs to encourage industrial growth while maintaining and enhancing water quality. The Island's development has centered around the recreation industry for which water quality is obviously vital. The plan will provide for the protection of these recreational amenities in the face of future industrial development.

The emphasis in the President's environmental message and Secretary Hickel's recently published regulations on conformance of waste treatment plant construction with basin programs and regional planning to ensure speedy and coordinated pollution abatement will require increased emphasis on the part of FWQA on implementing effective short-term planning and appraisals. The new regulations will require that, within a river basin, each treatment facility be part of a basinwide plan for pollution abatement and within a given city, each treatment plant be included in a metropolitan or regional waste treatment plan. The Agency's planners are developing a procedure to evaluate grant applications to help the States meet the requirements of the regulations and to better integrate planning and facilities construction, so that in the near future planning can be used to efficiently and effectively guide waste treatment installation. This will place additional and immediate demands on FWQA's planning capacity.

Towards this end, FWQA is increasing the emphasis upon quick appraisals of the status of comprehensive and coordinated programs in each river basin and preparing to make quick evaluations of the adequacy of and need for planning within metropolitan areas. In the latter regard, FWQA is working with the Department of Housing and Urban Development. Last year FWQA's Northeast Regional office, working with Housing and Urban Development, developed a joint set of comprehensive guidelines for regional sewerage systems. These guidelines can be used in preparing plans for metropolitan sewerage systems and are sanctioned by both Housing and Urban Development and the FWQA.

To assist in basin and regional planning, FWQA has developed a highly sophisticated systems analysis capability. Models have been developed to show the relationships between various stream flows, waste loads, water uses, and other factors that influence water quality. These models can handle up to fifteen sections of stream, fifteen reservoirs, ten discharge points and natural pollutants. Although relatively new, these models have been used successfully on the Sabine River, Texas; Skunk River, Iowa; Scioto River; Ohio; James River, Virginia; Broud River, South Carolina, and many others. For a given river basin, the models can provide information to determine how management practices influence water quality and what changes in management could be expected to provide a certain water quality and the cost of that quality.

Models have also been developed for the Delaware Estuary, and these are now being applied to the Potomac River–Chesapeake Bay system. These models help to relate tidal effects to water pollution. The Delaware model has provided a tool for determining the needed releases from upstream to protect Philadelphia's water intake from excessive salinity intrusion during periods of drought.

Another systems technique was applied in the San Joaquin Master Drain study in California. Here the model required inclusion of economic information, as well as the waste sources. The model provided the basis of measuring the impact of planned water resource development on an inland agricultural area, as well as on San Francisco Bay. Of major concern was the impact of pesticides and nutrients resulting from agricultural drainage. Through the use of this model, alternative locations of the drain outfall with consequent economic costs were determined, as were the costs of alternative treatment measures.

As planning for basinwide pollution abatement and regional waste treatment moves ahead in the future, the systems capability developed by FWQA will become increasingly important in the Nation's battle to achieve clean water.

Estuarine and Coastal Studies

For well over three and a half centuries, the estuarine and coastal waters of our Nation were thought of primarily as conveniences—places for the conduct of international commerce, locations for the residential and industrial development that resulted in our great cities, sites for mineral exploitation, and dumps for all kinds of wastes. Although this thinking is still commonplace, times are changing, and more and more people are becoming increasingly aware of the necessity to change our behavior with regard to these waters. We can no longer afford to treat our estuaries and the coastal waters over our Continental Shelf as endless sewers.

Because estuarine and Continental Shelf waters are so closely interrelated, pollution in one zone will affect the other. For example, hard pesticides, which are carried down rivers from the agricultural uplands and tend to accumulate in waters near the mouths of rivers, eventually spread into the surrounding oceanic waters. Conversely, an oil spill caused by the breakup of a tanker at sea will ultimately spread to the coastline, there to foul beaches and kill wildlife and waterfowl. Modification of the shoreline by dredging and filling will have an effect on life far out to sea. Ocean outfalls, while disposing of wastes at a distance from shore, are fre-

The polluted waters of upper New York Bay form the foreground for this view of Manhattan. Ocean outfalls and dumping of solid wastes imperil the fish and bottom fauna of these and other marine waters.

quently responsible for water conditions which make a shoreline area unfit for swimming or shellfishing. Sludge and solid wastes that are barged out to sea for dumping can return to shore on the currents and tides. Continental Shelf mineral development—ever increasing in importance—has the potential for major environmental damage.

The condition of the New York Bight area is a startling illustration of disposal wastes into coastal and ocean waters. The dumping of wastes near the New Jersey coast has recently come to the attention of a shocked public. Sewage sludge, treated and untreated, and various industrial wastes are a primary concern. A dumping area of approximately 14 square miles has been damaged and its bottom fauna severely impoverished. Even several species normally tolerant to pollution are absent from this area and evidence of pollution has been found on nearby beaches.

Not far away in the New York Harbor area, an outbreak of fish diseases has occurred over a three-year period. Large numbers of fish have neither tails nor fins, and there is some evidence that pollution may be at least partially responsible. Fish kills in the area are numerous, and there is growing concern about the contamination of shellfish—a threat both to the harvesting industry and to public health.

Federal Water Quality Administration (FWQA) over the last several years has strengthened its various programs, giving increasing emphasis to estuarine and coastal pollution. Increased enforcement activity along the coast—such as the recent conferences in Biscayne and Mobile Bays—has already been highlighted. Added emphasis is being given to oil clean-up activities. FWQA is now accelerating its work with the Coast Guard to prepare plans for a more speedy reaction to oil pollution incidents. Research and training programs that have a relationship to estuarine and coastal problems have also been increased. More emphasis is being given to studying pollution effects and ecological damages in the estuaries. More research chemists and marine biologists are being trained in FWQA-funded programs.

Because of the long-term cumulative impacts on the estuaries and coastal environment and because of the many interrelated actions affecting these waters—dredging and filling of marshes and construction of navigation facilities—considerable emphasis must be given to the overall planning and management of this valuable environment. Planning to protect our estuaries and coastal waters is a clear-cut example of the pressing need for environmental planning described previously.

As a result of increased public awareness of estuarine and coastal pollution problems, the Congress directed that a survey of estuarine pollution be made. FWQA, in November, 1969, submitted the report of this first comprehensive, definitive study of estuarine pollution to the Congress. With this report, the *National Estuarine Pollution Study,* proposed legislation for a comprehensive national management program for the estuaries and coastal zones, based on the report's recommendations, was submitted to the Congress.

The Study sought to obtain detailed information on the biophysicial, socio-economic, and institutional aspects of estuaries from a variety of sources. First, a series of 30 public meetings was held in the various coastal States to obtain information and opinion from the local citizens who are most directly affected by estuarine pollution. Second, information was collected from the coastal States concerning their laws and programs affecting estuarine uses and management. Third, studies were contracted to provide needed background on certain aspects of specific estuaries or on a restricted aspect of the Nation's estuarine areas. These include studies on ecology, economic and social values, sedimentation, and law. Reports on some of these studies are being published as the *Estuarine Pollution Study Series*. The first of these is entitled, *Legal Perspectives of Chesapeake Bay.* Others to be published will include *A Socioeconomic Analysis of Narragansett Bay,* and *The Social and Economic Values of Estuaries.*

A major part of the study was the development of the National Estuarine Inventory, an automated information system. This massive compilation of coastal zone information is the basis for the development of a continuing national Coastal Zone Management Information System to satisfy the information requirements of States, Federal agencies, and other entities for factual data on which to make decisions.

The recommendations which were presented in the Study were predicated upon the concept that the States should have the major responsibility for managing the estuarine and coastal zones and that the Federal role should be to provide coordination of the State programs within the national plan, to provide technical and financial assistance to the States and their subdivisions and to arbitrate conflicts between States.

Legislation to promote these aims is presently being considered by the Congress. The bill, if enacted, will provide for Federal grant support of State management programs. The prime objective will be the management of the estuarine zones in such a way as to permit maximum beneficial use with minimum damage.

Closely related to the National Estuarine Pollution Study (NEPS), the Fish and Wildlife Service has recently completed the *National Estuary Study* (NES). This study involved an intensive look at fish, wildlife, and recreational values of the coastal zone for the purpose of recommending a scheme for protection of extremely valuable areas. FWQA assisted in this latter study by making the data bank of NEPS available to the Fish and Wildlife Service as its base source of information. The NES is a complementary effort to FWQA's broader study of man's activities in the coastal zone and how pollution from these activities causes environmental damage to coastal resources.

FWQA has also sponsored a major study, conducted by the National Academy of Science and the National Academy of Engineering, to determine the state of knowledge on ocean waste disposal. The findings of the study are being used to help formulate approaches to the problem. A report—*Wastes Management Concepts for the Coastal Zone*—will be published later this year.

FWQA will continue to conduct and fund studies to increase our knowledge of the estuaries, their resources, the damages done to them by pollution, and their relationships to the surrounding land. In addition, direct technical and financial assistance will be provided to States for management and improvement of their estuaries.

Data and Information

Effective implementation and enforcement of water quality standards, development of regional and basin plans, administration of grants, and preparation of reports assessing costs of pollution control and abatement progress require up-to-date, accurate fact-finding and readily available data.

Several types of technical information are required to meet the various needs: specific data covering the status and effectiveness of municipal, industrial and Federal waste treatment and control facilities; current economic data associated with construction activities; and water quality data related to the water quality standards.

Collection, evaluation, and dissemination of data on chemical, physical, and biological water quality and other information relating to water pollution discharges is an essential element of the Federal Water Quality Administration (FWQA) program. Through effective coordination with other Federal and State agencies, such data and information are utilized at the national, regional and basin levels.

Collection and timely evaluation of reliable information on water quality is vital to the effective management of a dynamic national pollution control program. This has always been a requirement, but the need has intensified with the establishment and implementation of water quality standards and the resulting necessity of identifying priorities in waste treatment facility construction. Regardless of the number of treatment facilities constructed or the number of basin management plans completed, in the final analysis, program effectiveness can only be measured in terms of actual water quality improvements. And, this can be achieved only through adequate monitoring of water quality.

Thus, FWQA has been reorienting and expanding its data collection activities to identify compliance and noncompliance with water quality standards; improvements in water quality resulting from pollution abatement measures, such as waste treatment facility construction; and emerging water quality problems that should be corrected before crises arise.

Key steps required in the development of an adequate nationwide water quality surveillance system involve planning the system in close coordination with State and other Federal water data collection agencies and implementing the system by utilizing existing programs of FWQA, State pollution control agencies and other Federal water data collection agencies, principally the U.S. Geological Survey. During the past year, plans for integrated State-Federal water quality monitoring systems have been developed for six of the nine FWQA Regions and are now being implemented.

If a surveillance system is to be fully effective, thorough attention must be given to its design as well as its operation. In recognition of this, a systems analysis approach to the design of optimum water quality monitoring programs is under development. This approach will per-

These automatic monitors gather the accurate, up-to-date data required for effective implementation of water quality standards.

mit the monitoring subsystems, making up the nationwide integrated State-Federal surveillance system, to be designed and updated as necessary on a uniform basis in such a way as to ensure maximum program effectiveness.

To ensure the reliability of data collected by the coordinated program, an analytical quality control program, which is now under development, will become an integral part of the overall system. All cooperating agencies will be expected to participate in such a program. Thus far, a manual entitled, *Federal Water Pollution Control Administration Methods for Chemical Analysis of Water and Wastes, 1969*, has been published and distributed to all participating laboratories. Similar manuals covering standard biological and bacteriological laboratory procedures are under development. Quality control checks and procedures that will be employed on a routine basis in participating laboratories are also being developed.

A portion of the coordinated network is already in operation. It presently utilizes approximately 400 FWQA-funded and operated stations, 260 FWQA-funded and U.S. Geological Survey-operated stations, 200 stations jointly funded by the State and Federal agencies and 500 State-funded and operated stations. Ultimately, the network will encompass State and Federal stations numbering in the thousands. Network data will be supplemented by the findings of the many short-term intensive field studies of specific water quality problems that are conducted by FWQA.

In addition to water quality data, detailed knowledge of waste sources, treatment and discharges is also necessary to fulfill the needs of FWQA programs. Municipal sewage and industrial wastes are the two largest sources of pollutants. During 1969, the Pollution Surveillance Branch completed the processing and analysis of data on municipal waste facilities collected in a cooperative Federal-State inventory. This effort, the first since 1962, reflects conditions as of January 1, 1968. Because the need for timely and accurate data in this area is so critical, procedures have been developed for bringing the 1968 inventory up to date and for continually updating it to keep it current. In addition, data from the implementation plan portion of the water quality standards have been correlated to and integrated with the inventory to show schedules for providing additional municipal waste disposal facilities.

As for industrial wastes, plans have been made to initiate an inventory of industrial manufacturing and processing plants. Initially, this will be an in-house effort; eventually, it will be expanded to a joint FWQA-State cooperative project. Here, again, data from the implementation plans of the water quality standards will be valuable in planning and conducting the inventory. Once established, this inventory, like that of municipal facilities, will be continuously updated.

With the recent publication by the Secretary of new construction grants regulations, data on waste sources and discharges have become even more important. These regulations require the States to show that a proposed municipal facility is a part of, and in conformity with, a basin, regional or metropolitan pollution control plan before the project is declared eligible for a construction grant. In addition, regulations prescribe as a further condition for eligibility the provision of data on all waste discharges in the immediate proximity of a proposed plant which may affect its design and operation. For these reasons, additional data will be required on wastes characteristics and strengths.

In addition to the municipal and industrial inventories, a new collection effort has been planned and initiated to provide data on thermal discharges from electric power generating plants. By agreement with the Federal Power Commission, data for this inventory will be collected by that agency through a questionnaire on environmental control information.

To achieve the objectives of the coordinated data and information program, it is essential that the data collected be evaluated in an expeditious manner and made readily available to all users. Only in this way can appropriate follow-up actions be taken. FWQA's existing computerized data storage and retrieval system (STORET), coupled with additional computer programs, will meet these requirements. All data collected by FWQA will be placed in STORET to be available for analysis and use.

Using the most up-to-date computer technology, the data collected are entered in a central computer on a daily and weekly basis by remote terminals in all FWQA Regions. Similarly, questions can be asked of the central computer from the remote terminal and receive timely responses. This application is now being expanded to include several Federal and State agencies.

Evaluation and dissemination of the large amounts of data and information collected is assisted by the STORET system. Currently, this system is being expanded to include water quality standards and uses so that many questions can be asked, such as what facilities have inadequate treatment, what type of treatment is provided or waste contributed at any source, how many miles of streams are polluted, how many miles have been improved, what is the number of violations of water quality standards and where, and what uses have been affected and over how many miles. It will be several years before all of these questions can be accurately asked and answered for each individual basin, or region, or for the Nation as a whole. However, those questions must be answered to provide an effective overview of our rate of progress, and FWQA is beginning to build towards that capability now.

A quantitative analysis of changing trends in water quality and of the progress in abating pollution nationwide will be essential in guiding the course of the national water pollution control effort. It will also contribute to the assessment of national environmental conditions and trends required under the National Environmental Policy Act of 1969 by indicating whether we are gaining ground or falling behind in pollution control.

Economic Studies

For the first time, the Nation stands on the threshold of a major effort to reverse the heritage of neglect and to face the problems of a deteriorating environment. Massive investments will be required for environmental improvement and the expenditure of these funds will have major impacts on the economy. Formulation of sound public policy will require an increasing understanding of these costs, their distribution throughout the economy, and their economic impact, both on individual communities and firms and on the economy as a whole.

For three years the Federal Water Quality Administration (FWQA) has been conducting a series of economic studies aimed at gaining a deeper understanding of these factors and at assisting the Executive Branch and the Congress in formulating national policies and legislation. These studies have included: analysis of the national costs of treating municipal and industrial wastes and the impact of these costs on State and local governments; studies of sewer user charges as a means of financing local expenditures; studies of the need for economic incentives for industrial waste control; and studies to determine the extent of animal feedlot pollution and the costs of abating it. Collectively, these studies represent the most intensive and comprehensive effort ever made to understand the costs of water pollution control.

The findings of FWQA's economic studies are submitted to Congress annually. The first report projected municipal investment requirements for the period 1969 to 1973 and assessed the impact of funding required to meet municipal waste treatment needs on the municipal governments and bond markets. The second report examined the influences that determine investment levels and concluded that the critical factors were to be found in the dynamics of the situation—in the interaction of investment with time-conditioned growth, replacement, and demand for higher plant efficiencies. It was also found that regional cost differences, transmission costs, and the influence waste loadings were extremely important factors in analyzing the economics of water pollution control.

During the past year, these cost studies have concentrated on information needed to reshape the funding of the construction grants program to make it fully adequate to meet the Nation's needs. The Secretary of the Interior's legislative proposal for a Federal, State, and local investment of $10 billion over the next four years reflects the findings of these studies. The 1970 report, *The Economics of Clean Water,* defines the rate of investment needed to close the gap for municipal waste treatment in the years immediately ahead. This report provides the most thorough estimates ever developed for the Nation's municipal sewage treatment needs and costs. Detailed studies of the pollutional impact

Polluted rivers, such as this, present a challenge to people and governments at all levels—now is the time to institute urgently needed action programs.

of the inorganic chemicals industry and concentrated animal populations were also completed as separate sub-reports.

Various aspects of the socio-economic problems of water pollution are presented in the latest report. These include discussions and conclusions about investment trends and needs, Federal cost sharing, priority systems for grant funds, public treatment of industrial wastes, and regional waste handling systems. In addition, several estimates discussed in earlier reports were reviewed in view of the latest available information. These included investment estimates for collecting sewers, separation of storm sewers, industrial waste treatment and cooling facilities, and sediment control and acid mine drainage reduction.

The Water Quality Improvement Act of 1970 requires that a complete investigation and study of all methods of financing the cost of water pollution control, other than methods authorized by existing law, be made and the results submitted to Congress by December 31, 1970. To meet this requirement, FWQA has structured a study which will deal with pollution sources of all types including, but not limited to, municipal, industrial, agricultural, land and acid mine drainage, oil and accidental spills and debris. Questions of responsibility, ability to pay and equity will be addressed in allocating potential funding requirements among the private sector and the various levels of government. The study will examine the feasibility of a wide range of financing possibilities in the light of the analysis outlined above. Potential methods will include: conventional financing; loan arrangements; user, influent and effluent charges; taxation; insurance-type arrangements and others. In addition, potentials for reducing financing requirements by means of structural policy alternatives will be assessed.

Although considerable insight into and understanding of the economics of water pollution control has been gained through past studies, there are still many unanswered questions concerning the costs of pollution abatement and the impact that efforts to cleanse our environment will have on the national economy. The challenge is clear, however: if the Nation's water resources are to be enjoyed without the burden of increasing water pollution, *now* is the time to institute prudent action to clean up our streams. The people and their government have accepted the challenge. FWQA reflects their determination and will give continuing emphasis to devising policies and programs which will create a cleaner environment in the most expeditious and economical manner.

RESEARCH, DEVELOPMENT, AND DEMONSTRATION PROGRAMS

The search for new answers is an important part of the Federal pollution control mission. Federal Water Quality Administration (FWQA) is conducting a research, development, and demonstration program which is a coordinated, problem-solving program dedicated to exploratory research of new and imaginative pollution control methods; the engineering development of these methods to solve the practical problems associated with bringing an "idea" out of the laboratory and into the real world; and the demonstration of this new technology to go that extra, normally forgotten step of showing the decision-makers that new answers, new technology have really arrived and are available for use.

The program being conducted is highly mission-oriented. Each project responds to an identified need for an answer. These needs are specified and assigned priority primarily through imput from the non-research elements of FWQA. In short, responsiveness to the research needs of the Agency is a prime responsibility of this program.

There are really only two major categories of "answers" being sought. First, how are the water quality goals defined? Second, how are these goals reached with maximum effectiveness and at least cost? With regard to quality goals, research is required on the effects of pollution. What are they? How is the degree of effect related to the amount of pollution? And, how can the level and type of effect be predicted in advance? With this type of information we can improve and extend the water quality standards now being established and implemented for the Nation's waters. Simply knowing what water quality is required is not enough, of course. In those cases where we already have some ability to control pollution, new and improved means for control are needed in order to reduce the cost of pollution abatement to the very minimum possible. Beyond this, the need to develop and demonstrate means for controlling that pollution, which today is literally uncontrollable or untreatable at any cost, is assuming a high priority. Corollary to and, in fact, inseparable from this objective is the simultaneous upgrading of wastewater quality such that used water may be reused again—a concept of major significance in extending our relatively dwindling fresh water supply.

To assist in managing this program and in setting priorities and resource allocations, a problem-oriented project categorization is utilized. Eight major categories exist: the first five relate to single-source-related pollution problems from municipal, industrial, agricultural, mining, and from other sources. The last three categories relate to problems of a multiple-source nature, where the answers will be applicable broadly to many different sources of pollution. In the single-source category FWQA is working on such pollution problems as combined sewer discharges, pulp and paper wastes, agricultural runoff, acid mine drainage and oil pollution. In the multi-source categories FWQA has programs on eutrophication, thermal pollution, removal of nutrients and refractory organics, and effects of pesticides and other pollutants on fish and aquatic life. The mechanisms utilized in carrying out this program are three-fold:

(1) In-house research and development at eight laboratory locations and a number of associated field sites.

(2) Contract projects, primarily with industry.

(3) Grant projects with universities, industries, States and municipalities.

Contract projects are funded entirely with Federal dollars and are utilized primarily for laboratory investigations and pilot-scale research projects which involve a high degree of uncertainty and which are primarily aimed at determination of feasibility and development of design requirements. These are not the types of projects that municipalities and private corporations will readily sponsor with matching funds because of the large degree of risk involved. The work performed under contracts often requires highly-specialized personnel, equipment and facilities, having a high value over a short period of time, but limited value in the long term.

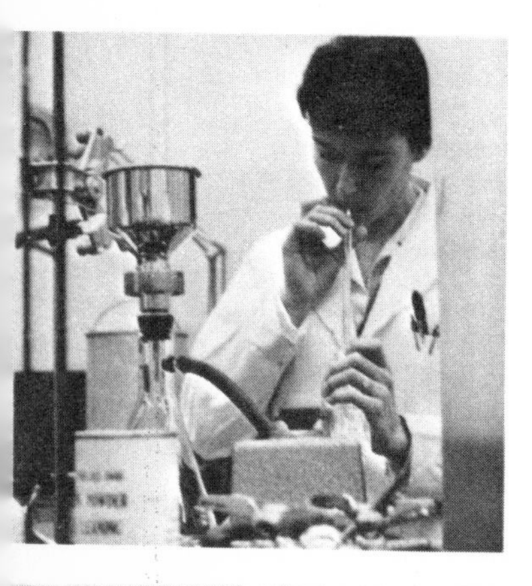

Research on advanced waste treatment technology is carried out at FWQA laboratories and through field evaluations.

Grant projects require some level of matching support from the grantee. Grants are employed in meeting objectives where it is desirable to utilize State, municipal, academic or industrial talents and expertise in carrying out research, development and, often, demonstration efforts on a cost-sharing basis to the mutual benefit of both the Federal government and the grantee.

FWQA's in-house activity forms the real foundation of an effective overall program. In-house researchers must establish objectives and plans of attack; they must review and evaluate the many, many project proposals received by this Agency; and they must be the ones to integrate the results of these efforts into a usable and applicable form. To do this most effectively, in-house staff must be involved in the work for which they are responsible.

FWQA's program is predicated on the assignment of specific areas of technical responsibility to each of eight laboratories. In this way, each laboratory functions as a national focal point for research on a given set of problems, and duplication of facilities, staff and effort among the various laboratories is avoided. Research laboratories are located in Cincinnati, Ohio; Athens, Gerogia; Ada, Oklahoma; Corvallis, Oregon; College, Alaska; Duluth, Minnesota; Narragansett, Rhode Island; and Edison, New Jersey. These laboratories are also responsible for operating a number of field sites to carrying out pilot plant work and necessary field studies. FWQA operates such field sites at Pomona and Firebaugh, California; Ely, Minnesota; Lebanon and Newtown, Ohio; Norton, West Virginia; and Washington, D.C.

In addition to the in-house efforts carried on at agency laboratories, FWQA is also involved in a number of joint efforts with the Bureau of Reclamation, Atomic Energy Commission, Office of Saline Water, Office of Water Resources Research, Public Health Service, and Tennessee Valley Authority.

Notable in the research, development and demonstration program are the special authorities to support both pilot-scale and full-scale demonstration projects on storm and combined sewer discharges, advanced waste treatment and wastewater renovation, and industrial waste treatment and control. These projects are particularly significant in permitting FWQA to carry on research and development findings into the demonstration phase, thereby literally showing what can be accomplished through the use of new technology and at what cost.

In order to effectively manage this program, communicate the results to users, and respond to special Administration, Congressional and public requests, a computerized management information system was instituted. As a result, up-to-date information is readily available on nearly 2,000 projects, on future needs, on priorities, on work plans, and on necessary planning, programming and budgeting data to effectively direct future efforts.

A supplementary project reports system has been established for the acquisition, filing, indexing and, most importantly, dissemination of research results. The final results in the form of reports and publications are indexed into a technical library, distributed, and made known to a wide range of users both inside and outside FWQA. In Executive Order 11514, President Nixon directed that the results of Federal research programs be made available for widespread use. FWQA will continue to emphasize this important aspect of the research, development and demonstration program.

The problems of water pollution, as previously described in the "Water Pollution and the Environment" chapter of this report, are so complex, so varied and so numerous that they have multiplied faster than solutions. To ensure that our technology is improving and to make existing control methods more effective in the overall effort to make America's waters clean and useable, FWQA has intensified its research programs. The Water Quality Improvement Act of 1970, enacted and signed into law recently, added emphasis to research programs in oil pollution, acid mine drainage, vessel pollution, and pollution in the Great Lakes, and FWQA is moving to meet these responsibilities.

The eight categories of research being conducted in FWQA's program are directed at solving the problems already discussed. These categories, the problems at which they are focused, and some of the recent accomplishments of the research are discussed below.

Municipal Pollution Control Technology

Municipal wastes, as indicated earlier, are a major source of pollution in the United States. Although a technology to treat these wastes has already been developed and is being applied, FWQA is continuing the search for better and more efficient ways of treating municipal wastes in conventional systems. For example, significant improvement and upgrading of treatment in overloaded plants has been demonstrated using synthetic organic polyelectrolytes.

Another major concern is research on methods to control the more complex municipal problems, such as combined sewer and urban sediment control. Combined sewers carry both sanitary sewage and urban runoff. During storms, the volumes in these sewers are often too much for local treatment plants and wastes are discharged untreated. Yet control of these discharges has largely been neglected until recent years because the only method of solving

Improved methods of controlling storm sewers are being developed.

the problem was separation of combined sewers, a costly and disruptive process. Through the efforts of FWQA's research program, a new technology for control of sewer discharges is being developed.

One of the alternatives being demonstrated is storage of excessive flows until they can be released to the treatment plants. Full-scale storage facilities under construction in Boston, Massachusetts; Milwaukee, Wisconsin; Dallas, Texas; and Shelbyville, Illinois, have determined the design criteria necessary for such facilities. Cost-effectiveness evaluation will allow other communities to economically design similar combined sewer pollution abatement facilities.

Another major alternative is treatment. Existing municipal and industrial treatment processes cannot be utilized for combined sewer overflow treatment because of the intermittent, widely fluctuating high-flow rates and the dynamic quality changes of combined sewer overflows. Screening and dissolved air flotation are two treatment methods which are amenable to the above constraints.

The demonstration of a novel, rotating collar, vibratory base screening treatment unit for combined sewer overflows was carried out in Portland, Oregon, in 1969. The unit provided primary treatment to normally bypassed sewage at a cost only slightly higher than the equivalent conventional treatment. The space utilization of the screens is one-tenth that of settling tanks.

Through these studies, a combination of control methods is being developed which will be applicable to the different combined and stormwater sewer problems throughout the country. Although determinations of the cost of controlling these discharges by the new methods being demonstrated are very preliminary, the total job may cost only about one-third of the earlier estimates based on separation.

Erosion and sediment from urban areas cloud rivers and impair their use. These waters generally are not confined to sewers, so the above methods cannot be applied to solve the problem. The National Association of Counties Reseach Foundation, in conjunction with FWQA, has therefore developed a *Community Action Guide for Erosion and Sediment Control.* This document will aid local officials in developing erosion and sediment control ordinances to control pollution from urban development construction projects. The control programs would be based on the establishment of control ordinances and on the use of present technology, such as vegetation control, mulching, sediment traps and other common erosion

Under water storage facilities for combined sewer ov flows have been developed with the help of Federal gra

control practices. Adoption of effective control programs based on this guide will substantially reduce the silt load to urban waters.

Of great importance is FWQA's research on joint treatment of municipal and industrial wastes. As has been pointed out, the benefits of joint treatment are considerable. Industry, while paying operating costs, is spared the burden of the capital costs; and regionalization of waste treatment and economies of scale help communities achieve more effective pollution control.

The benefits of joint treatment are recognized. Certain industrial wastes, however, have proved difficult to treat effectively in combination with domestic wastes. In this regard, our demonstration of the feasibility of joint treatment of domestic sewage and semi-chemical pulping waste from a paper mill in 1969 at Erie, Pennsylvania, was an encouraging breakthrough.

In addition, a joint municipal-industrial wastewater treatment engineering study of the Onondaga Lake watershed was also completed last year. Approximately 140 industries in the watershed participated in the study by assisting in characterizing their wastes, and it was recommended that a joint treatment system be implemented by Onondaga County during the remaining phases of the project.

The successful demonstration of joint treatment of industrial wastes in municipal treatment systems holds great promise for the future. FWQA is encouraging such joint treatment and numerous communities with significant industry within their jurisdiction are considering such treatment.

Industrial Pollution Control Technology

Industrial waste discharges, together with municipal wastes, comprise the two largest sources of pollution. Industrial wastes are complex—a result of the wide variety of products manufactured—and are discharged in enormous volumes. In order to effectively control pollution, industries must often face the heavy financial burden of installing waste treatment facilities. Current waste treatment methods, while sometimes adequate, are expensive and in many instances offer little hope of providing the type and degree of treatment which will be required in the future. Because of the competitive economic aspects, industries are continually searching for new means of reducing their wastes at lower costs.

An effective attack on industrial pollution—wastes from metal, chemical, petroleum, coal, paper and other product manufacture—requires a cooperative industry-government effort to conceive, develop, demonstrate, and install treatment processes, process modifications, and water conservation programs. Already, research funded by FWQA covers some industrial problems from almost all major sources of industrial pollution.

A grant project with the American Oil Company at its Mandan, North Dakota, refinery has demonstrated the feasibility of using a commercially available fluidized-bed incinerator for the disposition of refinery sludges. The project was initiated in May, 1968, and inquiries to date by others in the industry show a keen interest in the utilization of this technique to resolve their sludge disposal problems. Another oil company has indicated its desire to apply the method at its own refinery. The American Oil Company is presently considering the possible use of a much larger fluidized-bed unit at the Whiting, Indiana, refinery in the near future.

The color of pulp and paper mill wastes has long been an aesthetic nuisance, difficult to control. Interstate Paper Corporation at Riceboro, Georgia, has demonstrated the lime coagulation process for the removal of color from kraft pulping effluents. This installation is the first full-scale operation of its type and has obtained color reductions greater than 90 percent throughout the experimental program. The results of this grant have been utilized by both paper companies and State agencies in selection of effluent treatment processes to meet receiving water quality standards.

FWQA and the State of Vermont have jointly entered into a demonstration project which provides an excellent illustration of the side benefits of some industrial pollution control. A project initiated in late 1968 on the conversion of cottage cheese whey into an edible grade material has produced, on a pilot-scale, a high grade food powder for human consumption. A plant for the full-scale demonstration of the developed process has been completed and will be operable in 1970. The plant could ultimately have the capacity to produce 20 million pounds per year of dried edible whey. Cheese whey produced in this country represents pollution equivalent to that produced by a population of 16 million people.

FWQA research and development grants have been made to the pulp and paper industry to develop more effective and less costly treatment processes.

With the expansion of both the population and the industrial sector and the corresponding needs for water, conservation of water is becoming increasingly important. Much of today's research is directed toward finding easy-to-treat and re-use water effluents. A project with the Johns-Manville Products Corporation in Defiance, Ohio, demonstrates that a wastewater treatment system using diatomite filtration can effectively treat a waste stream, containing glass fibers, caustic and phenols, to a quality suitable for process reuse. The treatment facilities are operating on a 72,000 gallon-per-day basis with effective pollution control a demonstrated success.

Agricultural Pollution Control Technology

The most difficult sources of wastes to control are those that do not come out of pipes. Agricultural pollution is a good example of such "diffuse" wastes. Major forms of pollution associated with agriculture have already been identified as problems in earlier sections of the report. They include: nutrients; pesticides; salts and other materials in irrigation return flows; animal feedlot wastes; and silt and other solids from logging operations. Most of these wastes are not collectible and, therefore, cannot be treated in a conventional fashion. New and imaginative solutions are being sought for these problems.

Projects with Cornell University, South Carolina State, and South Dakota University are aimed at studying the addition of nutrients to streams from cropping practices as related to their respective geoagranomic areas. This is a precursor to the development of criteria for new management concepts that include considerations for waste management.

Thousands of beef cattle are concentrated in pens. Drainage from such feedlots is a major pollution problem.

The quality of irrigation return flows is a major problem in the arid sections of the country, primarily because of nutrients, silt, and salts. Treatment of such flows has long been considered impractical. A development program at Firebaugh, California, has developed two techniques for removing nitrates from irrigation return waters. These will be demonstrated on an engineering scale to obtain more definitive operating and cost data that will be applicable to a complete treatment system for the entire San Luis Drain.

Work is also under way with the Bureau of Reclamation to demonstrate a technique of forecasting the effects of irrigation practices on the quality of underground aquifers and surface streams before lands are irrigated. This method will enable us to make better provision for avoiding water quality damage in planning and developing new irrigation projects.

The tremendous load of animal wastes discharged from a rapidly growing number of animal feedlots is an area of particular concern in FWQA's research program. Projects have been initiated to demonstrate available techniques for treating runoff from animal feeding operations and for preventing its discharge to receiving waters. These include activated sludge, oxidation ditch, anaerobic-aerobic lagooning and management changes to control and collect the runoff. Cooperative projects with the Department of Agriculture have also been initiated to determine the quantity and pathways of nitrate addition to surface streams and underground water formations from excreta in beef feeding operations.

Mining Pollution Control Technology

Mine drainage, as noted in the discussions in "Water Pollution and the Environment" is a major pollution problem, particularly in the Appalachian Region. Past attempts to prevent or reduce such drainage have generally failed, and FWQA is emphasizing research to demonstrate the technology necessary to control such wastes.

A new method of preventing the formation of acid mine drainage has been proven through laboratory studies which have shown that an inert gas atmosphere which displaces oxygen will prevent acid mine drainage formation. This method is presently being field tested in an abandoned underground mine and is also being studied for use in operating underground mines. When applied to an operating mine this technique might also reduce the fire and explosion hazards to gassy mines.

Two methods of hydraulically sealing underground mines have also been demonstrated in the field. The first method used quick setting cementation materials placed near the mine portal; it was, however, relatively expensive. The second method used lime and limestone to eventually form an impermeable seal, also placed near the mine portal. This method was less expensive than the former.

The passage of the Water Quality Improvement Act of 1970 adds new emphasis to FWQA's program to demonstrate abatement techniques which will contribute substantially to effective and practical methods of acid or other mine water pollution control. As a result of the mandate of the new Act, the Agency will be stepping up its research in this area.

A sewage treatment system is installed aboard a Great Lakes freighter.

Control of Pollution from Other Sources

In addition to the pollutants already identified, there are a number of very significant waste sources for which improved technology is needed. These include recreational and commercial vessels, construction projects and impoundments, salt water intrusion, dredging, and oil pollution. Although some work has been done on all these problems, emphasis was given to vessel and oil pollution. The Water Quality Improvement Act of 1970 directs further attention to these sources of waste.

Increasing amounts of wastes are discharged from the ever-growing number of recreational and commercial vessels which use both inland and coastal waters. Suitable on-board equipment for properly treating or holding vessel wastes must be developed. In response to a request for proposals to demonstrate the feasibility of various control and/or treatment concepts for wastes generated on vessels, four projects were undertaken in 1969. One system demonstrated holding tanks on pleasure craft and an underwater storage bag for temporary storage of the pleasure craft waste prior to disposal by trucking to a sewage treatment plant. Other concepts are for holding tanks on large vessels and treatment utilizing an electro-chemical flocculating concept.

Closely related to vessel waste control, oil pollution has become a problem of major proportions and of increasing concern. The effects of drilling and tanker accidents, which release large quantities of crude oil into our coastal waters, have been described in detail elsewhere in the report. But the technology to avoid and to clean up such "spills" is woefully inadequate.

Primary program emphasis last year was placed on development of devices and techniques to restore oil contaminated beaches and to harvest oil from the water surface without the aid of additives. Fabrication of a unique centrifugal oil-water separator having high capacity and efficiency and relatively low power requirements was recently completed. An oil harvesting device for oil clean-up is also being designed and fabricated. The two units will be combined and tested at sea early in 1970.

Demonstration projects in progress under the direction of the Maine Port Authority in Portland, Maine, and the City of Buffalo, New York, developed valuable practical information on the effectiveness of a variety of oil containment and clean-up devices and techniques which were evaluated under actual conditions. In-sewer instrumentation for oil detection and oil traps was developed, demonstrated and evaluated. Modification of the inverted siphon is indicated to be an effective oil trap.

In order to use any of the above methods of treatment, the oil must be contained in the local area of the discharge or spill. A system of booms is generally used for containment, but the present systems have not been effective. Model studies were therefore initiated to develop criteria for effective design of booms for harbors, rivers and estuaries.

An increasing amount of attention is being given to methods of preventing oil pollution from tankers. For example, the purpose of one project started in 1969 was to determine the

Straw and booms are still two of the best methods of controlling and cleaning up oil spills. Research and development efforts are aimed at improving this technology.

feasibility of transporting oil in the form of a highly viscous emulsion, created by using ultrasonic techniques and certain additives. The thicker substance of the oil would prevent cargo loss in event of accidents. The same principle is being applied in the development of chemicals to rapidly gel oil within a tanker compartment after leaks are developed.

A joint American Petroleum Institute-FWQA Conference on Prevention and Control of Oil Spills was held in December, 1969. The meeting attracted over 1,200 registrants and 42 equipment exhibitors. Information developed in the course of industry and government programs in this country and the United Kingdom was exchanged, and reports were made on experience with the clean-up of recent large spills. The Conference summary pointed out some advances in oil pollution control technology but strongly emphasized the need for much greater effort in this area.

Water Quality Control Technology

This part of the research program includes all research, development and demonstration directed toward: the prevention and control of accelerated eutrophication and thermal pollution; the control of pollution by means other than waste treatment (e.g. industrial manufacturing process change to eliminate a waste); the socio-economic, legal and institutional aspects of pollution; the assessment and control of pollution in extremely cold climates; and the identification, source and fate of pollutants in surface, ground and coastal waters.

The accelerated aging (eutrophication) of our lakes, brought on by the increased discharge of nutrients (nitrogen and phosphorous) from municipal and industrial wastes and land runoff, has become a problem of major proportions. Technology has rapidly developed to effectively and economically control phosphorous discharges from municipal treatment plants to alleviate a portion of the problem and hopefully retard the aging process. Efforts are being made toward the replacement of the phosphates in laundry detergents with environmentally less harmful materials to eliminate this major source of nutrients. Many and varied approaches are being considered and new ideas sought to combat this extremely complex problem.

Another problem facing us results from the increased demand for electrical energy and the attendant requirement to dissipate waste heat to the aquatic environment. Significant effort is being expended to determine the actual temperature requirements of our surface waters and aquatic life and to discover means of preventing harmful effects of heat.

Development of water quality control technology will become of major and increasing importance as the pollution control payoff from waste treatment becomes increasingly marginal. This involves techniques other than conventional treatment systems, such as industrial process change or management of water resources to minimize the effect of waste discharges. These techniques are applicable in concert with or after high levels of waste treatment are provided.

Cold climate research has also proven to be of significant benefit. Many problems which have been solved elsewhere have required re-evaluation and investigation in Alaska because of the extremes of arctic climate. Efforts are focused on studying pollution problems specifically in regard to the arctic environment, such as determining the impact on Alaskan streams of sewage and other wastes resulting from a rapidly expanding population and industrial growth. An extended aeration system to stabilize wastewater has proven effective in arctic climates, and the use of physical-chemical techniques to provide reusable water for North Slope development camps offers promise.

The Water Quality Improvement Act of 1970 contains a special provision for demonstration of methods to provide central community facilities for safe water and pollution control in Alaskan villages. Today only eight percent of the native homes in Alaska have adequate sanitation facilities. FWQA's research and development staff will be working with the State of Alaska and the Department of Health, Education and Welfare to implement this provision of the new Act and to provide safe water and waste treatment for Alaskan natives, using both conventional and innovative methods.

Waste Treatment and Ultimate Disposal Technology

Waste treatment and ultimate disposal technology focuses on the development and demonstration of new processes and process modifications to control pollution from any source.

There are actually two corollary objectives to be attained through improved waste treatment technology. The obvious one is the alleviation of the Nation's increasing water pollution problems through removal of pollutants from waste effluents; the other is the renovation of wastewaters for deliberate reuse as industrial, agricultural, recreational, or, in some cases, even municipal supplies. These two objectives cannot really be separated, for as our ability to cleanse wastewaters increases, the resulting product water approaches closer and closer to, and may even exceed, the quality of a water supply. This concept, perhaps startling to the average citizen, will nonetheless play a larger and larger role in water resource management, especially in water-short areas.

The need for and the degree of advanced waste treatment will vary with the individual local needs for control of pollution and/or increased water supplies. To meet the spectrum of needs, almost 100 different processes and process variations for treatment and disposal of waterborne wastes have been considered. Some 85 of these processes are under active study at this time at almost 150 different locations throughout the United States. These studies are aimed at determining the efficacy and the cost of the various unit processes which may make up the advanced waste treatment systems of the future.

The fruits of this program have become apparent with the emergence of several advanced waste treatment systems into the demonstration plant phase. The methods being developed range across the spectrum of physical, chemical and biological techniques. They range from the "ordinary," such as filtration and gravity settling, through the "novel," such as biological denitrification, to the "exotic," such as reverse osmosis or ultrafiltration.

The government's investment in this effort has paid off handsomely. First generation process technology, capable of achieving greatly improved pollution control of municipal wastes, has already been brought to the stage of full-scale demonstration and is now available for use under many conditions.

An excellent example of the application of this technology was announced March 24, 1970, by Secretary Hickel and Mayor Walter E. Washington of Washington, D.C. The new process to be installed at the District of Columbia Blue Plains wastewater treatment plant will substantially reduce the pollution of the Potomac and is applicable to rivers and lakes throughout the Nation.

The new technique is the result of a series of research projects conducted jointly by FWQA and the District at the Blue Plains plant. Pilot plants have been testing the new system for two years. The process couples advanced biological techniques with a new physical-chemical treatment. The precipitation phase of the treatment process employs a greater use of chemicals than current processes, and pure oxygen, instead of air, is used in the biological phase of the treatment. The new process appears capable of removing nearly 100 percent of the biological impurities, 96 percent of the phosphates and 85 percent of the nitrogen in wastewater.

The results of this program have provided the necessary technology to reduce the pollution from municipal sources to essentially zero. The present cost is within economic feasibility, but further efforts are needed to optimize both processes and economics. This breakthrough will mean the development of effective, safe and economical wastewater systems, which, in effect, will amount to the same thing as creating a new water supply.

Water Quality Requirements Research

This program provides information on the effect of pollution needed to provide an improved scientific basis for determining the water quality necessary for municipal, industrial, agricultural, and recreational uses and for the propagation of fish and other aquatic life. This information is essential to the establishment and refinement of the Nation's water quality standards. Because of the tremendous number of new chemical compounds being synthesized and finding their way into our environment each year, intensive research investigations must be conducted to develop a predictive capability that will allow us to predict the potential pollutional impact of these compounds in advance.

Far too little is known about the effects of pollution. The drastic effects, such as the massive fish kill, can be easily recognized, but quite often the true cause of such events cannot be defined even with extensive investigation. To look ahead and to predict the occurrence of such events is, unfortunately, well beyond our current capability for any but the simplest stream systems under the least complicated set of environmental conditions and pollution loads. There is also the challenge of detecting, understanding and preventing the more subtle, long-term effects of pollution, which could, even now, be robbing us of valuable water resources. Such effects, as yet unknown, may be just as severe as the sudden fish kill, the unpalatable water supply or the condemned bathing beach. Because these problems are difficult to solve and the starting baseline inadequate, a rapidly accelerated program has been initiated.

Extensive, background data has been acquired and new test methods have been developed to better and more rapidly define the requirements for many uses. For example, a comprehensive research effort to develop sound information upon which to base temperature standards is underway. A temporary field site at a power plant has been established. A standard testing section to determine safe concentration of industrial waste in a natural waterway also continues to show promise. Our research on water quality requirements will continue its accelerated effort to provide the information necessary for the establishment of scientifically sound water quality bases.

Although there are monumental problems still facing the research program, the Agency and the Nation, there is much that is already known; there are problems that have economical solutions. In the future, considerable effort will be focused on putting the results of the research, development and demonstration program in the hands of those charged with implementing water pollution control in our Nation

THE HUMAN ELEMENT

In the final analysis, success or failure of the national pollution control effort will depend primarily upon the human element.

It will depend upon an informed public, which can express its voice intelligently and effectively in decisions affecting the quality of its environment. The President's March 7 Executive Order, issued in furtherance of the National Environmental Policy Act of 1969, placed great emphasis upon the need of the American people to know. He directed all Federal agencies to develop procedures for keeping the public fully informed on the environmental impact of Federal plans and programs and for enabling them to express their voice through public hearings on these issues.

Our success will also depend upon training and motivating a skilled work force to undertake the complex and technically demanding tasks of pollution control. People of many diverse skills and backgrounds will be needed to man the waste treatment plants, the laboratories, the offices of State and Federal regulatory agencies, industries, universities and local governments.

For the long run, the course of pollution control will be dependent most of all upon the attitudes and activities of the Nation's young people. As a group they have perceived—perhaps better than anyone else—that the quality of their lives in future years will depend on what we do about the environment today.

For all these reasons, FWQA is placing heavy emphasis upon the human element in pollution control—through informing the American public, through working with youth and through training and manpower development.

Informing the American Public

FWQA's public information program is founded on the firm conviction that our agency has a major responsibility to meet the American public's need and right to know.

Public information involves much more than mere voicing of official policy. It involves providing the public with full information on efforts to clean up the Nation's waterways, even if such disclosures may sometimes be controversial. This outlook recognizes that public information is often in opposition to public relations, and that its function is to serve the public first. As Commissioner Dominick recently told a group of FWQA information officers, they "are going to have to serve as the innovators, as the creative force, as the non-bureaucratic force, as the force in the Agency which gives us stimulation, new blood, new life, new challenges, new headaches—which gives us all of the things that a Federal bureaucracy could do without."

FWQA has received recognition for its information efforts. Senate Minority Leader Hugh Scott said in the *Congressional Record* of February 9, 1970: "President Nixon, in his State of the Union message, termed environment 'the great question of the 1970's.' It has become a matter of survival. Yet, despite some encouraging signs, too many Americans are still unaware of, or refuse to face up to, the danger. Clearly, there is an informational challenge as well.

'With this in mind, I was particularly gratified to learn that the Washington Chapter of the Public Relations Society of America has, for the second consecutive year, presented its Toth Award for professional excellence to FWQA's Public Information Office. With imagination, inspiration, and ingenuity, they have been alerting America to the multiplying dangers of pollution. Their message is crucial, and they richly deserve this recognition."

The message is being given to the American public by mail, by telephone, and in many other ways. Telephone requests from the news media, students, parents, service and fraternal organizations, and the general public have come from approximately 250 a week last year to nearly 600 a week at present. Correspondence requiring replies has risen from 4,000 a month last year to an average of 5,000 a month so far. Over the past 30 months, the Public Information Office has distributed over 2 million brochures, leaflets, and folders dealing with such subjects as water quality standards, estuaries, heat pollution, acid mine drainage, a primer on waste water treatment, fish kills, what citizens can do about water pollution, vessel pollution, and manpower and training needs. FWQA exhibits and posters have been used by the United States Post Office, the Water Pollution Control Federation, the Izaak Walton League of America, the Audubon Society, the National Rivers and Harbors Congress, the Boy Scouts of America and numerous State fairs and schools.

FWQA's efforts to inform the public have shown particularly gratifying results in television and radio campaigns. Eight film spots have been distributed to television networks and stations coast-to-coast. These spots were produced on what Variety Magazine described as a "shoestring" budget and were good enough to "make Madison Avenue shiver and shake." The Variety writeup continued: "The chiller is that the FWPCA (sic) division of the USD of I (United States Department of the Interior) did it on a production budget totaling $31,000, without an ad agency—and with a producer who had never turned out a blurb before." The International Broadcasting Awards and the American Television Commercial Awards—the advertising world's version of the Academy Awards—cited the "Clean Water" television spots as outstanding in the Public Service Category. Twenty-five radio "Clean Water" spot announcements were produced by FWQA's public information program. Some were interviews with prominent and average citizens, fishermen, conservationists and resort owners who had suffered as a result of water pollution. Another radio series provided a recording of New Orleans jazz by the Chicago Footwarmers, in which variations of popular songs were adapted to the theme of water pollution control.

The television and radio campaign has produced results. Mail addressed to "Clean Water, Washington, D.C.", solicited from viewers and listeners has shown a sharp rise. These letters are answered with literature which gives the correspondent an appreciation of the problem and of means to rectify it through community action.

Of course there is a temptation, in the midst of the ecological furor, to be overzealous. As a prominent columist observed, "The environment issue lends itself to grandstanding." It is a situation in which the fear words and the bright blue words come too easily. The public must not only be alerted to hazards, but also apprised of progress—progress being made in research, in clean-up agreements reached with industry, and in successful new approaches to the task at hand. Of the some 200 FWQA press releases issued since Secretary Hickel took office, many have dealt with new approaches for turning wastes into usable products, for using sludge as a fertilizer for crops, for new methods for controlling pollution from combined storm sewers—as well as with the oil disasters, the dying lakes, and the dangers posed by new contaminants.

The Water Quality Improvement Act of 1970 also points to the importance of adequately recognizing progress in pollution control. The Act authorized a program of official recognition by the Federal government to industrial organizations and local authorities which have demonstrated outstanding technological or innovative achievements in their pollution abatement programs.

Literature and radio campaigns inform the public of water pollution problems and solutions.

Looking to the future, the public information program of FWQA has produced a film entitled, *The Gifts,* which will be distributed to citizens groups and television. The movie on water pollution and its impact on the chain of life is narrated by Lorne Green, with original music by Skitch Henderson, and again sounds the theme that we must act—now.

In the publications field, a new booklet aimed at grade school children is being planned. The booklet may use drawings done by children because of their fresh charm and appeal.

Working with Youth

The quality of the environment is fast becoming the consuming issue on our campuses. At least 500 colleges and 1,500 high schools are expected to conduct environmental teach-ins on April 22, 1970. FWQA has been invited to participate in many of these events. Over 100 staff members are expected to serve as speakers and panel members, and a large volume of literature and other materials is being made available to individual campus sponsoring organizations.

As an agency whose mission is environmental protection and preservation, FWQA since 1969 has been deeply involved with students seeking to participate more effectively in the quest for environmental quality improvement.

SCOPE (Student Council on Pollution and the Environment) was created to serve as a two-way communication link between students and government on the issue of environmental quality. For the students it is an opportunity to obtain and apply governmental expertise and information to the process of formulating solutions to environmental problems and a chance to discuss their proposals for solving environmental problems with top-level government decision-makers. For the government it is a means of getting fresh viewpoints on environmental problems and solutions. Government agencies will be able to request student study and recommendations on specific points or issues.

SCOPE is composed of students at the college and high school levels interested in the issue of environmental quality. A SCOPE group was established in each of FWQA's nine Regions in

December, 1969. The first meeting of national representatives elected by each Regional organization was held in Washington on February 20–21, 1970. At the national meeting, Secretary Hickel committed a large amount of his time to listening to SCOPE representatives' proposals and answering penetrating questions that reflected their broad concern for all facets of the environment.

SCOPE was initiated by FWQA in response to Secretary Hickel's belief that improved communications would benefit both the Federal government and concerned students. SCOPE is an innovative experiment. Now that its basic feasibility has been demonstrated, the possibility of broadening its sponsorship both within and outside the Department of the Interior is being explored. Secretary Hickel recently announced the formation of a "Task Force on Environmental Education and Youth Activities" to act as a go-between for the Department and young people concerned about the environment. The Task Force's immediate projects include being the liaison group for SCOPE and making recommendations for the creation of a National Environmental Control Organization (ECO), proposed by the Secretary and modeled after the Peace Corps. The Task Force is also programmed to provide the focal point within the Department of the Interior for its participation in future national student teach-ins. Upon request, the group will provide assistance, information, and speakers to colleges, high schools, and private organizations.

A SCOPE meeting is conducted in Richmond, Virginia.

Perhaps the most basic point expressed by SCOPE members is that mankind will have to change many of its attitudes and aspects of its life-styles if we are to live within the earth's supply of natural and recreational resources over the long term. They see the need for general recognition that the earth and its inhabitants form a "closed system" and that actions by any segment of its population generally have an effect on other groups—or perhaps on the action-originating group at a later date. Further, they believe that remedial steps require changed attitudes and public acceptance and support for the expenditure of vast sums to improve the quality of our environment.

Public awareness and attitudes are at the heart of all of these broad concerns. In order to improve our understanding of the nature and magnitude of the public education task that lies ahead and to understand better what role organizations such as SCOPE can play, FWQA is seeking the help of the Institute for Creative Studies.

The Institute for Creative Studies is a private, nonprofit, educational corporation which attempts to use bright, imaginative, innovative high school and college students to apply modern research techniques and scientific methods to the resolution of policy problems. The institute began as a pilot project in the summer of 1967. The research projects are funded by government contracts and the Eugene and Agnes E. Meyer Foundation.

The only controls on the individual students' research projects are regular quality control review sessions, a formal interim report, and a thorough review of each project by a panel of experts at the end of the project period.

The Institute for Creative Studies will investigate in depth the role and nature of public attitudes on water pollution control problems. Additional topics may also be considered by the Institute for Creative Studies in connection with their work for FWQA.

Young men pick up debris along the banks of the Potomac River as part of "Operation Clean Waters."

FWQA has been involved in other work with young people. For example, a program called "Operation Clean Waters" has been conceived and organized by FWQA to involve youth directly in the clean-up of water. Pilot projects have demonstrated that teams of young men aged 16 to 21 can remove tremendous amounts of debris from waterways, thereby improving their aesthetic appearance and value for recreational use. These pilot projects have been carried out in the District of Columbia, Chicago, and Puerto Rico. This program will be expanded to a number of other cities. The new projects will be supervised entirely by local governments, with FWQA staff serving as advisors.

In another approach to young people, FWQA is developing a project with the Boy Scouts of America that will be known as "Conservation Good Turn." A Boy Scout *Leader's Guide* has been prepared outlining various projects which the Scouts can undertake, such as checking to see whether their community has a waste treatment plant; if the sources of pollution from industry are under control; and where other trouble spots are developing. The *Guide* gives directions for checking the quality of water in a stream or lake. We are anxious to enlist the support of the five million Boy Scouts in this country as another volunteer cadre for protecting the environment.

In addition to the involvement with these special youth programs, FWQA has a number of on-going programs which involve youth participation and offer young people an opportunity to work or study in the field of water pollution control. These programs—to be discussed in the following section on training and manpower—include traineeships and fellowships, grants to technical, professional, and secondary schools, in-house short-term training, and part-time or summer jobs.

Training and Manpower Development

Substantial expenditures for construction grants, research and development, technical assistance, and similar endeavors are outlined in various sections of this report. Effective utilization of these funds and achievement of clean water results will basically depend on adequate staffs of skilled and motivated people, from treatment plant operators to research scientists. We must very substantially increase both the number and proficiency of those employed in the water pollution control effort and, accordingly, manpower development has become a major program thrust within FWQA.

The objectives of manpower development programs are to assist in attracting and preparing new professionals, technicians and operators and to help prepare existing personnel to do a more effective job. To meet these objectives, FWQA is pursuing a number of approaches. These include support of and work with the universities to assure an adequate flow of engineers, scientists, and other professionals into the field; conduct of short-term training by FWQA staff, designed to upgrade the skills of those already in the field; and a variety of approaches to the training of sewage treatment plant operators.

FWQA is working to increase the flow of highly trained professionals through training grants awarded to academic institutions to establish or extend the scope of advanced training in water pollution control in their engineering, biological, physical and social science departments. Under this program, institutions are encouraged to develop the specialized and multidisciplinary training of scientists, engineers and administrators in water quality management. These grants support expansion and improvement of facilities and equipment, provide partial support of faculty salaries and offer stipends, dependency allowances and tuition to trainees. In 1969 training grants were awarded to 61 institutions. This type of grant will support 693 trainees in 1970, most of whom are working toward master's degrees.

Research fellowships are also awarded to individuals for specialized graduate and postgraduate research training involving investigations particularly related to FWQA's mission. These awards provide funds for institutional costs of education, stipends for the fellow and allowances for supplies. Fellowships are generally awarded to persons working towards the Ph.D. degree, the objective being to maintain the future supply of research scientists and engineers and university professors. A long training period is required to produce researchers and teachers, generally at least three years of full-time study after the bachelor's degree has been obtained. It is extremely important to maintain a steady flow of persons under training so that there are no major gaps in the supply of trained persons available to begin research and teaching careers. About three-quarters of the Ph.D. recipients who have received FWQA support through a fellowship or training grant embark on research and university teaching careers.

In 1969, approximately 300 students supported by FWQA training grants or fellowships received advanced degrees. They will make a significant contribution towards filling the demand for new professional talent in the field.

Other steps are being taken to increase this flow of talent. FWQA will be participating in intensified Federal efforts to improve the quality of education available at black institutions in accordance with declared Presidential support for a Black College improvement program. FWQA training grants have already been awarded to two such institutions. Predominantly Negro Delaware State College received support for development of an undergraduate water chemistry course to train baccalaureate candidates for pollution control-oriented jobs in industry. More recently, a grant was awarded to Howard University to support a Master of Science in Sanitary Engineering program. In 1970, we expect to consider a proposal for training pollution control microbiologists and biochemists at Tuskegee Institute. In the coming year, other black institutions will be investigated to determine opportunities for and means of developing professional training programs in water pollution control.

We are also exploring the need to encourage pollution control training at an earlier stage through increased emphasis in junior and senior high school science curricula. As a start in this direction, the Tilton School in New Hampshire was recently awarded a grant to provide for the modification and re-writing of a previously developed teacher's guide. The guide provides objectives, procedures and teacher's plans for scientific analysis of water pollution problems and consideration of social, legislative and historical factors. The revision will be performed in the summer of 1970 by a group of teams composed of a high school science or biology teacher and a student from each of forty different schools. These teams, during the regular school year, have gained experience in field and water laboratory testing techniques and will base their revisions on this experience. The teacher's guide is expected to become basic material for initiating secondary school courses emphasizing water pollution control at schools across the country.

We must not only attract and train new people for careers in pollution control; we must turn our attention to those already in the field. Water pollution control technology and techniques are developing rapidly. To be effective, pollution control personnel must be kept up-to-date on the latest developments. One of the best means of obtaining such updating is through attendance at short-term training courses. This type of training is also needed by the ever-increasing numbers of trained people shifting from related fields to water pollution control. They need to be acquainted rapidly with current knowledge and methods.

To meet these needs, specialized and advanced technical training is offered at FWQA laboratories to government employees and others working in pollution control. Special emphasis is given to training courses or programs which assist the State and local agencies in training their personnel, thus strengthening State and local effectiveness in the water pollution control effort.

Trainees are drawn from the professional, technical and treatment plant operator ranks across the Nation. In fiscal year 1969, more than 50 of these short-term courses were presented to approximately 1,300 persons at FWQA training facilities. The curricula included a variety of technical courses in water quality management of one or two weeks' duration. Also offered are orientation courses and short technical seminars to meet the special needs of particular Federal, State and local agencies or academic institutions. For example, in 1969 FWQA presented a two-week "Water Quality Studies" course in Harrisburg, Pennsylvania, to meet the needs of that State's employees. Also, two courses were offered to assist Federal agencies in meeting their increased responsibility to prevent water pollution: "Design and Management of Sewage Treatment and Disposal for Federal Installation," and "Water Pollution Control for Federal Installations." A special course was conducted for U.S. Geological Survey personnel to enable them to participate fully in the accelerated water quality monitoring program described elsewhere in the report.

Training of sewage treatment plant operators has been an area of special and increasing emphasis in the FWQA training program. The fastest and cheapest way to significantly improve water quality in the short run would be to operate existing treatment plants at reasonably efficient levels. Too often today, multi-million dollar plants produce unsatisfactory effluents which deny desired and obtainable water uses. Usually the reason is that these expensive plants are turned over to poorly trained personnel for operation and maintenance. Poor plant operation can result in undue pollution of the receiving waters with the resulting loss of water uses, such as closed swimming beaches. Poor plant maintenance can be extremely costly in yet another way. Most waste treatment plants are designed and constructed so as to have a useful life of at least twenty years. Improper plant maintenance can actually reduce that useful plant life to one or two years in extreme cases.

The need for competent, well-trained operators in the Nation's treatment plants is obvious. Traditionally, this has been viewed as a responsibility of State and local governments. The Federal government, and FWQA in particular, has taken a more active role in the past few years for very basic reasons. The job was not being adequately done at the State and local level: a large portion of existing treatment plants were, and are, being poorly operated and maintained. State and local governments often have had difficulty marshalling the financial and staff resources needed to conduct adequate training programs on their own. Therefore, FWQA has worked to provide advice, consultation and financial assistance to State and local governments to carry out operator training.

Recently, improved operation and maintenance of treatment plants has become more than a matter of Federal encouragement and assistance; it will be required in order for States and communities to receive construction grant assistance. It would make little sense for the Federal government to embark upon a major program to assist construction of treatment works without assuring that, once built, they will be adequately operated and maintained. Secretary Hickel's recently published regulations to this effect have been described elsewhere in this report.

FWQA is supporting operator training in several ways. First, and foremost, FWQA has assisted State and local governments in qualifying for funding for operator training under a variety of existing programs administered by other Federal agencies. This involves working with State and local governments to identify training needs, to formulate training programs to meet those needs, including assistance in such areas as curriculum development and instructor training, and to obtain Federal financial assistance. FWQA then works with Federal agencies to gain acceptance for Federal support of this training and to develop procedures to make funds available. Utilizing principally Manpower Development and Training Act (MDTA) funds which are administered by the Departments of Labor and Health, Education and Welfare, FWQA assisted projects that accomplished the training of 981 operators in fifteen States and in Puerto Rico in 1969. The number of operators trained under this mechanism in 1970 will total approximately 2,800 in 30 States.

The present use of MDTA funds illustrates the successful application of a multiple-purpose governmental program. FWQA-assisted projects utilizing Manpower Development and Training Act funds not only produce trained operators but also serve to enable persons classified as unemployed or under-employed to obtain better jobs and participate more fully in the economic life of the Nation. FWQA is further developing this approach through the Department of Defense's "Project Transition" which affords an opportunity to attract returning servicemen into the pollution control field. The "Project Transition" program provides enlisted military personnel with training for civilian jobs during their last six months of duty. Training is funded by the Manpower Development and Training Act and is administered by the Departments of Defense, Labor, and Health, Education and Welfare. FWQA is currently developing a pilot program to provide entry-level training in wastewater treatment plant operations for approximately 300 servicemen at Forts Belvoir, Virginia; Bragg, South Carolina; Hood and Bliss, Texas; and at the El Toro Marine Air Base in California. FWQA will use information gathered through a variety of programs to assist successful trainees in obtaining jobs across the country in waste treatment plants seeking qualified personnel.

We are moving forward in a number of other ways to upgrade operator training. Correspondence courses may prove the most practical method of reaching many operators of one-man plants—of which there are thousands. By late 1970 or early 1971 FWQA expects to have three correspondence courses available to help meet this need. The University of Michigan, under an FWQA grant, has developed a course utilizing programmed learning on chemistry of water and wastes for operators and technicians. Within FWQA's own short-term training teaching staff, a course on membrane filter methods in water microbiology has been developed. It will be aimed at operators. Under another grant, Sacramento State College has developed a course for improving the skills of operators in small and remote plants.

Efforts are also underway to better prepare those who will be responsible for training operators. FWQA developed and first offered a short-term training course for instructor development in April 1969. We co-sponsored with Clemson University the first large-scale national conference on operator training in Atlanta in November, 1969. This first-of-its kind meeting provided a forum for operator-trainers to meet together and listen to and discuss presentations on the latest instructional methods and teaching aids.

The President's February 4 Executive Order on control of Federally-caused pollution has established a vastly increased responsibility for FWQA to assist other Federal agencies in training operators of plants at Federal installations. The order requires Federal operators to meet levels of proficiency consistent with those being required of operators at the community level. To assist the Federal agencies, we will provide increased training opportunities, using FWQA training facilities and staff to present selective offerings of practical courses in waste treatment plant operation, methods and procedure—both for Federal operators and for personnel engaged in training Federal operators. This program will also provide FWQA with an opportunity to develop and test training techniques and materials which will ultimately be passed on to State and local governments for use in training large numbers of operators.

Enactment of the Water Quality Improvement Act of 1970 will further strengthen FWQA's activities and programs in training treatment plant operators. The new legislation authorizes a combination of grant, contract, and scholarship programs to attract and prepare students for careers in the design, operation and maintenance of waste treatment plants. Planning for implementation of new activities and approaches under this legislation is now underway.

In summary, FWQA is very substantially accelerating its training efforts, in concert with State, local and Federal agencies, with universities, and with others concerned. More effective manpower planning is needed to guide these efforts.

Trained operators are needed to assure more efficient waste treatment.

FWQA's last overall study of manpower needs, *Manpower and Training Needs in Water Pollution Control,* was submitted to the Congress in 1967. A much more specific appraisal of where and when job vacancies will occur and how they may best be met is now required. In 1969, FWQA initiated development of a manpower planning system which, when implemented, will define manpower demands, manpower supplies, and criteria for judging whether manpower resources are being effectively utilized. The system will provide carefully developed estimates of the total manpower needs in the water pollution control field and improve the identification of particularly severe manpower shortages. The system will also include more precise definition of occupations, manpower staffing guides, work force profiles, and industrial planners.

This manpower planning system will enable FWQA to formulate better action plans, through understanding the timing and nature of State, local, industrial and academic training needs. Rapid and effective implementation of this system will be needed to help us meet the training provisions of the Water Quality Improvement Act of 1970.

INTERNATIONAL ACTIVITIES

Public concern for environmental quality has reached international proportions in the last few years, and President Nixon has advanced the participation of the United States in efforts to solve global pollution problems.

The Federal Water Quality Administration (FWQA) is active on several major fronts of international activity in the environmental field. Efforts are moving ahead to meet the increasing pressures for an international leadership role in the environmental quality area.

The United States shares the North American continent with Canada and Mexico. A significant part of the water resources of the continent crosses or forms a part of the political boundaries between the United States and its two neighbors. This is especially true along the Canadian boundary where the Great Lakes system, constituting the largest source of fresh water in the world, is shared equally.

An important part of FWQA's involvement in international activities is the provision of technical support to the International Joint Commission (IJC). The latter was established pursuant to the Boundary Waters Treaty of 1909 between the United States and Canada. This activity includes membership on a number of international technical advisory boards which have been established by the IJC to investigate and report on specific boundary water problems referred to the Commission by the Governments of the two countries. At the present time, there are seven technical advisory boards working on the pollution problems of Lake Erie, Lake Ontario, the international section of the St. Lawrence River, St. Croix River (Maine), Niagara River, Detroit River, St. Clair River, St. Marys River, and Rainy River of the North.

Because of the serious acceleration of pollution in the highly industrialized areas of the Great Lakes, the work of the IJC and its advisory boards has assumed an increasingly important role in coordinating the remedial programs being carried on in the two countries to abate pollution. This coordination has resulted in significant agreement on the present levels of pollution in Lake Erie and Lake Ontario, the sources and amounts of pollutants reaching the Lakes and recommendations for an abatement program. In recent weeks, a comprehensive report on these agreements has been submitted to the IJC by its technical advisory board.

Other programs being coordinated through the IJC are oil contingency planning for boundary waters, vessel pollution control and review of off-shore drilling practices. In addition, the meeting of water pollution control technicians of the United States and Canada on boundary water problems has resulted in increasing cooperation in several areas which have not been referred to the IJC for consideration, such as Arctic pollution, exchange of scientific information, participation in pollution seminars and consultation on handling of oil spills.

Within the last year, meetings between higher levels of administrative personnel on matters of policy have developed as a result of the complexity of the pollution problems of the Great Lakes. Meetings were held between Secretary Hickel and Assistant Secretary Klein and their counterparts from Canada. As a result, the governments of both countries are moving closer together in a coordinated approach to pollution abatement in the Great Lakes. Additional meetings are being planned for FY 1970–71 involving White House level officials of the United States Government.

Although a Water and Boundary Treaty was established between the United States and Mexico in 1944, it contains no provision for formal institutions for dealing with pollution problems as is contained in the treaty with Canada. However, informal arrangements are established with the Water and Boundary Commission, and the FWQA does provide consultative services on border pollution problems when requested. Consultative services have been provided on border pollution problems stemming from domestic wastes in the Brownsville-Matamoros, El Paso-Juarez, Nogales, Yuma-Mexicali and Tijuana areas.

As the world's technicians turn to the task of controlling pollution of global waters, the development of a reliable mechanism for the exchange of existing and developing scientific information becomes increasingly necessary. As a result of this need the United States has established or explored bilateral agreements with other countries to exchange technical knowledge on water pollution control and research. Such agreements have been in operation with Germany and Japan for several years. Agreements to develop bilateral exchanges are presently being negotiated with the Soviet Union, France and Czechoslovakia. Requests for such agreements have been received from Sweden, the United Kingdom, Iceland, Poland and Romania. In negotiating these agreements, consideration is being given to including cooperation in specific research projects in problem areas of mutual interest, such as sludge disposal, the effects of pollution on fish and aquatic life, eutrophication of lakes, effluent standards and user charges.

The effect of detergent phosphate on the environment has been a matter of public and scientific discussion for several years and has recently come to the front as a major issue in the problem of accelerating lake eutrophication. As a result of the shared concern over the nutrient enrichment of Lake Erie, a joint United States-Canada team of scientists undertook a mission to Sweden in January, 1970, to investigate and study the use of low phosphate-content detergents in that country. Their findings will contribute to the development of policies for phosphate reduction in both nations.

During the past year, over 100 foreign water pollution technicians and scientists have visted the United States to study control programs and techniques which have been instituted in this country. A number of these visitors have participated in the short technical training courses offered in the FWQA Regional laboratories on various aspects of water pollution control technology. This represents a sharp increase over previous years and indications are that the number can be expected to double in the next 12 to 24 months. This increased number of foreign visitors is also expected to include higher-level government administrative officials than in past years.

With the establishment by President Nixon of the Environmental Quality Council and the concurrent structuring within the Department of State of an Office of Environmental Affairs, the Administration is gearing to meet increasing responsibilities in the international area. Most, if not all, of the international, multi-lateral organizations in the free world today are in some way engaged in carrying out programs in environmental protection. These programs consist mainly of establishing procedures and organizational arrangements for the exchange of technical and scientific information and of providing a platform for the discussion between government officials of member countries on environmental problems of general concern.

In recent months, however, increasing attention has been given to the development of international policies for environmental protection. Many international conferences and symposiums are scheduled for the next 12 to 24 months, including the international Water Pollution Control Research Conference in San Francisco in September 1970; the Environmental Safety Conference in 1971 in Prague, Czechoslovakia, sponsored by the Economic Commission for Europe; and the UN's major effort in this field in 1972 in Sweden. The conference will bring together the world's leading scientists and political leaders to discuss the environmental problems that beset the world.

FWQA has provided an increasing number of its technical and top administrative personnel to support these developing activities. This includes the appointment of agency representatives to environmental technical and planning panels which have been established in the North Atlantic Treaty Organization framework, the Economic Commission for Europe, the Organization for Economic and Cultural Development, and others. Co-sponsorship of the biennial International Congress on Water Pollution Research is a major undertaking of FWQA, and the Agency will be active in the planning and conduct of the next conference to be held in San Francisco in September, 1970.

The involvement of FWQA in the international field has been relatively small in the past and restricted to specialized technical fields. But sudden world concern for protection of the environment will thrust upon us an increasing pressure to share our knowledge, progress and technical capability with all Nations. This is especially true if the United States is to continue its present role as a leader in the free world.

ORGANIZATION, RESOURCES, AND FACILITIES

The capability of any agency to accomplish its mission is dependent upon its resources—budget, staff, and facilities—and upon how effectively those resources are organized and managed. During the past year, substantial efforts have been directed towards the improved organization and management of Federal Water Quality Administration (FWQA).

Major improvements have been made in FWQA's personnel systems and organization structure. Added emphasis is being given to systematic work planning—competing demands on the Agency's resources have made of prime importance the identification and maintenance of priorities, schedules, and objectives to guide our work. An Agency-wide accounting and management information system will be operational by July 1, 1970. This system will generate electronic data programs and develop reports which will aid top management in their decision-making process.

A formal directives system has been established to assure rapid and accurate communication of policy and instructions throughout the Agency. Better systems of delegation of authority and other management improvements are currently underway.

FWQA's mission is an increasingly complex one, and constant attention to modern management methods is an essential part of its overall job.

Organization

FWQA is organized along functional lines, as outlined on the attached organization chart.

During the past years, there have been a number of changes in FWQA's organizational structure, at both Headquarters and field levels, designed to marshall the Agency's resources most effectively to meet its changing mission.

With increasing emphasis placed on securing compliance with established water quality standards, the standards function has been transferred from the jurisdiction of the Assistant Commissioner for Operations to the jurisdiction of the Assistant Commissioner for Enforcement. The current emphasis on the environment as a whole is reflected by the proposed creation of the position of Assistant Commissioner for Environmental and Program Planning. The passage of the Water Quality Improvement Act of 1970 prompted the creation and staffing of an Office of Oil and Hazardous Materials.

The bulk of FWQA's activities is in the field. Of a present staff of 2,538 permanent and temporary employees, 592 are located in Headquarters, and 1,946 are assigned to the nine

FEDERAL WATER QUALITY ADMINISTRATION

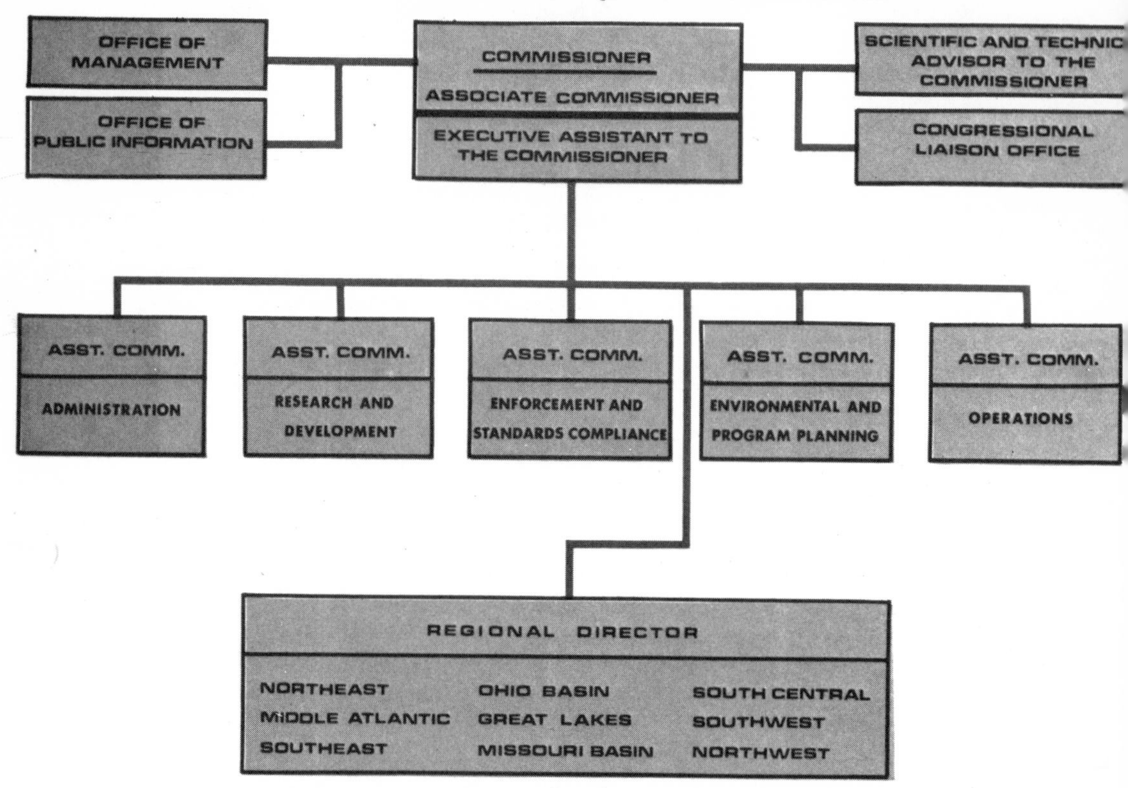

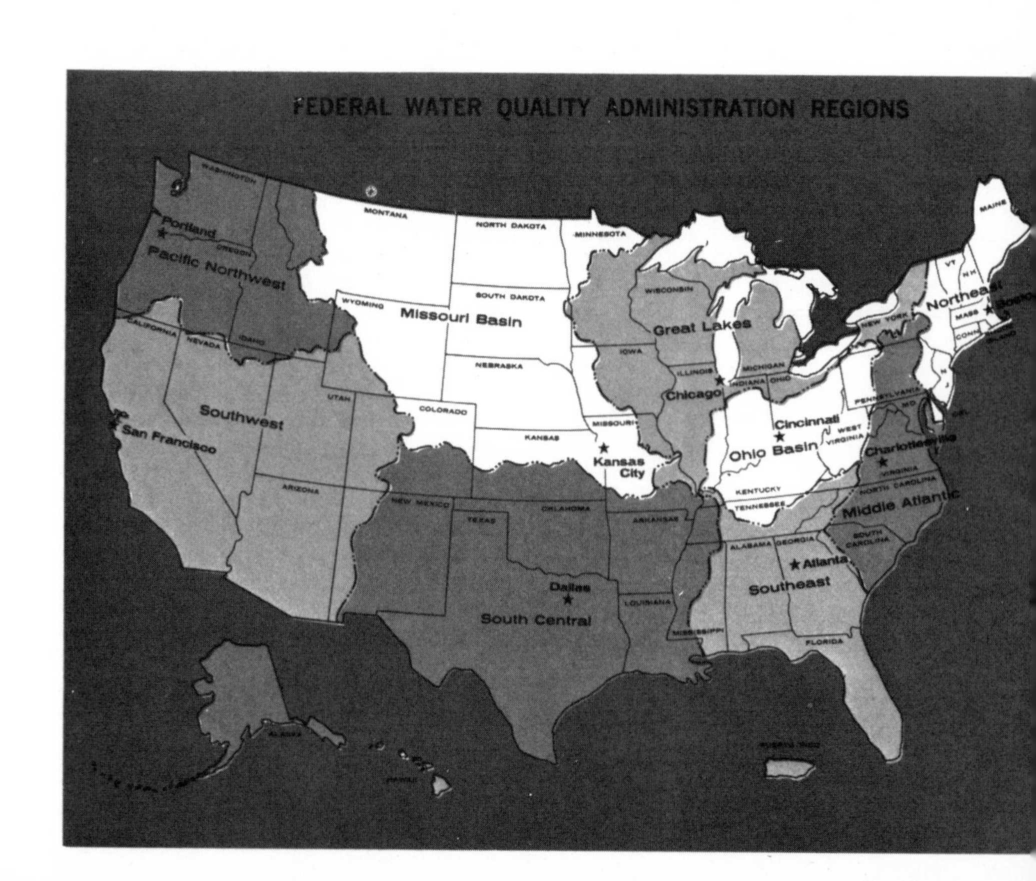

Regions. Regional boundaries are outlined on the attached map.

FWQA's Regions are organized along hydrologic lines to facilitate the planning and implementation of the clean-up of entire river basins and to aid our work with related water resource agencies. This method of organization is considered most effective in terms of the Agency's current program operations. It sometimes creates difficulties, however, for States whose boundaries fall within more than one Region. With the increased emphasis on effective working relationships with the States, major attention is being given to adjustments in responsibilities and lines of communication which will ease these problems.

The Department of the Interior's participation in the President's Federal Activities Review Program, which is designed to assure that services to State and local government are of maximum effectiveness, may lead to further adjustments in our Regional structure.

Personnel

FWQA's most valuable resource is its staff —a staff comprised of dedicated and experienced professionals, with backgrounds representing the many disciplines needed to operate an effective governmental agency. Heavily represented on the staff, because of the nature of the Agency's mission, are scientists and engineers with specialized experience in water pollution control, oceanography, and related fields. Lawyers, economists, public administrators, regional planners, and others provide the needed balance of skills.

A major management improvement during the past year has been initiation of an Agency career development system, designed to provide for planned intake of college graduates in entrance level positions; training and development for each careerist; a career counseling and appraisal system; and a centralized bank of data on all employees in an occupational field. This system will cover all scientific and engineering, technical support and administrative personnel by June 30, 1970. The Career Planning System will enable management to obtain, develop and retain a highly qualified workforce to meet mission goals and objectives in a timely and economical manner.

In addition to this system, a Graduate Fellowship Program was developed to provide a system to hire top quality graduate students who have completed all requirements for their advance degree but the thesis. They are hired on temporary appointments for one year and work on a special project selected by FWQA which can serve as the basis for their thesis. These employees form a pool of outstanding candidates for future employment with FWQA on a permanent basis.

Further in-house personnel management improvements were made by the implementation of a personnel program evaluation and management advisory service designed to measure the effectiveness of personnel management policies, practices and procedures. Lengthy interviews with managers at all levels, non-supervisory attitude questionnaire sessions, and discussions with Personnel Office staff members have provided the data for evaluation. At the conclusions of each survey, a report is made to management containing action items or recommendations for improving working conditions, employee morale, and supervisory performance.

Another innovation is the automated personnel system which results in statistical reports prepared by computer which greatly reduces the amount of time spent on this function at all levels of management. It also provides management with instant feedback of data needed for planning and other purposes. By the end of FY 1971, it is anticipated that all employee training records including FWQA-wide training needs will be fully automated. Also, the skills inventory file will be converted to an automated data bank to enable the instantaneous referral of outstanding candidates for vacant positions and to provide data needed for the manpower planning function.

Facilities

In addition to its Headquarters and Regional Office locations, FWQA conducts its work at 46 field stations and laboratories located in the field. These facilities range from complex laboratories, designed and operated to conduct sophisticated research, to small field stations, studying special problems. A variety of physical facilities is needed. At the Southeast Water Laboratory on the University of Georgia campus at Athens, controlled environmental chambers, designed to simulate varying conditions in the natural environment, have been constructed. Work with these chambers is shedding new light on basic pollution relationships in streams. In Newtown, Ohio, an entire tributary has been protected and controlled with weirs and other devices to test the long-term effects of low level toxic wastes on biota under natural conditions. This unique facility has already attracted the attention of scientists across the Nation. A small laboratory on a floating barge provides a

base for a team of investigators studying pollution along the Florida coast. The National Water Quality Laboratory at Duluth, Minnesota, provides special facilities to conduct a wide range of studies designed to determine environmental requirements of fresh water organisms.

During the past year, the Bears Bluff Laboratory on the South Carolina coast was leased to FWQA by a non-profit educational institution. This facility will provide an invaluable opportunity to conduct work on environmental requirements of southern waters marine life—an important need in the establishment of improved water quality criteria.

Currently, FWQA is completing a comprehensive review of the need for additional facilities. A 5-year proposed facilities program has been developed. It is designed to provide necessary facilities and laboratory space for the future.

Scientists in FWQA laboratories conduct research on pollution and its effects on aquatic life.

Budgetary Resources

FWQA's budgetary resources for the past, current and coming fiscal years are shown below. These figures show a significant increase for water pollution control, reflecting the high priority this program is receiving from the President and Congress during a period of overall budgetary stringency.

Appropriations to Federal Water Quality Administration (in thousands of dollars)

	FY 1969	FY 1970	1971 President's Budget
Research, development and demonstration	$43,611	$37,260	$44,092
Comprehensive planning			
Federal planning studies	4,936	5,214	5,143
Planning grants	1,250	1,782	2,900
Control of Pollution from Federal activities	858	1,031	1,158
Technical support	5,732	6,181	6,188
Pollution surveillance	2,690	4,012	4,286
Training			
Federal activities	1,274	1,573	1,801
Training grants	4,000	4,620	5,250
State and interstate program grants			
Grants	10,000	10,000	10,000
Grants administration	329	344	394
Construction grants for waste treatment works			
Grants	214,000	800,000	[a]
Grants administration	2,674	4,198	5,883
Enforcement	4,042	4,381	5,256
Executive direction and support	5,279	5,528	5,667
Total, new obligational authority	300,675	*886,124	*[a]98,018

[a] For 1971, $4 billion is provided in proposed legislation which would provide contract authority for use over four years for grants to localities for construction of waste treatment works, of which $1 billion will be allocated in 1971 and in each of the next three fiscal years.

*Amounts shown exclude any consideration for Water Quality Improvement Act of 1970.

B. Ocean Waters

Ocean Dumping, A National Policy. A Report to the President Prepared by the Council on Environmental Quality, October 1970. Pages 1-37, 39-41.

On April 15, 1970 President Nixon directed the Chairman of the Council on Environmental Quality to work with federal, state and local governments in preparing a comprehensive study of ocean dumping. The study was to recommend research needs and legislative and administrative actions. Only a few months after this Presidential message the need for a national policy became evident. A Department of Defense plan to dump large quantities of lethal gasses off the Atlantic coast aroused a heated public outcry. Despite the outcry the dumping proceeded as scheduled.

The Council on Environmental Quality has previously been discussed in Chapter I of this work. In preparing the report the Council worked closely with the following entities: Atomic Energy Commission, Department of the Army, Department of Commerce, Department of Defense, Department of Health Education and Welfare, Department of the Interior, Department of State, Department of Transportation, Executive Office of the President, National Council on Marine Resources and Engineering Development, National Science Foundation, and the Smithsonian Institute. The final report consisted of 45 pages, plus ten pages of introductory matter. The excerpt reproduced below consist of 37 pages plus six pages of introductory matter. The import of the study is quite clear: current international, federal, state and local controls are grossly inadequate to cope with the problem. The Ludicrous nature of our current controls is (ironically?) revealed by the following quotation from page 35 of the Council's report:

> "The international law governing the high seas . . . is codified in the 1958 Geneva Convention on the High Seas. This Convention provides for freedom of navigation . . . and other freedoms recognized by international law, such as dumping."

CHAPTER I

Ocean Dumping: Location, Quantities, Composition, and Trends

About 48 million tons of wastes were dumped at sea in 1968. These wastes included dredge spoils, industrial wastes, sewage sludge, construction and demolition debris, solid waste, explosives, chemical munitions, radioactive wastes, and miscellaneous materials. This chapter indicates rapid increases in ocean dumping activity over the last two decades and the potential for great increases in the future. At the same time, ocean dumping of wastes from other sources should decrease through implementation of water quality standards and new Federal laws dealing with control of sewage from vessels and with oil pollution.

DISPOSAL SITE LOCATIONS

Data on disposal sites are still incomplete, with little definitive information on sites off Alaska and Hawaii and outside the U.S. contiguous zone (more than 12 miles offshore). There are almost 250 disposal sites off U.S. coasts. Fifty percent are located off the Atlantic Coast, 28 percent off the Pacific Coast, and 22 percent in the Gulf of Mexico. Table 1 summarizes the number of sites for each major area and the number of permits issued for their use. The locations of the disposal sites are indicated in Figure 1.

Table 1.—*Ocean Dumping: Site Location Summary (22, 66)*

Coastal area	Number of sites	Active Corps disposal permits
Atlantic Coast	122	136
Gulf Coast	56	50
Pacific Coast	68	71
Total	246	257

Not included in Table 1 are some 100 artificial reefs constructed by private concerns under permits issued by the U.S. Army Corps of Engineers. (66) These reefs, sometimes formed of old car hulks or tires, are intended to provide artificial shelters for fish.

QUANTITIES AND TYPES OF WASTES

The categories of wastes covered in this report are used because of the large quantities of materials currently dumped, their potential for increase, or their special characteristics, such as toxicity. The quantities for each category are summarized by coastal region in Table 2. Radioactive wastes and chemical munitions are not included in the table because weight is not a meaningful descriptor. Each, however, will be discussed later.

The Bureau of Solid Waste Management estimates that the data in Table 2 represent about 90 percent of ocean dumping. However, the data undoubtedly underestimate the size and scope of the problem because of the time lapse and the possibility of many small community operations or illicit operations by private firms. Also not included in the table are those wastes that are piped to sea.

Each major category of ocean dumping sources is now discussed and the possible chemical composition of the wastes delineated as an aid in evaluating their present and potential effects on the marine environment.

Dredge Spoils

A large percentage of dredging is done directly by the Corps. The remainder is done by private contractor under Corps permit. Spoils are generally disposed of in open coastal waters less than 100 feet deep.

LEGEND

D DREDGE SPOILS
I INDUSTRIAL WASTES
S SEWAGE SLUDGE
E EXPLOSIVES
R RADIOACTIVE WASTES
W SOLID WASTE
X INACTIVE SITE

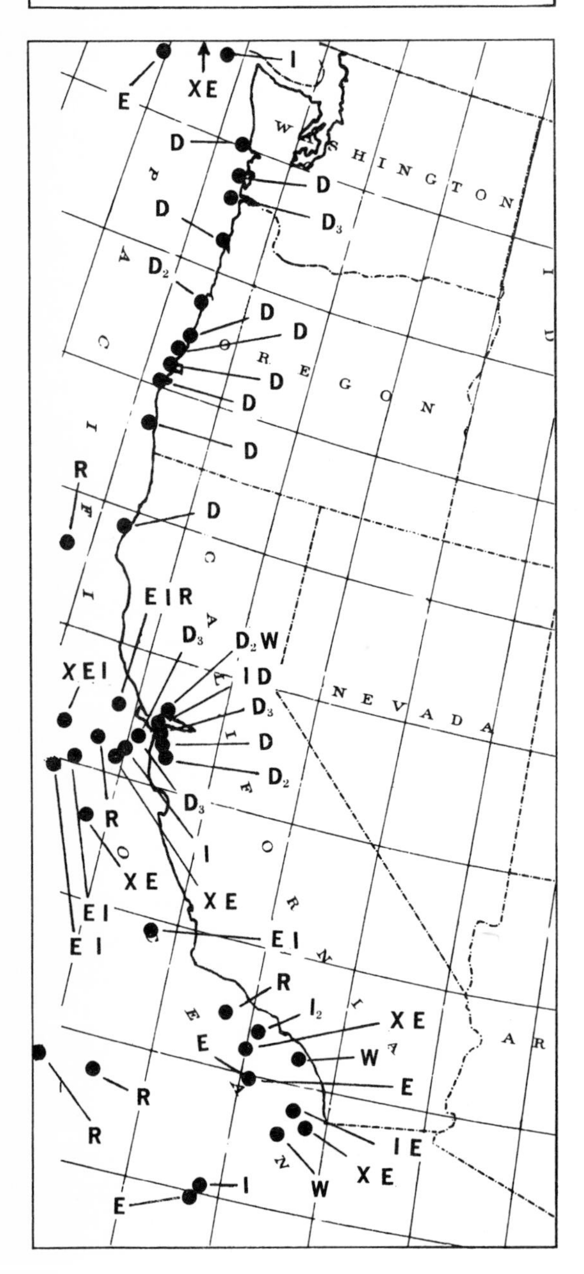

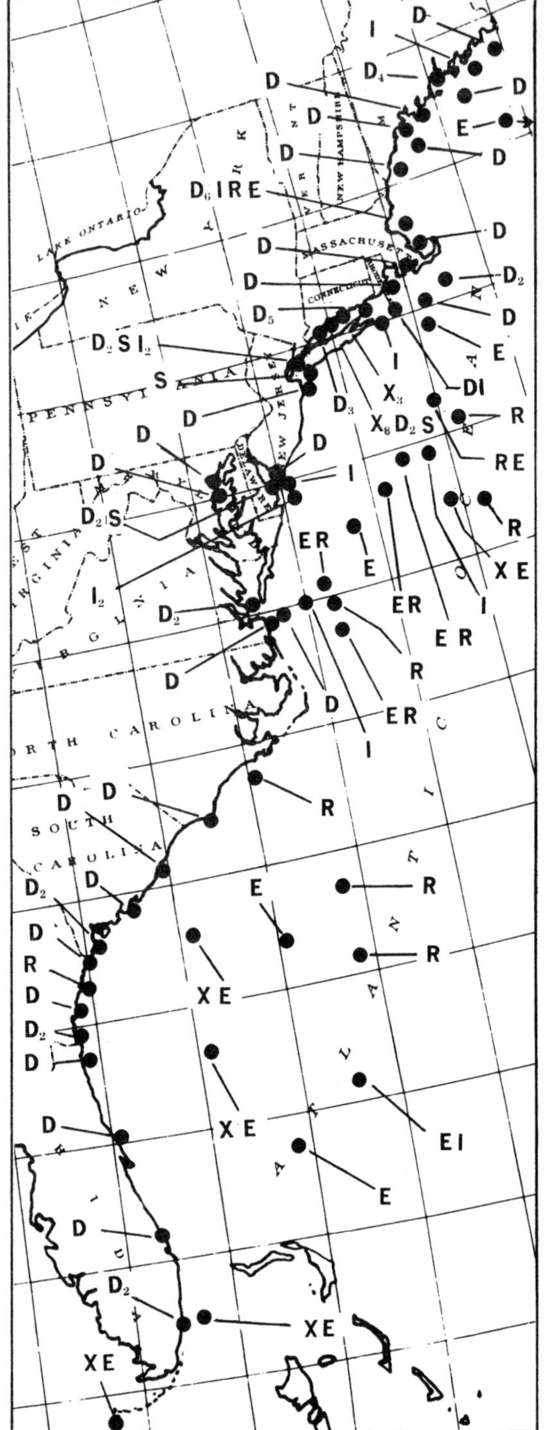

Figure 1.—Known Dumping Sites Off U.S. Coasts (22, 66)

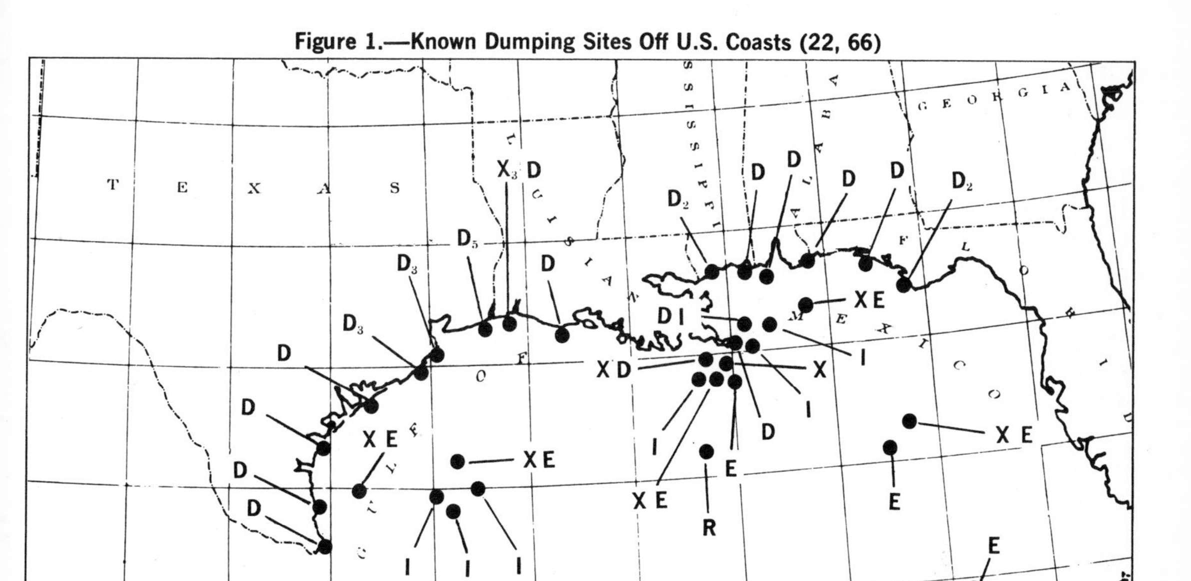

TABLE 2.—*Ocean Dumping: Types and Amounts, 1968* (66)

(In tons)

Waste type	Atlantic	Gulf	Pacific	Total	Percent of total
Dredge spoils	15,808,000	15,300,000	7,320,000	38,428,000	80
Industrial wastes	3,013,200	696,000	981,300	4,690,500	10
Sewage sludge	4,477,000	0	0	4,477,000	9
Construction and demolition debris	574,000	0	0	574,000	<1
Solid waste	0	0	26,000	26,000	<1
Explosives	15,200	0	0	15,200	<1
Total	23,887,400	15,966,000	8,327,300	48,210,700	100

Dredge spoils account for 80 percent by weight of all ocean dumping. The Corps of Engineers estimates that about 34 percent (13 million tons) of this material is polluted. Contamination occurs from deposition of pollutants from industrial, municipal, agricultural, and other sources on the bottom of water bodies. The quantities of polluted dredge spoils are shown in Table 3.

Polluted dredge spoils vary at every location according to the land-based sources of pollution. Detailed quantitative analyses of the pollutants in dredge spoils in the coastal

TABLE 3.—*Estimated Polluted Dredge Spoils* (22)

Coastal area	Total spoils (in tons)	Estimated percent of total polluted spoils [1]	Total polluted spoils (in tons)
Atlantic Coast	15,808,000	45	7,120,000
Gulf Coast	15,300,000	31	4,740,000
Pacific Coast	7,320,000	19	1,390,000
Total	38,428,000	34	13,250,000

[1] Estimates of polluted dredge spoils consider chlorine demand; BOD; COD; volatile solids; oil and grease; concentrations of phosphorous, nitrogen, and iron; silica content; and color and odor of the spoils.

areas are not available. An analysis by the Federal Water Quality Administration (FWQA) of polluted spoils from Lake Erie indicates that a total of 82,091 tons of spoils created 10,500 tons of chemical oxygen demand (COD). (23) These large quantities of oxygen-demanding materials can reduce the oxygen in the receiving waters to levels at which certain fish and other aquatic populations cannot survive. Also present were toxic heavy metals. Even with substantial dilution, the levels of heavy metals in the spoils may deleteriously affect marine life, as shown in Table 4.

TABLE 4.—*Heavy Metals Concentrations in Dredge Spoils (23, 36)*

(In parts per million)

Metal	Concentrations in dredge spoils	Natural concentrations in sea water	Concentrations toxic to marine life
Cadmium	130	.08	.01–10.0
Chromium	150	.00005	1.0
Lead	310	.00003	.1
Nickel	610	.0054	.1

Industrial Wastes

Industrial wastes were the second largest category of pollutants dumped at sea in 1968 (4.7 million tons, or 10 percent of the total). (66)

Most industrial wastes are commonly transported to sea in 1,000- to 5,000-ton-capacity barges. Sites are 4 to 125 miles off the Atlantic Coast, from 25 to 125 miles off the coast of the Gulf of Mexico, and from 5 to 75 miles off the Pacific Coast. Most of the sites are at the nearshore end of the range.

Highly toxic industrial wastes are sometimes contained in 55-gallon drums and are jettisoned from either merchant ships or disposal vessels at least 300 miles from shore. The containers are sometimes weighted and sunk. More frequently, they are ruptured at the surface, either manually with axes or by small arms or rifle fire. (66)

The breakdown for disposal methods by geographic area is shown below.

TABLE 5.—*Industrial Wastes by Method of Disposal (66)*

(In tons)

Coastal area	Number of sites	Bulk wastes	Containerized wastes	Total
Atlantic Coast	10	3,011,000	2,200	3,013,200
Gulf Coast	6	690,000	6,000	696,000
Pacific Coast	7	981,000	300	981,300
Total	23	4,682,000	8,500	4,690,500

Table 6 shows the relative quantities of major industrial wastes found in a survey of 50 producers in 20 cities.

TABLE 6.—*Industrial Wastes by Manufacturing Process (66)*

Type of waste	Estimated tonnage	Percent
Waste acids	2,720,500	58
Refinery wastes	562,900	12
Pesticide wastes	328,300	7
Paper mill wastes	140,700	3
Other wastes	938,100	20

The types of contaminants in industrial wastes dumped at sea vary greatly because of the diversity of industries and production processes involved. Many of the wastes are toxic—some highly toxic. For example, refinery wastes, which are 12 percent of the total ocean-disposed industrial wastes, can include cyanides, heavy metals, mercaptides, and chlorinated hydrocarbons. Pulp and paper mill wastes may contain "black liquor" and various organic constituents which are toxic to the marine environment. Chemical manufacturing and laboratory wastes that are dumped include arsenical and mercuric compounds and other toxic chemicals. (66)

Sewage Sludge

Sewage sludge is the waste solid byproduct of municipal waste water treatment processes. These solids can be further treated by digestion, a process which allows accelerated decomposition of the sludge to control odors and pathogens. Most sewage sludge is disposed of on land or is incinerated. Relatively small amounts (4.5 million tons on a wet basis) are currently dumped at sea, of which almost 4.0 million tons are dumped off New York harbor. (66) As of 1968, there were no similar operations on either the Gulf or Pacific Coasts, although sludge is being discharged from Los Angeles by pipeline.

Sewage sludge in digested or undigested form contains significant quantities of heavy metals. A study by the FWQA indicated that copper, zinc, barium, manganese, and molybdenum are present in sewage sludge. (9) The concentrations and types of toxic materials vary because sludge is the residual of waste water treatment and contains whatever domestic and industrial contaminants have entered the system. Table 7 shows the minimum, average, and maximum values for three heavy metals found in one analysis of sewage sludge.

TABLE 7.—*Heavy Metals Concentrations in Sewage Sludge (8, 9, 36)*

(In parts per million)

Metal	Concentrations in sewage slud~e			Natural concentrations in sea water	Concentrations toxic to marine life
	Min.	Avg.	Max.		
Copper	315	64?	1,980	.003	.1
Zinc	1,350	2,459	3,700	.01	10.0
Manganese	30	262	790	.002	-------------

Sewage sludge also contains significant amounts of oxygen demanding materials. In 1969, sludge dumped in the New York Bight, encompassing the New York harbor and some adjacent coastal areas, had an oxygen demand of about 70,000 tons. (15) These wastes also include some bacteria that cause diseases in man.

Construction and Demolition Debris

Only New York City disposes of debris at sea in significant quantities because of the lack of nearby available landfill. Sea disposal is conducted with 3,000- to 5,000-ton capacity barges that are towed some 9 miles offshore. These materials are generally inert and nontoxic.

Solid Waste

Solid waste, the byproducts and discards of our society, amounts to approximately 5.5 pounds per capita per day collected by municipal and private agencies. (28) Although these wastes total approximately 190 million tons per year, ocean disposal accounted for only about 26,000 tons. (66) Ocean dumping of solid waste occurred exclusively on the Pacific Coast, where they were generated by cannery operations and commercial and naval shipping operations. Other sources no doubt exist, but the overall magnitude of the current problem is minor.

The composition of solid waste, ascertained by sampling, is shown in Table 8. It is presented here to indicate the materials that would be introduced into the marine environment if ocean dumping of solid waste becomes a common practice.)

Solid waste disposed of in the ocean interacts with the water, but the resultant chemical products are difficult to determine. Studies have been done on the interaction between solid waste and fresh water in sanitary landfills as the water percolates through the waste materials. (The resultant mixture of water and chemicals is called leachate.)

TABLE 8.—*Composition of Solid Waste* (28)

Type of waste	Average (percent)
Paper products	43.8
Food wastes	18.2
Metals	9.1
Glass and ceramics	9.0
Garden wastes	7.9
Rock, dirt, and ash	3.7
Plastics, rubber, and leather	3.1
Textiles	2.7
Wood	2.5
Total	100.0

The percentage of pollutants in solid waste is not nearly as high as in sewage sludge or dredge spoils, but it does contain nutrients, oxygen-demanding materials, and heavy metals. Laboratory studies of water contaminated by solid waste have shown significant quantities of heavy metals, with zinc, nickel, and magnesium present in concentrations of 13, .27, and 378 parts per million respectively. (29) These concentrations are well above toxic levels for marine life.

Up to 50 percent of solid waste is usually paper, wood, plastics, and rubber, all of which can float to the surface. Particularly significant are the plastics which will not become water soaked and will not degrade for many, perhaps even hundreds, of years. Even if baled before ocean disposal, it is almost certain that over time the bales will disintegrate and the floatables will rise to the surface. The potential esthetic problems of large quantities of solid wastes floating to the surface and then being carried to shore are staggering.

Explosives and Chemical Munitions

Unserviceable or obsolete shells, mines, solid rocket fuels, and chemical warfare agents have been disposed of in deep water for many years. In 1963, the Navy initiated Operation "CHASE," in which munitions were disposed of by sinking them in obsolete hulks. Since then, 19 gutted World War II Liberty ships containing munitions have been scuttled. In the last six operations, the weapons were to detonate, but the S.S. ROBERT LOUIS STEVENSON failed to do so as planned and is located on the continental shelf near Alaska in 2,200 feet of water.

Since 1964 at least 18,342 tons of ammunition and explosives have been dumped in this manner. Additional cargoes of approximately 35,000 tons containing an unknown proportion of net explosives were also scuttled. A detailed listing of the ships scuttled, their cargoes, and disposition are shown in Table 9.

Detonation of explosives can result in trace amounts of lead, nickel, bronze, and other metals in the water, depending on corrosion processes and the materials used in the munitions.

Radioactive Wastes

Most nuclear waste products are liquid and of low radioactivity. They consist mostly of decontaminated process and cooling waters from reactors, fuel processing, and other operations. Small amounts of liquid wastes are highly radioactive; they result from the reprocessing of reactor fuel elements.

Solid radioactive wastes are produced by contamination of equipment and other materials during nuclear power plant operations, from medical use, and by research and development activities.

Solid radioactive wastes have been buried in carefully controlled landfill sites. Low-level liquid nuclear wastes are treated and/or stored to reduce radioactivity before disposal. High-level liquid wastes are stored exclusively in tanks at land-based sites.

TABLE 9.—*Explosives and Chemical Munitions, 1964–1970 (30)*

Year	Name	Total cargo (in tons)	Nature of cargo	Net explosives (in tons)	Disposition
1964	S.S. John F. Shafroth	9,799	A&E	Unknown	SDW
	S.S. Village	7,535	A&E	Unknown	SDW
1965	M.V. Coastal Mariner	4,040	A&E	512	D at 1,000′
	S.S. Santiago Iglesia	8,715	A&E	408	D at 1,000′
1966	S.S. Issac Van Zandt	7,500	A&E	1,625	D at 4,000′
	S.S. Horace Greely	6,033	A&E	442	D at 4,000′
1967	S.S. Robt. L. Stevenson	6,600	A&E	2,327	S
	S.S. Corporal Eric G. Gibson	9,005	Chem.	None	SDW
	S.S. Monahan	833	A&E	Unknown	SDW
1968	S.S. Mormactern	7,763	Chem.	N.A.	SDW
	S.S. Richardson	7,437	A&C	138	SDW
1969	S.S. Cape Tryon	7,626	A&E	1,145	DU
	S.S. Cape Catoche	6,348	A&E	1,359	DU
	S.S. Cardinal O'Connell	6,431	A&E	2,144	DU
1970	S.S. Frederick E. Williamson	5,245	A&E	478	DU
	S.S. Cape Comfort	6,200	A&E	N.A.	DU
	S.S. Walker D. Hines	6,500	A&E	N.A.	DU
	S.S. David Hughes	5,000	A&E	N.A.	DU
	S.S. LeBaron Russell Briggs	2,664	Chem.	N.A.	SDW

Definitions: A&E=ammunition and explosives; N.A.=not available; DU=Detonated unintentionally; SDW=sunk in deep water; D=detonated; S=sunk at less than 4,000 feet and did not detonate as planned; A&C=ammunition and cylinders contaminated with residues of GB nerve gas.

Liquid and solid radioactive wastes which have been dumped in the ocean are usually in concrete-filled metal drums or containers. Table 10 summarizes the amounts of these wastes disposed of at sea.

The quantities of radioactive materials disposed of at sea have decreased dramatically for several reasons. First, in 1960 the Atomic Energy Commission placed a moratorium on new licenses for disposal of radioactive wastes in the ocean. Only one commercial organization (which has never conducted any sea disposal), two Government agencies, and one university are still authorized to dispose of radioactive wastes in the ocean. Second, the major contractors of the AEC have not disposed of any wastes at sea since 1962. And for economic reasons, those firms with licenses are phasing out sea disposal of radioactive wastes in favor of land disposal.

TABLE 10.—*Radioactive Wastes: Historical Trends, 1946–1970 (70)*

Year	Number of containers	Estimated activity at time of disposal (in curies)
1946-1960	76,201	93,690
1961	4,087	275
1962	6,120	478
1963	129	9
1964	114	20
1965	24	5
1966	43	105
1967	12	62
1968	0	0
1969	26	26
1970	2	3
Total	86,758	94,673

Two sites have been used for disposal of most of the wastes in the Pacific Ocean. These sites are approximately 48 nautical miles west of the Golden Gate Bridge. One commercial firm has disposed of wastes in the Pacific Ocean farther than 150 miles from the U.S. coast; these disposals, 11 in number, were at depths greater than 6,000 feet. In the Atlantic Ocean, the major sites for disposal were in the area of Massachusetts Bay, approximately 12 to 15 miles from the coast; approximately 150 miles southeast of Sandy Hook, N.J.; and approximately 105 miles from Cape Henry, Va. With the exception of the Massachusetts Bay site, disposal was at depths greater than 6,000 feet. The Massachusetts Bay site was in 300 feet of water.

PAST TRENDS

Figure 2 shows significant increases in ocean dumping activities during the years 1951–1968. These data do not include dredge spoils or explosives because historical data could not be readily reconstructed. Radioactive wastes are also excluded because of their negligible weight contribution.

Table 11, on which Figure 2 is based, shows a fourfold increase in tonnage dumped at sea from 1949 to 1968. The 28 percent increase between the 1959–1963 period and the 1964–1968 period is largely attributable to dramatic increases in industrial wastes and sewage sludge disposal. In 1959, industrial wastes disposed of at sea approximated 2.2 million tons. By 1968, the amount had increased to over 4.7 million tons, a 114 percent increase in 9 years. The amount of sewage sludge disposed of at sea increased by 61 percent in the same period, from 2.8 million tons to 4.5 million tons. (66)

FUTURE TRENDS

Assessing future trends in ocean dumping requires analysis of basic population trends. Population growth is accompanied not only by increased amounts of wastes but also by decreased space available for their disposal.

Between 1930 and 1960 the coastal population increased by 78 percent, compared with a 48 percent increase nationwide. (36) The figures below (25) indicate the population growth in the coastal region projected through the year 2000:

1960	57, 946, 000
1970	68, 397, 000
1980	76, 607, 000
1990	92, 940, 000
2000	106, 900, 000

TABLE 11.—*Ocean Dumping: Historical Trends, 1949–1968* [1] *(66)*

Coastal area	1949–1953		1954–1958		1959–1963		1964–1968	
	Total	Avg./Yr.	Total	Avg./Yr.	Total	Avg./Yr.	Total	Avg./Yr.
Atlantic Coast	8, 000, 000	1, 600, 000	[2] 16, 000, 000	3, 200, 000	27, 270, 000	5, 454, 000	31, 100, 000	6, 200, 000
Gulf Coast	[3] 40, 000	8, 000	283, 000	56, 000	860, 000	172, 000	2, 600, 000	520, 000
Pacific Coast	487, 000	97, 000	850, 000	170, 000	940, 000	188, 000	3, 410, 000	682, 000
Total	8, 527, 000	1, 705, 000	17, 133, 000	3, 426, 000	29, 070, 000	5, 814, 000	37, 110, 000	7, 422, 000

[1] Figures do not include dredge spoils, radioactive wastes, and military explosives.

[2] Estimated by fitting a linear trend line between data for preceding period and data for succeeding period.

[3] Disposal operations in the Gulf of Mexico began in 1952.

Figure 2.—Average Annual Tonnage Dumped at Sea—
by Coastal Area (66)

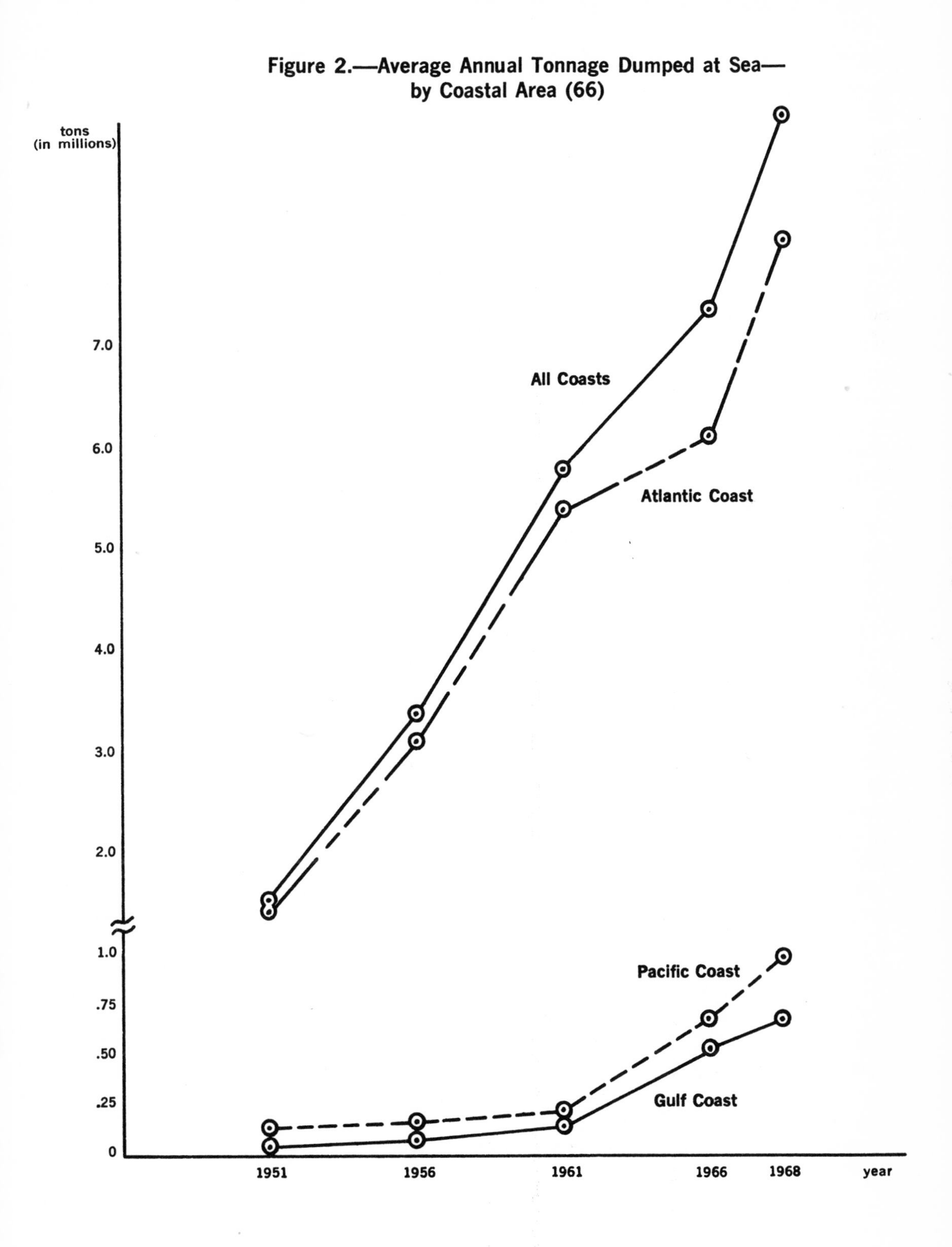

Solid Waste

About 65 million tons of solid waste are generated annually in the coastal region. Based on a conservative estimate of 8 pounds of waste generated per person per day in the year 2000—the generation rate which will be reached by 1980—over 150 million tons will need to be disposed of for that one year. (28) If 10 pounds per person per day are generated, total wastes in the coastal area will be close to 200 million tons, more than triple current levels. The pressure to use the ocean for waste disposal will increase as land disposal sites become more scarce, costs increase, and metropolitan areas face political problems in obtaining new land disposal sites. Several cities are currently exploring the use of the ocean as a solid waste disposal site, and this interest is expected to increase. In some cases operations may begin within a year. If even a small percentage of the solid waste annually generated in the coastal area were disposed of at sea, the quantities entering the marine environment would be many orders of magnitude greater than all solid waste disposed of at sea to date.

Sewage Sludge

Based on an average of .119 pounds of sludge generated per person per day, potential sludge disposal quantities for the coastal region can be roughly estimated. (37) In 1970, approximately 1.4 million tons of sludge will be disposed of in the coastal areas, and in the year 2000, approximately 2.1 million tons will be generated, an increase of 50 percent in 30 years. If anything, these figures may underestimate future quantities of sludge. For example, between 1960 and 1980, 20-year period, the sludge generated by the Baltimore-Washington area is expected to increase from 70,000 tons to 166,000 tons, or about a 140 percent increase. New York City's sludge barged to sea is expected to increase from 99,000 tons in 1960 to about 220,000 tons in 1980, a 120 percent increase in 20 years. (66)

Industrial Wastes

The volume of industrial production, which gives rise to waste production, is increasing at a rate of 4.5 percent annually, or three times the population growth rate. Additionally, the FWQA estimates that the manufacturing industry is responsible for three times as much waste as that produced by the Nation's population. And about 40 percent of the Nation's industrial activity is concentrated in the estuarine economic region. (36) Given increasingly stringent water quality standards and the ever expanding level of industrial waste generation in the coastal zone, the potential for increased industrial waste dumping at sea is great.

Radioactive Wastes

The amount of liquid and solid radioactive wastes will rise with projected increases in nuclear power generation. The amount of high-level liquid radioactive wastes will increase from 100,000 gallons in 1970 to 6,000,000 gallons by the year 2000 and radioactive solid wastes, from approximately 1 million cubic feet in 1970 to 3 million cubic feet by 1980. (70) As mentioned earlier, however, ocean dumping has been virtually nonexistent since the early 1960's because of the AEC moratorium and the economic advantage of land disposal.

Large radioactive structures, an additional source of radiation, are not yet a significant problem. In the past, the few that became obsolete have been decontaminated, dismantled, and kept under surveillance on land—with the exception of parts of one nuclear submarine, which were disposed of in the ocean. Currently, however, there are 16 nuclear power plants in operation, 55 under construction, and 25 for which construction permit applications are pending with the Atomic Energy Commission. (70) If current forecasts are realized, by the year 2000, the equivalent of up to 1,000 nuclear power units, each with a capacity of some 1,000 megawatts, may be operating. In addition, the Navy has about 90 nuclear-powered submarines and surface ships, and many more may be built in the next 30 years as a large portion of the current naval fleet is replaced. Commercial nuclear ships—currently the N.S. SAVANNAH is the only one—may become economically feasible in the future.

A lifetime of 10 to 30 years for the power plants' and ships' reactor vessels is reasonable in terms of physical or technological obsolescense. Their radiation levels vary considerably, up to 50,000 curies of induced radiation in each structure. (70)

Individually none of these sources adds significant amounts of radioactivity to the ocean. Taken together, however, the increases could be of significant concern.

Dredge Spoils

In the long run, the reduction of polluted discharge from municipal and industrial sources, brought about by water quality standards, will lessen the problem from dredge spoils. However, they will remain a problem for at least the next 5 to 10 years. During this period, there will be pressures for more dredging to deal with increasing marine commerce, to meet the desire of cities for new deep-water harbors, and to provide draft for larger vessels (including the supertankers used to transport oil). These needs will all increase total dredging and hence dredge spoils.

Explosives and Chemical Munitions

The following are Department of Defense estimates of conventional munitions planned for disposal: in 1970, 103,777 tons; in 1971, 88,835 tons; and in 1972, 80,000 tons. (26) These quantities are several times larger than the total volume of these wastes disposed of at sea in the last two decades. They indicate the quantities which would enter the marine environment if no other disposal technique were employed.

Chemical munitions have also been disposed of at sea in three deep-water disposal operations, but actual quantities involved are not known. No future ocean disposal operations are planned. Biological agents have not previously been disposed of at sea, and no future disposal is projected.

SUMMARY

The data indicate that the volume of wastes dumped in the ocean is increasing rapidly. Many are harmful or toxic to marine life, hazardous to human health, and esthetically unattractive. In all likelihood, the volume of ocean-dumped wastes will increase greatly due to decreasing capacity of existing disposal facilities, lack of nearby land sites, higher costs, and political problems in acquiring new sites.

Ocean Pollution

CHAPTER II deals with the effects of ocean dumping in terms of the broader problem of ocean pollution. This view is necessary because wastes affect marine ecosystems no matter where or how the pollutants originate and because pollutants tend to interact, sometimes synergistically, in the environment.

Marine pollution has seriously damaged the environment and endangered humans in some areas. Shellfish have been found to contain hepatitis, polio virus, and other pathogens; pollution has closed at least one-fifth of the Nation's commercial shellfish beds; beaches and bays have been closed to swimming and other recreational use; lifeless zones have been created in the marine environment; there have been heavy kills of fish and other organisms; and identifiable portions of the marine ecosystem have been profoundly changed.

THE PATHWAYS OF POLLUTION

In order to understand the effects of pollutants on marine ecosystems, one needs to understand how pollutants are dispersed and concentrated. The dispersal of wastes depends on the material involved. Most wastes, but far from all, sink to the bottom. Others, such as solid waste, oil, and garbage, contain many floatable materials. Floating wastes can be transported great distances by current and wind. Early in 1970, the Heyerdahl expedition encountered wastes over large areas of water in mid-ocean, reporting that the ocean was "visibly polluted by human activity." (55)

Suspended materials, such as fine particles, are also transported by currents over great distances. For example, horizontal currents flush the 500 square miles of the New York Bight, completely exchanging the water in less than 1 week. (42) Vertical movement is considerably slower, and pollutants may remain in layers of water for quite some time.

Pollutants enter living systems through biological concentration. Billions of tiny phytoplankton organisms act as a great biological blotter, picking up nutrients, trace metals, and other materials. Organisms feed on the phytoplankton and successively pass the pollutants on to higher organisms. As this process moves through the food chain, concentrations reach their highest levels in predators such as marine mammals, birds, and man. An example of the food chain may be seen in the North Atlantic—1,000 pounds of phytoplankton produces:

- 100 pounds of zooplankton or shellfish
- 50 pounds of anchovies and other small fish
- 10 pounds of the smaller carnivores
- 1 pound of the carnivores harvested by man. (41)

The concentration of chemicals by phytoplankton and subsequent further concentration within the food chain have lethal and sublethal effects on organisms.

Heavy metals have been found in toxic concentrations in plankton, seaweed, and shellfish, although levels of concentration in the surrounding water were not high. The ability of biota to concentrate materials varies from a few hundred to several hundred thousand times the concentrations in the surrounding environment. (8, 42, 48) Table 1 shows phytoplankton concentration factors for selected metals.

EFFECTS ON MARINE LIFE

Pollution affects marine life directly through toxicity, oxygen depletion, biostimulation, and habitat changes.

TABLE 1.—*Phytoplankton Concentration of Some Heavy Metals.* (45)

Metal	Concentration factor
Aluminum	100,000
Cobalt	1,500
Copper	30,000
Iron	45,000
Lead	40,000
Radium	12,000
Zinc	26,000

Toxicity

Although plants and animals are sometimes killed by toxic wastes, organisms may be affected by concentrations far below the lethal level. Sublethal effects include reduced vitality or growth, reproductive failure, and interference with sensory functions.

Copper was found in the waters of the New York Bight in concentrations greater than 0.120 milligrams per liter. (8) These concentrations, found throughout the water column, indicate widespread copper contamination.

With even lower concentrations of copper, laboratory experiments have shown that:

- Concentrations of 0.1 milligrams per liter killed soft clams in 10–12 days. (62)
- Concentrations of 0.05 milligrams per liter killed polychaete worms in 4 days. (63)
- Concentrations of 0.1 milligrams per liter inhibited photosynthesis in kelp 70 percent in 9 days. (16, 17)

Pesticides and other toxic materials are a major cause of fish kills in fresh water. Although there are few recorded fish kills in the ocean resulting from pesticides, pesticide concentrations are rising every year. They reduce the size and strength of mollusk shells. Reduced growth rate and reproductive activity in fishes exposed to sublethal doses of pesticides and copper have also been shown. (54)

Pesticides endanger higher predators because of biological concentration. For example, pesticides amplified through the food chain damage birds' reproductive capability and in some cases seriously reduce their populations. The peregrine falcon is the most dramatic example; pesticide accumulation through the food chain has led to drastic reduction and projected extinction in the coterminous United States.

Oil introduced into the marine environment produces several adverse effects: Reproduction and other behavior is altered. Direct contact with respiratory organs weakens or kills animals. And oil clogs their filtering mechanisms. (67) Experiments with oysters have shown that when water-soluble fractions of oil were introduced into water, the amount of water filtered by the oysters decreased from between 207 and 310 liters per day to between 2.9 and 1.0 liters after 8 to 14 days. (13)

Cancer in fishes is very likely a result of contact with certain waste products. Cancerous growths on the lips of croakers have been found in areas of the Pacific Ocean polluted by oil refinery wastes. (65) Growths on several species including White Seabass and Dover Sole caught in oil polluted areas have been reported. (72) Oysters and barnacles are also known to concentrate cancer-producing agents.

Laboratory tests with "black liquor" from a paper mill showed that 0.05 grams per liter affected photosynthesis and 1 gram per liter killed the four species of phytoplankton tested. (66)

In laboratory experiments with polluted sediments from the New York Bight disposal area, the following sublethal effects were shown:

- Serious infections were found in native species.

- Bottom waters inhibited phytoplankton cell growth and division. (34)

Lethal and sublethal effects from toxic wastes are complex and not well understood. But evidence is mounting that these effects may be widespread and very harmful to the marine environment. Their potential for deferred and long-range ecological damage must be taken into account in any program to control ocean dumping.

Oxygen Depletion

Oxygen supports marine and aquatic life and is necessary to the biological degradation of organic materials. Organic wastes dumped or discharged into water bodies demand oxygen to decompose. If waste loads are too heavy, the oxygen levels become depleted and the diversity of marine organisms is altered.

Many of the Nation's rivers, estuaries, and harbors are in this condition. In the Potomac estuary, severely polluted by municipal wastes, dissolved oxygen levels approach zero in some reaches during low flow periods of warm summer months. (33)

When all the oxygen is depleted, organisms die, and anaerobic bacteria produce hydrogen sulfide and methane gas, which are malodorous. Large amounts of oxygen are required to decompose some materials. The dissolved oxygen in 320,000 gallons of air-saturated sea water is required to oxidize 1 gallon of crude oil completely. (64) If the oxygen level is already low, damage from oil spills may increase.

Dumping undigested sewage sludge in the ocean can create a significant demand on the dissolved oxygen. And oxygen depletion can develop rapidly. In the New York Bight waste disposal area, where sludge has been dumped for 40 years, the oxygen concentration as a percent of saturation declined from 61 percent in 1949 to 59 percent in 1964. It then dropped to 29 percent in 1969 and was as low as 10 percent in the center of the dump. (42) This may indicate that a threshold was reached and that the water quality then deteriorated rapidly.

Oxygen levels fell below those necessary to sustain life in species of lobster and crab normally found in the area. Researchers have noted that:

> the most striking effect observed was the *extreme depletion of dissolved oxygen* in the bottom waters over the disposal areas during the summer months. Levels frequently fell below 2 parts per million during the period from July to mid-September . . . This condition is undoubtedly caused by the heavy oxygen demand of the organic-rich waste materials coupled with the reduced mixing rates normally found during the summer. (43)

Oxygen deficit in a waste disposal area may be self-perpetuating. The accumulation of organic matter, sulfides, and some metals can act as a reservoir of future oxygen demand. Even after the disposal of the organic matter is stopped, it may be a long time before the area recovers.

Biostimulation

Some wastes, such as sewage sludge, are particularly rich in nutrients, such as phosphates and nitrates. These nutrients can cause biostimulation—the accelerated fertilization of plant life. When the plants die, oxygen necessary to support marine life is used in their decomposition. And when dead algae are carried to beaches, they rot and produce unpleasant odors.

By creating excessive blooms of algae, biostimulation indirectly changes the nature of bottom sediments and thus whole communities of bottom organisms. For example, areas which formerly supported surf clams in sand may become covered with an algal mud to which the surf clams cannot adapt. Sediments adjacent to disposal areas show greatly increased concentrations of organic matter. Some come directly from the wastes, but other material filters down from algal blooms. (2)

In the past, biostimulation has been recognized as a major problem of fresh waters, but not of the oceans. Increasingly, however, biostimulation is affecting estuaries and bays and even some portions of the continental shelf.

Shock

Explosions from dumping of munitions cause death in marine organisms surrounding the explosion point. The Department of Defense calculates that detonation of 1,000 tons of explosives—the approximate amount contained in the September 4, 1970, "Deep Water Dump" off Washington State—generates a shock wave that will kill most marine animals within 1 mile of the explosion and will probably kill those fish with swim bladders [1] out to 4 miles from the explosion.

Habitat Changes

Evidence indicates that waste disposal practices drastically alter certain marine communities. Habitat changes are the most common change that can affect entire ecosystems.

The most pronounced ecological changes, caused by dumping sewage sludge and polluted dredge spoils, have been found in the New York Bight. The consistency of bottom sediments changed from sand or hard mud to muddy ooze. Nematode worms, normally tolerant of pollution, were completely absent from the center of the dredge spoil dump and were found in very low numbers in the center of the sewage sludge dump. (2)

Changes in the kinds and quantities of sediments deposited may alter ecosystems. The plague of starfish in the Pacific may be an example of this effect. In recent years, the numbers of Crown of Thorns starfish have multiplied. This coral-eating starfish has devastated large areas of the coral reefs off many Pacific islands and the Great Barrier Reef of Australia. The population explosion may be linked to sediment protecting the larval starfish from their predators, which normally keep the population in balance. The sediment results from blasting, dredging, and dumping.

Significant changes in the benthic ecology of the Southern California coast have been caused by wastes from several municipalities. (11) These wastes brought about a shift in the marine population. Large numbers of sea urchins replaced other organisms and grazed off most of the giant kelp beds near the sewer outfalls. Because of the commercial value of giant kelp and the habitat it provides for many marine animals, the changes were an economic and an ecologic loss.

Habitat changes may be quite subtle. Near a sewer outfall off San Diego, species variety declined an average of 30 percent. Populations of remaining species sometimes overran their food supply. The loss of species diversity made the ecosystem less stable. (71)

HUMAN IMPACTS

Public health problems are created by toxic agents and pathogens that find their way into the human food chain through seafood. Floating refuse and surface films reduce recreation opportunities and damage esthetic values. Economic losses are incurred when seafood

[1] A large group of fish with respiratory organs that adjust to different depths.

species are killed or are rendered inedible by pollution.

Public Health

The standard method for determining the potential public health hazard of fish is the coliform bacteria count. (These harmless bacteria are rough indicators of pathogens.) If the count exceeds Food and Drug Administration (FDA) standards, shellfish beds are closed to harvesting.

Effluents from land-based sewage outfalls are the major source of coliform bacteria, but ocean dumping of sewage sludge is also significant. The FDA found that ocean bottom sediments up to 6 miles from the New York Bight sludge dump contained coliform counts that exceeded permissible levels. On May 1, 1970, this area, 12 miles in diameter, and a similar area off Delaware Bay were closed to shellfishing. Clams harvested for sale in the New York Bight contained coliform bacteria 50 to 80 times higher than the standards set by FDA. (2)

Hepatitis virus are carried by shellfish. A 1961 outbreak of infectious hepatitis was traced to raw shellfish taken from Raritan Bay, N.J. (36) Shellfish have been collected with polio virus concentrated to at least 60 times that of surrounding waters. (52)

White perch have become actively infected with human pathogens by exposure to human wastes, and they may transmit these pathogens over considerable distances. Exposure is sufficient for them to develop antibodies to such human diseases as pseudo-tuberculosis, paratyphoid fever, bacillary dysentery, and a variety of chronic infections. (40)

Aquatic and marine organisms are capable of concentrating radioactivity to high levels (45). In a study near Oak Ridge National Laboratory, dead embryos and abnormalities appeared in irradiated broods of killifish. This is the only example of a natural marine or aquatic population subjected to high-level irradiation over many generations. (68)

Hydrocarbons of the type known to cause cancer in man and animals are concentrated by oysters and mussels in polluted areas. These substances remain invisible and odorless in seafood tissues, even after frying. (28) Cancer in humans has not yet been traced to consumption of carcinogens from seafood, but public health officials do not discount the possibility.

Between 1953 and 1960, 111 persons were reported to have been killed or to have suffered serious neurological damage near Minamata, Japan, as a result of eating fish and shellfish caught in areas contaminated by mercury. Among these were 19 congenitally defective babies whose mothers had eaten the fish and shellfish. Subsequently, at Niigata 26 more cases of mercury poisoning were noted. (1) The fish eaten by the affected Japanese contained from 5 to 20 parts per million of methyl mercury.

Mercury pollution recently discovered in 33 States and in Canada caused many fishing areas to be closed. Concentrations of as high as 5 parts per million have been found in fish in the Great Lakes. (1)

Loss of Amenities

The coastal zones provide recreation and beauty for the 60 percent of the Nation's people dwelling there. Oceans afford swimming, boating, water skiing, sport fishing, and wildlife viewing opportunities,[2] and they are some of the most scenic areas of the United States.

Many beaches have been closed to swimming because of the high coliform content of the water. Most closed beaches are near large

[2] The Bureau of Sport Fisheries and Wildlife estimates that as many as 100 million people observe the wildlife of the U.S. estuarine zones.

metropolitan areas, such as San Francisco and New York. Floating materials, such as solid waste and oil, pose a major threat to amenity values. Rotting algae and anaerobic waters cause unpleasant odors and visual pollution. And debris are often a hazard to small boats.

Economic Loss

Significant economic losses result from ocean pollution. A major loss is the commercially valuable fish or other seafood species killed directly or indirectly or rendered inedible. They represent serious social and financial losses because of the near subsistence level of many fishermen.

In 1969, the total catch of crabs, lobsters, shrimp, oysters, clams, and scallops was 729 million pounds. Because one-fifth of the Nation's 10 million acres of shellfish beds are closed due to contamination, it can be estimated that the total catch would have been 181 million pounds higher. This estimate is probably low, since the closed areas are particularly productive—in lush estuarine systems in close proximity to large cities where they would have been harvested intensively. Figure 1 indicates the financial impact assuming a loss of one-fifth the potential catch.

The loss is well documented in San Francisco Bay. (36) Prior to 1935, the annual commercial harvest of soft shell clams was between 100,000 and 300,000 pounds. Today clam-digging is virtually nonexistent because of pollution. The annual commercial landings of the shrimp fishery prior to 1936 were as high as 6.5 million pounds; landings in 1965 were only 10,000 pounds.

Contamination by pesticides or mercury has rendered nine species of fish unfit for consumption by humans. Many States have banned fishing and impounded fish because of mercury poisoning, and the FDA impounded coho salmon due to high levels of DDT.

Figure 1.
Potential Value of U.S. Shellfish Catch, 1969
$320 million

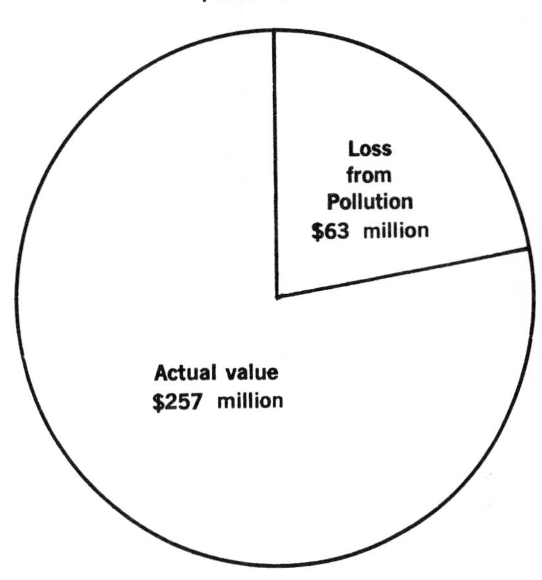

Even where contaminant levels do not prevent safe consumption, the food may be discolored or tainted. Sludge decay can result in the production of hydrogen sulfide, which blackens the shells of clams and oysters and affects their taste and odor. (36) In even very small amounts, oil can taint the flesh of fish. The discharge residue from burning 2.6 gallons of a gasoline-oil mixture in an outboard motor was sufficient to taint fish in 1 acre-foot of water. (67)

A further ocean dumping cost is that of cleaning up or rehabilitating polluted beaches and other shores. If projected increases in solid waste are dumped at sea, continuous and expensive clean-up operations will be required.

SUMMARY

The information presented in this chapter is necessarily incomplete. Knowledge of ocean pollution is rudimentary, and generally it has not been possible to separate the effects of ocean dumping from the broader issue of ocean pollution. Yet one general conclusion is apparent: There is reason for significant concern. Dealing with ocean pollution requires that all sources be greatly reduced. If no action is taken and ocean dumping continues to increase, the long-term damage to the marine environment will be great.

CHAPTER III

Alternatives to Ocean Dumping

THE critical or potentially critical sources of ocean pollution and their effects on the marine environment are described in Chapters I and II. Based on these findings, a strong national policy has been recommended to stop or limit ocean dumping substantially. The extent to which the recommended policy can now be implemented depends on existing alternatives for handling wastes.

This chapter sets forth alternatives, both interim and longer term. The interim alternatives discussed are practical, available disposal techniques which can be used now to reduce or prevent damage to the marine environment without shifting the problem to another part of the environment. Long-term alternatives look toward recycling, resource conservation, and more economic and environmentally safe techniques of waste management. Costs and capacity are estimated to indicate the impact of the alternatives.

The types of wastes for which alternatives are presented include: solid waste, sewage sludge, dredge spoils, industrial wastes, construction and demolition debris, radioactive wastes, and explosive and chemical munitions.

Although dredge spoils and industrial wastes are the two largest sources of ocean dumping, solid waste is discussed first because the alternatives are largely applicable to the other wastes dumped in the ocean.

SOLID WASTE[1]

The amount of solid waste dumped in the ocean is not yet significant, less than 1 percent of all wastes disposed of in the ocean. Only about 26,000 tons were dumped in the ocean in 1968, (66) compared to the 190 million tons of municipal solid waste collected and disposed of on land. (28) However, many communities are beginning to look to the ocean as a place to dispose of solid waste in light of increasing population; increasing per capita rates of solid waste generation; and the declining capacity, increasing costs, and lack of nearby land disposal sites. If many coastal cities were to dump solid waste in the ocean, many millions of tons would be introduced annually into the marine environment. Although little research has been done on how solid waste affects marine ecology, it is known that improper disposal of solid waste on land seriously contaminates ground water. Further, floating materials from the solid waste dumped in the ocean would be unattractive, especially when carried to shore. Accordingly, the policy recommended would prohibit new sources of solid waste in the ocean and call for phasing out existing sources.

Interim Alternatives

Nationwide, landfill capacity is generally adequate. The average time remaining for currently used landfills in all metropolitan areas is 16 years, although some large metropolitan areas will soon exhaust their current sites. (28) Only 10 percent of land disposal operations are sanitary landfills, in which the wastes are covered daily by soil. The other 90 percent are open dumps, which create many health and esthetic problems. Rodents and insects breed and carry infectious diseases, and ground water often becomes polluted. Esthetically, open dumps are unattractive and malodrous. Converting open dumps to sanitary landfills can be accomplished relatively quickly and inexpensively.

There are two alternatives to ocean dumping of solid waste. New sites can be developed, but often at a considerably increased distance. Or incinerators can be constructed. By reducing the volume, possibly up to 90

[1] Includes residential, commercial, industrial, institutional, and agricultural solid wastes.

percent, they can prolong the use of existing sites by many years.

The barriers to acquiring new sites are political and financial. Communities are reluctant to be the dumping ground for the wastes of large metropolitan areas, and transport to distant sites increases costs. Transfer stations and rail or transfer truck operations make these longer hauls more costly than collection vehicles' traveling only a few miles to the disposal area. But they provide more flexibility in site selection. The barriers to the construction of new incinerators are largely financial. They are expensive to build and to operate. More stringent air pollution standards will add to both capital and operating costs.

Comparative costs for various alternative methods of disposal are shown in Table 1. As it indicates, the additional costs for use of rail haul and land disposal instead of ocean dumping are not so high when the distances are comparable. For example, when the wastes are transported 50 or 100 miles by either method, the costs of land disposal are less than 10 percent higher.

If conducted correctly, rail haul and land disposal offer an economically attractive method of disposing of solid waste. However, the political problems are a significant barrier to a good economic and environmental solution. A stronger regional approach to waste management, better disposal operations, and adequate payment for the use of land could well overcome these barriers.

One possible alternative deals with the problems of both solid waste disposal and abandoned strip mines. Because of the small incremental costs involved in rail haul, large coastal cities could haul their wastes to these mines economically.

Available acreage within range of the three coastal areas has been estimated. In the mid-Atlantic States of Ohio, Pennsylvania, West Virginia, Virginia, New York, and New Jersey, over 660,000 acres of unreclaimed surface-mined land are available. Over 300,000 additional unreclaimed acres are available in the Gulf Coast States, Texas, Alabama, Mississippi, Louisiana, and Florida. On the West Coast, California and Nevada have approximately 150,000 acres of available, unreclaimed surface-mined land.

Nationwide, surface mining has disturbed over 3.2 million acres of land. The Department of the Interior estimates that over two-thirds of this acreage is completely unreclaimed. This 2 million acres represents 3,300 square miles of potential solid waste disposal sites. (31)

TABLE 1.—*Comparison of Estimated Solid Waste Disposal Costs (28, 47)*

[On a cost-per-ton basis]

Unit process	Sanitary landfill at nearby site	Incineration at central city site	Rail haul and landfill			Baling and ocean dumping			Incineration ship-based
			50 mi.	100 mi.	150 mi.	20 mi.	50 mi.	100 mi.	
Collection [1]	$15.00	$14.00	$14.00	$14.00	$14.00	$14.00	$14.00	$14.00	$14.00
Transfer operation [2]	0	0	4.05	4.05	4.05	4.20	4.20	4.20	0
Haul	0	0	2.65	3.00	3.45	.60	1.30	2.25	0
Disposal [3]	1.25	10.50	.65	.65	.65	0	0	0	10.89
Total	16.25	24.50	21.35	21.70	22.15	18.80	19.50	20.45	24.89

[1] Higher cost of collection for nearby landfill due to lack of central city site.

[2] Higher cost of ocean baling due to higher density requirements.

[3] Lower cost of landfill operation due to baling.

These figures do not consider suitability of terrain, amount of cover material, volume in need of fill, or other limiting factors. Nevertheless, there are access roads and rail lines to almost all this land, and if legal and social barriers can be removed, the problems both of providing large disposal areas and of reclaiming the land would be solved.

Containerizing wastes—that is, enclosing them in plastic or other material to prevent interaction with the sea—raises a number of potential problems. First, any containment system will still allow leaching of the wastes, some of which are toxic. Second, containment systems will probably not isolate the wastes from the ocean environment indefinitely. Plastics and other floatables are likely to be released eventually. As indicated in Table 1, the economics of containerizing wastes are not significantly better than for land disposal, assuming that solid waste would have to be dumped some distance from shore.

Ship-based incineration has also been suggested as an alternative disposal technique. It appears, however, to have little economic or environmental advantage. As Table 1 indicates, the costs are higher than for rail haul or land-based incineration. And difficulties of systematically locating and using sea dump sites may be a problem compounded by the difficulties of operating during bad weather. Further, many of the materials are noncombustible, and the effects of large amounts of ash residue on the ocean environment are not clearly known.

Longer-Term Alternatives

Although ship-based incineration may not be practical, other advances in incineration may have long-term benefits for solid waste management. A new type of incinerator, the CPU-400, is being developed under a Bureau of Solid Waste Management contract. Shredded and dried refuse is burned in a fluidized bed reactor to produce gas for turboelectric power generation. A 400-ton-per-day modular unit will produce up to 15,000 kilowatts of electric power. Total annual cost is projected at between $4.27 per ton for a municipal utility and $5.99 per ton for private ownership; the difference is a function of the interest rate. (18) (Current incineration costs are $10.50 per ton.) Depending on revenues from the sale of electricity and residue byproducts, the net cost could be reduced. Soon in the pilot plant stage, this incinerator may provide a low-cost, environmentally sound method of dealing with solid waste.

Recycling may also become general practice. Technology exists to recycle many types of paper, glass, aluminum, and ferrous metals, among others. Currently, 19 percent of the materials used to manufacture paper products in the United States are recycled rather than virgin materials. (28) Eighty-five percent of all automobiles taken out of service are recycled and used in steelmaking, and tires and aluminium cans are beginning to be recycled. (28) The problems and associated costs of separation; transportation; poor secondary markets; and other legal, economic, and social barriers have limited recycling. However, with new approaches to these barriers, new technology, and the need to conserve resources, recycling may become practical on a broad scale in the future. And as more materials are reused, disposal needs will lessen. It is important to note that inexpensive but environmentally unsound practices such as ocean dumping discourage waste reuse and recycling, which are desirable in the long term.

SEWAGE SLUDGE

In 1968, about 200,000 tons of sewage sludge on a dry basis were disposed of at sea, compared to about 3 million tons disposed of by other means. Increasing population and the higher levels of treatment required to meet water quality standards will generate even more sludge. Given the difficulties of sludge disposal and the high costs involved, pressures to use the oceans will necessarily increase. The environmental problems from sludge disposal in the ocean are significant, in terms both of volume and of the toxic and sometimes pathogenic materials involved. Accordingly, the policy recommended would phase out ocean disposal of sewage sludge and prevent new sources.

Alternatives (Interim and Longer Term)

Sewage sludge is primarily disposed of by using it as a soil conditioner or landfill and, to a much lesser degree, by incineration. The costs of present ocean disposal operations are generally far below costs for land-based disposal. Ocean disposal a few miles from shore costs an average $1 per ton. (66) Table 2 contains more detailed data on the per-ton-mile costs for longer hauls.

TABLE 2.—*Barge Haul Costs for Sewage Sludge Disposal (37)*

City	Distance (miles)	Cost per-ton-mile	Cost per ton
New York City	25	$0.30	$7.50
Elizabeth, Md	30	.23	6.90
Baltimore, Md	230	.08	18.40
Philadelphia, Pa	300	.04	12.00

Depending on distance, actual barge haul costs range from $1 to $12 per ton. Thickening, a process preparatory to barging, can add $2 to $6. Digestion can raise total ocean disposal costs by $5 to $18 per ton. Total ocean dumping costs can range from $3 for undigested sludge deposited nearshore to perhaps $40 per ton for digested sludge dumped several hundred miles offshore. The current average is low because most communities that use the ocean for disposal dump undigested sludge nearshore. Table 3 summarizes costs for land and ocean disposal of sewage sludge.

TABLE 3.—*Estimated Costs of Land-Based Sewage Sludge Disposal (37, 50)*

Location	Method	Cost per ton
Land	Digestion and lagoon storage (Chicago)	$45
	Digestion and land disposal [1]	22
	Composting	35–45
	Processing into granular fertilizer (net cost)	35–50
	High temperature incineration	35–60
Ocean	Barging undigested sludge	3–18
	Barging digested sludge	8–36
	Piping disposal	12–30

[1] At Chicago, with a 7-mile pipeline to the land disposal site.

These data indicate that land-based sewage sludge disposal is more expensive than nearshore ocean disposal. But when sewage is digested and barged a distance from shore, the costs become comparable, and land-based disposal may even be cheaper. As indicated in the discussion on solid waste disposal alternatives, the capacity does exist to handle more sewage sludge. But current land-based operations are often not adequate to protect the environment.

Pipeline disposal of treated sewage sludge, used by Los Angeles, has been proposed for other areas. Because piped and barged sludge materials are the same, the same policy is recommended. Further, the potential savings for piping are not significant in light of the potential environmental impact.

Piping digested sewage sludge 7 miles from Los Angeles costs an estimated $1.55

per ton. (37) FWQA estimates that current costs on the East Coast would double the net cost—a function of both increasing costs since the Los Angeles pipeline was constructed and the higher construction costs on the East Coast. Costs for longer pipelines to limit environmental damage would increase at a linear rate, and perhaps even faster, as the distance increased because of construction and pumping difficulties. A 30-mile pipeline might raise the cost to $12 per ton and a 50-mile pipeline to perhaps $20 to $30 per ton.

More promising is the use of digested sludge for land and strip mine reclamation and for a supplemental crop fertilizer. As discussed earlier, many strip mines are in need of reclamation. Sewage sludge is high in nutrient value and can be used to improve lands low in organic matter.

The Metropolitan Sanitation District of Chicago has intensively researched the environmental impact and potential of using digested sewage sludge as a crop fertilizer and in land reclamation. Their studies document the nutrient value, lack of odor, and safety when used on all types of land, including clay, sand, and acid strip mine tailings. Depending on crops and soil condition, other nutrients may be needed, but the sludge can supply much of the needed nutrients and moisture. Chicago now spends over $20 million annually to dispose of 900 tons (on a dry weight basis) of sewage sludge per day, using incineration, lagoon storage, and other methods. (50) The District is prepared to initiate a program of rail or barge haul for sludge disposal and land reclamation within a year. The program should cost approximately the same amount as current operations and has potential for large savings if pipe transport becomes feasible. Use of sludge for land reclamation looks promising, but it must be carefully controlled and monitored to assure no environmental harm.

In this discussion of land-based sewage sludge disposal, the alternatives to ocean dumping do not involve significantly greater costs. However, a phase-out period is required because of substantial commitments by some communities and the lead time necessary to develop the alternatives.

DREDGE SPOILS

Disposal of dredge spoils—38 million tons—represents 80 percent of all ocean dumping in 1968. (66) Removed primarily to improve navigation, spoils are usually redeposited only a few miles away. About one-third is highly polluted from industrial and municipal wastes deposited on the bottom. (22) Their disposal at sea can be a serious source of ocean pollution. The recommended policy to phase out ocean disposal of polluted dredge spoils recognizes that the speed of implementation depends almost entirely on available alternatives.

Interim Alternatives

Disposing of all dredge spoils on land is not possible simply because of the vast tonnage. The Corps of Engineers estimates that of the total dredge spoils removed from each coastal region, 45 percent, or approximately 7,120,000 tons, on the Atlantic Coast are polluted; 31 percent, or 4,740,000 tons, on the Gulf Coast, are polluted; and 19 percent, or 1,390,000 tons, on the Pacific Coast are polluted.

Until land-based disposal facilities can handle these quantities, the following interim operational techniques are recommended: First, the pollutant level of dredge spoils should be determined by sampling and analysis for such key factors as BOD and concentration of heavy metals. If the spoils are not polluted, they can be disposed of in the ocean.

However, care must be taken in the location of disposal sites and in the method of disposal in order to minimize turbidity and to protect marine life.

For polluted dredge spoils, current disposal practices are not adequate, but mitigation of damage to the environment is possible without recourse to sophisticated and/or expensive processing techniques. The estimated cost increases for hauling polluted spoils farther from the dredging site are presented in Table 4.

TABLE 4.—*Estimated Dredging Costs Per Cubic Yard (24)*

Method	1 mile	3 miles	10 miles	20 miles	50 miles
Hydraulic pipeline dredging	$0.95	$1.30	(1)	(1)	(1)
Dipper dredging and dump scows	1.10	1.25	$1.50	$1.80	$3.60
Hopper dredging	0.28	0.34	0.54	0.81	1.66

[1] Pipeline dredging operations beyond 3 miles are usually not practical because of problems in handling long floating pipelines and the extra pumping equipment involved.

Most spoils are now deposited within a few miles from shore in less than 100 feet of water. Table 5 summarizes the additional costs for disposing of polluted dredge spoils farther out to sea using a hopper dredge.

As the table indicates, the additional cost for dumping polluted dredge spoils 10 miles rather than 3 miles out is $2.7 million annually. For 20 miles, the additional cost is $6.2 million; for 50 miles, it is $17.5 million.

Diking is another interim alternative for disposing of polluted dredge spoils. Briefly, a dike is constructed to hold the dredge spoils nearshore or at the shoreline. Its effectiveness depends on the prevention of contaminated spoils' interaction with surrounding waters. At Cleveland, diking was successful in containing over 99 percent of the contaminants in dredge spoils removed from Lake Erie. (23)

Estimates for 35 dike projects on the Great Lakes indicated that the costs of diking and depositing dredge spoils vary greatly—from $0.35 to over $6 per cubic yard. (23) The increased cost for disposal by diking over open-lake disposal ranged from $0.03 to almost $5.50 per cubic yard, with an average increase of $1.50 per cubic yard.

Diking is not without environmental problems. Dredge spoils would not provide fill of sufficient strength to allow use of the diked area for many years. Hence, areas of the coastal zone, already in high demand, would be unusable. Further, diking is unattractive and may cause greater environmental problems than controlled dispersal of pollutants.

Longer-Term Alternatives

Reduction in the volume of sediments requiring dredging and higher levels of treatment of wastes will both lessen the problem of polluted dredge spoils. Erosion control through improved construction, highway, forest, and farm planning and management will reduce future dredging needs. One example is the recently completed stream bank stabilization project on the Buffalo River,

TABLE 5.—*Estimated Costs for Disposal of Polluted Spoils Using Hopper Dredge*

Coastal area	Tons	3 miles	10 miles	20 miles	50 miles
Atlantic Coast	7,120,000	$2,421,000	$3,845,000	$5,767,000	$11,819,000
Gulf Coast	4,740,000	1,612,000	2,560,000	3,839,000	7,868,000
Pacific Coast	1,390,000	473,000	751,000	1,126,000	2,307,000
Total	13,250,000	4,506,000	7,156,000	10,732,000	21,994,000

which reduced maintenance dredging requirements 40 percent. (23) The level of pollution in dredge spoils will be reduced by the higher levels of treatment of municipal and industrial wastes required by Federal-State water quality standards within a few years.

High-temperature incineration of contaminated dredge spoils is a longer-term alternative requiring further development and testing. Such incineration can render spoils an inert ash, safe for land disposal. Processing costs are a function of the size of the plant, the percent of total solids, and the percent of volatile solids. Figure 1 illustrates disposal costs per cubic yard for incinerating dredge spoils whose total solid content ranges between 30 percent and 45 percent (a normal range) and volatile solids between 10 percent and 20 percent (a normal range). Also shown are costs for aerobic stabilization, a process similar to that used for sewage treatment. These costs can range from $2 to $12 per cubic yard or roughly 4 to 24 times current ocean disposal costs. Compared to disposal 20 miles out to sea, however, incineration is 3 to 15 times as costly. But compared to disposal at 50 miles, incineration may cost the same or it may be as much as 8 times more costly.

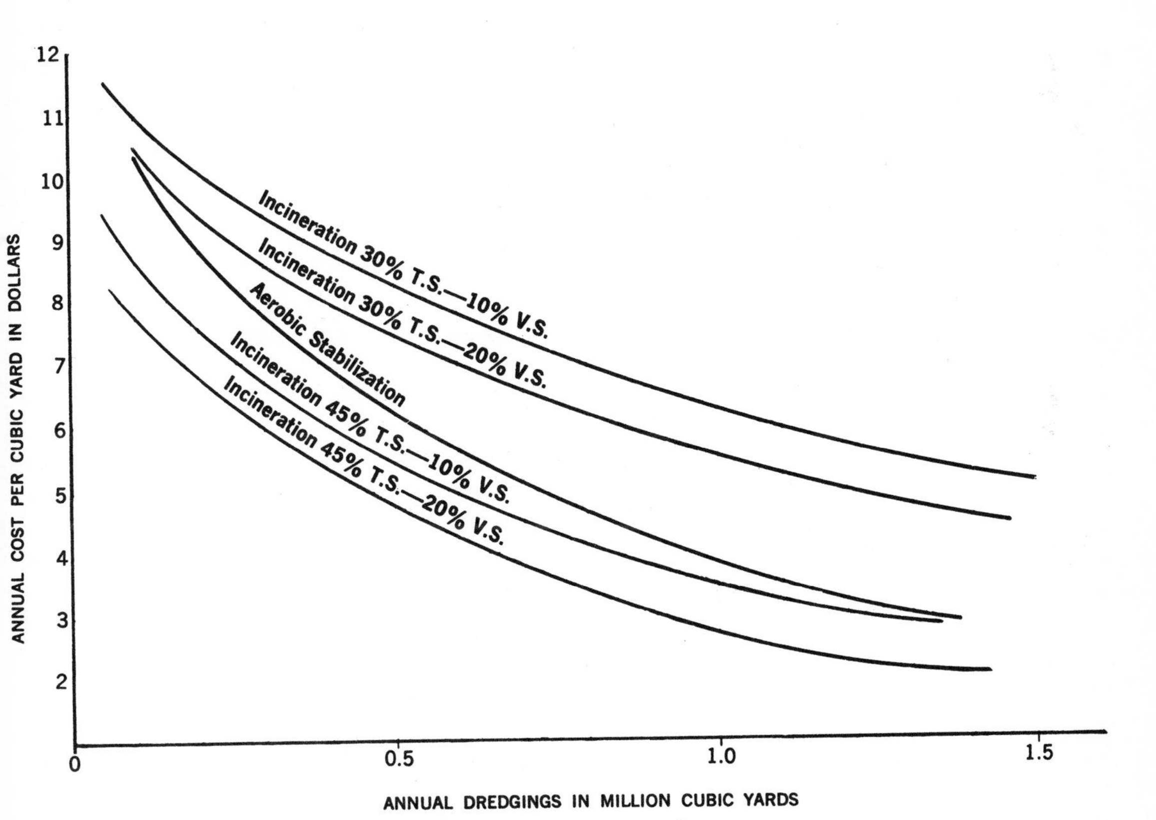

Figure 1.—Total Annual Cost Per Cubic Yard for Complete Treatment Using Incineration and Aerobic Stabilization (23)

T.S.=TOTAL SOLIDS

V.S.=VOLATILE SOLIDS

Special treatment to remove toxic materials so that the sludge may be used as a fertilizer either on arid lands or for ocean farming is possible. An approach similar to that discussed for use of digested sewage sludge as a fertilizer may be feasible.

INDUSTRIAL WASTES

Industrial wastes vary widely, but they usually contain nutrients, heavy metals, and/or other substances toxic to marine biota. Although the volume of industrial wastes is 10 percent of all wastes disposed of in the ocean, it is minor compared to the quantities of industrial wastes treated at land-based facilities.

The policy recommended would call for termination of ocean dumping of industrial wastes as soon as posssible. Ocean dumping of toxic industrial wastes should be terminated immediately, except in those cases in which no alternative offers less harm to man or the environment.

Interim Alternatives

Many industries utilize ocean disposal because it is cheaper and easier than other disposal processes. Table 6 shows costs for bulk and containerized wastes.

TABLE 6.—*Industrial Wastes Disposal Costs* (*66*)

Method	Average cost/ton	Range of cost/ton
Bulk wastes	$1.70	$0.60-$9.50
Containerized wastes	24.00	$5-$130

The costs of discharging bulk wastes directly into the sea are significantly lower than for other disposal techniques. Containerization, used mainly for toxic materials, is much more costly than dumping bulk wastes.

Industrial wastes can be treated and disposed of on land, or they can be incinerated. Whichever technique is used, it is necessary to assure that the environment is protected. Treatment of wastes should not add to stream pollution, and incineration should not add to air pollution. Deep-well disposal of toxic wastes is generally undesirable because of the danger of ground water pollution.

Unlike the other categories discussed, industrial wastes are not homogeneous. Hence, interim disposal methods will vary not only among the different types of wastes but also according to process, location, local practices, and other factors. The costs of using some alternatives will be significantly higher than for ocean dumping, but as a portion of total production costs, generally they will not be great. Total industrial pollution control costs, as a percentage of gross sales, are well under 1 percent, although costs for some industries are much higher.

Longer-Term Alternatives

In the long term, changes in industrial production processes and recycling offer great promise for reducing or reusing industrial wastes. For example, the average waste from modern sulfate paper plants is only 7 percent of wastes in the older sulfite process. In some cases, recycling will be an alternative to ocean disposal. Two West Coast refineries are now recycling oil wastes instead of disposing of them at sea.

Toxic wastes present a more difficult problem. They cannot be stored indefinitely, but allowing ocean disposal is a disincentive to development of adequate detoxification and recycling techniques and of production processes with fewer toxic byproducts. But highly toxic wastes will continue to be produced, and many will not be amenable to land disposal.

One alternative worthy of further study is the establishment of regional disposal, treatment, and control facilities. Federally or privately operated, the facilities could conduct research on and provide for waste detoxification and storage. Complicated disposal processes that are too expensive or complex for a single company could be used jointly to dispose of wastes. Fees would need to be sufficiently high to encourage development of private solutions, except in the most troublesome cases or when significant economies would result from shared use of facilities.

CONSTRUCTION AND DEMOLITION DEBRIS

Construction and demolition debris, less than 1 percent of all wastes dumped in the ocean, (66) are composed mainly of dense and inert materials. Because of the small amounts dumped and their character, these wastes are not a threat to the marine environment. Moreover, amounts dumped in the ocean are not expected to increase significantly because of their high value as landfill. The recommended policy assumes continued ocean dumping, but with care to prevent damage to the marine ecosystem.

RADIOACTIVE WASTES

Since 1962, no significant quantities of radioactive wastes have been dumped at sea. Rather, they have been stored at several sites operated or regulated by the Atomic Energy Commission or at sites regulated by the States. Increasing demands for electricity and for use of nuclear power portend a dramatic increase in the amounts and kinds of nuclear wastes produced. Hence, it is important to develop policy to prevent contamination of the ocean.

The policy recommended would continue the practice of prohibiting high-level radioactive wastes in the ocean. Dumping other radioactive materials would be prohibited, except in a very few cases for which no practical alternative offers less risk to man and his environment.

Alternatives (Interim and Longer Term)

The quantity of nuclear wastes is not large, and the technology for storing and treating them is well developed. However, the AEC estimates that the amount of high-level liquid radioactive wastes will increase approximately sixtyfold between 1970 and the year 2000. High-level wastes, usually liquid, are now stored on an interim basis in large, well-shielded tanks. In the long run, the wastes will be solidified, reducing their volume by a factor of ten, for eventual storage in special geological formations, such as salt mines. As new nuclear facilities are constructed, provision is being made for parallel construction of storage tanks and treatment facilities to handle the wastes.

Solid radioactive wastes have been buried in carefully controlled landfill sites. In 1970, about 40,000 cubic yards of solid radioactive wastes will be buried in approximately 15 acres. (70) The increase in the amount of these wastes in the next decade will require about 300 acres. This figure could be reduced with compaction and incineration, which are currently being used or planned.

Low-level liquid wastes from nuclear power generation, medical facilities, etc. are treated and/or stored to reduce radioactivity. A small amount is eventually released to the environment under controlled conditions.

Large radioactive structures, chiefly reactor vessels and associated parts, have heretofore not presented a significant problem. With the exception of ocean disposal of the SEAWOLF submarine reactor vessel, obsolete reactor vessels and associated parts have been decontaminated, dismantled, and stored on land. Sixteen nuclear power plants are now operating, and 80 are either under construction or permit applications are pending. There may be as many as 1,000 plants by the year 2000. When reactor vessels are taken out of service, each used structure is a source of high-level induced radiation.

There are three alternative ways to dispose of these vessels and associated parts: ocean disposal; entombment in place, with final disposition after radioactive decay; and dismantling and burial. Ocean disposal is the cheapest method when the facility is on the coast or when waterborne transportation is available. Entombment provides an opportunity to monitor disposal operations carefully but occupies valuable land during the period of radioactive decay. Dismantling and burial is the most expensive of the alternatives.

Because of the need to keep all sources of radioactivity at the lowest possible level, ocean disposal of the wastes should be avoided except when no alternative offers less harm to man or the environment. These cases should be carefully examined to assure that no safe and practical alternatives do exist. If ocean disposal is necessary, it should be carefully controlled.

EXPLOSIVES AND CHEMICAL MUNITIONS

Large quantities of explosives and some chemical warfare agents have been disposed of at sea. No biological warfare agents have been disposed of at sea. The policy recommended would prohibit ocean disposal of chemical and biological warfare agents and phase out disposal of explosive munitions.

Alternatives (Interim and Longer Term)

Ocean disposal of munitions was developed as an alternative to burning them in the open. That practice is often hazardous, is noisy, and creates air pollution.

Other alternatives to ocean dumping are available and should be used. In some cases weapons can be dismantled and critical components, such as gunpowder, lead, etc., either disposed of safely or sold for reuse. Centralizing the disposal of obsolete munitions may be desirable to provide efficient dismantling. Alternatively, portable disposal facilities, under development by the Department of Defense, offer promise. When salvage value is significant, commercial contracting for disposal services may be possible. Mass underground burial or detonation is another alternative.

The alternatives used for disposal of munitions will depend on ability to train people for disposal operations, relative costs, available sites, and their environmental impact. Dismantling and recycling the materials is the preferable alternative from an environmental point of view, but facility and manpower constraints may dictate the use of other alternatives to ocean dumping.

For chemical warfare agents and munitions, the alternatives to ocean disposal are neutralization and incineration. Toxic chemical warfare agents can be separated from munitions or containers and then treated. Facilities are currently being modified at the Rocky Mountain Arsenal near Denver, Colo., for disposal of toxins. Similar facilities for treatment of chemical warfare agents are needed elsewhere. (26)

SUMMARY

Interim alternatives exist to mitigate the environmental damage of ocean dumping. Land capacity can be expanded by use of rail haul, and strip mines and other lands can be reclaimed. In the long run, technological advances and new methods of recycling should help reduce pressures for ocean disposal. The major conclusion is that a program of phasing out all harmful forms of ocean dumping and prohibiting new sources is feasible without greatly increased costs.

CHAPTER IV

Legislative Control of Ocean Dumping

THE previous chapters indicate the need for a national policy to control ocean dumping. This chapter examines the adequacy of State and Federal regulatory authorities to implement that policy.

STATE CONTROL ACTIVITIES

Although by tradition and Federal law the States have primary responsibility for water pollution control, the response of the coastal States to ocean dumping has not been extensive. Where the Federal Government has assumed authority over ocean dumping—in New York, Baltimore, Boston, and Hampton Roads, Va.—States have subordinated their activities to Federal control.

In some circumstances States exercise regulatory authority. California, for example, through State and regional agencies, has provided the leading role in control of ocean dumping of such materials as municipal garbage and industrial chemicals and solid waste. In the San Francisco Bay area and in the San Diego area, regional water quality control boards regulate ocean dumping operations and provide for monitoring and surveillance to enforce the regulations. Disposal operators are required to file detailed trip reports and a monthly summary of the volume and types of wastes dumped. In the San Diego area, prior notification of ocean dumping is required so that a board staff member can accompany the dumping vessel. In the Los Angeles area, the California Department of Fish and Game is the lead agency. In Oregon, the State Board of Health regulates ocean dumping, with special emphasis on chemicals. No other States regulate ocean dumping to a greater extent than California and Oregon.

State regulation has not established a basis for an extensive and comprehensive method of controlling ocean dumping. Besides general lack of authority and programs, State jurisdiction would generally be limited to the 3-mile territorial sea.

FEDERAL CONTROL ACTIVITIES

Four Federal agencies have some responsibilities for ocean dumping: the Corps of Engineers, the Federal Water Quality Administration, the Atomic Energy Commission, and the Coast Guard.

Corps of Engineers

The Corps of Engineers is the only agency with regulatory authority to control dumping of a broad class of materials. This authority stems from Corps responsibility for maintaining navigation in U.S. territorial waters. In general, the Corps has no power other than in internal navigable waters and in the territorial sea.

Special authority for the port areas of New York, Baltimore, and Hampton Roads, Va., was given to the Corps of Engineers under the Supervisory Harbors Act of 1888 (33 U.S.C. 441–451b). Under that Act, the Corps exerts jurisdiction over ocean dumping beyond the territorial sea by controlling transit through the territorial sea. The Act provides for the appointment of a harbor supervisor to control ocean dumping, authorizing him to issue permits for the transportation and dumping of materials into the ocean. For ocean dumping in territorial seas, the Corps relies on both section 4 of the Rivers and Harbors Act of 1905 (33 U.S.C. 419) and section 13 of the Rivers and Harbors Act of 1899 (33 U.S.C. 407). Through the regulatory and permit authority conferred by the Supervisory Act, logs and fathometer charts are required of tugboat operators transporting material for dumping to provide surveillance of their operations. Infrequent ship and aircraft patrols are made for the same purpose. The permit operation has three steps: application by the prospective dumper according to the type of waste, issuance or rejection of a permit by the Corps after review, and monitoring of operations by the Corps as waste materials are transported to the designated dumping grounds.

The Corps has cautiously exercised its power under the 1899 and 1905 Acts. Its policy on enforcing these authorities can be attributed largely to emphasis on navigation in the enabling statutes. Until recently there was considerable doubt whether the Corps could deny a permit to a prospective waste disposal applicant for any reason other than obstruction to navigation. These doubts were dispelled only on July 16, 1970, when, in *Zabel* v. *Tabb*, —— F. 2d —— (5th Cir.), a Federal circuit court reversed a district court ruling. The district court disputed Corps authority to consider environmental as well as navigational factors in denying a permit and directed that the permit be granted. The circuit court, relying on the Fish and Wildlife Coordination Act (16 U.S.C. 661–666c) and the National Environmental Policy Act of 1969 (42 U.S.C. 4331–4347), held that the Corps does have this authority and could deny the permit.

Despite jurisdictional limitations, the Corps has occasionally concurred in ocean dumping outside the territorial seas when its direction was requested. For example, dumping areas have been established off Boston Harbor by the Corps, but with full recognition that authority was lacking. In such instances the action is taken at the request of the user. Often when the Corps receives a request to dump in areas beyond the territorial sea, it simply issues a letter of no objection. Prior to issuing such a letter, the Corps consults other governmental agencies such as the Fish and Wildlife Service of the Department of the Interior and the fish and game department of the affected State.

In the New York Bight area, the Corps has designated areas for the deposit of rock, dredged material other than rock, cellar dirt, sewage sludge, chemicals, and other substances. Specific regulations define the areas in which dumping can take place. Special permits, usually of 3 months' duration, are issued for the transit of material to the dumping areas.

Criminal penalties are authorized to punish violations of the various Corps authorities. Fines of up to $2,500 may be levied, or imprisonment up to 1 year may be imposed. Under the Supervisory Harbors Act, when dredged matter is illegally dumped, a fine of $5 per cubic yard of material can be prescribed.

Corps authority over ocean dumping has several limitations: First, with the exception of three harbors, it is restricted to the 3-mile territorial sea; yet most waste disposal sites lie outside the territorial sea. Second, its authority originates from responsibility for the navigability of waterways, not for their ecology. Third, while operational authority is lodged in an agency with responsibility to promote navigation, the water quality agency has no direct control over actions of the operating agency. In fact, the Corps could conceivably issue permits for activities that FWQA believes damage the quality of marine waters. Fourth, to a large extent the Corps regulates itself because it is a major producer of dredge spoils, the material most commonly dumped at sea. This is the type of conflict of interest that the creation of the Environmental Protection Agency was designed to prevent. Nonethless, the Corps has capabilities which could be effectively used to implement the recommended policy on ocean dumping. It possesses a large field organization strategically located in areas where ocean dumping regulatory action is important.

Federal Water Quality Administration

The Federal Water Quality Administration (FWQA), in the Department of the Interior, administers section 10 of the Federal Water Pollution Control Act, as amended (33 U.S.C. 466g). Under this section, States develop water quality standards for interstate and coastal waters within their jurisdiction. The standards require Federal approval, thus becoming joint Federal-State standards.

These standards consist of water quality criteria (e.g., 5 parts per million of dissolved oxygen) to meet designated water uses (e.g., water supply, recreation, etc.). The standards must also include an enforcement and implementation plan in which remedial measures are to be taken in accordance with a schedule for achieving the water quality levels established. The Federal Water Pollution Control Act provides procedures for abating pollution which violates water quality standards, endangers health or welfare, or interferes with the marketing of shellfish in interstate commerce.

The Administration has proposed amendments to the Federal Water Pollution Control Act (S. 3471) that would authorize the Secretary of the Interior to establish water quality standards for the contiguous zone when pollution in these waters is likely to cause pollution in the territorial sea and to set standards for discharge beyond the contiguous zone of substances transported from territory under U.S. jurisdiction. The legislation would also call for specific effluent discharge requirements for all discharges into waters covered under the Act.

The authority of FWQA under the Federal Water Pollution Control Act, even with the proposed new amendments, would not be adequate to control ocean dumping. First, there is no authority for requiring permits to dump wastes in the oceans—authority essential to enforcement of any effective control program. Second, the Act's general thrust is control of continuous discharges that clearly violate the water quality standards, rather than control of intermittent dumping.

Other sections of the Federal Water Pollution Control Act deal with ocean disposal of specific materials or classes of materials. Section 11 of the Act prohibits discharge of harmful quantities of oil into the navigable waters of the United States and the contiguous zone, but it deals only with oil and is aimed chiefly at spills, rather than at purposeful dumping.

Section 12 of the Act provides authority for Federal agencies to clean up and to prevent discharge of hazardous substances into the navigable waters of the United States and the contiguous zone. Hazardous substances are those that present an imminent and substantial danger to the public health and welfare. Many materials now dumped in the oceans could be classified as hazardous: solid waste containing heavy metals, DDT, or other persistent pesticides and sewage sludge from limited-treatment facilities. But regulating intentional ocean disposal of materials is beyond the scope of section 12.

Section 13 of the Act provides for control of sewage from vessels, chiefly by requiring the installation of marine sanitation devices.

Although FWQA lacks authority for issuing permits to control ocean dumping, it has several related responsibilities. These include approval, and in some circumstances establishment, of water quality standards in interstate and coastal waters; enforcement; research; technical assistance; monitoring; and other water quality functions.

Atomic Energy Commission

The Atomic Energy Act of 1954 authorizes the AEC to regulate the receipt, transfer, and possession of nuclear source, byproduct, and special materials (42 U.S.C. 2077, 2092, 2111); these include most radioactive substances. In addition, the AEC has authority to regulate and control contractually the use of radioactive materials for its own activities, such as AEC-supported research and development programs. These authorities cover ocean disposal of radioactive materials but not other wastes.

Coast Guard

The Coast Guard is the principal maritime law enforcement agency. It enforces or assists in the enforcement of all Federal laws on the high seas and waters subject to the jurisdiction of the United States and has authority to make inspections, searches, seizures, and arrests. In addition, the Coast Guard can assist other Federal agencies and State and local governments in carrying out their responsibilities. The Coast Guard's law enforcement capability can be an effective means of enforcing controls and standards set by other agencies, but it has no independent authority to control ocean dumping.

RECOMMENDATIONS

Authority to control ocean dumping is currently dispersed among several agencies. Jurisdiction is generally confined to the territorial sea, where most material is currently not dumped. Authority that is now used for control is not lodged in agencies responsible for environmental control. Conflicts of interest exist in that some regulatory powers are exercised by agencies with operational responsibilities in the same area.

These problems must be resolved before a national policy on ocean dumping can be implemented. Full regulatory responsibility—involving both setting standards and issuing permits—should be placed in one organization. The Council recommends that this agency be the Environmental Protection Agency.

The organization charged with implementation of the national policy should have as its chief purpose the protection of the environment. It should also command sufficient research and monitoring resources for evaluating the environmental effects of the broad spectrum of materials currently dumped in the oceans.

Authority to control ocean dumping must be tied closely to efforts to abate other sources of pollution in the marine environment. Municipal and industrial discharge in rivers and harbors, urban and rural runoff, and other sources are important components of marine pollution. A regulatory program for ocean dumping should be defined to complement the efforts in these other areas.

Most of the wastes now dumped in the oceans originate in the United States and are transported to sea for dumping. Accordingly, primary jurisdictional emphasis should shift from a territorial basis to regulation of the transportation of materials from the United States for dumping.

The Environmental Protection Agency will have the broad responsibility as well as the necessary supporting programs to protect the marine environment. To give it the power to regulate ocean dumping, legislation is required.

CHAPTER V International Aspects of Ocean Disposal

THE oceans of the world are a truly international resource, forming a vast environmental system through which its components circulate or are dispersed by currents and the migrations of organisms. They are critical to maintaining the world's environment, contributing to the oxygen-carbon dioxide balance in the atmosphere, affecting global climate, and providing the base for the world's hydrologic system.

Within the oceans, fish may travel great distances during their lifetimes. Although the oceans are important to all nations, they are particularly significant for many developing countries, which increasingly depend on fisheries for essential protein. A disturbance in the chemistry of the oceans which could be multiplied in the food chains would have a major impact on food-deficient nations. Hence, pollutants from one country may ultimately affect the interests of many other nations.

WORLDWIDE CHEMISTRY OF THE OCEANS

Of the materials entering the oceans through natural processes, the amounts of two, mercury and lead, have probably been doubled by man's activities. In addition, man has introduced new chemical compounds, such as chlorinated hydrocarbons (including DDT), gasoline, dry cleaning solvents, and other organic materials, whose biological significance is unknown.

The rate of transfer of mercury from land to oceans by natural weathering is estimated at 5,000 tons per day. (38) This amount, about one-half the total world production of mercury, is used by agriculture and industry in such a way that it eventually enters the oceans. As yet, this approximate doubling has not been chemically measured, but it is thought responsible for the 10 to 20 times increase in mercury found in sea birds off Sweden between prewar years and the 1950's (5) and for additions to the high mercury content of fish off Japan.

Natural weathering introduces into the oceans about 150,000 tons of lead each year. Man introduces about 250,000 tons in the Northern Hemisphere alone (69). Most of this lead is derived from the washout into the oceans of atmospheric lead produced by burning gasoline enriched with tetraethyl lead. Industrial waste products further contribute lead. Over the last 45 years these additions have raised the average lead content of ocean surface waters from 0.01–0.02 to 0.07 micrograms per kilogram of sea weater. (19) Slow mixing within the oceans keeps the lead within the upper layers, the region where biological productivity is greatest and the chances of biological enrichment highest. However, the biological effects of this changing lead concentration remain unknown.

Industrial wastes and sewage sludge also introduce large quantities of such metals as vanadium, cadmium, zinc, and arsenic. Man's contribution relative to nature's is not known, but civilization may well be close to matching nature's contribution of these materials to the oceans.

The fact that man is changing the chemical composition of the oceans focuses attention on the need for international action to control the introduction of wastes into the ocean.

INTERNATIONAL LAW ON WASTE DISPOSAL

In an environmental sense there are no subdivisions within the oceans. The highly productive coastal waters are continuous with and contribute to the biologic activity of the deepest trenches. Legally, the oceans are divided into the seabed and the superjacent waters, and further subdivided into distinct zones with particular legal characteristics. International law governing ocean waste disposal must take into account these legal characteristics and the material to be dumped.

Four conventions, referred to as The Law of the Sea Conventions, were adopted at Geneva in 1958 codifying existing international law and establishing new rules governing the law of the sea. The Convention on the Territorial Sea and the Contiguous Zone sets out three zones—the territorial sea, the high seas, and the contiguous zone between them.

Narrow bays, estuaries, and other semi-enclosed areas are classed as internal waters. Seaward of the internal waters and of the low-water line along uninterrupted coasts is the territorial sea, extending for 3 miles. Between 3 and 12 miles from the shore is the contiguous zone. The contiguous zone, together with the waters lying seaward of it, comprise the high seas. Each has distinct legal characteristics affecting rights to dispose of materials in it and to control such disposal.

A coastal state (nation) has exclusive control over its internal waters and its territorial sea. In these areas, the coastal state has exclusive power to determine dumping sites and to enact necessary sanitary and pollution laws to protect its citizens and their property. These laws can be enforced against ships of both the coastal state and of foreign registry. In addition, a coastal state may control the transport of waste products from its ports. However, in its territorial sea, the coastal state must permit the innocent passage of foreign vessels that do not prejudice its peace, good order, or security. As discussed in Chapter IV, Congress has enacted legislation that covers ocean disposal of oil and sewage wastes from vessels.

Within the contiguous zone, 3 to 12 miles out to sea, the coastal state may exercise some control necessary to prevent pollution. The right to exercise these controls in the contiguous zone, however, does not change the high seas status of those waters. Under the terms of the Convention on the Territorial Sea and the Contiguous Zone, a coastal state cannot act to prevent dumping in the contiguous zone unless such action is necessary to prevent infringement of sanitary regulations within its territorial sea.

The international law governing the high seas, the largest jurisdictional zone, is codified in the 1958 Geneva Convention on the High Seas. This Convention provides for freedom of navigation and of fishing, freedom to lay submarine cables and pipelines, freedom to fly over the high seas, and other freedoms recognized by international law, such as dumping.

The Convention sets forth two fundamental concepts: It declares the high seas as an area not subject to sovereignty, and it states that the freedoms of the seas which are recognized in international law must be exercised by states with reasonable regard to the interests of all other states in their exercise of freedom of the high seas. Inasmuch as one use may interfere with another current or potential use of the high seas, the reasonable regard standard holds that there must be an accommodation of the various and possibly conflicting uses of the high seas.

The right to dispose of waste materials in the high seas is a traditional freedom of the seas. However, under the standards set out in the Geneva Convention on the High Seas, this freedom—like all other freedoms of the seas—must be exercised with reasonable regard to other states' use of the oceans. It is not possible to say that any particular waste disposal or dumping project will meet the requirements of international law. Only after careful consideration can it be determined that a particular ocean dumping proposal meets the reasonable regard standard set out in the Convention. For example, a project for disposal of unpolluted dredge spoil may be suitable for an area of the high seas in which disposal of chemical waste would neither be suitable nor legal.

Unfortunately, the law of the sea conventions do not establish a hierarchy of ocean uses. However, international law places paramount importance on the protection of human life. It allows destruction of property to save human life or to prevent greater property damage. Clearly, any dumping activity that threatens life or directly damages property violates international law.

It is important to recognize that the law of the sea is based primarily on conventions or other agreements which were concluded prior to current understanding of the actual and potential impacts of dumping on the marine environment. Consequently, present international law appears inadequate to deal with possible long-term environmental effects of various actions.

INTERNATIONAL ACTIVITIES

Many international organizations engage in activities related in some way to marine pollution. Most of these activities are designed to exchange ideas and/or to coordinate national efforts. It is important to recognize, however, that in most cases, their concern with ocean pollution and particularly with ocean dumping is only incidental or peripheral. Although efforts such as the International Decade of Ocean Exploration will provide useful data, the IDOE does not give the highest priority to ocean pollution. Combined annual expenditures on activities designed to improve environmental quality, of which ocean waste disposal problems constitute but a small part, probably do not exceed $5 million, a small sum compared with the $100 million of the FWQA in fiscal year 1970 for water pollution control and research alone.

Research concerned with ocean pollution and establishment of controls on waste disposal is undertaken mainly through national efforts, rather than by the intergovernmental agencies. Even national efforts are limited. Basic studies of the character of the oceans and the seabeds have dominated U.S. oceanographic research. There has been little or no emphasis on such questions as the capacity of the oceans to absorb wastes.

Several countries have begun to search for solutions. Canada is developing regulations governing the disposal of garbage and sewage from vessels. As now drafted, the regulations would apply to non-pleasure craft within the territorial sea and inland waters of Canada and would require new vessels in Canadian inland waters to carry sewage treatment equipment. The regulation would also prohibit discharge of garbage in all Canadian waters. Israeli scientists have been studying pollution of the Mediterranean coast off Tel Aviv since 1963. All new vessels constructed for the Argentine Merchant Marine are required to meet international standards on waste disposal, including holding tanks and oil-water separation tanks. Argentinian law also requires all foreign ships to be similarly equipped or access to Argentina ports will be denied. Similar legislation is contemplated for pleasure craft.

NEED FOR INTERNATIONAL ACTION

International cooperation is essential to preservation of the oceans. The quantities of wastes dumped in the oceans are increasing rapidly in this country and will increase internationally as other countries experience similar waste disposal pressures. Consequently, control of ocean dumping necessitates action.

Recognition of the need for international cooperation is an initial step toward reaching worldwide agreements to control ocean pollution. There will be obstacles. Nations' interests in the oceans vary, as do their ideas on the controls that may be required.

RECOMMENDATIONS

The United States should assist in finding a solution to the international problem of ocean dumping through a twofold approach. First, it must systematically attack its own problems. As a significant polluter of the ocean and at the same time a technologically advanced nation, the United States must show its serious intention to meet its responsibility as a matter of urgent national priority. In demonstrating determination to preserve the marine environment, the Nation will develop valuable information on costs, effects, and technology associated with ocean dumping and its alternatives.

Second, the U.S. should take the initiative to achieve international cooperation on ocean dumping. The Council on Environmental Quality recommends that at the outset the Federal Government develop proposals to control ocean dumping for consideration at international forums such as the 1972 U.N. Conference on the Human Environment at Stockholm. U.S. initiative should suggest a basis for international control over ocean dumping similar to the policy recommended in this report. Provision should be made for:

- Cooperative research on the marine environment and on the impacts of ocean dumping of materials;
- Development of a worldwide monitoring capability to provide continuing information on the state of the world's marine environment;
- Development of technological and economic data on alternatives to ocean disposal.

Domestic and international action is necessary if ocean dumping is to be controlled. The United States must show its concern by strong domestic action through implementation of recommended policy. But unilateral action alone will not solve a global problem. International controls, supported by global monitoring and coordinated research, will be necessary to deal effectively and comprehensively with pollution caused by ocean dumping.

References

1. Abelson, P. H. 1970. Methyl Mercury. Editorial. Science 169:3942.
2. Ad Hoc Committee. 1970. Evaluation of Influence of Dumping in the New York Bight. Report to the Secretary of the Interior. (mimeograph)
3. Anonymous. Panama American. 1970. Contamination Threatens RP Marine Life. January 17.
4. Armstrong, N. E., and Storrs, P. N. 1969. Biological Effects of Waste Discharges on Coastal Receiving Waters. Background paper prepared for the NAS–NAE Steering Committee for Coastal Waste Disposal Workshop. (mimeograph)
5. Berg, W., Johnels, A., Sjostrand, B., and Westermark, T. 1966. Mercury Content in Feathers of Swedish Birds for the Past 100 Years. Journal Oikos 17:71–83.
6. Blaycock, G. G. 1969. The Fecundity of a *Gambusia affinis affinis* Population Exposed to Chronic Environmental Radiation. Radiation Research 37:108–117.
7. Blumer, M., Souza, G., and Sass, J. 1970. Hydrocarbon Pollution of Edible Shellfish by an Oil Spill. Woods Hole Oceanographic Institution Ref. No. 70–1. (mimeograph)
8. Buelow, R. W., Pringle, B. H., and Verber, J. L. 1968. A Preliminary Investigation of Waste Disposal in the New York Bight. Department of Health, Education, and Welfare, Public Health Service, Northeast Marine Health Sciences Laboratory. (mimeograph)
9. Burd, R. S. 1968. A Study of Sludge Handling and Disposal. Supported in part by grant No. PH 86–66–32 of Department of the Interior, Federal Water Pollution Control Administration.
10. Butler, P. A. 1961. Effects of Pesticides on Commercial Fisheries. pp. 168–171. *In* Proceedings of Gulf and Caribbean Fisheries Institute, 13th Annual Session.
11. California State Water Quality Control Board. 1964. An Investigation of the Effects of Discharged Wastes on Kelp. Pub. No. 26.
12. Chamblin, J. 1969. Rumblings from the Deep. Science News 96:213–214.
13. Chipman, W., and Galtsoff, P. S. 1949. Effects of Oil Mixed with Carbonized Sand on Aquatic Animals. Department of the Interior, Bureau of Commercial Fisheries Spec. Scientific Rep. No. 1.
14. Chow, T., and Patterson, C. 1966. Concentration Profiles of Barium and Lead in Atlantic Waters off Bermuda. Earth Planetary Science Letters 1:397–400.
15. City of New York Environmental Protection Administration, Department of Water Resources. 1970. Information provided to the Council.
16. Clendenning, K. A., and North, W. J. 1960. Effects of Wastes on the Giant Kelp, *Macrocystis pyrifera*. p. 82. *In* Proceedings of the First International Conference on Waste Disposal in the Marine Environment.
17. Clendenning, K. A., and North, W. J. 1958. The Effects of Waste Discharges on Kelp, Quarterly Progress Report of the Institute of Marine Resources, University of California, La Jolla. IMR Ref. No. 58–6.
18. Combustion Power Company, Inc. 1969. Combustion Power Unit—400. Prepared for Department of Health, Education, and Welfare, Public Health Service, Bureau of Solid Waste Management under contract No. PH 86–67–259. (mimeograph)
19. Cronin, L. E. 1969. Biological Aspects of Coastal Waste Disposal. Background paper prepared for the NAS–NAE Steering Committee for Coastal Waste Disposal Workshop. University of Maryland, Natural Resources Institute Ref. No. 69–99. (mimeograph)
20. Davis, W. B. 1969. Monitoring of the Marine Environment for Waste Components and for Effects of the Introduced Wastes. Background paper prepared for the NAS–NAE Steering Committee for Coastal Waste Disposal Workshop. (mimeograph)
21. Darnell, R. M. 1967. Organic Detritus in Relation to the Estuarine Ecosystem. pp. 376–382. *In* Lauff, G. H. (ed.). Estuaries 1967. American Association for the Advancement of Science Pub. No. 83.
22. Department of the Army, Corps of Engineers. 1970. Information supplied to the Council.
23. Department of the Army, Corps of Engineers, Buffalo District. 1969. Dredging and Water Quality Problems in the Great Lakes. Vol. 1.
24. Department of the Army, Corps of Engineers, New York District. 1970. Information supplied to the Council.
25. Department of Commerce, Office of Business Economics, Regional Economics Division. 1970.

26. Department of Defense, Office of the Assistant Secretary of Defense for Health and Environment. 1970. Information supplied to the Council.
27. Florida Department of Air and Water Pollution Control. 1970. Escambia Bay Menhaden Kill. Smithsonian Institution Center for Short-Lived Phenomena Event Notification Rep. No. 76–70.
28. Department of Health, Education, and Welfare, Bureau of Solid Waste Management. 1970. Information supplied to the Council.
29. Department of Health, Education, and Welfare, Bureau of Solid Waste Management. 1970. Sanitary Landfill Guidelines. Review draft.
30. Department of the Navy. 1970. Information supplied to the Council.
31. Department of the Interior. 1967. Surface Mining and the Environment.
32. Department of the Interior. Bureau of Outdoor Recreation and National Park Service. 1969. Gateway National Recreation Area: A Proposal.
33. Department of the Interior, Bureau of Sport Fisheries and Wildlife. 1970. National Estuary Study.
34. Department of the Interior, Bureau of Sport Fisheries and Wildlife, Sandy Hook Marine Laboratory. 1969. The Effects of Waste Disposal in the New York Bight—Interim Report for January 1, 1970. Prepared for Department of the Army, Corps of Engineers, Coastal Engineering Research Center.
35. Department of the Interior, Bureau of Sport Fisheries and Wildlife, Sandy Hook Marine Laboratory. Undated. Statement on the Effect of Sewage Pollution on Fisheries of the Middle-Atlantic Coastal Zone. (mimeograph)
36. Department of the Interior, Federal Water Pollution Control Administration. 1970. The National Estuarine Pollution Study.
37. Department of the Interior, Federal Water Quality Administration. 1970. Information provided to the Council.
38. Goldberg, E. 1970. The Chemical Invasion of the Oceans. pp. 178–185. *In* Singer, S. F. (ed.). Global Effects of Environmental Pollution.
39. Gunnerson, C. G. 1970. An Appraisal of Marine Disposal of Solid Wastes off the West Coast: A Preliminary Review and Results of a Survey. Department of Health, Education, and Welfare, Bureau of Solid Waste Management.
40. Janssen, W. A., and Meyers, C. D. 1968. Fish: Serologic Evidence of Infection with Human Pathogens. Science 159:547–548.
41. Ketchum, B. H. 1970. Biological Implications of Global Marine Pollution. pp. 190–194. *In* Singer, S. F. (ed.). Global Effects of Environmental Pollution.
42. Ketchum, B. H. 1970. Testimony Before the Subcommittee on Air and Water Pollution of the Senate Committee on Public Works. March 5.
43. King, K. 1970. The New York Bight, A Case Study of Ocean Pollution. (mimeograph)
44. Korringa, P. 1968. Biological Consequences of Marine Pollution with Special Reference to the North Sea Fisheries. *Helgolaender Wissenschaftliche Meeresuntersuchungen* 17:126–140.
45. Lowman, F. G., Rice, T. R., and Richards, F. A. 1969. Accumulation and Redistribution of Radionuclides by Marine Organisms. *In* National Research Council-National Academy of Sciences. 1970. Radioactivity in the Marine Environment. (mimeograph)
46. Macek, K. J. 1968. Growth and Resistance to Stress in Brook Trout Fed Sublethal Levels of DDT. Journal of the Fisheries Research Board of Canada 25:2443–2451.
47. Massachusetts Institute of Technology. 1970. Economic Aspects of Ocean Activities. Vol. III.
48. Massachusetts Institute of Technology. 1970. Study of Critical Environmental Problems, Summary of Major Findings and Recommendations. (mimeograph)
49. MacKenthun, K. M. 1969. The Practice of Water Pollution Biology. Department of the Interior, Federal Water Pollution Control Administration.
50. Metropolitan Sanitary District of Greater Chicago. 1970. Information provided to the Council.
51. Mihursky, J. A. 1969. Patuxent Thermal Studies, Summary and Recommendations. University of Maryland, Natural Resources Institute Spec. Rep. No. 1.
52. Mitchell, J. R., Presnell, M. W., Akin, E. W., Cummins, J. M., and Liu, O. C. 1966. Accumulation and Elimination of Polio Virus by the Eastern Oysters. American Journal of Epidemiology 86:40–50.

53. Morgan, J. and Pomeroy, R. D. 1969. Chemical and Geochemical Processes Which Interact with and Influence the Distribution of Wastes Introduced into the Marine Environment and Chemical and Geochemical Effects on the Receiving Waters. Background paper prepared for the NAS-NAE Steering Committee for Coastal Waste Disposal Workshop. (mimeograph)
54. Mount, D. C. 1968. Chronic Toxicity of Copper to Rathead Minnows (*Pimephales promelas, Rafinesque*). pp. 215–223. *In* Water Research.
55. National Council on Marine Resources and Engineering Development. 1970. Marine Science Affairs—Selecting Priority Programs. *In* Annual Report of the President to the Congress on Marine Resources and Engineering Development.
56. National Academy of Sciences Committee on Oceanography-National Academy of Engineering Committee on Ocean Engineering. 1970. Wastes Management Concepts for the Coastal Zone—Requirements for Research and Investigation. (mimeograph)
57. National Research Council-National Academy of Science. 1970. Radioactivity in the Marine Environment. (mimeograph)
58. Nichols, G., Curl, H., and Bowen, U. 1960. Spectrographic Analysis of Marine Plankton. Journal of Limnology and Oceanography 4:472–478.
59. North, W. J. 1970. A Survey of Southern San Diego Bay. (mimeograph)
60. Parrish, L. P., and Mackenthun, K. M. 1968. San Diego Bay: An Evaluation of the Benthic Environment. Department of the Interior, Federal Water Pollution Control Administration.
61. Pearson, E. A., Storrs, P. N., and Selleck, R. E. 1970. A Comprehensive Study of San Francisco Bay. University of California at Berkeley, Sanitary Engineering Research Laboratory Rep. No. 67–5.
62. Pringle, B. H. 1970. Trace Metal Accumulation by Estuarine Molluscs. (in press)
63. Raymont, J. E. G., and Shields, J. 1962. Toxicity of Copper and Chromium in the Marine Environment. pp. 275–290. *In* American Public Health Association. Recommended Procedures for the Bacteriological Examination of Sea Water and Shellfish.
64. Revelle, R., Wenk, E., Corino, E. R., and Ketchum, B. H. 1970. Ocean Pollution by Hydrocarbons. (mimeograph)
65. Russell, F. E., and Kotin, P. 1956. Squamous Papilloma in the White Croaker. Journal of the National Cancer Institute 18:857–860.
66. Smith, D. D., and Brown, R. P. 1970. An Appraisal of Oceanic Disposal of Barge-Delivered Liquid and Solid Wastes from U.S. Coastal Cities. Prepared by Dillingham Corporation for Department of Health, Education, and Welfare, Bureau of Solid Waste Management under contract No. PH 86–68–203. (mimeograph)
67. Tarzwell, C. M. 1970. Toxicity of Oil and Oil Dispersant Mixtures to Aquatic Life. Presented at the International Seminar on Water Pollution by Oil. (mimeograph)
68. Templeton, W., Nakatani, R., and Held, E. 1969. Radiation Effects. *In* National Research Council-National Academy of Sciences. 1970. Radioactivity in the Marine Environment. (mimeograph).
69. Tutsumoto, M., and Patterson, C. 1963. The Concentration of Common Lead in Sea Water. pp. 74–89. *In* Geiss, J. and Goldberg, E. (eds.) Earth Science and Meteorites.
70. U.S. Atomic Energy Commission. 1970. Information provided to the Council.
71. Water Resources Engineers, Inc. 1970. Ecologic Responses to Ocean Waste Discharge: Results from San Diego's Monitoring Program. Prepared for California State Water Resources Control Board and San Diego Regional Water Quality Control Board.
72. Young, P. H. 1970. Some Effects of Sewer Effluent on Marine Life. California Fish and Game 50:33–41.

C. Estuaries

National Estuary Study, January 1970. Volume 1, Main Report. United States Department of the Interior, Fish and Wildlife Service, Bureau of Sport Fisheries and Wildlife and Bureau of Commercial Fisheries. Pages 18-79.

The National Estuary Study was undertaken pursuant to section 2(a) of the Estuary Protection Act. The Act, approved on August 3, 1968, can be found in 82 United States at Large 625 and 16 United States Code Annotated section 1222. Section 2(a) of the Act directed the Secretary of the Interior to conduct a comprehensive study of the Nation's estuaries, including the land and waters of the Great Lakes. The National Estuary Study utilized, synthesized, and added to two major previous studies:

> Our Nation and the Sea, January 1969. Commission on Marine Science, Engineering and Resources. This study was undertaken pursuant to the Marine Resources, Engineering and Development Act of 1966, approved on June 17, 1966 and found in 80 United States Statutes at Large 203 and 33 United States Code Annotated section 1101.
>
> The National Estuarine Pollution Study, Submitted to the Congress November 3, 1969. U.S. Department of the Interior, Federal Water Pollution Control Administration. This study was undertaken pursuant to the Clean Water Restoration Act of 1966, approved on November 3, 1966 and found in 80 United States Statutes at Large and 33 United States Code Annotated section 1151.

Thus our estuarine resources have been amply studied. Unfortunately, ample study has not resulted in adequate protection of this unique and finite national resource. About 73 percent of the nation's estuaries have been either severely or moderately modified by man, and almost all of the remaining 27 percent has been slightly modified.

The National Estuary Study consists of seven volumes. Volume 1 contains the "Main Report." Volumes 2-7 contain the supporting staff and consultants' studies. The excerpts which are reprinted below are taken from the "Main Report." The overall message is painfully clear: Unless massive reforms are instituted the estuaries will be destroyed within a few decades. Immediate action is imperative.

NATIONAL INTEREST IN ESTUARIES

Congress has several times found that there is a National interest in the varied and diverse resources of the Nation's coastal areas, including estuaries. The facts support a finding that protection, conservation, and restoration of the Nation's estuaries is in the National interest.

The reasons are:

> There is a widespread physical and economic disparity between the factors which effect estuary production and the distribution of benefits;
>
> Many of the basic resources of estuaries are common properties which are interstate if not international in character and extend far beyond the effective jurisdiction of governments below the National level;
>
> Estuaries, even though used by Man for centuries, are essentially wild, natural systems which have diversity of conditions directly proportional to the range of natural variation found within an estuary and which support a variety of fish, wildlife, and esthetic values which should be retained for coming generations;
>
> Estuaries have relevance to urban regions.

This study, although only able to generally anticipate the principal mechanisms for planning and implementing management of coastal resources, concludes such mechanisms should be at the State level. Therefore, the National interest in estuaries can best be accomplished by reinforcing State ability to better manage estuary resources; by authorizing Federal participation in establishing goals and objectives for the protection, conservation, and restoration of estuary resources, and by undertaking certain specific activities in estuaries in keeping with the Federal role in the spectrum of management applied to other resource systems.

Physical and Economic Disparities

Physically, estuary systems respond more than other natural systems to forces which occur at places geographically removed from the identified estuary. The most pronounced is the relationship of fresh water inflow. Physical changes in the quantity, quality, and timing of surface and/or sub-surface fresh water inflows will have a marked effect on the estuary. And there are others, for example littoral deposition which alter the movement of tides. Many of the physical characteristics which effect the biologic operation of estuary systems are subject to manipulation in response to other National interests; water supply, flood protection, and shore erosion control, for example.

There are also major separations between the incurring of cost and the realization of benefits from estuary resource use. That is, the total cost of developing estuary wetland for housing or using estuary water for cheap waste disposal is not paid by the person or firm which takes the action. Many of the values of estuaries lost through such development can only be perceived by a National viewpoint.

Resources are Interstate or International in Character

Clearly estuarine dependent fish species are important in meeting the National and international protein requirements, both directly and indirectly. It is true that realization of the full benefit is dependent on improved harvesting and marketing technology, and regulation of marine banks, but the maintenance of stocks is directly related to the present or improved levels of biologic productivity of estuaries. Moreover, the National interest and responsibility in the management of certain fish species and migratory waterfowl dependent on estuaries is already recognized by treaty.

Maintenance of Natural Systems

The total natural heritage of the Nation consists of a broad array of conditions each supporting characteristic natural communities of plants and animals. As the Nation expanded geographically and in numbers and affluence, recognition has grown that the maintenance of portions of each plant and animal community is important to the health, well-being, and enjoyment of all citizens. And the interest in and support of public agencies' efforts to manage land and water resources entrusted to their care in ways to maintain remnant natural systems transcends all political boundaries. The public interest in wilderness, wild, natural and scenic rivers, refuges, and parks extends beyond those who directly benefit from the existence of such management to those who derive satisfaction in knowing that portions of the Nation's legacy of plants, animals, and scenery will be available to provide contrast to the increasingly pervasive changes wrought by Man.

The estuaries have largely been excluded from such interest. Only to the extent that some portion of the total estuary environment was important for some specific use have public management schemes at any level of government been implemented in an estuary. Yet an estuary, as a geographical and hydrologic concept, contains within it a vast diversity of habitats, and each is utilized by various species of plant and animal life. After all, on a gross scale an estuary is an edge between land and sea and in biologic terms an edge will provide for greater diversity of species than climax communities. Edges in the terrestrial environment depend on change; estuaries are in a constant state of change. Taken together, the Nation's estuaries support a larger number of plant and animal communities than any other natural type.

It is as or more important that representative examples of natural estuary systems be accorded comparable recognition as terrestrial communities now protected, conserved, and restored in the National interest, such as wilderness areas.

Estuaries and Urban Conditions

Many estuaries in and adjacent to major metropolitan areas have remained undeveloped and constitute great open space resources. There is a swelling National interest in the condition of life in these urban areas. Maintenance of estuary systems can be a significant factor in shaping the growth characteristics of urban regions and in providing for close-in natural contrasts for the residents of the cities.

III. PHYSICAL AND BIOLOGICAL CHARACTERISTICS OF ESTUARIES

Estuaries have unique biological and physical characteristics that make them an especially valuable part of the natural environment. Their form, beauty, and richness derive from the interaction of the land and the water, the wind, the tides, the sun, the rocks, the soil, and the rivers--all in a relatively small part of the Earth.

Elevated, resistant-rock coasts like those on the Pacific Ocean tend to have small, open estuaries. Downwarped coasts overlain by coastal erosional plains like those on the Atlantic Ocean and Gulf of Mexico usually have broad, shallow estuaries with inlets restricted by outer banks, barrier islands, or spits.

The shaping never stops. Great earth movements and glaciation have occurred, and the form of the land is being gradually modified by the ceaseless tides and runoff, by sun and wind, and by ocean currents.

Estuaries and the Great Lakes trap nutrients and sediments as well as the man-made wastes conveyed by runoff. The sediments, nutrients, and wastes tend to gather in their shallow and quiet fringes. The sediments aggrade into marshes, then into dry land over geological time spans, finally replacing the estuary with a delta.

The net result is a system of shallow, rich, and productive estuarine areas, within the larger coastal system, in which fresh and sea waters mix. They have unique and particular characteristics of special importance to Man because of their esthetic qualities and the living marine resources which they support.

Estuarine Ecology

The estuaries of the United States and the Great Lakes include a great variety of tremendously fertile, vibrantly productive ecological systems. They range from the stable, normally placid, coral-enclosed lagoons of tropical Hawaii to the ice-scoured, high latitude estuaries of North-slope Alaska. In between are systems of every kind and gradation. In total they include nearly every combination of natural productivity and limitation as modified by impacts of Man's activities.

These combinations result from the sum of natural and imposed conditions: of wind, tide, and runoff; of nutrient inflow and salinity; of light, heat, and cold; and of dredging, filling, pollution, and diversion of freshwater inflow. Marshes and shallows are eliminated by dredging and filling, and estuarine areas are degraded by heat, chemical, and other pollution and by diversion of inflows.

The lightly stressed systems support a great diversity of poorly adaptable species; the heavily stressed systems support fewer, more-adaptable species, frequently in great abundance but not necessarily of great value to Man. The systems of estuaries have a variety of life niches that typically support widely adaptable marine species with contingents of freshwater forms in the tidal heads of tributaries and fresh marshes and of narrowly adaptable marine species at their seaward inlets, and in the saline marshes.

The Great Lakes, being entirely fresh and without significant tides, have wind, light, and cold-stressed systems as well as those caused by human use. They have a variety of life niches including many within the Lakes themselves.

The productivity of the zone depends on favorable balances of tidal and river forces, precipitation and evaporation, enrichment and depletion of nutrients, and Man's conservation and use as well as many other factors. A less complex, but similar system of areas of the Great Lakes also has comparably important values to Man.

The geographic variety of the Nation's estuarine and Coastal Zone is more fully described and defined by regional sections, with maps, at the end of this chapter.

Estuarine Dynamics

The tidal surge and flow and the seasonal cycles of runoff, and climate, highlight the rhythmic, ceaseless change--the dynamic nature of the estuarine zone. The tides and runoff combine to accelerate the circulation--the mixing, flushing, and enriching of estuaries. These unique processes invigorate estuarine waters with oxygen from the sea, recycle nutrients from the rivers and marshes and from the sea, and exchange them with the marshes and coastal marine waters. In sum, they maintain a rich variety of highly productive waters and wetlands.

The rhythmic tidal changes give the Coastal Zone waters a more prominent time dimension than that of inland waters. This tidal rhythm, the power of the surf, and the broad vistas of sea, beach, and headlands, are the charm of the Coastal Zone of which estuaries are a part.

The Great Lakes also are unique. Because of their size, they exhibit patterns of circulation and wave development more akin to the sea than to ordinary lakes. They have but minor astronomical tides and their inflowing rivers are generally small. But they are invigorated by winds and have rich shallows and coastal marshes. They too have dynamic and esthetically pleasing vistas.

Modifications of the Nation's Estuaries

Most estuarine areas of the United States have been modified more or less severely by Man's activities. The degree of modification is indicated by regional zones in the following table:

Biogeographic Zone 2/	Degree of Modification of Estuaries 1/		
	Slight	Moderate	Severe
		Per Cent	
North Atlantic	44	48	8
Middle Atlantic	5	68	27
Chesapeake Bay	44	50	6
South Atlantic	36	60	4
Biscayne and Florida Bay	50	50	0
Gulf of Mexico	15	51	34
S.W. Pacific	19	19	62
N.W. Pacific	13	50	37
Alaska	80	20	0
Hawaii	54	15	31
Great Lakes	35	46	19
United States	27	50	23

Source: Field evaluation carried out by Fish and Wildlife Service's personnel during course of Estuary Protection Act study.

1/ All estuaries and subestuaries were individually rated for each zone. The percentage refers to the proportion of these individual areas that were rated as indicated.

2/ Also see the maps and descriptions of each zone at end of this chapter.

Two views of marsh and tidelands habitat near Juneau, Alaska. Estuaries are wild, natural systems which support fish and wildlife that are interstate and international in character. Such values are enjoyed by people from coastal and inland areas and from all walks of life.

The degree of estuarine modification tabulated above is displayed graphically on the maps which follow each of the descriptions of the 11 identified biophysical estuarine zones, presented subsequently in this chapter.

Estuarine Life and Esthetics

The incessant flux and rich nutrient supply of estuaries supports a wealth of aquatic and semi-aquatic life, both in number and varieties. This wealth of life--particularly the fish, wildlife, and higher plants--is of tremendous value and interest to Man for its inspiring beauty and grace, its intriguing life histories, its sport value, and its value as food. The microscopic and other supporting life is of direct interest, also, to the scientist, the student, and it is of life and death interest to higher animals that feed on it.

The multitude of waterfowl, shorebirds, and marsh birds frequenting the estuarine flats and tidal marshes, and the ubiquitous seabirds of the beaches, rocky shoreline and coastal islands contribute in great measure to the public's outdoor enjoyment. Birds are as much a part of the landscape as is the pounding surf, the rocky headlands, sandy beaches, mud flats, and tidal marshlands. They add the spice and sparkle to the coastal seascapes and panoramas for the millions of coastal urban dwellers and inland people attracted to the Nation's shorelines.

Marine mammals also add zest to coastal vistas. Sea lions, seals, sea otters, and whales entice and entrance thousands of viewers at certain locations, particularly along the Pacific Coast. Porpoises re charming and entertaining companions to the multitude of coastal oaters on all sea coasts.

Trees, shrubs, and grasses are a backdrop to many coastal vistas nd provide character to the landscapes of the estuarine and Great akes areas that attract many of our vacationing citizens.

Literally hundreds of estuarine dependent species of shellfish, ther invertebrates, and fishes, including freshwater and anadromous ishes, are avidly sought by sport fishermen. Most of these species re also harvested by commercial fishermen. Without the rich habitat f estuaries, their production would be reduced to insignificant levels.

The Great Lakes support a great variety of freshwater fishes, both ative and introduced, that are highly valued by sport and commercial ishermen. Many of these fish use the rivers, shoals, and weedy bays or feeding and spawning.

The great continental flocks of waterfowl, rails, and other birds hat are popular with hunters depend heavily on the Coastal Zone for uch of their wintering, feeding, and nesting habitat. They nest in he Zone to a great extent in Alaska and to some extent in most stuarine zones and the Great Lakes.

Estuarine areas are important also for sport hunting of terrestrial ildlife, for commercial trapping of fur animals, and for harvest of elp and other plants.

Coastal vistas and seascapes draw millions of visitors to glory in their beauty and solace. The power and rhythm of the breakers, the astness of the sea, the endless strands of beach or, equally, the intricate rocky promontories and indentures, and the triumph of bent

Estuaries differ physically as well as in the animals they support. The upper photo is of McNeil Creek where it enters Kamishak Bay, Alaska (BSFW Photo by Dick Chace), and the lower photo is of Laguna Atascosa National Wildlife Refuge, San Benito, Texas (BSFW Photo by L.C. Goldman).

grass or gnarled trees are all ages-old attractions to youngster and oldster alike, to both coastal residents and inlanders. The estuarine zone also has rewarding attractions for outdoor recreationists, including boaters, swimmers, and many, many more.

The texts and individual maps of the 11 estuarine zones of the Nation are presented in order in the following sections, beginning with the North Atlantic. The maps indicate the degree and extent of estuarine modification caused by Man's activities.

A more detailed discussion of the biophysical environment of the Nation's estuaries and the Great Lakes is presented in Chapter I of the staff report in Vol. 2, App. A.

A special report entitled: Estuary Landscape Survey and Analysis included in Vol. 4, App. D, more fully discusses the landscape values of the Nation and presents pertinent maps of landscape and cultural values of the estuarine and Great Lakes regional zones.

Another special report entitled: Securing Additional Data Supporting the Biophysical Estuary Evaluation is to be found in Vol. 4, App. C. It presents the status of information on sedimentation of estuaries and the value of such areas in hurricane and other storm protection.

North Atlantic Estuarine Zone

This zone extends from Canada around the Gulf of Maine to Cape Cod as shown on the map following this description. It is glaciated throughout. The coast is rugged on the north, reducing gradually to nearly a plain from Boston south around Cape Cod Bay. The estuaries of Maine are narrow with deep, open inlets, steep fringing marshes and flats, and rocky coastal islands. Southward the coastline is more even; its marshes and beaches more gently sloped and extensive.

Great tidal range, cold winters, and encroachment onto the limited estuarine wetlands are the major factors that limit fish and wildlife production in the North Atlantic Zone. Pollution is a locally limiting factor. Scenic resources are threatened by summer-home and other encroachments. The degree of estuarine modification is shown on the zone map.

The zone is noted for its lobsters, clams and scallops, but much of its great fisheries is not heavily dependent on estuaries. Its estuarine-dependent Atlantic salmon and sturgeon are rare or endangered. Many species of waterfowl and shorebirds stop over in the zone to feed on the abundant food in the shallows. The marshes and islands are nesting grounds for black duck, eider and scaup.

The intricate, rocky shores of Maine with their splotches of marsh, tide pools, and mud flats, backed by forests and fronted by the surf, provide beautiful seascapes of great variety and interest. Beaches of the southern part of the zone are broader, sandy, and inviting. The hardwoods in New England are a brilliant colorful backdrop for coastal landscapes in the fall.

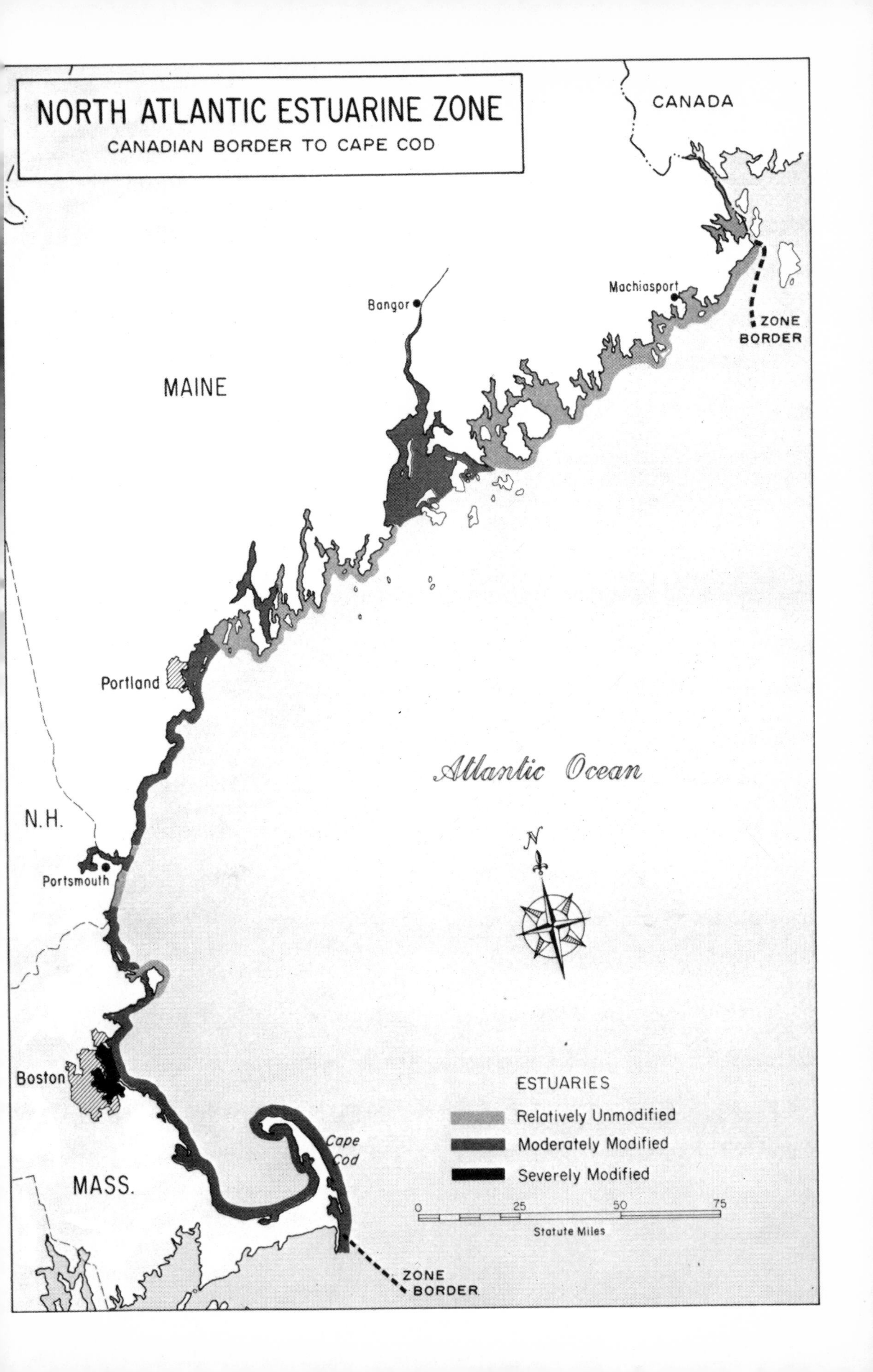
NORTH ATLANTIC ESTUARINE ZONE
CANADIAN BORDER TO CAPE COD
CANADA
Machiasport
ZONE
BORDER
Bangor
MAINE
Portland
Atlantic Ocean
N.H.
Portsmouth
N
Boston
ESTUARIES
Relatively Unmodified
Moderately Modified
Severely Modified
Cape
Cod
MASS.
0
25
50
75
Statute Miles
ZONE
BORDER

Middle Atlantic Estuarine Zone

The map following this description delineates this zone which extends from Cape Cod south to Cape Charles, opposite Norfolk. It is glaciated and of rugged to low relief as far south as the New York bight; further south it is coastal plain.

Small marsh-lined estuaries occur at intervals along the glaciated part of the coast. Along the coastal-plain segment, estuaries and marshes are nearly continuous behind outer banks, and marshes fringe Delaware Bay. The outer-bank beaches are wide and sandly with gently deepening offshore waters.

Tidal and other natural stresses to estuarine life are moderate in this zone. Pollution and land occupation for port cities and industry are the causes of greatest estuary loss and degradation. The severity and extent of estuarine modification is shown on the accompanying map.

Sport and commercial fish and shellfish of the zone include a great number of estuarine-dependent species of which striped bass, bluefish, and clams are best known. The Atlantic sturgeon of the zone is rare and deserves protection.

Myriad numbers of many species of waterfowl and shorebirds flock to the Middle Atlantic for wintering and to rest there during migrations. They depend greatly on its extensive estuarine marshes, mud flats, waters, and outer beaches. Black duck, wood duck, gadwall, and a few blue-winged teal nest in the zone.

From the intricate bays and islands on the north to the southern coastal lagoons and windswept outer banks, the Middle Atlantic offers accessible and attractive seascapes, sailing waters, and extensive, popular beaches. The headlands, islands, rugged shores, sounds, and challenging channels of the glaciated coast form a scenic, sailing, and boating resource that has few equals. The outer-beach resource is outstanding for its great extent and proximity to metropolitan centers.

Great South Bay on Long Island, New York, part of this zone, was the subject of one of the eight special case studies prepared during this study. A report giving full detail on Great South Bay is in Vol. 3, App. B.

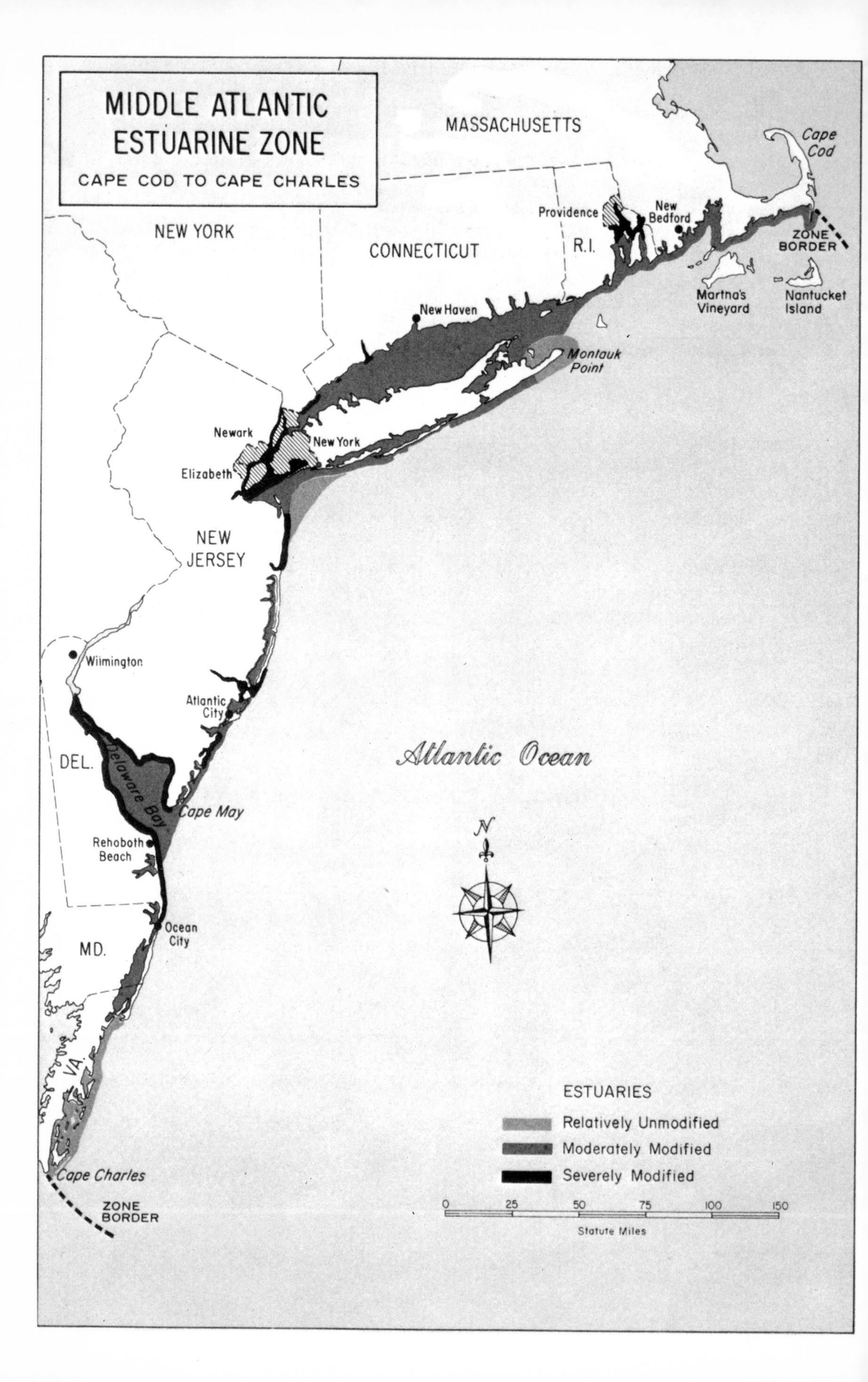

MIDDLE ATLANTIC
ESTUARINE ZONE
CAPE COD TO CAPE CHARLES
MASSACHUSETTS
Cape Cod
NEW YORK
CONNECTICUT
R.I.
Providence
New Bedford
ZONE BORDER
Martha's Vineyard
Nantucket Island
New Haven
Montauk Point
Newark
New York
Elizabeth
NEW JERSEY
Wilmington
Atlantic City
DEL.
Delaware Bay
Cape May
Rehoboth Beach
Ocean City
MD.
VA.
Cape Charles
ZONE BORDER
Atlantic Ocean
N
ESTUARIES
Relatively Unmodified
Moderately Modified
Severely Modified
0
25
50
75
100
150
Statute Miles

Chesapeake Bay Estuarine Zone

This, the estuary of the drowned lower Susquehanna River and tributaries, extends inland nearly 200 miles due north from its inlet, between Cape Henry at Norfolk and Cape Charles. It is 5 to 40 miles wide. Chesapeake Bay is unique among estuaries of the World for its size, complexity, and productivity. A map of the Bay follows this description.

The Bay's tidal-river reaches and baylets are fringed with marshes, those on the eastern-shore plain being most extensive. Bluffs and narrow beaches edge some of its western shores.

Tides and other natural factors are generally favorable to life of the Bay. Pollution and land occupation are locally limiting, but the scenic beauty and life of the Bay has not been greatly degraded. The accompanying map shows the degree and extent of modification by Man.

The Bay is popularly known for its abundant oysters and blue crabs, but it yields greatly of a variety of other marine, anadromous, and freshwater food, industrial and sport fish, such as the popular and commercially important striped bass and the high yielding menhaden.

The Bay is a key area for wintering and resting of migratory waterfowl and shorebirds. Great flocks of ducks of many species, geese and whistling swan home to it annually. Many State and Federal refuges have been established around the Bay to accommodate these birds. Some species, notably black duck, wood duck, gadwall, and blue-winged teal nest in the Bay wetlands.

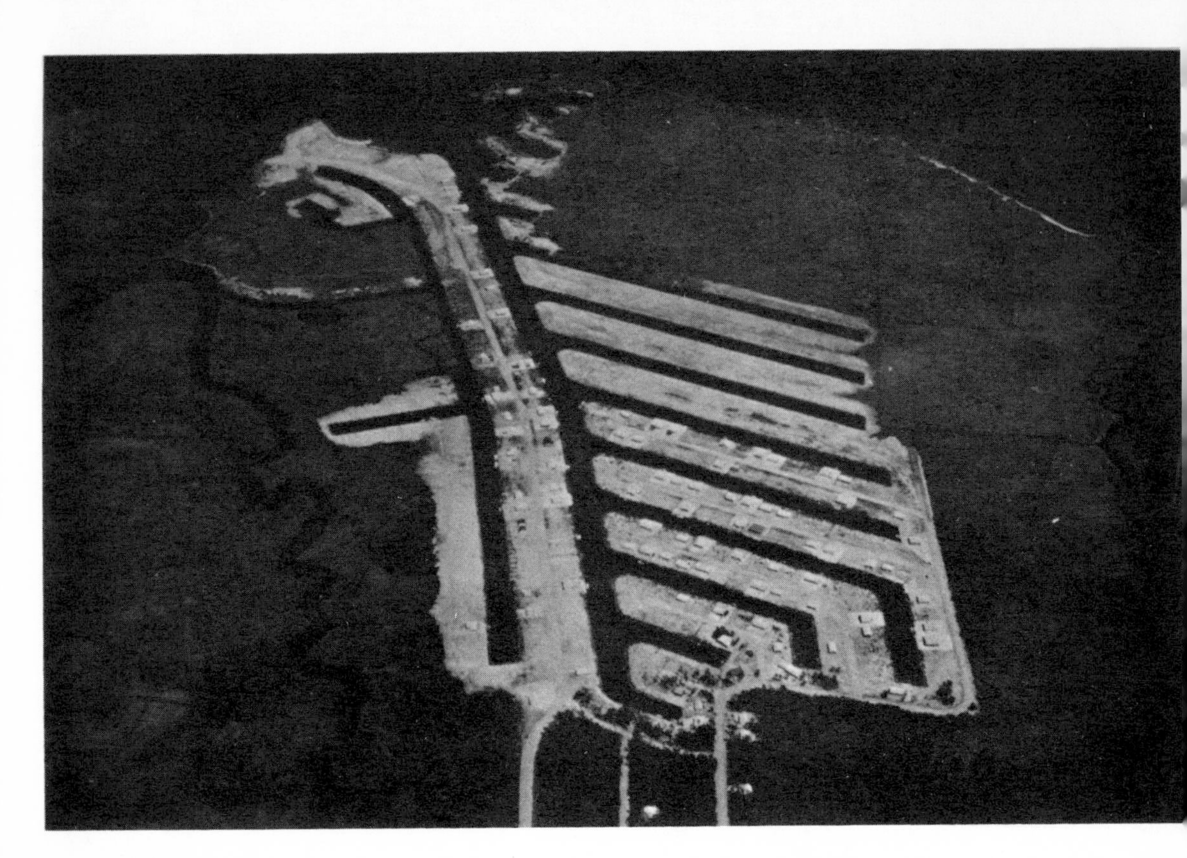

Man can change a coastal marsh, but it is difficult for him to restore one if he builds houses on it. These two photos were taken in the coastal marsh of New Jersey near Atlantic City. (BSFW Photos by C.D. Evans)

Chesapeake Bay offers a wealth of scenic, historic, and recreational attractions, including many of the Nation's shrines near its shores. Its vistas of marshes, chalky bluffs, and verdant hills are notably scenic.

The Bay's variety of wide waters, placid coves, and long western arms are the delight of boaters. The small, timbered swamps of its lower tributaries are the most northern sites of the southern bald cypress. The endangered southern bald eagle maintains a remnant population on its shores.

A special, much more detailed, report on Chesapeake Bay is in Vol. 3, App. B.

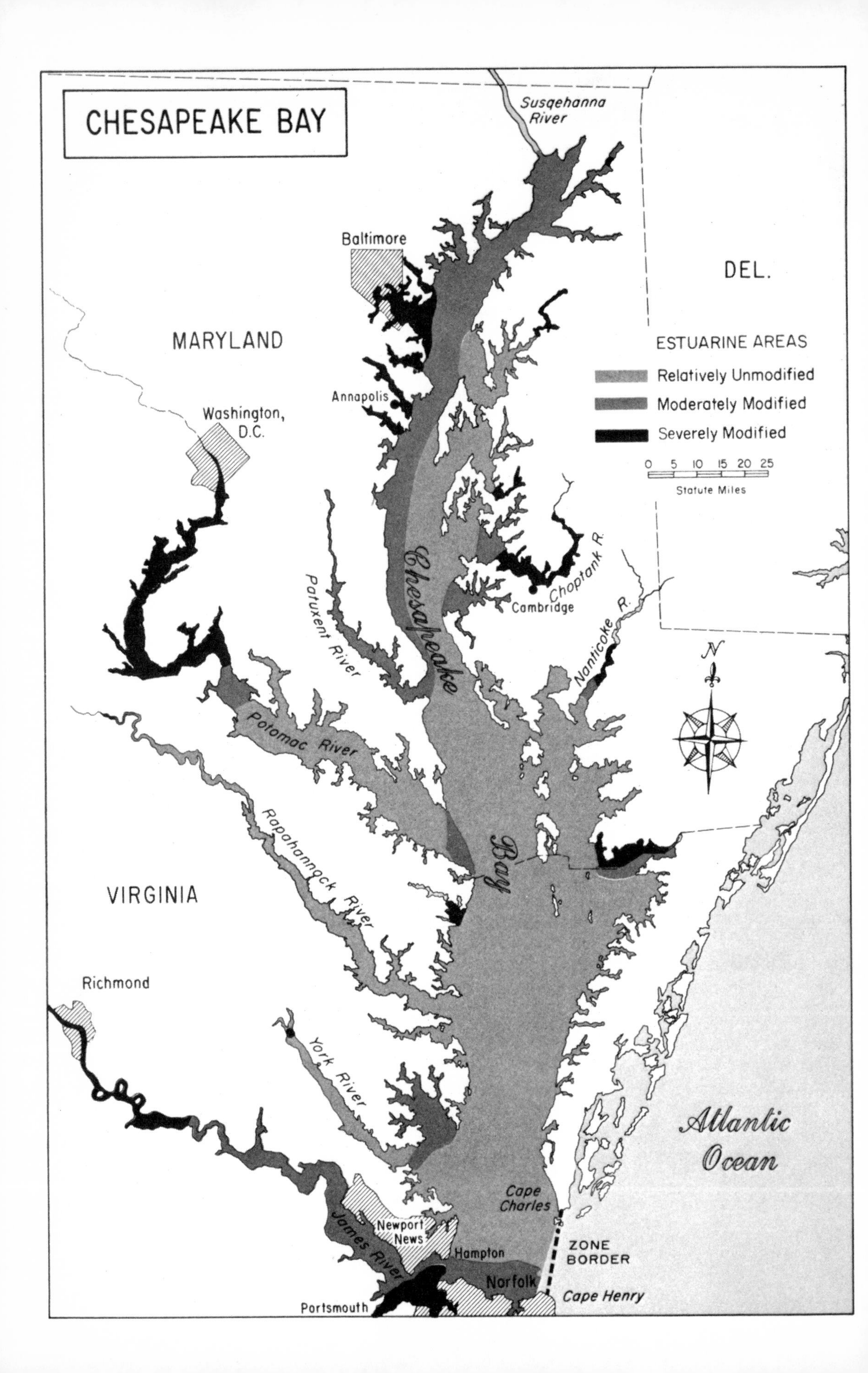

CHESAPEAKE BAY
Susqehanna River
Baltimore
DEL.
MARYLAND
ESTUARINE AREAS
Relatively Unmodified
Moderately Modified
Severely Modified
0 5 10 15 20 25
Statute Miles
Annapolis
Washington, D.C.
Chesapeake
Choptank R.
Cambridge
Patuxent River
Nanticoke R.
N
Potomac River
Bay
Rapahannock River
VIRGINIA
Richmond
York River
Atlantic Ocean
Cape Charles
Newport News
James River
Hampton
ZONE BORDER
Norfolk
Cape Henry
Portsmouth

South Atlantic Estuarine Zone

This zone extends from Cape Henry southward to Fort Lauderdale, Florida, as shown on the map following this description. On the north it is bordered by a 100-mile wide plain, and the Florida Peninsula is of extremely low relief.

Laterally joined estuaries behind nearly continuous barrier islands are typical of the South Atlantic Coast. The complex estuary of the Pamlico-Albemarle-Currituck Sound behind the outer banks of Cape Hatteras is a particularly noteworthy feature of the region.[1/] Between Cape Fear and Jacksonville the estuaries are more open; the barrier islands more fragmented.

Tidal marshes, riverine swamps, and wetlands extend inland almost the width of the plain along some streams, and almost continuously along the landward sides of the estuaries. Great swamps such as Okefenokee, Dismal, and many lesser ones are located inland from the extensive tidal marshes.

Tides are moderate in this zone, and other natural factors are generally favorable to living resources. Pollution and land occupation have degraded a few estuaries, but most are in good condition. The degree and extent of modification is shown on the map of the zone.

[1/] A special report on the Pamlico-Albemarle-Currituck Sound estuary complex, one of eight areas selected for special case studies, can be found in Vol. 3, App. B.

Estuarine-dependent fish yield abundantly of great variety of species in the South Atlantic Zone. Spotted sea trout, bluefish, striped bass, and freshwater largemouth bass are popular with sport fishermen, and shrimp, menhaden, crabs, and oysters provide large commercial yields.

Waterfowl, shorebirds, and marsh birds of many species find a winter haven in the great expanses of large and small estuaries and wetlands in this zone. The endangered dusky seaside sparrow occurs in Florida sections of the zone. Ducks that nest in the zone include mallard, gadwall, and wood duck.

The South Atlantic zone is well known for its salubrious climate gained from the proximity of the Gulf Stream. The extensive seascapes, outer beaches, and estuarine waterways of Florida and the islands off Georgia are valued for their attractions. The northern coast is more varied with numerous islands, intricate estuaries, outer banks, and small but fine beaches. Magnificant pines and moss-draped oaks vie with palms and great expanses of marsh and swamp to please the coastal viewer. Historical sites of early Europeans abound in the zone.

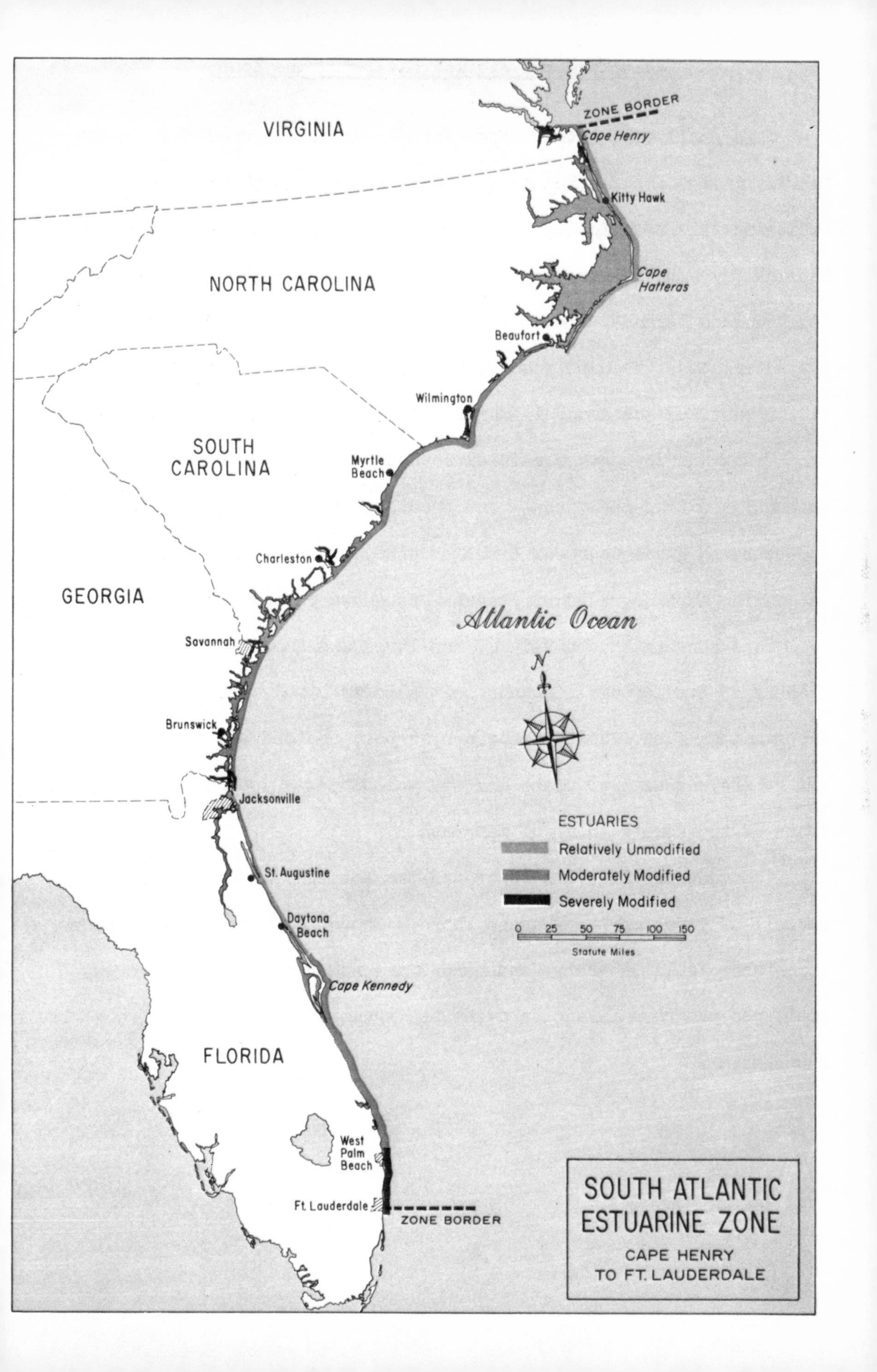
VIRGINIA
ZONE BORDER
Cape Henry
Kitty Hawk
NORTH CAROLINA
Cape Hatteras
Beaufort
Wilmington
SOUTH CAROLINA
Myrtle Beach
Charleston
GEORGIA
Atlantic Ocean
Savannah
N
Brunswick
Jacksonville
ESTUARIES
Relatively Unmodified
Moderately Modified
Severely Modified
St. Augustine
Daytona Beach
0 25 50 75 100 150
Statute Miles
Cape Kennedy
FLORIDA
West Palm Beach
Ft. Lauderdale
ZONE BORDER
SOUTH ATLANTIC ESTUARINE ZONE
CAPE HENRY TO FT. LAUDERDALE

Biscayne and Florida Bay Estuarine Zone

This small zone extends from Fort Lauderdale around the tip of Florida inside the Florida Keys to Cape Romano, as shown on the map following this description. It includes Biscayne Bay, Barnes Sound, Florida Bay, the western edge of the Everglades, Whitewater Bay, and Ten Thousand Islands. Small islands abound offshore and in the bays. The entire coast is barely above sea level. It is enclosed by the Keys on the east and south and is mainly within Everglades National Park.

Tides and most other natural conditions favor estuarine life, but seasonal and long-term drouth and tropical storms cause occasional stresses. Man has aggravated the drouths, and pollution and wetland occupation have degraded some areas, as shown on the map of the zone.

This zone is the nursery of the large and important pink shrimp fishery of Dry Tortugas Grounds. It also is highly productive of many important finfish, including tarpon, snook, spotted sea trout, pompano, and others, as well as crabs and spiny lobster. Oysters grow on the roots of the zone's abundant mangrove.

The zone provides important habitat for many of the wading birds, as well as birds of the sea and shore. Brown pelicans, water turkeys, cormorants, gulls, egrets, and herons are numerous. The many, scattered mangrove islands provide the required seclusion for rookeries and resting.

The zone is the home of the rare and endangered Florida great white heron, Florida manatee (sea cow), southern bald eagle, Caribbean monk seal, Cape Sable sparrow, Key deer, and the peripherally endangered wood ibis, roseate spoonbill, eastern reddish egret, the American crocodile and alligator, and possibly the Florida mangrove cuckoo.

The esthetic attractions of the zone are widely known. The coral reef underwater gardens in the Gulf Stream off Barnes Sound are notable. From Biscayne Bay through Florida Bay to the Ten Thousand Islands, the tropical waters, bordering mangrove forests, and inshore wetlands all add up to a uniquely attractive setting centered on Everglades National Park.

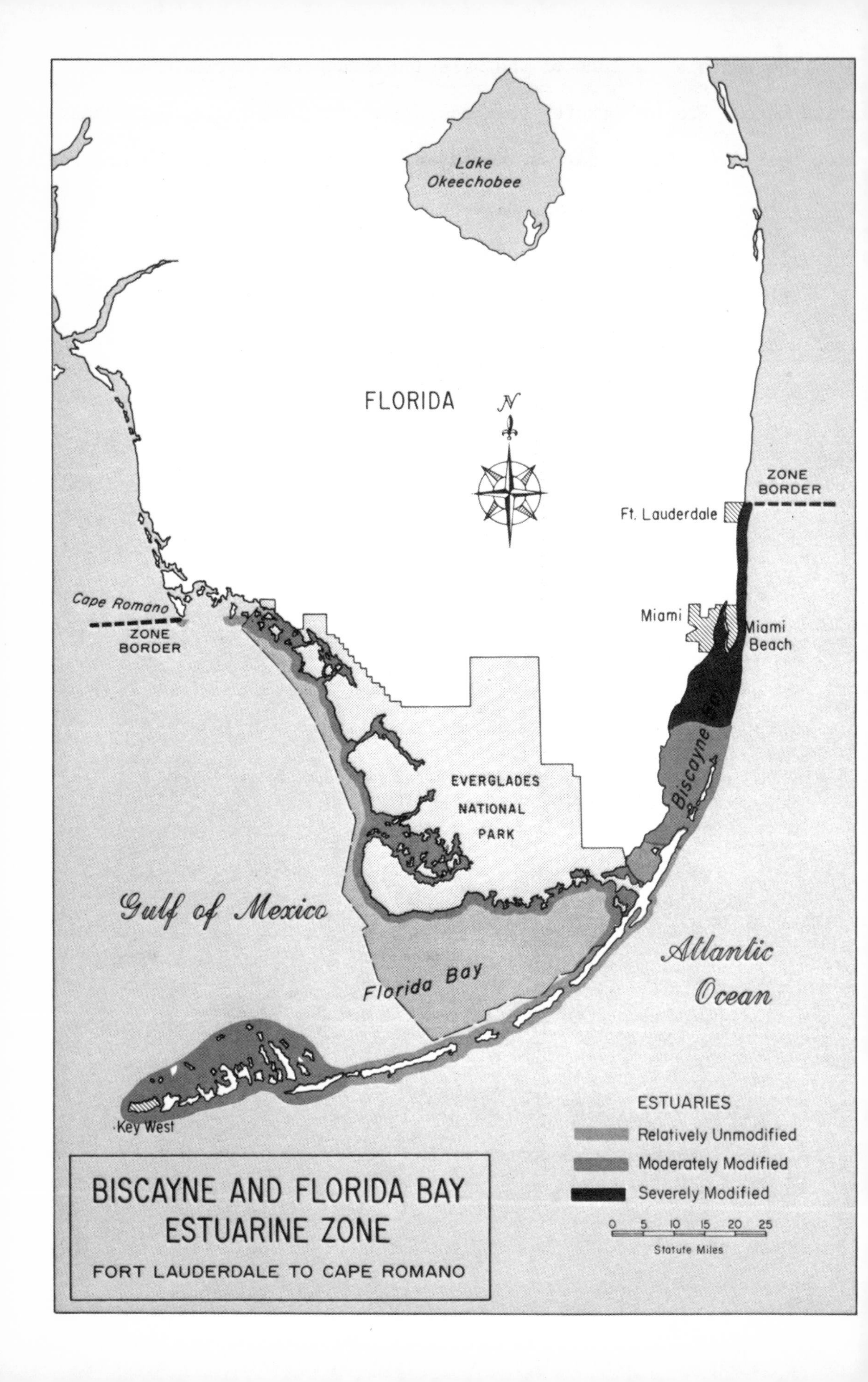

BISCAYNE AND FLORIDA BAY ESTUARINE ZONE

FORT LAUDERDALE TO CAPE ROMANO

Gulf of Mexico Estuarine Zone

This zone extends around the United States Coast of the Gulf of Mexico from Cape Romano in southwest Florida to the Texas-Mexico border near Brownsville as shown on the accompanying map. The entire zone borders a wide coastal plain of generally low relief.

The dominant and unique estuarine feature is the Mississippi River Delta with its vast area, great river channel and flow, highly intricate network of lesser channels and bayous, and its numerous lakes, bays, marshes, estuaries, and islands. Notable also are Tampa, Mobile and Galveston Bays; the elongate complex of bays, marshes, and islands of the western Texas coast; and the long estuaries and barrier islands of the Florida, Alabama, Mississippi, and Texas coasts.

Although natural conditions of tide, runoff and climate are generally favorable to estuarine life in the Gulf zone, the western estuaries suffer drouths, and the entire zone is subject to storm calamities on occasion. Man has degraded many estuaries of the zone by pollution, land occupation, drainage of wetlands, and diversion of runoff, as indicated on the zone map.

The Gulf Coast is highly productive of a great variety of estuarine-dependent fish and shellfish, notably red snapper, spotted sea trout, menhaden, shrimp, crabs, and oysters.

Birds are outstanding resources of the Gulf Coast. Waterfowl and other birds of the Central, Mississippi, and Atlantic flyways all winter in the extensive and rich marshes and estuaries of the Gulf Coast.

Southern Florida is the home of many sub-tropical and tropical birds including the roseate spoonbill. Nesting rookeries of egrets and herons are a great attraction at many Gulf marshes. The whooping crane, southern bald eagle, greater sandhill crane, and Cape Sable sparrow, as well as the American alligator and Florida manatee, occur in the zone but are rare or endangered. The brown pelican, once abundant, has rapidly diminished in numbers. The endangered whooping crane winters only on Aransas Bay, Texas.

Gulf Coast scenic attractions relate importantly to the wintering birds and to the life of swamps and marshes. The moss-draped swamp, riverine, and coastal trees add beauty and character to the zone. Scenic vistas, fine beaches, and seascapes occur on the Coasts of Florida, Alabama, Mississippi, and Texas.

Special reports on Apalachicola Bay, Florida, and on Aransas Bay, Texas, two of eight areas selected for special case studies, are included in Vol. 3, App. B.

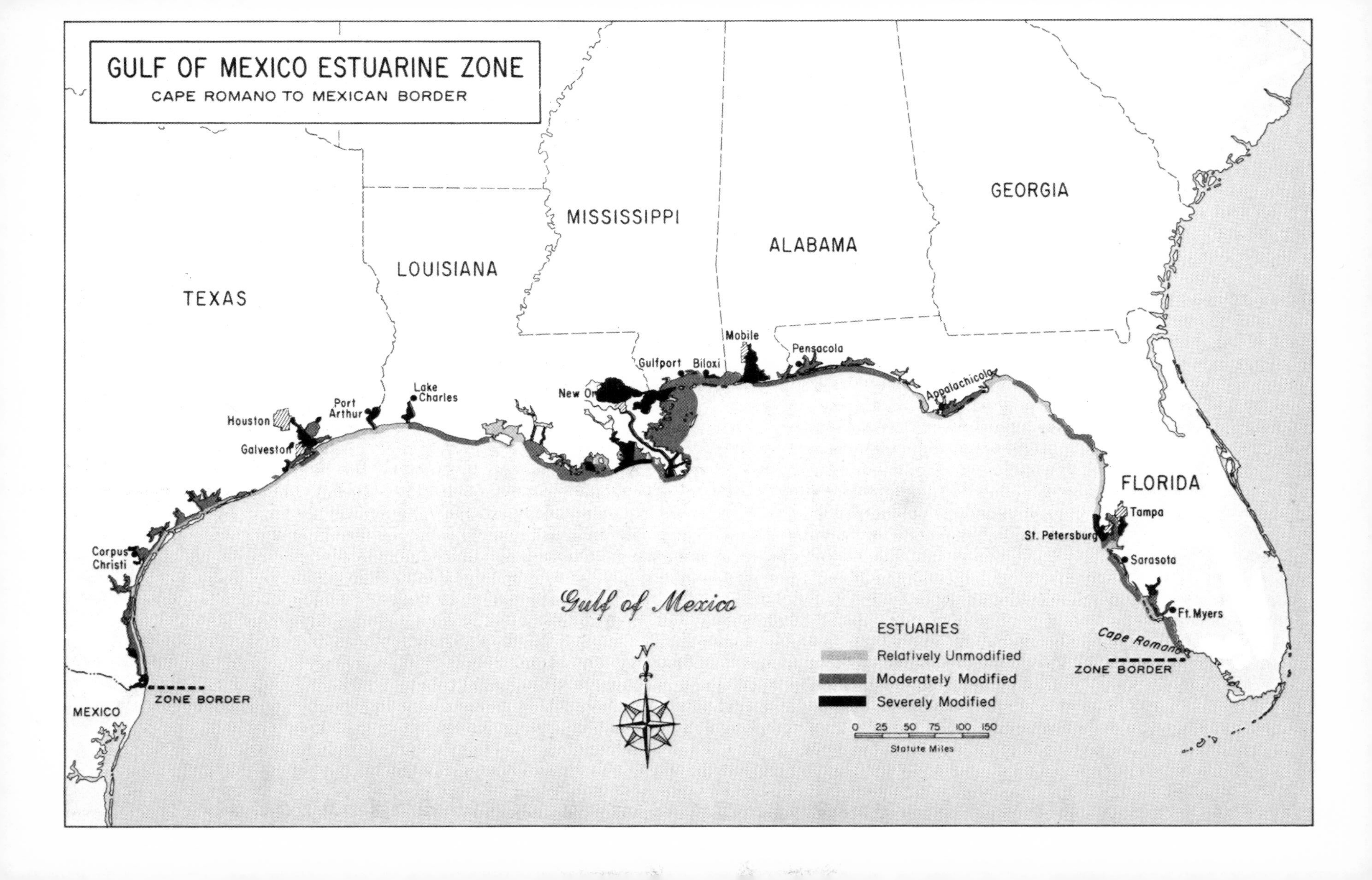
GULF OF MEXICO ESTUARINE ZONE
CAPE ROMANO TO MEXICAN BORDER
TEXAS
LOUISIANA
MISSISSIPPI
ALABAMA
GEORGIA
FLORIDA
MEXICO
Houston
Galveston
Port Arthur
Lake Charles
New Or
Gulfport
Biloxi
Mobile
Pensacola
Appalachicola
Tampa
St. Petersburg
Sarasota
Ft. Myers
Cape Romano
Corpus Christi
ZONE BORDER
ZONE BORDER
Gulf of Mexico
N
ESTUARIES
Relatively Unmodified
Moderately Modified
Severely Modified
0 25 50 75 100 150
Statute Miles

Southwest Pacific Estuarine Zone

This zone, extending from the border of Mexico north to Cape Mendocino, is shown on the map following this description. This coast is uniformly of resistant-rock. Its estuaries are widely spaced and small except San Francisco Bay, San Diego Bay, and Tomales Bay. Most of them have fringing marshes. Spits nearly enclose San Diego Bay and some others are partially enclosed or enclosed seasonally by inlet bars.

Although the southern part of the zone is favored by sunshine and warmth, it is naturally deficient in precipitation and runoff, and its tides are of moderately high range resulting in seasonally closed estuaries. Man has aggravated the natural conditions by encroachment on the estuarine wetlands and shallows, and by diversion of streamflow. Pollution has degraded most of the estuarine ports and waters near cities, as indicated on the zone map.

Important estuarine-related fishes of the zone include the anadromous chinook and coho salmon, steelhead trout, striped bass and others as well as several species of marine finfish, shellfish, and freshwater fish. Other species of marine fish frequent the outer estuaries and bays.

Great flocks of a variety of waterfowl winter in the estuaries of the zone. The few eel-grass beds of Bodega, Tomales, Drakes, and Morro Bays are very important to black brant which winter in Mexico and nest in Alaska. San Pablo Bay, in the San Francisco Bay complex, is a favored wintering site of the scarce canvasback. Mallard, gadwall, and cinnamon teal breed in the zone.

Continental populations of shorebirds of some 33 species, numbering in thousands, use the dwindling feeding areas of the zone on their migrations and join resident birds to winter and to nest in the zone.

Huge numbers of sea birds frequent the shore, and many nest on offshore rocks. California brown pelicans nest in Mexico and winter in the zone. The only U.S. colony of elegant terns is on San Diego Bay. Three species of rails of the zone are rare or endangered as is the California least tern.

Observation of marine mammals from coastal vantage points is very popular.

The sheer coastal front and jewel-like estuaries and bays of much of the coast in this zone offer spectacular scenic vistas of rugged cliffs and headlands with tenacious wind-swept cypress and pines above a pounding surf. The redwood groves of this zone are excelled only by the larger trees and groves further north.

San Francisco Bay and the semi-tropical, palm-fringed bays of Southern California such as Mission Bay and San Diego Bay with their old Spanish Missions are of great scenic and historic interest.

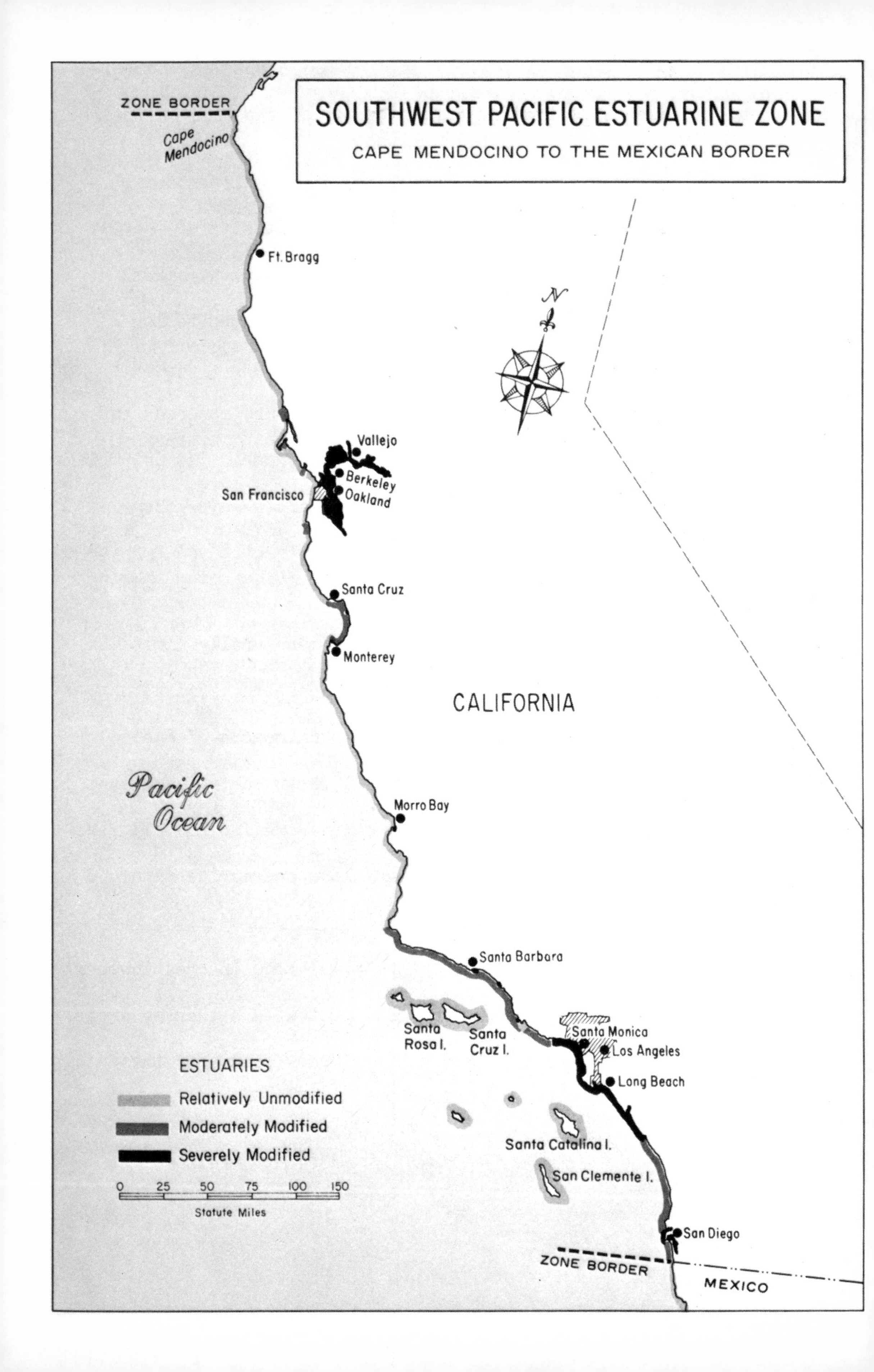

SOUTHWEST PACIFIC ESTUARINE ZONE
CAPE MENDOCINO TO THE MEXICAN BORDER
ZONE BORDER
Cape Mendocino
Ft. Bragg
Vallejo
Berkeley
Oakland
San Francisco
Santa Cruz
Monterey
CALIFORNIA
Pacific Ocean
Morro Bay
Santa Barbara
Santa Rosa I.
Santa Cruz I.
Santa Monica
Los Angeles
Long Beach
Santa Catalina I.
San Clemente I.
San Diego
ZONE BORDER
MEXICO
ESTUARIES
Relatively Unmodified
Moderately Modified
Severely Modified
0 25 50 75 100 150
Statute Miles

Northwest Pacific Estuarine Zone

This estuarine zone extends from Cape Mendocino, California, to the Canadian line in the Straits of Juan de Fuca and Straits of Georgia as shown on the following map. The southern part of the zone has a resistant-rock coast, but in Washington the coast is reduced to low coastal flats and islands by erosion of sedimentary rocks. The marsh-bordered estuaries of the zone are scattered but are more numerous and generally larger than those in the Southwest Pacific zone. The Columbia River mouth and Puget Sound are unique in their separate ways--the Columbia for the quantity and force of its outflow; Puget Sound for its intricate channels and islands.

Moderately high tides and strong onshore winds naturally limit biological production, but other conditions are favorable. Pollution and encroachment onto estuarine lands are the major impacts of Man's works. The degree and extent of modification is shown on the zone map.

Fishes of importance in the Northwest Pacific zone include all five species of Pacific salmon and many other anadromous and marine fish and shellfish.

Waterfowl and shorebirds frequent this coastal zone in great numbers during migrations, and many winter in the more protected estuarine areas. Nesting waterfowl include mallard, wood duck, widgeon, cinnamon teal, and blue-winged teal.

Heron rookeries are common in the zone and many sea birds nest on near-shore islands along the zone. The rare California clapper rail occurs in the southern part of the zone.

Harbor porpoises, sea lions, and harbor seals are the common marine mammals of the zone. Sea otter have been recently reintroduced to coastal waters of Washington. The rare humpback whale frequents bays and estuaries of the zone.

Scenic resources of the zone are outstanding. The famed coast redwood groves of the Redwood National Park and California State Parks, are unique in the World. The Olympic National Park rain forests are equally inspiring and wonderful.

Spectacular panoramas of breakers and rock cliffs, surf-cut rocks and islands, sea-lion and sea-bird rocks, rocky tide pools with gardens of sea life, and everywhere, green-mantled mountains are the nearly universal expectation along the coast in this zone.

A report on a special study of Willapa Bay, Washington, is included in Vol. 3, App. B.

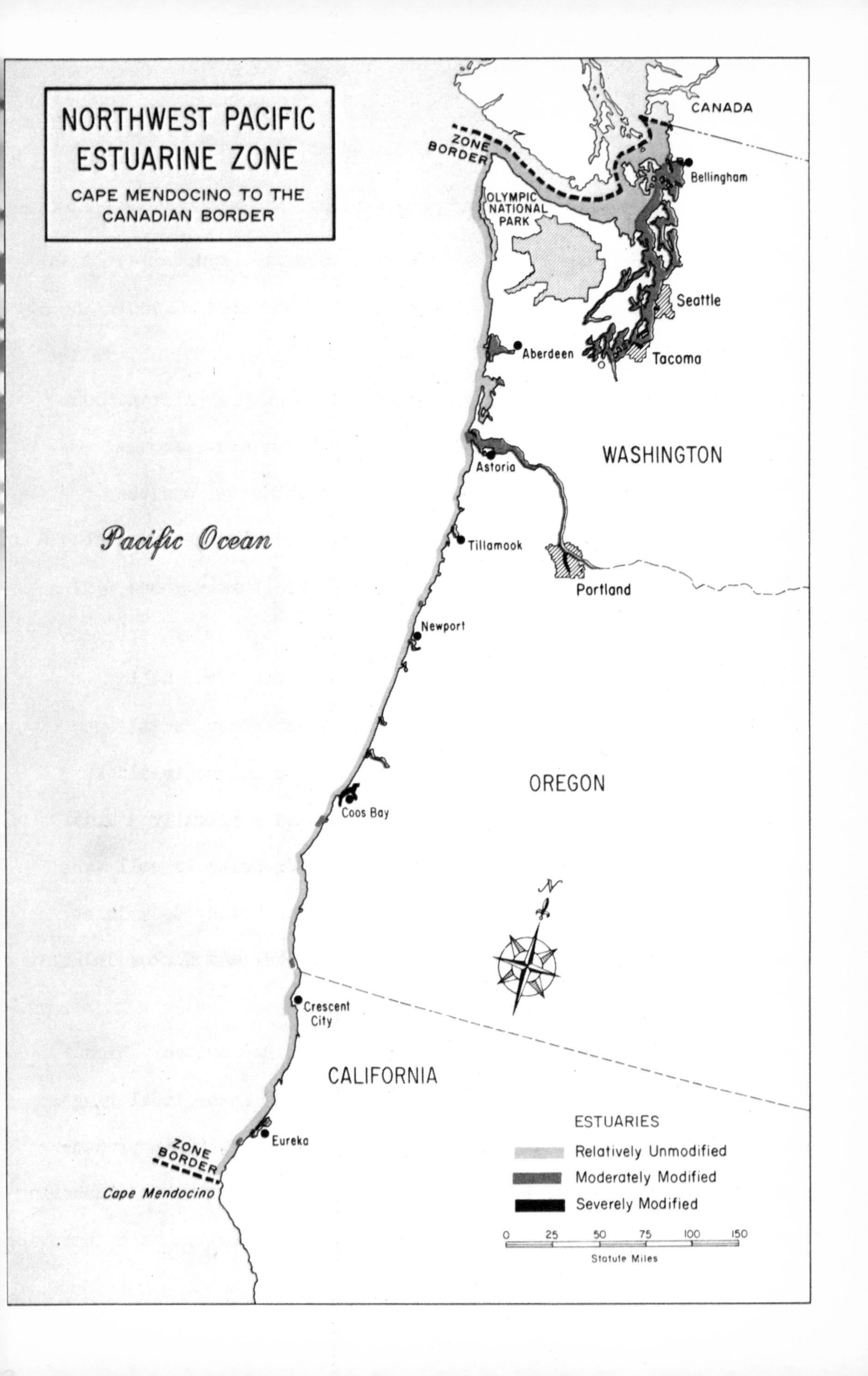

NORTHWEST PACIFIC ESTUARINE ZONE
CAPE MENDOCINO TO THE CANADIAN BORDER
CANADA
ZONE BORDER
Bellingham
OLYMPIC NATIONAL PARK
Seattle
Aberdeen
Tacoma
WASHINGTON
Astoria
Pacific Ocean
Tillamook
Portland
Newport
OREGON
Coos Bay
N
Crescent City
CALIFORNIA
ESTUARIES
Relatively Unmodified
Moderately Modified
Severely Modified
Eureka
ZONE BORDER
Cape Mendocino
0 25 50 75 100 150
Statute Miles

Alaska Estuarine Zone

This zone includes all of Alaska as indicated on the map following this description. It is a vast and diverse zone with a detailed tidal shoreline of 33,900 miles or about one-third of the coastal shoreline of the United States including the Great Lakes.

All of the coastline around the Gulf of Alaska is resistant rock as are sections of the western coast, but extensive areas around Bristol Bay, on the Yukon River Dèlta, Seward Peninsula, and the northwest and north slopes are coastal plain. From Southeastern Alaska to the bight of the Gulf of Alaska, the coast is glaciated, and many active glaciers persist.

The estuarine areas of Alaska include the Yukon River Delta, glacier-fed fjords, large open estuaries, and numberless, small open bays, coves, and inlets. The Yukon River Delta is unique in Alaska and rivals the Mississippi River Delta in size and complexity. Tidal and freshwater marshes cover the broad Yukon River Delta as well as extensive areas near Yakutat Bay, the Copper River Delta, Cook Inlet, the northern shores of the Alaska Peninsula, Bristol and Kuskokwim Bays, Norton and Kotzebue Sounds, and the North Slope.

Tides of Cook Inlet and estuaries of Bristol Bay are very high and they are generally high in the Gulf of Alaska. These tidal ranges combined with the cold of Alaska cause ice scouring and limit production, and degradation of estuarine areas by oil exploration and lumbering are the major impacts of Man's activities in Alaska. To date, the modifications are minor as shown on the zone map.

Many Pacific and Central Flyway waterfowl nest in Alaska, and great numbers that nest in Alaska also winter there. The waterfowl numbers of Alaska swell dramatically during spring and fall migration times when nearly the entire population of birds from Alaska, Siberia, and northwest Canada rest and feed in its estuaries.

Many species of marine mammals occur in Alaskan coastal waters. They include the harbor porpoise, several seals, northern fur seal, sea otter, walrus, beluga whale, and humpback whale. The ribbon seal and humpback whale are endangered species.

The estuarine scenery of the central Gulf of Alaska is unexcelled. Also, Southeastern Alaska has intricate forested shorelines, myriad emerald lakes, and white-water streams--fronted by icy fjords and backed by mainland alpine peaks and hanging glaciers.

A special study of Bristol Bay is reported in Vol. 3, App. B.

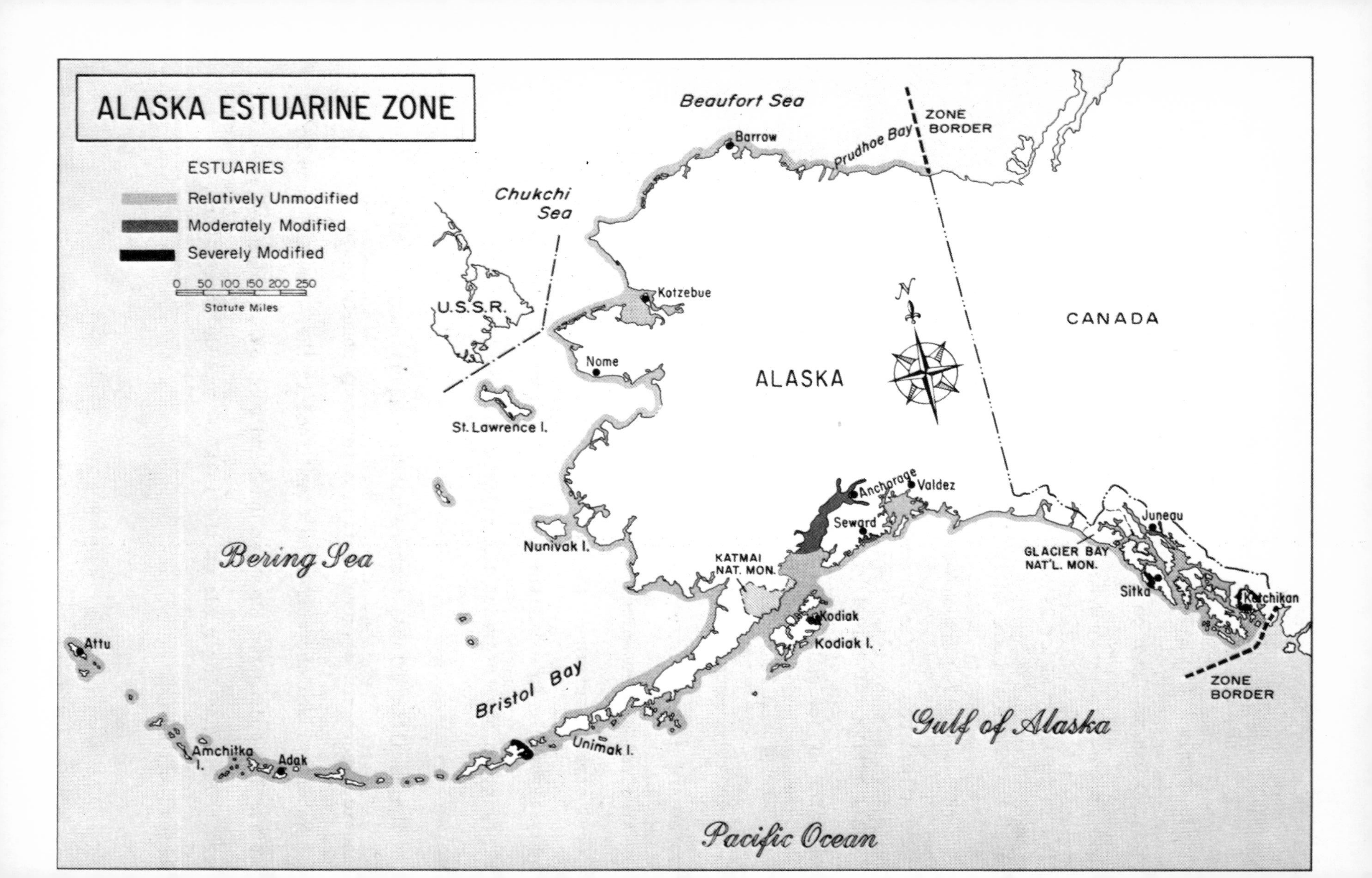
ALASKA ESTUARINE ZONE
ESTUARIES
Relatively Unmodified
Moderately Modified
Severely Modified
0 50 100 150 200 250
Statute Miles
Beaufort Sea
Barrow
Prudhoe Bay
ZONE BORDER
Chukchi Sea
U.S.S.R.
Kotzebue
Nome
St. Lawrence I.
ALASKA
CANADA
N
Anchorage
Valdez
Seward
Juneau
GLACIER BAY NAT'L. MON.
Sitka
Ketchikan
ZONE BORDER
KATMAI NAT. MON.
Kodiak
Kodiak I.
Nunivak I.
Bering Sea
Bristol Bay
Unimak I.
Attu
Amchitka I.
Adak
Gulf of Alaska
Pacific Ocean

Hawaii Estuarine Zone

This zone, the State of Hawaii, is a chain of volcanic mountain islands nearly 1,600 miles long located near the center of the North Pacific Ocean as shown on a following map. Of the approximately 6,450 square-mile total area of the islands, all but about 6 square miles is in the 8 major islands. These extend 375 miles at the east end of the chain and are entirely within the Tropics.

Hawaii, the youngest, highest and largest island (about 4,000 square miles) has only a narrow, peripheral shelf of lowlands and is still active volcanically. The other major islands have a greater proportion of lowlands. The greatest extent of estuarine lowlands is on Oahu around Pearl Harbor, but small bays on the windward sides of the islands, usually partly enclosed by coral reefs, are estuarine. Marshes and mangrove swamps occur but are not extensive.

Tropical storms, seismic waves, and leeside drouths are the only natural stress factors of the Islands; all other natural factors are highly favorable to estuarine life. Occupation of estuarine lands and pollution have degraded the major port areas, as shown on the zone map.

The Hawaiian Islands inshore waters support many tropical esthetic, sport and commercial fishes in great abundance.

Waterfowl and shorebirds nest and feed in the estuarine zone in Hawaii. The endangered Hawaiian duck and some other waterfowl nest there. Other nesting birds of the islands include shearwaters, boobies, and terns.

The rare Hawaiian monk seal and Pacific bottlenose dolphin are important marine mammals in the Islands.

The beauty of Hawaii is both well advertised and deserved. The coral-reef enclosed bays of the windward sides of the islands with their garden-like settings of tropical vegetation are a special delight. The highly colored and intricately marked coral-reef fishes also deserve special note as scenic attractions as do the albatrosses and other sea birds that frequent Hawaiian shores.

HAWAII ESTUARINE ZONE

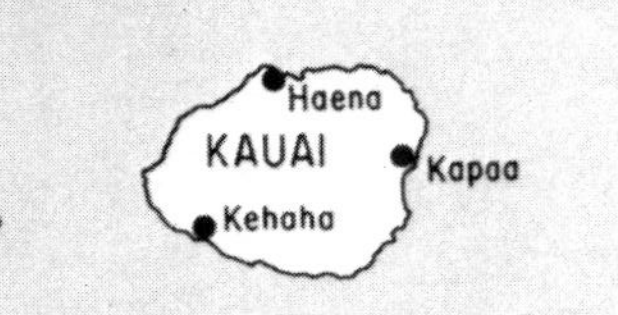

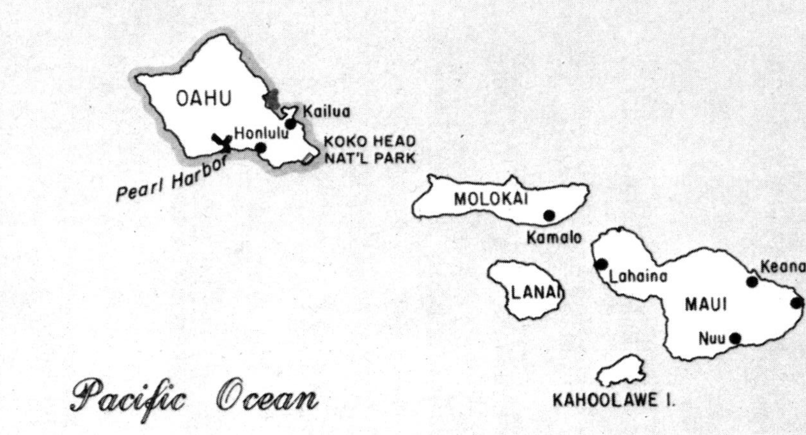

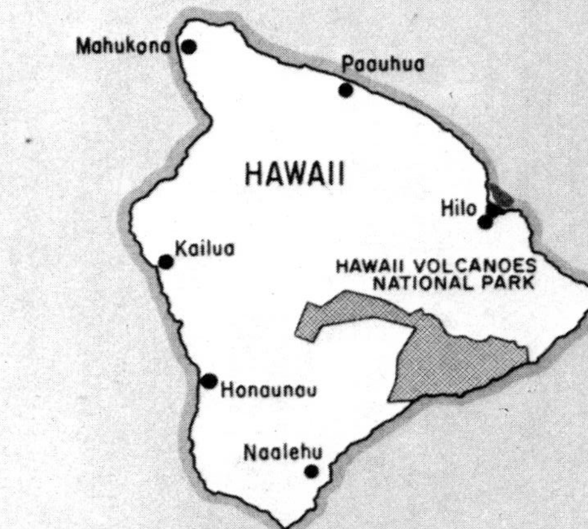

Great Lakes Estuarine Zone

This regional zone includes Lake Michigan and the United States shores of Lakes Superior, Huron, Erie, and Ontario as well as of Lake St. Clair between Lakes Huron and Erie as appears on the map following this description. The lakes occupy basins carved and deepened by Pleistocene glaciers from pre-glacial erosional basins and valleys of the St. Lawrence River system.

No true estuaries occur in the Great Lakes, but many of their shallows and marshes are similar to those of estuaries. The prominent areas of this type include Green Bay and Bay de Noc in Lake Michigan; Saginaw and Thunder Bays in Lake Huron; Anchor Bay in Lake St. Clair; and the southeastern shores of Lake Ontario. In Lake Superior, Keweenaw and Whitefish Bays, and the Apostle Islands are notable fish and wildlife and scenic resources. The major natural limitations to productivity in the Great Lakes are winter cold and high winds. Man has degraded many shallow bay and marsh areas by dredging and filling and by pollution as shown on the zone map.

The fish and wildlife resources of the Great Lakes include a wide variety of freshwater fishes. Some of the native fishes, including lake sturgeon, longjaw cisco, deepwater cisco, blackfin cisco, and blue pike are endangered.

The Lakes and the Lakes basin are migration stops for many species of migratory waterfowl and other birds of the Central, Mississippi, and Atlantic Flyways. Nesting waterfowl of the Lakes include wood duck, mallard, black duck, blue-winged teal, and ring-necked duck.

The marsh and upland, bay and river habitats of the lakes also are important for many birds and mammals as are the aquatic habitats of the Lakes.

The scenic bluffs and forested lakeshores, shoreline headlands, tree-studded rocky islands and peninsulas, and great expanses of water and beaches of the Great Lakes provide a mid-Continent wonderland of sea-like beauty and recreational utility.

A special study of Saginaw Bay, Michigan, is included in Vol. 3, App. B.

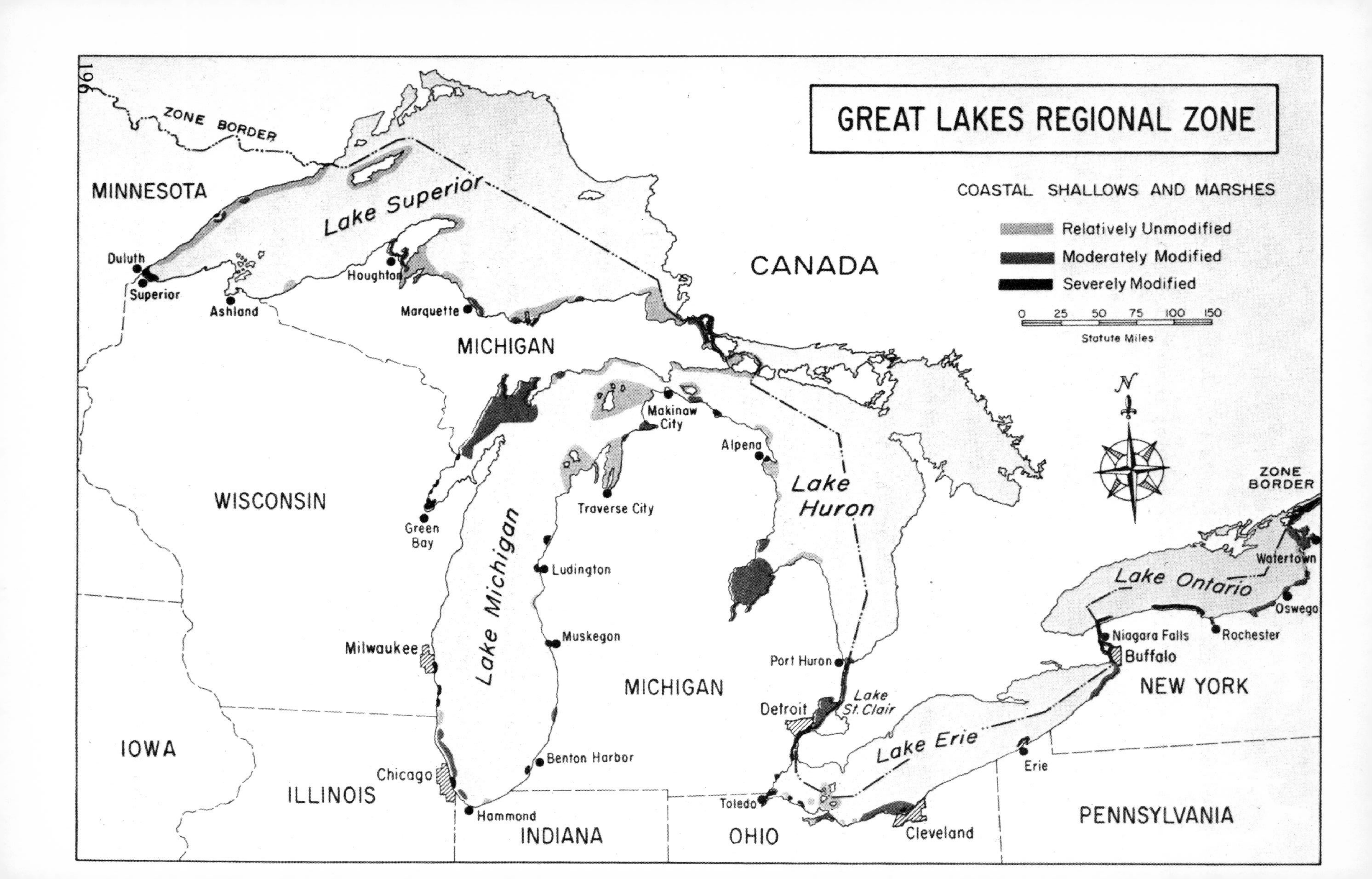
GREAT LAKES REGIONAL ZONE
COASTAL SHALLOWS AND MARSHES
Relatively Unmodified
Moderately Modified
Severely Modified
0 25 50 75 100 150
Statute Miles
N
ZONE BORDER
ZONE BORDER
MINNESOTA
CANADA
MICHIGAN
WISCONSIN
MICHIGAN
IOWA
ILLINOIS
INDIANA
OHIO
PENNSYLVANIA
NEW YORK
Lake Superior
Lake Michigan
Lake Huron
Lake St. Clair
Lake Erie
Lake Ontario
Duluth
Superior
Ashland
Houghton
Marquette
Makinaw City
Alpena
Traverse City
Green Bay
Ludington
Muskegon
Milwaukee
Benton Harbor
Chicago
Hammond
Port Huron
Detroit
Toledo
Cleveland
Erie
Buffalo
Niagara Falls
Rochester
Oswego
Watertown

IV. ESTUARY PROBLEMS

Ever since man discovered the seas and their rich bounty he has depended particularly on the estuaries for their readily harvested shellfish and finfish. And since he learned to travel the seas he has been focusing his uses ever more intensively on the estuaries. Here he first found embarkation sites and refuge for his sallies upon the seas. From these initial discoveries of utility, man has continued to center his commerce, industry, populations, and culture about estuaries. Today a preponderance of the World's major cities are located on them, and in the United States our eight most populous metropolitan areas are located on estuaries on the Great Lakes as are 15 of our 20 largest cities.

This concentration of human activity combined with the distinctive natural qualities of estuaries and similar Great Lakes areas intensifies the interactions and conflicts among uses and magnifies the use impacts on the environment. Many of the uses of estuaries modify or affect the land or water environment more or less seriously and more or less permanently. Some of them preclude other uses; others are compatible with many uses.

Certain uses such as those associated with extensive dredging and filling of marshes and shallows or with the discharge of toxic material can result in a complete disruption of the delicate ecological balance and even destruction of all plant and animal life. This was pointed out graphically in the Delaware Estuary Comprehensive Study of the Federal Water Pollution Control Administration in 1966. According to this report, waters of many estuaries have turned toxic and anaerobic; fish yields have declined to a few percent of the production capacity; water contact sports have been restricted or forbidden because of health hazards; boating has been hampered by floating debris; and expanding metropolitan and industrial developments have encroached upon the few remaining wild areas in the metropolitan vicinity and thereby destroyed the marshes and the wildlife which they support. In short, important estuarine resources upon which society depends are being destroyed indiscriminately to satisfy urban or industrial demands which may have a social value only in the immediate locality.

There are numerous Federal, State, and local laws and regulations related to management of the estuaries and the Great Lakes. The Federal, State, and local governments carry out a wide variety of programs which affect these regions. However, most of the laws and the programs concern much broader areas and relate to the estuaries and the Great Lakes only to the extent that they are a part of the whole. The government programs are administered by individual agencies with variously prescribed interests and responsibilities. There often is inadequate coordination among the programs of the various agencies, and duplication and conflicts are common.

Use Conflicts

All of the present uses of the estuaries and the Great Lakes satisfy human needs. The Nation's estuaries and the Great Lakes areas are ecologically complex; extremely varied, physically and biologically; and heavily used by people for many purposes. When not modified unduly, they are a buffer against storm and hurricane damage. They produce abundantly a great variety of fish, birds, and mammals of commercial and social importance to the people. Their broad vistas of land and open water are a form of natural beauty found nowhere else.

Because of their location at river mouths they collect and trap water-borne wastes, as well as nutrients from the lands and river sources. When not overtaxed they have the capacity to assimilate and dissipate wastes through natural processes.

As noted, they are all locationally appropriate centers of human activity. This concentration of people makes them desirable sites for outdoor recreation. People from the interior as well as those from the coastal States flock to these estuarine and coastal sites. This greatly aggravates the problem of maintaining environmental quality while increasing the demand and need for such quality.

Fishing and Recreation

Commercial fishing and recreational use, when not excessive, has the least damaging effect on the natural environment and on other estuarine uses. Although the dedication of certain portions of estuarine areas as wildlife refuges or public parks preempts their use for other purposes, this use is not accompanied by irreversible changes in the environment. Intensive developments such as small boat marinas and commercial fishing facilities may damage the natural environment if not planned and operated properly.

Navigation

This was one of the earliest uses of estuarine waters and continues to be one of the most important. The estuaries and the sheltered bays of the Great Lakes provide ports which serve as important links in oceanic, coastal and inland commerce. Hundreds of millions of tons of cargo pass through these ports each year. Between 1950 and 1965, foreign and domestic shipping increased at an annual growth rate of about 3 percent, and the foreign commerce portion increased at an annual rate of 6.7 percent. The Department of Transportation expects the foreign sector of water-borne commerce to triple in volume by the year 2000. Clearly, navigation channels, harbor works, and docks are essential to our National welfare. Yet, new navigation channels can be located so as to prevent much of the environmental damage which has occurred in the past and actually improve ecological conditions in some cases. Similarly new ports can be located and designed so as to reduce damaging impacts.

Although the recent trend has been toward larger vessels and deeper channels, new technologies of shipping, transfer of cargoes, and warehousing give promise of lessening the concentration in estuaries and in the bays of the Great Lakes. Advantage should be taken of these technologies to free many existing harbors of blight and to replace the blight with public parks and recreation areas central to our cities.

Landfill

Much of the most damaging use of estuaries--to create land--is not an essential on-site use. In many cases, the shore and adjoining shallow waters have been filled with material dredged from the productive bottoms together with sludge and garbage from the cities to create land for homes, motels, industries, roads, airports, and various service establishments. Dredging and filling is often carried out to create a series of alternate fingers and channels with a maximum number of waterfront lots for residential housing. Such dredging and filling for permanent or seasonal housing can destroy the natural environment which is responsible for the high value of the home sites. Although each home has a waterfront, the water itself may become polluted and unfit for fish life or swimming. The filling eliminates public use of the shore and access to the water, including enjoyment of the shore and water landscape. It also eliminates the habitat and basic nutrient sources of the quiet tidal edges and marshes, and the dredging of submerged material deepens the water and thereby decreases the most productive portion of estuarine bottom. Yet, with the high demand for waterfront property, developers can realize substantial profits by buying submerged or marshland at a low price and selling the filled land for much more.

Landfill for residential housing is not an essential use of estuarine areas. The houses could be located at sites away from the immediate shorelines where they would still have scenic outlooks but would not eliminate public access to the estuaries or destroy natural productivity of the waters.

Landfill is commonly placed in these water areas for airports and connecting roads. They often offer the only open space available for this purpose within a convenient distance of metropolitan centers. Although not estuarine dependent, airports may have to be located in these waters in the absence of alternative sites. However, they could be located and designed to minimize damaging environmental impacts.

Industry

Industrial developments have often been located in the shallow waters and marshes of these areas because of the convenience to water transportation, low land prices, and ease of waste disposal. Most of the industries are not estuarine-dependent and could perform the same function at an upland location. Obviously land, transportation, and waste treatment costs would be higher and this could result in higher prices of products to the consumer.

Industries which are estuarine-dependent could be located, constructed, and operated so as to eliminate or at least reduce damaging effects on the environment. Some form of environmental quality standards might be imposed by government on these activities.

Water Pollution

Because of their location, estuaries and the bays of the Great Lakes are particularly susceptible to water-borne pollution. Part of this represents releases from municipal and industrial sources including thermal power plants. Another part represents runoff from uncontrolled non-point sources such as city streets, farms, animal feed lots, lands under development, natural erosion, and accidental spills of various types. Federal-State pollution control programs are expected to decrease the harmful aspects of planned point-source waste disposal, but it is certain to be a long uphill process. There is a need for a speed-up in the development of improved technological processes for treatment of domestic and industrial wastes and a need for additional facilities. There is also a need for additional facilities and action by public bodies to reduce pollution from non-point source wastes. Rigorous enforcement of water-quality standards in estuaries and the bays of the Great Lakes should help prevent degradation below established standards. If the estuarine environment and its living resources are to be protected, conserved, and developed, it is essential that adequate water quality be maintained.

Mining

Mining and extractive uses of the estuaries and the Great Lakes include petroleum exploration and development; salt and sulphur mining; extracting of sand, gravel, shell, coral and fill material from water bottoms; mining of phosphate rock, and extraction of minerals from seawater. Other types of mining are not now economic. By far the most significant mining with respect to both value and environmental maintenance is that of petroleum. Sand, gravel, shell, and fill-material extraction and phosphate mining are of great significance in some areas. Also extraction of salts from seawater is a significant estuarine use in San Francisco Bay.

The principal impacts on the environment from petroleum activities are those associated with careless exploration and operation and from accidents. Extraction of sand, gravel, shell, and fill-material removes productive bottoms together with their plants and animals and spreads silt and fines which smother other bottom habitats.

Most of the effects of mining are more or less transitory although some may be relatively permanent such as in the fragile environment of Arctic, North Slope Alaska. Extensive dredging can cause permanent changes in the degree of sea-water incursions into some marshes and can result in long-term changes in currents, circulation, mixing, salinity, dissolved oxygen, and consequently of the productivity of the estuarine area.

Mining and extractive activities which cause substantial environmental damage need not be carried out in the estuaries and the Great Lakes if alternative sites are readily available. However, the cost may be increased by shifting the activities to other areas. Such activities which must be carried out in these waters should be planned and conducted so as to keep the environmental damage to a minimum.

Water Developments

Water conservation, flood control, and other developments in the watershed of an estuary or Great Lakes area affect the volume, rate, and timing of runoff that reaches the area. The effects on an estuarine system depend on the magnitude of the changes in runoff and on the character of the estuary. The reduction of freshwater inflow through diversion of water to another watershed, direct consumption, or evaporation can affect the salinity of the estuary and disrupt the delicate balance needed to maintain the living organisms. On the Great Lakes, flow-through flushing is reduced.

The Florida Everglades and the Texas estuaries provide much of the principal rearing areas of the extremely valuable Gulf shrimp. A substantial reduction in the runoff to these areas could well result in the complete elimination of the Gulf shrimp industry. The Nation's shrimp catch had a dockside value of more than $100 milion in 1967 and 1968 and represented the Nation's most valuable commercial fishery product.

The needs of the living organisms of the estuaries must be provided for in designing water development projects on the streams providing the freshwater inflow or the viability of the estuaries will be destroyed.

Economic and Social Values

Estuarine uses serve important human needs. The estuarine ports handle hundreds of millions of tons of foreign, coastal, and inland commerce each year. The commerce associated with water transportation, together with the industries dependent upon estuarine resources, produce necessary goods and services and generate many hundreds of millions of dollars of income. These industries provide jobs for local residents and provide tax revenue to the local communities.

Although the estuarine-based recreational activities provide large-scale, social benefits to both local and non-local people, these benefits are not priced in the market place. There are direct tangible benefits from the service establishments which cater to these recreationists. However, these are only part of the value of recreation; they do not include, nor do they relate to, the value of the recreational experience itself. The natural environment and its living natural resources, as long as they are kept viable, are available for recreational and esthetic enjoyment by all of the people of the Nation.

It should be possible to determine that commercial activity which would provide the greatest dollar return for each dollar invested in a particular estuary. From a strictly economic or cost-effectiveness standpoint, it might seem logical to assign exclusive use to that purpose alone. However, this would not be desirable from a public or social standpoint. In the past, estuarine uses have been determined largely by the private developers who have overwhelmed local communities with promises of the economic benefits which they would receive. What is needed is a far stronger action by government on the side of the environmental public values of estuaries to counteract the inducements offered by private developers.

In summary, a degree of control by some public body above the local government level over the uses of estuaries and the bays of the Great Lakes seems mandatory if the environment and the living resources are to be protected, conserved, and developed. This would be provided by the Department of the Interior's Coastal Zone Management System. Uses not estuarine-dependent should be located in other areas. Uses which must be carried out in the waters should be conducted so as to have the least damaging effects possible on the natural environment. Government must take appropriate action if these areas are to be preserved for all of the people both today and tomorrow.

The various uses of the estuaries and the Great Lakes area, together with their effects on the environment and its living resources, are discussed fully in Chapter II of the Staff Report: Estuarine Uses and Conflicts, in Vol. 2, App. A. Special reports entitled: Some Economic Factors Affecting the Estuarine Zone, Including Market Outlooks for Selected Products; Technological Impacts on Estuary Resource Use; and Conflicts and Problems in Specific Estuaries are included in Vol. 5 as Appendices E, F, and G, respectively.

Legal Problems

Except for the laws authorizing the National estuarine pollution study and the current National estuarine study, there are no Federal statutes directed specifically toward the estuaries. Excepting Constitutional reservations related to commerce and defense, State law is generally prevailing in the estuaries and Great Lake areas. Several States have recently enacted laws for the protection of coastal wetlands and intertidal areas and others are considering them.

Federal Laws

Some of the broadest powers of the Federal Government relate to the Constitutional power to regulate commerce among the several States and with foreign nations. In the estuarine zone, this is directly applicable to the navigable waters and the navigation thereon. However, the powers are much broader, and have been interpreted to include a variety of activities related to and affecting such commerce. The war or defense powers of the Federal Government also are broad in these waters. They relate to the installation of structures, navigation improvements, and any activities which may be construed as affecting the defense capability of the Nation. Federal powers of taxation, other than the income tax, are essentially limited to indirect ones which have little utility for regulating or managing private activities in these waters.

Some have suggested that development in the estuaries and the Great Lakes considered to be against the public interest could be greatly discouraged by enacting a Federal law requiring the taxing as income, rather than capital gain, any income derived from development of estuarine lands which caused significant loss of public resources.

In the estuarine zone, our philosophy that individuals and local interests be assured, as fully as possible, freedom of choice and freedom from central control is tested against as highly vulnerable and as highly valuable an array of public resources, renewable and non-renewable, as are found anywhere. To complicate the matter further, responsibility in this zone is more sharply divided between the Federal and State sovereignty than elsewhere.

State Laws

The States hold in trust the submerged lands of coastal and navigable waters within their territorial boundaries, including all estuaries, but excluding those lands in some States which have passed to private hands and such other lands and waters ceded to the United States for military or park purposes. Yet the Federal Government exercises primary control of the overlying water under the Constitutional commerce and defense powers. In earlier times when the Federal interest was narrowly construed, the waters and their inherent values were easily neglected by the Federal government. Also, the submerged lands having unknown intrinsic values were considered worthless and were easily neglected by the States. Further, the shared responsibility in the Coastal Zone proves the adage of everybody's responsibility being nobody's; it has encouraged neglect of ill-defined responsibilities.

Ownership of intertidal lands varies among the States. Generally they are part of the submerged lands held by the States, but in some States they are the property of the riparian owner. In the Great Lakes the States own or have jurisdiction over all submerged lands.

The States hold Constitutionally reserved police powers and thus have more direct ability to manage estuarine development and use than does the Federal government. They can act and regulate for purposes of health, safety, and general welfare. Excepting powers limited by Federal law or forbidden by the Federal Constitution, the limitations on State action are self-imposed by their own constitutions and laws which may be amended.

Both Federal and State agencies have carried out or encouraged programs of drainage and reclamation of private wetlands in the estuarine zone for mosquito control, timber production, agriculture, industry, and urban development. Much of such drainage has been ill-advised and uneconomic. Substantial public interests in fish, wildlife and scenic beauty have been sacrificed under these governmental programs to further private development and gain. However, the situation is changing. Several of the States, notably Massachusetts, Connecticut, North Carolina, and California (on San Francisco Bay), have moved to control unwarranted drainage and filling of marsh lands in the estuarine zone. Florida recently set aside 26 estuarine areas as aquatic preserves, and is establishing bulkhead lines in coastal counties at or above mean high tide. Other States and the Federal agencies involved should follow these forward-looking examples.

Several States encouraged filling and loss of State submerged and tidal lands by disposing of them to private developers for nominal or non-monetary consideration. Also, some States gave such lands to local governments with stipulations that they be developed. Other such lands have been filled and used for public purposes such as airports, roads, docks, parks, etc. But such policies are being abandoned in many States in favor of protection of such lands from destructive development. The public is demanding environmental quality and ecologically sound policies. It is no longer willing to countenance profligate gifts of its heritage in estuarine submerged and tidal lands.

Real Property Taxes

Real property tax laws and policies of assessment presumably can encourage or discourage development of estuarine lands. Taxing on the highest potential use rather than present use often forces owners to sell, particularly if the owner is financially unable to support preservation of a primitive or conservation use. However, this effect is largely one of timing; if the demand exists, the sale and development probably would eventuate regardless of the taxing policy.

Some have suggested that the Federal government impose a charge on persons engaging in developments which would damage the estuarine environment. Revenue from such charges could be used to establish a fund for use by the State in acquiring control of other estuarine lands requiring protection.

Zoning

Zoning of land use is potentially a powerful tool for fostering rational and balanced use in the estuarine zone, but it has practical limitations. It has been a power delegated almost entirely to local governments. Also, it has been generally accomplished through readily changed administrative action rather than by statute. This has often permitted local pressures to influence decisions in the direction of private gain at the expense of public interest. It is simply a reflection of values perceived at the community level rather than at a State or National level.

There is a growing tendency for States to reclaim zoning powers expecially in relation to flood plains. Hawaii has adopted statewide zoning and Alaska and Oregon have taken recent steps in this direction. Effective and adequately enforced zoning can regulate uses of estuarine and Great Lakes lands, and prevent unnecessary, destructive dredging and filling of submerged and intertidal lands as has been demonstrated in Massachusetts and in the San Francisco Bay area of California. Wisconsin requires its counties to enforce shoreline zoning satisfactory to the State, and Massachusetts has invoked State level zoning in estuarine and coastal areas.

The Federal government has no direct zoning power, but it has in the case of certain recent national seashore statutes (Cape Cod, Point Reyes, and Fire Island) imposed Federal zoning through the expedient of the property power. In these instances, the threat of Federal condemnation of lands under the eminent domain authority is encouraging the local governments to establish adequate zoning regulations for protection of the natural resource values of areas contiguous to the seashores.

Control Over Dredging and Filling

Federal authority for the protection of estuarine and Great Lakes areas is contained in the 1899 Navigation Act. Under this authority, permits are required from the Corps of Engineers of the Department of the Army for any dredging, filling or other modification in these and other navigable waters of the United States. The Coast Guard now administers this Act in relation to causeways and bridges. Yet the 1899 Act is single-purpose, designed to protect navigation alone. It took the 1958 amendments to the Fish and Wildlife Coordination Act to broaden this authority to cover the protection of fish and wildlife resources as well. Under a 1967 agreement between the Secretary of the Interior and the Secretary of the Army, the Corps of Engineers now considers effects on fish and wildlife and environmental quality along with effects on navigation in determining whether or not to issue a permit for dredging and filling. However, private developers have questioned this authority of the Federal Government to deny permits on grounds other than effects on navigation, and the matter is now before the courts. Additional legislation may be necessary to clarify this authority of the Federal Government.

It is often difficult for the Corps of Engineers and the Department of the Army to determine whether or not dredging and filling should be permitted in estuarine areas even when the Fish and Wildlife Service and the Department of the Interior oppose the activities because of damage to the estuarine environment and the living resources. The local communities generally favor proposed developments because of the economic activity they will generate and the additional taxes they will provide. The State economic development agencies generally favor such works while the State conservation agencies oppose them. The Governor of the State seldom takes a position.

A fuller discussion of legal aspects of estuarine management is included in Chapter III of the staff report Vol. 2, App. A. A special more-detailed report entitled: Federal, State, and Local Laws and Tax Policies Affecting the Use of Estuarine Resources is in Vol. 6, App. I.

Institutional Problems

The views of different levels of government differ by proximity to the estuary problems. As long as the local community is constrained only by internal desire for growth, it is likely to allocate land within its jurisdictional boundaries to a complete mix of economic uses including industry, commercial, and residential. Viewed from a regional perspective, however, it might be more desirable for the particular community to not allocate any area to industry. Similarly there is no reason to assume that a mosaic of State development plans would correspond to a plan that would best serve the National interest.

Each level of government serves a particular constituency. The higher the level of government, the more diverse is the constituency and the more remote the concern of the constituency will be from the influence of individual profit and loss situations. Thus, the decisions of local government are likely to reflect the system of values dictated by the profit and loss situation of the residents of that jurisdiction. The fact is that for reasons that are viewed as legitimate by residents, local jurisdictions will not--perhaps cannot--commit substantial land and water resources to the production of values realized by society at large.

Local allocation of resources is also influenced by the prevailing financial structure. For example, a community which is totally dependent upon self-generated revenue to support all essential public services is less likely to dedicate land for social uses than a community which receives financial assistance from regional or State governments.

In contrast, action of the Federal government should reflect a value structure of all the people in the National society. It is at the National level that decisions to commit specific resources to wilderness areas and wild rivers can be made. The value expression involved is political and transcends even the proxies for total market values, such as X dollars per recreation visitor day.

Federal Government

There is no agency which has overall responsibility at the Federal level for the management of estuaries and similar areas of the Great Lakes. There are a number of line agencies which carry out programs directly or provide financial or technical assistance to others in programs which affect these lands and waters. At times these programs may be duplicative or in conflict with one another.

Each line agency operates under laws developed by particular Congressional committees to serve specified needs. They are often supported by non-Federal clientele groups representing the interests they serve. Certain programs of these agencies are financed by earmarked funds derived from special taxes on the goods or services provided.

One Federal agency may be assisting a counterpart State agency in constructing a highway or landfill in an estuary while another Federal agency may be assisting its counterpart State agency in establishing a wildlife refuge or park in the same area to preserve its natural values. In estuaries, one Federal agency is responsible for developing navigation channels and ports, a second is responsible for assisting industrial development, a third for assisting urban development, a fourth for water pollution control, and so on, while another Federal agency is responsible for preservation of the natural resources. Often there is little coordination.

The Fish and Wildlife Coordination Act does require the Corps of Engineers of the Department of the Army to consult with the Fish and Wildlife Service of the Department of Interior as well as with the counterpart State agency and to consider the recommendations of the Secretary of the Interior in planning navigational developments in estuaries and the Great Lakes. This Act together with the earlier mentioned agreement between the Secretary of the Army and the Secretary of the Interior provides for consideration of effects on fish and wildlife and environmental quality of non-Federal work proposed in navigable waters under Department of the Army permit. The Fish and Wildlife Coordination Act provides for similar coordination in connection with water-use projects proposed or licensed by various

Federal agencies. Many of these projects can affect estuaries. Section 4 of the Estuary Protection Act, P.L. 90-454, gives the Secretary of the Interior specific review authority over Federal water and land development projects affecting estuaries.

The Water Resources Council directs and coordinates the Federal-State comprehensive planning for the use and development of the water and related land resources of the Nation's river basins. It also administers programs of grant-in-aid to the States for such planning. The Council has insisted that planning under its guidance give balanced consideration to all relevant public objectives including those of environmental quality, recreation, and fish and wildlife. It has specifically included estuaries and estuary-related resources in its comprehension.

The four river basin commissions and the Federal-State interagency committees operating under the Council cover much of the coast line of the United States. The comprehensive planning which they carry out involves estuarine and Great Lakes resources. All concerned Federal and State agencies participate. Neither the Council nor the commissions have development authority of their own.

The Delaware River Basin Commission, established by Federal-interstate compact, has been effective in coordinating water development and water quality control among the party States. Its broad powers make it a particularly important coordinating authority in Delaware Bay estuary of New Jersey, Delaware, and Pennsylvania.

The Economic Development Administration of the Department of Commerce administers coordinated government planning in the estuarine zone under the Public Works and Economic Development Act of 1965. The New England, Coastal Plain, and Upper Great Lakes Regional Commissions established under the Act provide Federal assistance to the States in developing programs to improve economic conditions in their areas.

The above described programs of Federal agencies may have marked influence on estuarine and Great Lakes areas. These programs give various degrees of consideration to environmental and living natural resource aspects. The primary emphasis has been on developments which increase the goods and services valued and readily exchanged in the market place. The non-market values realized by society as a whole have often been neglected.

The recently authorized Council on Environmental Quality (P.L. 91-190, approved January 1, 1970) will continually assess the status and condition of the Nation's environmental resources including those of the estuaries and the Great Lakes. It will provide a board of review for all aspects of environmental concerns of the Nation and report annually through the President to the Congress on its findings and recommendations arrived at through its own and other sources as well as through consultation with the Citizens' Advisory Committee on Environmental Quality established by Executive Order numbered 11472, dated May 29, 1969.

The Bureau of the Budget, in the Office of the President, encourages coordination among Federal, State, regional and local programs through the Federal planning, programming, and budgeting processes, and it normally implements Presidential directives through these processes, preparation of Executive Orders, and issuance of Bureau Circulars.

Adequately controlled commercial fishing and recreational use are compatible wit
maintaining estuaries in their natural state as well as with many other uses of
estuaries by man.

State Governments

As in the case of the Federal Government, the States do not have lead agencies responsible for the overall management of their estuarine and Great Lakes areas. They have the same mixture of line agencies--each having responsibility over specific program affecting these lands and waters. Again the programs may be duplicative or in conflict at times. A State lands agency may be responsible for selling or leasing submerged lands in an estuary while a State fishery agency may be responsible for protecting and developing fisheries which are dependent upon these lands and the associated waters. The various agencies work closely with their Federal counterparts and are supported by the special interest groups on the outside. In some States there is coordination of the various programs in the office of the Governor.

In other States, there is little or no coordination of the various programs. As mentioned earlier, a limited number of coastal States have enacted legislation recently to provide better control over the use of estuarine lands and water. However, in the majority of the Coastal States there is inadequate control over the development and use of these areas and insufficient effort to protect the environmental values.

Local Government

The majority of local communities are unable or unwilling to provide adequate protection to the environment and the associated living natural resources of the estuaries and the areas of the Great Lakes within their jurisdiction even though broad social values to the people of the State and the Nation are involved. Most of the communities are dependent upon local revenue to finance the services which their residents require. They tend to favor those residential or industrial developments of land and water areas which provide the highest direct return. They are often subject to strong pressures from private developers and from the majority of their residents to encourage developments and often cannot afford the luxury of preserving fish, wildlife, and esthetic values.

See Chapter III, Vol. 2, App. A (staff report) for a fuller discussion of institutional problems. A special report in Vol. 6, App. H entitled: Evaluation of Existing Methods of Coordinating State and Federal Actions explores these matters in detail, and a special staff report entitled: Inventory of State Development Plans and Policies for Uses of Estuarine Resources, is to be found in Vol. 7, App. J. Another special report, Evaluation of Existing Public Management Schemes Relevant to Estuarine Resource Use, also is included in Vol. 7, App. K.

Summary

The record of environmentally destructive use of the estuaries and the Great Lakes clearly shows the need for greater control by government. As demonstrated in this study and the earlier studies of estuaries and the Coastal Zone, present institutional arrangements are not providing orderly development or environmental protection.

stuaries and tidelands offer many advantages to industry. Tidelands were the
nly site available for the airport at Juneau, Alaska. (upper photo). At Bay City,
ichigan, petroleum and chemical industries border the Saginaw River, and foreign
nd Great Lakes ships dock along the banks of the dredged channel (lower photo).

It is apparent that primary responsibility for management of the estuaries and comparable areas of the Great Lakes should be vested in the States. The local governments are not in a position to exercise this responsibility adequately. The primary authority of the Federal Government under the commerce clause of the Constitution relates to the navigability of these waters, and it should not have primary responsibility of overall management. However, there is a broad National interest in the protection and development of the living natural resource and environmental values of these lands and waters for all of the people. There is a need to designate a single Federal Agency to coordinate the many Federal programs which affect these areas and to provide financial and technical assistance to the States in carrying out their management responsibilities.

V. NATIONWIDE SYSTEM OF ESTUARINE AREAS

The fundamental thrust of the Estuary Protection Act -- the provision that made it different from the other authorizations calling for Coastal and Estuary studies -- was its call for an examination of the desirability and feasibility of establishing a Nationwide system of estuarine areas. No comparable direction has been contained in any other enactment.

In approaching this assignment, consideration was given to other systems now in use for Federal administration. These include the National Park System, the National Wildlife Refuge System, (both operated by the Department of the Interior), and the National Forest System, (operated by the Department of Agriculture).

In each of these systems, a special piece of the terrain with prescribed boundaries is identified for inclusion because of its unique characteristics that make it especially suitable.

Thus the Park System components need to have special qualities for scenic, historic, and recreation purposes. Units included in the National Wildlife Refuge System must have existing or potential capabilities for the creation of outstanding wildlife habitat. Forests must be manageable natural units and have substantial timber, watershed, and/or grassland values before they can be considered for inclusion in the National Forest System. Moreover, they must be all or nearly all in Federal ownership.

In every case, these are exclusive concepts. Not all of the scenic or historic areas of the Nation qualify for inclusion in the

National Park System. Not all of the good wildlife areas of the Nation qualify for inclusion in the National Wildlife Refuge System. Obviously, not all of the forests, watershed or grassland areas of the country are in the National Forest System.

These concepts of systems applied to estuarine areas would necessarily call for choosing a limited number of the estuarine areas of the Nation for the special status of being incorporated in the Nationwide system of estuarine areas. The criteria for incorporation in such a system might include:

a. pristine qualities

b. possibilities of effective management action to protect and restore natural values

c. need for rejuvenation because of degraded condition

d. preempting an estuary so that it would not be damaged by a real estate or industrial development

e. a combination of two or more of the above criteria.

It is not possible to make a choice of only a limited number of the Nation's estuarine areas for inclusion in a Nationwide system using these criteria because every estuarine area in the Nation has at least one of them. The alternative, obviously, is to include all of the estuarine areas of the Nation in such a system, and that is the conclusion of this report.

This does not mean, of course, that places like New York Harbor, an estuarine area, should be subject to special measures by State and Federal governments for the restoration and protection of natural values. On the other hand, this most assuredly does not mean that a relative few of the Nation's estuaries would be selected for annointment by inclusion in the system, with the rest excluded.

All of the Nation's estuaries have actual or potential natural values that should be safeguarded for the benefit of the people.

As in the Coastal Zone management proposal submitted by the Department of the Interior to the Congress on November 13, 1969, the Coastal and Great Lakes States are the keys to successful estuary preservation and restoration.

This means that the Conservation or Fish and Game Departments of the State governments are the ones to take the lead in identifying estuaries for management and measures to protect and restore their natural values. The States generally own the waters and submerged lands of estuarine and inshore Great Lakes areas. Also the States, of course, have jurisdiction over the fish and wildlife of the estuaries, except for migratory birds.

Benefits to bird watchers at Anahuac National Wildlife Refuge, Texas (upper photo
and clam diggers on the tideflats of Puget Sound, Washington (lower photo) cannot
be priced accurately in the market place.

VI. SUMMARY OF STATE COMMENTS

The majority of the Coastal and Great Lakes States responding to a request from the Bureau of Sport Fisheries and Wildlife for an expression of their views on the feasibility and desirability of establishing a nationwide system of estuaries or estuarine areas indicated that such a system was both feasible and desirable. However, their agreement with such a proposal was qualified by insistence that such a system remain under State jurisdiction and management. Additionally, the States expressed a need and intention to seek financial assistance so they can acquire, develop, and manage any estuarine areas of the system. Several of the States also would like assistance from the Federal government other than financial to help them to manage and protect estuarine areas.

Almost 75 percent of the States have experienced a need for protection through acquisition in estuary situations where legal controls were not adequate. Even with this experience only a little over half of the States favored Federal acquisition in such situations, and then mainly with qualifications. Some would limit Federal acquisition to situations where the State recognized the need but could not accomplish the acquisition. Others would accept Federal help in acquisition but would insist on State management.

Estuary development or management funds are available to the Coastal and Great Lakes States through general funds appropriated by the State legislature, fishing and hunting license fees, Pittman-Robertson and Dingell-Johnson funds, and other Federal aid monies.

Generally there is a shortage of available funds in most of the concerned States to adequately protect, develop, or manage their estuarine areas. Sources of additional funds proposed by the States include: recreational user fees or licenses; industrial use fees assessed to compensate damage resulting from industrial development and use; State general funds; Federal grant or cost-share funds; earmarked Federal funds derived from oil and mineral Outer Continental Shelf leases; and a national saltwater license with the fees derived therefrom to be made available to the States.

Viewpoints of the Coastal and Great Lakes States are quite varied regarding placement of emphasis on the protection of estuaries. Their opinions are about equally divided between unharmed and both unharmed and degraded areas. Several States favor a priority system which would include such things as giving special consideration to Federal lands under military control, emphasis on all areas with priority to pristine areas, or special emphasis to those areas where the economic input will result in the greatest return to users. Others feel the entire system of the Nation's estuaries and Great Lakes areas should be considered.

Environmental quality standards for estuaries are favored by most of the States, but a few believe that existing water quality standards and air pollution laws are sufficient protection for the estuaries. Since specific considerations are felt necessary by some of the States, close consultation with all States will be necessary if such standards are proposed. There is an awareness on the part of the States that there has been and continues to be a loss of valuable estuarine habitat. The consensus is that something must be done immediately to safeguard what is left.

Most of the States commented on the preliminary drafts of report materials sent to them. Many of them reaffirmed the need for Federal assistance in the protection of the Nation's estuaries. All were very helpful in their suggestions for improvement of the report.

III

Air Pollution

A. The Pollutants

Environmental Pollution: A Challenge to Science and Technology. Report of the Subcommittee on Science, Research, and Development to the Committee on Science and Astronautics, U.S. House of Representatives, Eighty-Ninth Congress, Second Session, Serial S, Revised August 1968. Pages 19-25.

Congressional interest and concern with air pollution has been relatively recent. Although Congress evidenced serious concern with air pollution in 1963 when it passed the Clean Air Act, approved December 17, 1963 and found in 77 United States Statutes at Large 392 and 42 United States Code Annotated section 1857, it was not until 1970 that Congress directed the federal government to itself establish national air quality standards and directly regulate certain types of pollution. This was accomplished by the Clean Air Amendments of 1970, approved December 31, 1970, and found in 84 United States Statutes at Large 1676 and 42 United States Code Annotated sections 1857a-1858. It is of course proper for the federal government to assume an active role since the "ambient air" is not confined with state boundaries.

Part of the welcome recent trend toward greater federal activity in this field is attributable to congressional staff studies like that which is reprinted below. The Subcommittee on Science, Research, and Development, chaired by Emilio Q. Daddario, heard eleven days of testimony from twenty eight witnesses and consulted with numerous experts in preparation for the writing of the report. The numbers in parentheses in the following excerpt from the report are references to pages in the printed transcript of this testimony, Hearings before the Subcommittee on Science, Research, and Development, of the Committee on Science and Astronautics, U.S. House of Representatives, 89th Congress, 2d Session, Volumes I and II, July 20, 21, 26, 27, 28, Aug. 3, 4, 9, 10, 11, 17, and September 19, 1966.

The report itself was prepared under the direction of Richard A. Carpenter, senior specialist in science and technology, of the Science Policy Research Division of the Library of Congress. Its concern was with the entire spectrum of environmental problems rather than with air pollution. Perhaps for this reason the excerpt reproduced below is unusually succinct in identifying and describing the major pollutants and their characteristics.

Needed Research—Air

The atmosphere has evolved from prehistoric times through a number of changes to its present, apparently stable, composition of essentially nitrogen and oxygen. Carbon dioxide is one of several minor constituents. There is no evidence of any contaminant remaining in air indefinitely. Rainfall and settling are rather efficient cleansing mechanisms. However, locally, and for brief periods of time, episodes of air pollution occur which are quite disturbing. In addition, air carries corrosive gases and dirt which do costly damage over wide areas.

Finally, polluted air is difficult to clean up at the user site; i.e. engine intakes, human gas masks, automobile or building filters. We live in the air much like marine life in the ocean. While it is possible to assure the purity of drinking water in the midst of pollution, it is not easy to breathe clean air unless the entire atmosphere in the vicinity is uncontaminated.

Goals for air pollution abatement are at present of the trend type. Being unable to set ambient air standards as yet, the objectives are expressed in terms of percent reduction from present emissions (see pp. 48, 586). Electric power generation and automobiles account for 83 percent of the air pollution due to combustion, so that air pollution is mainly an urban problem (see p. 276).

1. Internal combustion engine

The automobile engine uses air as a "working fluid" or expansion gas, and also some of the oxygen combines with fuel to produce heat. The cool walls of the combustion cylinder keep some of the fuel from burning. What does burn produces carbon monoxide as wel as carbon dioxide and water. The high flame temperature causes some of the nitrogen and oxygen in the air to combine into nitrogen oxides. Thus, a large volume of air is contaminated in passing through the engine and unless the pollutants are removed before going out the tailpipe, the moving car scatters wastes into the atmosphere. Once there, they may disperse but still maintain a substantial concentration in central cities, especially when low wind conditions or temperature inversions prevent circulation. Photochemical smog is a product of the action of sunlight on undispersed air pollutants.

The increasing concentration of automobiles means that city dwellers have no real hope of regaining clean air as long as the internal combustion engine is used for personal transportation devices. Even if advanced abatement devices (by present standards) are used, the total pollution load will return to present levels in 1980 (see p. 348). Nevertheless, there was agreement that no effort should be spared in improving this powerplant as much as possible, because alternative propulsion means are far off in time and quite expensive (see pp. 261, 328). Electric power for cars, trucks and buses should be encouraged but appears to be applicable only for a small portion of the vehicle population (see pp. 800, 847, 853). Gas turbine engines are now in development which may yield higher combustion efficiency and will not require lead additives in the fuel.

A systems approach to the internal combustion engine problem is underway by the Bureau of Mines (with important assistance from the American Petroleum Institute). All factors affecting the end results from photochemical smog will be examined (see p. 304). Cleaner air is possible, if not clean air.

The challenge to research is that devices or engine modifications costing as much as $50 per car may be required on a nationwide basis, totaling several hundred million dollars per year in investment. This annual outlay should be a great motivation for R. & D. to cut the cost. A second challenge from automobile pollution is to pin down the hazards to humans so that judgments can be made clearly on esthetic, annoyance, or health bases. All are sufficient to justify some abatement activity in some locales, but judgment will soon be necessary on nationwide, expensive programs and the scientific facts are lacking (see p. 359).

2. Diesel engine

The repugnant and annoying odor from diesel engines is, ironically, accompanied by lower emissions of the smog components than obtained from the internal combustion engine (see p. 309). The odor problem is particularly bad, however, because large-sized diesel engines are used in buses and trucks which concentrate on city streets. The pedestrian or passenger cannot get away from such pollution. Buses of municipal transit companies should be especially monitored for they add pollution unnecessarily at the very worst location—the crowded city street. Strict enforcement of existing smoke and fume laws would lead diesel operators to better maintenance which would accomplish some improvement. A recent suggestion of perfuming the exhaust to mask the odor seems a step in the wrong direction as this merely introduces a new contaminant to the air. Fundamental combustion research is necessary to gain full benefit from the diesel engine.

3. Nitrogen oxides

Whenever air encounters high temperatures (in electrical arcs, internal combustion engines, combustion, steel making, etc.) the nitrogen and oxygen combine to a small extent into oxides of nitrogen (see pp. 341, 448). These compounds are also produced in nitric acid and other chemical manufacturing processes. The brownish fumes of nitrogen dioxide may be seen at chemical plants even though extremely efficient control techniques are used. This is because of the very high optical absorbtivity of the gas—a small concentration stains the air. More importantly, oxides of nitrogen are irritants in breathing and have a chemical activity in smog, combining with hydrocarbons and oxygen to produce the eye irritating effects.

Whereas completing the oxidation of hydrocarbons and carbon monoxide in automobile exhaust holds some promise for their amelioration, such treatment does not effect oxides of nitrogen.

It is likely that the irritating factor alone will require nitrogen oxide abatement in the future for some cities and industrial plants. The reduction of hydrocarbons from automobiles might actually increase the concentration and possibly the net annoyance of nitrogen oxides. Intensive research on the prevention of their formation and on means of removal from gas streams is needed.

4. Lead

Lead is a toxic heavy metal, of course. Much has been done to avoid its ingestion by humans while making use of the properties of lead and its compounds. It is natural that there is concern for the lead which is emitted into the air from combustion of gasoline containing tetraethyl lead as an antiknock agent. One hypothesis is that a substantial increase in environmental lead levels will lead to poisoning since the human system would be accommodated only to some lesser natural concentration. Apart from acute lead poisoning (from painted toys, pipes, etc.) which is well controlled and understood, there may be insidious effects of long-term, low-level exposure to lead (see p. 322).

This line of thought has stimulated some considerable research and surveys of lead in the air. There are yet insufficient data to resolve the issue, and unfortunately, many scientists who should remain objective are becoming polarized as to their opinion and thus less useful to policymakers.

Stopping the use of tetraethyl lead in gasoline would be one resolution of the problem. Reduction in amount might be helpful as would modification of engines to use lower octane fuels. Unless an equally cost-effective substitute were available (and competitors have tried to find one without success for 40 years) the additional cost to the public would be $700 million per year for each additional cent per gallon required to otherwise obtain present gasoline quality (see p. 798).

So the challenge to science and technology is again twofold: clearly identify the tbreat, and find a low cost solution if it is needed.

5. Carbon dioxide

The amount of carbon dioxide (CO_2) in the atmosphere is about 300 parts per million (0.03 percent). The concentration varies from place to place particularly because of the local production of varying amounts of CO_2 by oxidizing carbonaceous materials via animal respiration or fuel combustion. Green plants use carbon dioxide in photosynthesis. The ocean absorbs carbon dioxide and it is eventually deposited in carbonate minerals. The amount in the air is increasing over past values because civilization, in a few centuries, is buring gas, petroleum, and coal which were formed over many eons of time. Counteracting forces which might restore a lower equilibrium concentration are too slow.

Even the combustion of all the fossil fuels in the earth would not raise the concentration of carbon dioxide to a toxic level. But it is considered a pollutant because of its absorptive effect on the transmission of heat radiated upward from the earth by the atmosphere. The more carbon dioxide, the less heat is passed back into space and the warmer the earth—thus the term "greenhouse effect" (see pp. 288, 353).

There is some evidence that smoke, dust, and haze are also increasing with civilized activity. These particles in the air might increase the reflectivity (albedo) of the earth. Less energy would reach the surface and a counteracting trend to the greenhouse effect would be obtained.

The net result is that we do know the carbon dioxide concentration has increased and is increasing. Once again we do not know what this means or what to do about it if action is called for. Studies underway may give a more complete picture on a global basis by 1975 (see pp. 108, 451). Research on these long-term atmospheric problems will be valuable to weather technology as well as pollution.

6. Episode prediction

Since air most often does move and become cleansed by diffusion and rain, the most damaging air pollution often occurs in brief periods of time when the pollutant load is high and the atmosphere is stagnant.

As an alternative to expensive abatement efforts to maintain very low contamination levels every day, it might be satisfactory to cut down emissions only when meteorological conditions are unfavorable (see pp. 312, 368). The ability to predict conditions of weather and pollutant combinations which produce severe episodes depends on careful determination of alarm levels as well as weather forecasting. If successfully demonstrated, episode prediction would save a lot of money and damage in certain urban areas; providing of course, that the willingness to sharply curtail activities on occasion was well established in the public mind.

7. Coal—Electric power generation

Urban air pollution is often more of a problem because of fossil fuel burning steam-electric generating plants located in or nearby many cities. Present investment and the costs of long-distance transmission mitigate against removing powerplants from the cities. There is insufficient low sulfur fuel in the eastern part of the United States to substantially reduce the sulfur dioxide problems. The rise in temperature of water and air which absorbs the waste heat of powerplants is often called thermal pollution.

Electricity consumption is growing faster than the population. In some ways this is an asset to pollution abatement; for example, its use in place of small gasoline engines or space heaters can decrease the pollution load to the air. But the net effect may be lessened if the powerplant increases in size and pollution emissions mount at its location. Hydroelectric power should be expanded to the fullest extent possible, but there is not enough additional potential to effect the pollution problem.

Location of coal-fired plants away from urban areas is possible for new installations. Very high stacks and high gas velocities may be used to place pollutants above inversion layers and far enough from the ground so that diffusion can keep concentrations low. (See p. 881.) Such stacks are no more a hazard to air traffic than TV towers, but they are a rather costly abatement method. The Tennessee Valley Authority provides an opportunity for Federal funds to lead the way in experimentation and in demonstrating new techniques for reducing pollution from electricity generation. (See p. 434.)

8. Sulfur dioxide

Sulfur is a chemically bound component of most fossil fuels—it cannot be removed short of destroying the fuel molecule (see pp. 303, 454, 268, 911). When the carbonaceous fuel oil or coal is oxidized the sulfur is converted to gaseous sulfur dioxide which subsequently forms droplets of sulfuric acid in moist air. These droplets are very damaging to the respiratory system. In particular when combined with small particle pollution and stagnant air, severe damage of the type noticed in the Donora, Pa., New York, and London episodes may occur (see pp. 304, 522). It is worth noting that these episodes (only a few in a century) have a perhaps exaggerated impact on thinking in air pollution.

Research in the past to remove sulfur dioxide from stack gases has not resulted in economically feasible processes (see pp. 48, 414–419). The objective was to recover a valuable sulfate or elemental sulfur, the sale of which would offset the cost of recovery. This economic analysis should include some subsidy factor corresponding to the decreased damage in the environment when a sulfur dioxide free atmosphere was obtained. Then the recovery process might become a substantial source of this valuable element (see p. 827). The TVA alone emits sulfur equivalent to one-eighth of the annual national production. Pollution control can be coupled with resource conservation. The industrial firm which develops a workable process should find a ready market. But economics will be demonstrated only on a large-scale installation. A possibility for Government-industry cooperation in this phase is now under consideration.

9. Nuclear power

The Atomic Energy Commission has done an excellent job of restricting waste products from nuclear-electric power generation (see pp. 452, 899). But there still are wastes and the further consideration of how to handle radioactive materials over a long period of time in increasing quantities is not yet solved. Nuclear powerplants reject somewhat more waste heat to cooling water than do coal plants but this is a matter of engineering design choice.

Burial, retention in containers, or deep sea disposal of nuclear wastes with very long half lives has so far been effective. Reprocessing of spent fuel elements takes place away from the powerplant so that central city location of nuclear electric generators is possible. The operating safety of the plants is predicated on a "zero defect," or no-accident, statistical concept which is not subject to verification except by experience. It has been calculated, for whatever it is worth, that natural radionuclides in coal would result in more radioactivity in the stack gases than the amount escaping from nuclear plants (see pp. 330, 332). Nuclear energy does not result in carbon dioxide production.

If the technology for handling spent fuel element wastes (in greatly increased amounts of low-level, long half-life character) can satisfy the strict criteria for protecting the environment from radioactivity, nuclear-electric energy seems a very promising means of reducing urban air pollution (see pp. 818, 900).

10. Fossil fuels

There are enormous reserves of coal and even with an increasing rate of usage they will not be exhausted for hundreds of years. The present investment in coal-fired electric generating plants, and the realistic cost differential between this and other energy sources, precludes any rapid demise of the coal industry (see p. 908). Further elucidation of the threat from carbon dioxide might alter this picture.

Petroleum is being discovered at about the rate at which it is consumed, but estimates of reserves indicate that supplies could be exhausted in 50 to 200 years. Natural gas, tar sands, and oil shales also seem to be more limited than coal.

There is some suggestion in the testimony that fossil fuels, particularly petroleum hydrocarbons, should be preserved for a higher use in petrochemical manufacturing or for portable powerplants rather than be burned up for electric power (see pp. 349, 796, 822, 824). To consider this concept, some assumption must be made as to how far into the future the responsibilities of today's society can logically be extended. The answer would seem to depend on how clearly and precisely the natural resource, energy, and population picture can be predicted. In fact, of course, the situation is so complicated with political and social forces that even a few decades ahead there are great uncertainties.

The best course of action today is to try to avoid precipitating irreversible environmental changes or those which might take hundreds of years to straighten out if the need arises. In the meantime the same two courses of scientific activity noted before should be expedited: learn much more about global processes in air and water; and acquire more technological tools and alternatives to deal with consequences. Such technology would include controlled nuclear fusion, solar power, gravitational or tidal energy, weather modification, etc.

11. Particulates

The example of Pittsburgh, Pa., shows that smoke or soot (unburned carbon particles) fly ash, and cement plant dusts can be substantially controlled. However, there are several problems which remain. First, more imaginative uses could be found for the collected solids, as they still must be disposed of some way. Extraction of valuable metals or the use of fly ash in concrete are examples.

In gas cleaning techniques, there are still improvements to be made in removing extremely small particles (which also need research as to their physiological effect) and in achieving very high percentage removal (important for large installations) (see pp. 326, 516).

Particulates distributed high in the atmosphere may have significant effects on the weather, changing the reflectivity of the earth and serving as nuclei for condensation. Small particles may rise from industrial and urban air pollution, or they may be injected at high altitude by rocket exhausts and jet aircraft (the supersonic transport will operate at 70,000 feet). Air pollution is thus not confined just to the biosphere but extends to the fringes of the atmosphere.

B. Health Effects

<u>Air Quality Criteria</u>. Staff Report Prepared for the Use of the Subcommittee on Air and Water Pollution, Committee on Public Works, United States Senate, July 1968. 90th Congress, Second Session. Pages 33-55, 59-63.

Clean air is sometimes thought of as an amenity. It should be seen for what it is: a necessity of life. The selection reproduced below clearly indicates that polluted air can and is causing serious health problems, including death. Perhaps most disturbing are the indications that the continuous low level pollutants which do not cause acute episodes may nevertheless have a serious cumulative impact far exceeding that caused by the isolated but more dramatic episodes of air pollution.

Senator Edmund S. Muskie, the Chairman of the Subcommittee on Air and Water Pollution, commissioned this report as part of the continuing program of research and investigation of this Subcommittee. The original report, written by the Subcommittee staff--especially Mr. Richard D. Grundy, consists of sixty nine pages, plus introductory material. The material omitted from the selection reprinted below concerned mainly the methodology to be used in developing air quality criteria.

III. HEALTH EFFECTS OF AIR POLLUTION

3.1 Introduction

Studies of air pollution episodes have demonstrated the immediate effects of changes in air pollution. Daily levels of air pollution measured in terms of levels of black suspended matter and sulfur dioxide have been correlated with indices of mortality and morbidity. The effects, which are most noticeable in individuals suffering from cardiac or respiratory conditions, are observed within a few hours of a deterioration in atmospheric conditions, and are frequently evident even with quite small increases of pollution.

Concurrently, there exists the condition of chronic exposure of the general population to low levels of contaminants which may have carcinogenic, mutagenic, or tetratogenic effects. At the present time, however, too little is known about this subject to permit precise classification of such contaminants.

Limited studies suggest that indirect effects occur because of air pollution. (Graham 1932.) First, air pollution obscures natural light which apparently reduces resistance to infection and retards recovery from illness.

Second, the resultant reduction of light and sunshine may cause physiological effects which are no less serious than the direct physical effects.

When reviewing the types of effects which have resulted from air pollution exposures it has been helpful to classify the extent to which greater morbidity and mortality are a direct consequence of:

(*a*) Short-term massive exposure during air pollution episodes;
(*b*) Continued exposure involving nonepisodic exposures;
(*c*) Combinations of chronic and episodic exposures.

When considering the resultant indices of health and welfare effects it is requisite that the observed effects be associated with particular contaminants or combinations of contaminants. Many types of effects have been postulated under the above classifications.

This section is intended to serve as a brief review of current thinking on air pollution as it contributes to disease mortality and morbidity. No attempt has been made to develop air quality criteria. The development of air quality criteria in accordance with the provisions of the Clean Air Act is the responsibility of the Department of Health, Education, and Welfare. This section outlines a framework for evaluation of the completeness of the scientific evidence on health and welfare effects.

3.2 Health Effects During "Air Pollution Episodes"

Mortality during air pollution episodes during the last two decades provides ample evidence of air pollution exposures which result in the death of susceptible individuals in the general population. While the results of these exposures to individual peaks are dramatic, the medical effects of continued exposure to polluted urban atmospheres are probably more important (Martin 1964).

Air pollution episodes are characterized by high levels of air pollution over relatively short periods of time. The resultant increased levels of exposure are accompanied by death or acute illness. It should be recognized, however, that mortality during air pollution episodes tends to reflect the effects of the environment in the terminal stages of the disease rather than on its initiation or evolution (Holland and Reid 1965.)

The most widely publicized of these episodes have occurred in the heavily industrialized Meuse Valley of Belgium in 1930; in Donora, Pa., in 1948; in New York City in 1953 and 1966; and in London in 1952 and 1962. During the London smog of 1952, 4,000 more deaths occurred in that city than would normally have happened during a similar period of time. These episodes, and others which have been studied, are summarized in table 2.

TABLE 2.—REPORTED DISEASE MORBIDITY AND MORTALITY OCCURING DURING AIR POLLUTION EPISODES

Year and month		Location	Reported excess deaths	Reported illness	References
1873	Dec. 9–11	London, England			Logan (1953), Scott (1953).
1880	Jan. 26–29	do			Logan (1953).
1892	Dec. 28–30	do			Logan (1953).
1930	December	Meuse Valley, Belgium	63	6, 000	M. Firket (1931), J. Firket (1936).
1948	October	Donora, Pa	20	6, 000	PHS (1949).
1948	Nov. 26–Dec. 1	London, England	700–800		Logan (1949, 1953).
1952	Dec. 5–9	do	4, 000		Ministry of Health (1954).
1953	November	New York, N.Y			Greenburg, et al. (1962A).
1956	Jan. 3–6	London, England	1, 000		Martin (1961).
1957	Dec. 2–5	do	700–800		Bradley, et al. (1958).
1958		New York, N.Y			
1959	Jan. 26–31	London, England	200–250		Martin & Bradley (1960).
1962	Dec. 5–10	do	700		Scott (1963).
1963	Jan. 7–22	do	700		Do.
1963	Jan. 9–Feb. 12	New York, N.Y	200–400		Greenburg, et. al. (1967).
1966	Nov. 23–25	do			Glasser, et al. (1967).

Typically, air pollution episodes are characterized by an immediate rise in mortality. While there is some disagreement among investigators as to the onset of the observed health effect in general, it occurs between the first and third days of the episode. A comparison between the Meuse Valley, 1930; Donora, 1948; and London, 1952, showed some similarities and some differences (WHO 1961):

"In all three episodes the induced illnesses involved chemical irritation of the eyes, nose, throat, and respiratory tract, with a consequent effect upon the heart in susceptible persons as a result of the respiratory tract irritation. There were two important differences between the health effects in London as compared to those in the Meuse Valley and Donora. These were (a) the time during the episode that deaths occurred, and (b) the ratio of the number who became ill to the number who died. In the Meuse Valley and Donora outbreaks, deaths did not begin until after the pollution had been collecting for 2 days. In the London episode, however, the excess deaths began on the first day, and even within the first 12 hours. In regard to the ratio of the number of persons who became ill to the number who died, it is to be noted that if one considers the total number of persons who were ill during the episode in Donora, the proportion of illness frequency to deaths becomes about 300 to 1. If only the very severely ill are considered the proportion turns out to be about 75 to 1. For the Meuse Valley incident the data do not permit an estimation of a corresponding ratio, since numerical data on sickness frequency were not reported, although the number of deaths was known. Competent authorities on the London episode of 1952 noted that morbidity rates did not increase in a proportion which one might have expected with such a known mortality (Ministry of Health 1954)."

Air pollution episodes are further characterized by an abrupt fall in death rate following the termination of the event. During the 1952 London episode cardiovascular deaths fell abruptly as soon as the fog cleared and respiratory death rate followed a few days later. This suddenness of the deterioration in atmospheric conditions is remarkable (Martin and Bradley 1960).

Past studies of air pollution have tended to seek effects among those population members suffering from bronchitis and emphysema. Discrepancies have been noted that suggest that future investigations should consider a wider population base and the relative importance of cardiovascular and respiratory cause of death.

Studies of the effects of air pollution on the respiratory tract must consider the fact that chronic respiratory diseases, such as acute obstructive ventilatory disease, chronic bronchitis, pulmonary emphysema, and bronchial asthma, all may involve the heart. The cardiorespiratory system functions as a unit with one part of the system compensating for failure of other parts.

In order to understand fully the impact of air pollution episodes on health it is necessary to study simultaneous stresses such as cold weather or endemic diseases, which often seem to accompany air pollution episodes. Such biological stresses have been reported in connection with selected episodes. Ultimate epidemiological investigation will need to delineate the relative effects of these accompanying stresses. Air pollution episodes have been noted only during the winter, and with the onset of spring this association with sickness and death disappears (Lawther 1958, Martin and Bradley 1960, Martin 1961). Nevertheless, the pollution of the atmosphere by the biologically harmful substances which precipitated the episode should be avoided to the maximum extent possible.

Generalization on the age groups particularly affected by episodes is not possible. Age specific death rates have been determined for selected episodes, but it is obvious that work needs to be done to improved reporting practices in future studies. Emphasis should be placed in particular on infant death rates where data is severely lacking. It has been reported that the age groups particularly affected by episodes are usually the very elderly, and to a lesser extent the later middle ages and infants under 1 year. Detailed analyses of the London episode of 1952, however, showed that percentage increases in mortality were approximately the same in all age groups. This apparent inconsistency deserves further study.

Both the clinical and post mortem findings in connection with the December 1952 fog were consistent, with pollution having a nonspecific effect on persons already suffering from serious respiratory or cardiac lesions. These individuals represent the susceptible group as far as death is concerned. Whether the disease has resulted from chronic air pollution exposure or not, the exposure during the episode represents the final intolerable stress on already impaired organs.

Analysis of several London air pollution episodes supports the hypothesis that the critical air pollution level in London is reached when daily concentrations of black smoke reach 2.0 milligrams per cubic meter of air and 0.40 ppm (1.14 milligrams per cubic meter) of sulfur dioxide (Scott 1956, 1958, 1959, and 1963). While this level represents an unusually high atmospheric concentration, there is little proof that the presence of sulfur dioxide at this level is harmful to humans (Greenburg et al. 1967). Scott (1959) stated that evidence so far does not preclude smoke being the determining factor rather than a combination of the two pollutants.

While it has not been established that either particulates or sulfur dioxide are the causative agents of the adverse effects on health associated with increased pollution, it is reasonable to assume that the concentrations of the causative agents are related to one or both of these contaminants; their levels thus serve as an index, which may be related more precisely to the level of an identified active agent (Clifton 1967).

Martin (1961) was more explicit when he stated that from the chemical point of view both levels of black suspended matter and sulfur dioxide provide a crude, convenient index of general air pollution until more is known of the actual toxic substances.

Air pollution episodes offer the ultimate possibility of differentiating the role of individual contaminants in the cause-effect relationship involving man. Episodes could be studied in those cities that succeed in cleaning up their air in whole or in part, and these results compared with those areas in which little or no improvement is made. However, discretion should be exercised in comparing the results from different cities directly, as London is as different from New York, as New York is from Los Angeles, or London is from Sheffield. It is unrealistic to assume that comparable levels of particulate and sulfur dioxide result in comparable total air pollution exposures under any of these conditions.

A number of air pollution episodes have been studied in retrospect; those of particular interest are summarized herein.

1930 MEUSE VALLEY, BELGIUM

Between December 1 and 5, 1930, there was a heavy fog which covered the Meuse Valley, Belgium. Several hundred individuals were severely attacked with respiratory symptoms and 63 died (J. Firket 1936). Results were such that "there is no reason to believe that only the ill and the aged were seriously affected by the toxic attacks" (M. Firket 1931). All the sick people felt retrosternal pain, and fits of coughing and dyspnea, and some had asthmatic attacks. Autopsy reports indicated that the noxious substance was inhaled during the last few hours of life. Respiratory symptoms disappeared on the sixth day when the fog abated. Chemical analyses found 30 substances polluting the atmosphere, with maximum concentrations being reached by carbon monoxide, carbon dioxide, nitrous gases, sulfur dioxide and hydrogen fluoride. The investigator reported that only by the action of sulfur pollution could the simultaneous illnesses occurring be explained. Pathological finding also favored sulfur dioxide and sulfur trioxide as the offending gases (J. Firket 1936).

1948 DONORA, PA.

During the latter part of October 1948, Donora, Pa., a town of 13,000 population, experienced a large number of acute illnesses and 20 deaths during a heavy smog. A detailed study (Schrenk, et al. 1949) showed that the cause of the episode was an accumulation in the atmosphere of chemical irritants; this accumulation developed under conditions involving a prolonged weather inversion. The investigators deduced that no one agent was responsible. It was concluded that the observed effects were due to a chemical agent (possibly sulfur dioxide) plus particulate matter. Because of lack of knowledge about the toxic effects of low concentrations of the irritant gases, this could not be said with certainty. Detailed studies of disease morbidity, however, were conducted 10 years after the episode. (Sec. 3.3.)

1952 LONDON, ENGLAND

Possibly the most extensively studied air pollution episode to date is the great London fog of December, 1962. A sharp increase in both respiratory and cardiovascular deaths was observed during the first day of the fog. While cardiovascular deaths fell abruptly as soon as the fog cleared, respiratory deaths continued at a high level for a few days before falling (Ministry of Health 1954).

A detailed analysis of the excess deaths occurring for the first week of the episode (December 7 to 13, 1952) showed a three-fold increase in deaths for babies (4 to 52 weeks) and adults (greater than 55 years of age) as compared with the previous 3 weeks. For all other ages there was an approximately two-fold increase in deaths (Scott 1953).

During the first week of the episode death rates from bronchitis were 10 times as high as for the previous 3 weeks; for pulmonary tuberculosis, 4.5 times as high; for other respiratory disease, 6 times as high; for lung cancer, 2 times as high; and for disorders of the heart and circulatory system, 3 times as high (Scott 1953).

Detailed investigations were made of 1,280 post mortem reports of persons who had died suddenly either before, during, or immediately after the episode. In these post mortem reports no cases of sudden death were found which could not have been explained by previous respiratory or cardiovascular lesions.

The need for large population groups in order to study the acute effects of air pollution is apparent when considering this episode. During this episode in the greater London area (population 8,205,000) the excess death rates were equivalent to 12.2 per 100,000 population per day of fog. During the fog of January 1956 they were 3.0 per 100,000 (Martin 1961). While it is not requisite to have a large population group in order to exhibit an air pollution episode (e.g., Donora), it is obvious that large population groups exhibit effects for smaller changes in air pollution levels. Further, large population groups do provide detail useful in determining specific causative agents (e.g., contaminants) through comparison of successive episodes.

It is important to note that the maximum daily mean sulfur dioxide concentration during the London 1952 episode was 1.3 parts per million. The generally accepted toxic limit for an exposure of several hours is 10 parts per million (Wilkins 1954). Therefore, it was unreasonable to assume that healthy persons were injured solely by sulfur dioxide exposure. On the other hand it is possible that those individuals with previous histories of respiratory illnesses who were affected during air pollution episodes may have been affected solely by sulfur dioxide.

It has been established that during air pollution episodes both sulfur dioxide and particulate concentrations in the atmosphere reflect comparable increases. It can be assumed that a similar relationship exists between sulfur dioxide and other contaminants. Estimated concentrations of carbon monoxide, carbon dioxide, and possibly sulfuric acid (although the evidence here is very conjectural) may each have approached or reached levels at which slight physiological effects are said to occur (Wilkins 1954). The possibility of a synergistic effect between sulfur dioxide and sulfuric acid, or particulates and sulfur oxides was suggested, also.

In the limited investigation of morbidity associated with the London fog of December 1952 (Ministry of Health 1954), information on sickness was collected from as many sources as possible. Sources included national insurance sickness claims, applications for admission to hospitals, pneumonia notification, and observations of general practitioners. Analysis of this information did little more than confirm many of the conclusions drawn from the mortality results. There was some indication that the proportionate increase in numbers of cases of illness (morbidity) was not as large as the increase in deaths (mortality). Also, the morbidity rates did not rise as suddenly as the mortality rates in the early days of the fog.

1953 NEW YORK CITY

In the New York air pollution episode of November 15 to 24, 1953, the observed increase in the number of deaths was distributed generally among all age groups (Greenburg, et al. 1962A). An analysis of emergency clinic visits to four major city hospitals served as an index of morbidity. Greenburg, and colleagues, (1962B) studied case records of upper respiratory, cardiac, and asthma clinic visits. Statistically significant increases were observed for upper respiratory and cardiac illnesses; however, the study failed to reveal any effects of air pollution on the number of asthma clinic visits. Measurement of air pollution levels was limited to sulfur dioxide and smoke shade determinations as the two most reliable measures of atmospheric contaminants (Greenburg, et al. 1962A).

1957 LONDON, ENGLAND

In the air pollution episode of December 2 to 5, 1967, an estimated 800 to 1,000 excess deaths occurred (Bradley, et al. 1958). As in previous episodes, deaths occurred primarily in the first day. There was also a marked early increase in the number of acute illnesses necessitating hospital admission. The observed increase in deaths and cases of sicknesses occurred before the ambient air levels reached sulfur dioxide and particulate concentrations which were previously regarded as critical. The investigators suggested the need for further investigations into critical levels of air pollution.

1958–59 LONDON, ENGLAND

During the winter of 1958–59 there were six or seven severe air pollution episodes in London accounting for 200 to 250 excess deaths (Martin and Bradley 1960). The sudden rises in mortality during the episodes suggested that many of the deaths occurred among people who were already seriously ill with respiratory conditions or suffering from cardiovascular lesions. Analysis of the deaths showed significant correlations between levels of black particulate matter and mortality (all causes), and mortality *(bronchitis)*. Unfortunately, data on deaths from cardiac diseases were not available. Among the more important observations was the fact that the number of deaths was found to fall on the second day of the episode even though the air pollution levels remained high (Martin 1961). The possible stress of viral infection as a contributing factor was noted.

A significant positive association was seen between the daily number of deaths and black particulate matter, a slightly lower association with sulfur dioxide, and a negative association with visibility. Martin mentioned that from the chemical point of view both suspended matter and sulfur dioxide are but crude indices of air pollution and until more is known of the actual toxic substances in the air they but provide a convenient and ready method of measuring air pollution.

1962 LONDON, ENGLAND

In December 1962, a particularly severe London fog occurred, comparable in many respects with that of 1952 (Min. of Health 1963, Scott 1963). The fog did not appear to be so dense, so uniformly persistent, or so irritant to the nose and throat. It lasted approximately 80 hours as compared with 96 in the earlier episode. Sulfur dioxide air pollution levels were comparable, but smoke pollution levels were associated with the episode as compared to 4,000 in 1952.

Caution has been stressed in attempting to draw conclusions from comparison of these two events (Min. of Health 1963, Scott 1963, Lawther 1965). The temptation to the epidemiologist to ascribe this improvement to the fall in smoke is great. However, factors other than air pollution may have played a part in determining the observed level of mortality. In particular, newspaper and radio publicity had alerted the population so that susceptible people suffering from cardiac and respiratory diseases could take precautions. Lawther (1965) emphasized that there is no assurance that the structure of the population at risk was the same. Before the relative effect of the decrease in smoke can be evaluated these factors must be considered. In addition, the effects of antecedent epidemics of infectious disease must be studied and lung series of stimuli and responses observed before firm conclusions can be reached.

1963 NEW YORK, NEW YORK

The New York City episode of January and February 1963, afforded an opportunity to study the effects of three stresses at the same time: air pollution, viral infection, and cold weather.

During the entire month of January the average sulfur dioxide level was 0.34 parts per million, and 0.46 parts per million from January 29 to February 12. A maximum average hourly reading of 1.50 parts per million was reached on February 6 and sustained for a period of 4 hours. While little proof exists that the presence of sulfur dioxide at these levels is harmful to humans, these levels nevertheless represent unusually high concentration for urban atmospheres (Greenburg, et al. 1967). During the critical period, 7 out of 15 days, or 47 percent of the days, had "dark smoke shade readings" of 4 Coh units or higher.

During the critical period 809 excess deaths took place from all causes among individuals 28 years of age and older. Although children and infants represent possible susceptible groups they were not considered. The statistically significant increases were attributed to influenza, pneumonia, vascular lesions affecting the central nervous system, diseases of the heart, and "all others," and excluded accidents, homicides, and suicides.

When the effects of other biological stresses than air pollution were removed, an estimated 200 to 400 of the 809 excess deaths experienced were attributed to air pollution. This increase in mortality centered in the older age groups (45 to 64, and 65 and over).

3.3 Health Effect Following Air Pollution Episodes

Theoretically, disease morbidity provides a more sensitive index of air pollution effects than mortality. However, as discussed in the previous section of this report, limited data are available for air pollution episode periods. Further review of the literature reveals a single study of disease morbidity following an air pollution episode.

1948 DONORA, PA.

A very interesting study of disease morbidity was carried out 10 years after the October 1948, air pollution episode in Donora, Pa. (Ciocco, et al. 1961). Early in 1957, mailed questionnaires and personal interviews contacted more than 99 percent of the 4,092 individuals who were residents in 1948. The question was posed as to the extent to which subsequent health experience was related to complaints during the episode.

Questions were asked on the prevalence and incidence of certain chronic conditions as well as certain cardiorespiratory symptoms. The findings were essentially that persons reporting acute illness at the time of the episode subsequently demonstrated higher mortality and morbidity rates than persons with mild complaints at the time of the episode. The differentials in subsequent illness were narrowed considerably when persons who reported chronic cardiorespiratory illness which antedated the 1948 episode were removed from the comparisons. This suggests that the episode aggravated an existing respiratory infection producing a more complex or critical form of disease.

3.4 Chronic Exposure and Disease

The medical effects of chronic exposure to polluted urban atmospheres are more important to the community than the immediate results of air pollution episodes. These effects are manifested mainly on the respiratory system. There is ample evidence that air pollution is an etiologic agent in the occurrence or aggravation of acute nonspecific respiratory disease, chronic bronchitis, lung cancer, pulmonary emphysema, and bronchial asthma. Further, there are indications that the severe cardiorespiratory disease seen in middle life is the end result of exposure to an environmental background or to personal habits that began to have an effect even in childhood (Reid 1964A, 1966).

The most intriguing evidence of the possible serious long-term adverse effects of environmental exposure during childhood is from observations on the lung cancer death rates among British migrants. Quite consistently, such studies have shown that emigrants from the United Kingdom experience death rates from lung cancer above those of their contemporaries in the land of adoption and below those in their home country. This excess over the prevailing rates in the new country cannot be explained by the number of cigarettes smoked (Haenszel 1961). These findings suggest that exposure to some factor or factors in the British environment in early life, perhaps air pollution, predisposes the individual to the later development of a more serious form of respiratory disease. The differences cannot be accounted for on the basis of diagnostic habit. The simplest explanation is that their removal to the United States is followed by a reduction in symptoms and an arrest in the progression of bronchitis and its allied diseases. On the other hand, it is also possible that structural lung damage received early in life, particularly in the British urban environment, may remain to cause the residual excess observed in lung cancer above the level found in the native born (Reid, et al. 1966).

D. D. Reid (1964A) has generalized from British experience that the ill-effects of air pollution seen in children must be followed by or be associated with an epidemic increase of chronic lung disease in adults. The question is whether, in the industrial areas of Western Europe and the United States, a similar situation exists but is as yet unrecognized.

CAUSATION

The variability of response of the "normal" individual renders quantitative investigation into disease causation of air pollution time consuming and complex. Not all healthy individuals react to the same concentrations of irritants and this variation may be ascribed to "normal" factors such as age, sex, etc. Also there are population groups which may be considered susceptible to air pollution. These groups may include the very young and the very old, also the bronchitis patient whose disease is in part an attempt by his body to protect his lungs against further irritation by secreting more mucous than is "normal." In some phases of his disease he might therefore be able to cope with more irritant than his counterpart. In the advanced stages of the disease, however, a strain may be placed on his heart and lungs to the extent that death occurs. Whether the effect is indirect or direct, the end result is the same and air pollution abatement and control is indicated. The requirement to take action will not ordinarily allow us time to unravel the whole causation chain. A few links may have to suffice, depending upon the circumstances.

The investigator considering causation must inquire how and by what mechanism these effects are produced. In this investigation of causation consideration must be given to several aspects of association which include strength, consistency, specificity, temporality, biological gradient, plausibility, coherence, and analogy (Hill 1965).

Strength

By far the most significant aspect to be considered is the strength of the association. While emphasis must be placed upon the strength of an association, care must be exercised not to dismiss a cause-effect hypothesis merely on the grounds that the observed association appears to be slight. An appropriate example is the prospective studies into air pollution as it relates to lung cancer. The fact that lung cancer morbidity and mortality rates show only a small correlation with air pollution, but are quite closely correlated with smoking (Reid 1958), does not in any sense preclude air pollution as a causative factor in the disease etiology. Rather the implication is that air pollution adds to cigarette smoking as a biological stress or insult to the respiratory system. Although smoking is the dominant factor there remains enough room for an "urban-factor" which includes air pollution.

Of course there does not have to be one urban factor any more than there is one cause of lung cancer or chronic bronchitis. The association between lung cancer and air pollution has not yet been proven. Also, the specific mechanism is still to be elucidated.

Consistency

Consistency of observed associations was the second feature singled out by Hill for consideration. Namely, is an observed association repeatedly observed by different persons, in different places, circumstances and times? This requirement is of particular import today when many alert minds are at work postulating environmental associations between air pollution and disease.

In some cases a hypothesis may appear conclusive on the basis of customary tests of significance, but whether a true hazard has been revealed or chance is the explanation may sometimes be answered only by a repetition of the circumstances and observation. The same results from precisely the same form of inquiry invariably will not greatly strengthen the original evidence, although a great deal of weight can be placed upon similar results reached in quite different ways (e.g., prospectively and retrospectively).

Take, for instance, the question of the role of sulfur dioxide during air pollution episodes. Air pollution levels have traditionally been measured in terms of concentrations of particulates (smoke) and sulfur dioxide. During successive air pollution episodes a correlation has been observed between disease mortality and morbidity and levels of particulate and sulfur dioxide. Research studies have not provided evidence that sulfur dioxide and sulfuric acid at the concentrations present in the atmosphere have a harmful effect on man. Repeated laboratory studies have failed to show any deleterious effects even at comparable levels to those found during air pollution episodes. Nevertheless the possibility, or even probability, exists that sulfur contaminants may have a harmful effect when associated with other contaminants (e.g., particulate). Although no clear evidence of this exists, it is readily apparent that sulfur dioxide correlates with mortality and morbidity during air pollution episodes.

While these sulfur compounds undoubtedly merit the attention they receive, an approach based upon quantities of pollutants without consideration of toxicologic properties is not the full story. Indeed, such an approach may seriously mislead those who seek an association between air pollution and disease. There is without question a statistical association between particulates, sulfur dioxide and the incidence of bronchitis, but their concentrations also serve as indices of many other combustion products. The physical form and chemical composition of these other combustion products may separately, or in combination, be of greater relevance to chronic disease etiology. Final resolution of the role of particulates and sulfur dioxide in the causation of chronic bronchitis and other diseases will be reached only when prospective studies are conducted. These studies must center on comparison of the incidence of disease between areas with different levels and types of air pollution. Relevant factors such as smoking and socioeconomic status must also be taken into consideration. An example of the type of study required was conducted by Douglas and Waller (1966) in Great Britain. Four broad pollution levels were chosen and the health status of children up to the age of 15 determined.

The data were limited to 3,131 families who remained at the same address for the first 11 years of the study or moved to other areas in the same pollution-level group. Socioeconomic status was determined and there was no observed tendency for the poorer families to be concentrated in the more heavily polluted areas. Morbidity information was obtained from two major sources. First, interviews between health visitors and mothers when the children were aged 2 and 4 years. Second, reports of examination conducted by school doctors when the children were aged 6, 7, 11, and 15 years. Hospital admissions were also recorded and special absence records were kept at school when the children were 6½ to 10½ years old.

The results consistently showed that upper respiratory tract infections were not related to the amount of air pollution, but lower respiratory tract infections were related. The frequency and severity of lower respiratory tract infections were observed to increase with the amount or degree of air pollution with both boys and girls being similarly affected. No difference was found between children in middle class and those in working class families. An association was found between lower respiratory infection and air pollution for each age group examined.

Meadows (1961) has suggested that part of the association between bronchitis incidence and pollution might be due to the fact that people with chronic bronchitis tend to descend the social scale. Therefore they tend to move into a poorer and cheaper type of housing in the more polluted areas of the cities. Although resolution of the problem is difficult, there is good evidence that even when allowance is made for smoking, occupation, social class, overcrowding and the many other undesirable features of urban life there remains an urban factor which can be attributed to air pollution.

On the reverse side of the coin there are those occasions when repetition is absent or impossible and yet we should not hesitate to draw conclusions. A unique example can be drawn from the occupational experience of nickel refiners in South Wales, England, as presented by Sir Austin B. Hill (1962) in the Alfred Watson Memorial Lecture to the Institute of Actuaries:

> "The population at risk, workers and pensioners, numbered about 1,000. During the 10 years 1929 to 1938, 16 of them had died from cancer of the lung, 11 of them had died from cancer of the nasal sinuses. At the age specific death rates of England and Wales at that time, one might have anticipates one death from cancer of the lung (to compare with the 16), and a fraction of death from cancer of the nose (to compare with the 11). In all other bodily sites cancer had appeared on the death certificate 11 times and one would have expected it to do so 10–11 times. There had been 67 deaths from all other causes of mortality and over the 10 years' period 72 would have been expected at the national death rates. Finally, division of the population at risk in relation to their jobs showed that the excess of cancer of the lung and nose had fallen wholly upon the workers employed in the chemical processes.
>
> "More recently my colleague, Dr. Richard Doll, has brought this story a stage further. In the nine years 1948 to 1956 there had been, he found, 48 deaths from cancer of the lung and 13 deaths from cancer of the nose. He assessed the numbers expected at normal rates of mortality as, respectively 10 and 0–1.
>
> "In 1923, long before any special hazard had been recognized, certain changes in the refinery took place. No case of cancer of the nose has been observed in any man who first entered the works after that year, and in these men there has been no excess of cancer of the lung. In other words, the excess in both cites is uniquely a feature in men who entered the refinery, in, roughly, the first 23 years of the present century.
>
> "No casual agent of these neoplasms has been identified. Until recently no animal experimentation had given any clue or any support to this wholly statistical evidence. Yet I wonder if any of us would hesitate to accept it as proof of a grave industrial hazard?"

A parallel example can be drawn from environmental experience with asbestos and has not resulted in equivalent remedial action. The first case of malignancy (diffuse mesothelioma of the pleura) associated with asbestos exposure was diagnosed in South Africa in 1956 (Wagner, et al. 1960). By 1959, Sleggs and Marchand had been able to establish an association between the Cape Asbestos fields or the industrial use of asbestos and histologically proven malignancy in 32 or 33 patients. The majority of these patients had not actually worked with asbestos but had lived in the vicinity of the mines and mills, and some had left the asbestos fields as young children, or in their teens. The average time between first exposure and the development of the tumors was 40 years. A summary of the findings as of 1961 was presented by Wagner (1965) and is as follows:

"At the end of 1961 we had diagnosed a total of 87 pleural and two peritoneal mestheliomas. In only two cases had it not been possible to establish a history of exposure to asbestos dust and only in one of the cases was there no definite exposure to crocidolite. This was a man who had made fireproof clothing while serving in the South African Air Force during the last war. The youngest case diagnosed, aged 21, gave a history of a very brief exposure as an infant when he was taken to a 'cobbing' site by his mother from the age of 6 weeks until he was weaned. Of the 87 cases, 12 had had industrial exposure and the rest came from the region of the Cape Asbestos fields. More than half of these people had not been employed by the asbestos industry. A few of them had had occupational exposure—two men had been road workers in an area where before 1940 many of the roads were surfaced with material from the asbestos dumps, and one man had worked in a railway yard where the same material had been used as ballast for the rails. Another man had been employed digging wells in the vicinity of the asbestos mines.

"The remainder of the people had had environmental exposure due to living in the vicinity of the mills and dumps in an arid, windswept region. An example of this was a woman who was born in a town on the asbestos fields in 1900. She left at the age of 5, and was in good health until she was 55 when she developed a pleural effusion. She died from a pleural mesothelioma 18 months later. Detailed questioning on her childhood revealed that she attended an infants school near to an asbestos dump. The children used to enjoy sliding down this on their way home. Two of her playmates at this school have subsequently died of the same condition.

"Even among those working in the asbestos mines, the actual exposure seems to have been slight—a number of them cobbed under relatively dust-free conditions. In these cases, there was no evidence of asbestosis in their lungs, but only a few asbestos bodies and fibers in the air spaces. Therefore, the association appeared to be with exposure to asbestos dust rather than a sequelae of asbestosis."

In the confirmed cases attempts have been made to determine the type of asbestos to which the patients were exposed. Of the known types of asbestos crocidolite shows the greatest association with the mesothelial tumor of the pleura and peritonem, but not uniquely with this fiber. The risk of mesothelial tumor appears least with chrysotile and anthophyllite fiber types. The other type of malignancy associated with asbestos exposure, the bronchial tumor, is not specifically associated with crocidolite. Studies are now being conducted to obtain quantitative information about the risk of developing these tumors in groups exposed to only one fiber type. No one disputes the cause-effect relationship here although complete knowledge of the pathogenesis is unknown. With a 20-to-60-year incubation period and increasing evidence of higher than expected prevalence of asbestos bodies and fibers in the lungs serious consideration should be given to a preventive medicine action before levels in the environment create a general health hazard.

Specificity

The specificity of an association was the third characteristic mentioned by Sir Austin B. Hill for consideration. This characteristic is most useful in the study of short-term effects. This is particularly true in occupational environments where exposures are limited usually to specific contaminants and specific work groups. Under these conditions specific diseases have been observed that cannot be attributed to other exposures. These effects are very specific and there is a strong agrument in favor of causation. When the same diseases are observed following environmental exposures, then the causative agent can be assumed. The difficulty occurs when considering the long-term effect of air pollution and isolating the influence of other factors. Nevertheless epidemiological studies on the long-term effects of specific air contaminants have shown that exposures for years to high levels of dust containing free silica can result in silicotic pulmonary sclerosis, exposures to asbestos can result in asbestosis, etc. The criterion of strict specificity is most useful in the study of short-term exposures; for example, evaluation of the biological effect of exposure to carbon monoxide by the measurement of the carboxyhaemoglobin (Wilson et al. 1926).

Emphasis on the need for specificity before drawing conclusions is unrealistic. Generally diseases may have more than one cause, and hypotheses of causations based upon 1-to-1 relationships are infrequent, although, if we knew more of the answers we might get back to a single factor. In short, where specificity exists it is possible to draw conclusions without hesitation; however, where it is not apparent, we are not thereby necessarily left "sitting on the fence."

Temporality

The temporality of an association is the fourth factor which must be considered. This factor is particularly relevant to chronic diseases. This is a question of whether a particular environment promotes disease or whether the individuals who select the environment are more likely to contract the disease.

Although the temporal problem does not arise frequently, it needs to be remembered as contributing to the "urban factor." Inherent in social class differences in disease prevalence is the question of whether those individuals who because of social class differences live in areas with higher air pollution are more likely to contract respiratory disease, or whether the disease results from their environment. Unfortunately complete resolution of this question is not forseeable and the temporal factor will continue to plague epidemiological investigations.

Biological gradient

Clearly the existence of a biological gradient, "exposure dose-response" curve, rather than a threshold effect, is an important factor in disease causation. Such evidence is of particular value in studies of disease causation. For instance, the fact that the death rate from cancer of the lung rises linearly with the number of cigarettes smoked daily, adds a very great deal to the simpler evidence that cigarette smokers have a higher death rate than nonsmokers. An assumption of a threshold effect would have oversimplified the cause-effect hypothesis. Although the "exposure dose-response" curve also simplifies the complex relationship, it puts the case in perspective.

Often when dealing with environmental exposures the difficulty is in securing satisfactory quantitative measurement of either exposure or effect which will permit definition of the "exposure dose-response" curve, but we should seek such parameters of measurement. For smoking, unlike air pollution, it is possible to determine exposure dose which is directly related to tobacco consumption (e.g., cigarettes per day). Therefore, a biological gradient can be determined for smoking.

Plausibility and coherence

Plausibility and coherence are two other factors of association which must be considered. In short, when an association is observed that is new to science or medicine we must not dismiss it too lightheartedly as what is biologically plausible depends upon the biological knowledge of the day. On the other hand, the cause-effect interpretation of our data should not seriously conflict with the generally known facts of the natural history and biology of the disease.

Thus, in the discussion of lung cancer the coherence of its association with cigarette smoking does not distract from the known urban/rural ratio of lung cancer. Laboratory evidence would enormously strengthen the hypothesis and possibly could determine the actual causative agent. On the other hand, lack of such laboratory evidence cannot nullify the epidemiological observations in man.

Experimental evidence can provide the strongest support for a causation hypothesis, and occasionally on the basis of experimental evidence alone preventive action is taken. In turn, lack of experimental evidence is not grounds for rejection of a hypothesis. Experiments may only have failed to reproduce environmental conditions. Experiments done in laboratories all over the world have failed to produce effects from sulfur dioxide at comparable levels to those found in urban atmospheres. Lawther (1965) has been careful to point out that the effect of sulfur oxide may be due to combination with other contaminants and that their number and the variability of time responses are so legion that laboratory studies have simply failed to reproduce the responsive combination. Under any situation, levels of sulfur dioxide are indicative of the causative agent.

Particulate levels in the atmosphere are another case in point. Their measurement serves as an index of a host of substances of both organic and inorganic natures. Particular interest centers on these substances because of laboratory evidence on the presense of carcinogens.

Analogy

In some cases it is fair to judge by analogy. Where occupational and environmental health effects have been observed from one chemical agent, it is reasonable to assume that similar cause-effect relationships exist for similar chemical contaminants. A typical example is that of chemical carcinogens.

A number of different viewpoints were presented by Sir A. B. Hill as a basis for determining disease causation. None of these viewpoints can be expected to provide indisputable evidence for or against a cause-effect hypothesis. However, they do, with greater or lesser strength, provide support for the fundamental question—is there any other way of explaining the observed exposure-effect relationship? No formal tests of significance can answer this question. Rather, such tests can, and should, remind us of the role that the play of chance can create and they do instruct us in the likely magnitude of the effects.

Evidence on disease causation can be derived from both prospective and retrospective studies. Powerful additional evidence is to be gained from prospective epidemiological studies which show a change in the incidence of a disease following removal of a suspected agent (e.g., the reduction in lung cancer in those who have stopped smoking). This technique is not always applicable, but when associations are observed the results present compelling evidence. In the case of the nickel, and probably the asbestos workers, the timelag between exposure and the manifestation of disease is too great to make this technique useful.

Protection of public health requires action based upon the best available evidence, and while strong evidence is requisite this does not imply crossing every "t." All scientific work is incomplete and is liable to be upset or modified by advancing knowledge. This does not confer upon us a freedom to ignore the knowledge we already have, or to postpone the action that appears demanded at a given time.

CONTRIBUTING FACTORS

Air pollution has been shown to contribute to the etiology of respiratory diseases. Other factors also are involved, with smoking predominating over all factors including air pollution. However, only restricted groups are exposed intermittently to high levels of cigarette smoke, while air pollution affects the whole population fairly continuously throughout life. Under conditions of prolonged exposure even small concentrations of toxic substances may have surprisingly large cumulative effects (Reid 1964C).

Bronchitis

Epidemiological studies of chronic bronchitis have shown that many factors are involved in its causation. Social environment and cigarette smoking have been clearly shown to dominate the background of this disease. Before reviewing the contributing factors it is useful to characterize the medical symptoms termed bronchitis.

There are differences in the terminology and definitions used in the United States and Great Britain, but the first phase of chronic bronchitis is characterized by a proliferation and hypertrophy of the mucous-secreting elements in the airways, presumably in response to inhaled irritants. This phase, which is almost certainly reversible, is manifested clinically by chronic coughing to expel the excess mucous. When later infection supervenes the sputum becomes purulent and is frequently followed by destruction of lung substance which leads to emphysema. Cigarette smoke is certainly a powerful enough irritant to produce early changes; these changes frequently regress when smoking is stopped (Lawther 1965).

Studies have shown a relationship between bronchitic mortality rates and environmental factors which include air pollution, population density, social index, smoking, and so forth. (Buck and Brown 1964, Wicken and Buck 1964). Correlations were observed between bronchitis morbidity and levels of particulates for both sexes and all social-class areas. However, the correlations did not reach significance in the urban areas until the social index was included. Although some correlation was observed between bronchitis morbidity and sulfur dioxide the correlation was greater after the inclusion of social index.

Significant positive correlations were observed between bronchitis mortality and particulates, sulfur dioxide, and social index. Sulfur dioxide levels and social index accounted for area differences. In the areas covered bronchitis mortality rates appeared to be more strongly associated with levels of air pollution than smoking habits.

While the exact causation mechanism has not been defined, epidemiological studies have shown that many factors are involved, with the social environment and cigarette smoking clearly dominating the background to this disease. A number of factors have been incriminated in the etiology of chronic bronchitis which must be considered in interpreting area differences (Fairbairn and Reid 1958, Holland 1966B).

First, statistically significant correlations have been made between standardized mortality ratios from bronchitis and various indices of air pollution including smoke and smoke deposition (Stocks 1959), and sulfur dioxide (Pemberton and Goldberg 1954).

An indication of the importance of the association between air pollution and respiratory and cardiac disease as a cause of premature death was observed in studies of chronic bronchitis mortality and morbidity in postmen (Reid and Fairbairn 1958, Holland and Reid 1965). A highly significant correlation was found between bronchitis mortality and morbidity in areas where thick, polluted fog was frequent. A more recent confirmation and amplification of this earlier work by Buck and Brown (1964) confirms the earlier positive correlation between air pollution exposure levels and male bronchitis death rates. This correlation was independent of differences in social characteristics such as population density and overcrowding.

While an analysis of loss in productive capacity among the postmen was not made, among the older bronchitics, a positive correlation was observed between absence from work due to minor respiratory infections and air pollution levels. Also, it was observed that as levels of air pollution exposure increase, the number of minor respiratory infections leading to serious complications increased.

Second, demographic factors have been shown to correlate with chronic bronchitis morbidity and mortality rates. While mortality rates rise steeply with age, both mortality and morbidity rates are higher in males than in females (Reid 1956, 1964; Holland 1966A). Thus area rates, which will be affected by population makeup, must be adjusted accordingly. The need for consideration of relevant data on age and sex is obvious (Reid 1964C).

Third, smoking has been clearly incriminated in the etiology of chronic bronchitis (PHS 1964, 1967). Holland (1966A) in a rough calculation of the roles of smoking and air pollution indicates that smoking is a factor of 5 or 6 in symptons compared to a factor of 1.2 for environment and air pollution. When comparison is made between the concentrations of various chemical agents inspired in cigarette smoke and levels found in urban air, the difficulty of detecting the effect of air pollution exposures is hardly surprising (Reid 1964C).

Fourth, there is a distinct effect of social class seen in bronchitis mortality rates. The mortality rates are highest in those individuals in the lower social classes and lowest in those belonging to higher social classes (Goodman, et al. 1963, Winkelstein, et al. 1967, Zeiberg, et al. 1967). That this gradient is unlikely to be due to occupation of the males is shown by the fact that a similar gradient is present for their wives. However, the social class effect can also be in part the effect of the environment, and exactly which part of the social class factor is to blame is very difficult to disentangle. A social class gradient, however, has been found in bronchitis mortality and morbidity in the United Kingdom (Edwards 1959, Buck and Brown 1964, and Wilken and Buck 1964).

Fifth, climatic factors have been shown to correlate with London morbidity and mortality rates (Holland, et al. 1961). Adult admissions of respiratory patients to London hospitals show a negative correlation with temperature which is independent of levels of air pollution, humidity, rainfall, sun hours, barometric pressure and season of year. The only other meteorological variable to have an effect independent of the others was found to be atmospheric pollution.

The sixth factor related to the development of chronic bronchitis is respiratory infection in both children and adults. Little is known of the relationship of childhood infection and the development of chronic bronchitis in later life. Studies by Rosenbaum (1961) showed that Army recruits coming from polluted industrial towns suffered from more respiratory infections while in the Army than those from rural areas. Lee (1957) showed that the prevalence of chronic otitis media in youths presenting themselves for the British national service medical examination was highest in those coming from large urban areas.

It is essential that epidemiological studies of air pollution and bronchitis mortality and morbidity rates take into consideration the above factors although it may be difficult, if not impossible, to disentangle their separate effects. Some of the confusion can be minimized by comparing only individuals of the same occupations and similar smoking habits. An example is the study by Holland and Stone (1965) comparing the prevalence of respiratory symptoms in U.S. East Coast telephone men with similar employee groups in the United Kingdom.

Lung cancer

The association between lung cancer and air pollution is, on superficial examination, the same as for bronchitis with the accompanying "urban factor." Epidemiological investigations have shown that lung cancer rates are higher among urban than rural populations. Superimposed on the urban factor is the effect of smoking, the dominant factor in the case of lung cancer (PHS 1964, 1967; Reid 1958). The role of air pollution in the disease etiology is small in comparison and difficult to assess quantitatively. It has been estimated that the urban-rural ratio for lung cancer deaths is 1.32 to 1.67 for males smoking less than one pack of cigarettes per day (Haenszel, et al. 1964).

At the same time it has been noted that a consistently higher lung-cancer risk exists for smokers migrating to metropolitan areas. This risk is lower among the least mobile and highest among mobile populations with three or more residences of exposure. (Haenzel, et al. 1962, 1964). It has been suggested that the higher lung cancer risk might be related to the abrupt imposition of additional particulate matter present in urban atmospheres with no intervening period of gradual adaptation to the increasing exposure. Additional epidemiological inquires are required to elaborate on this migrant effect with particular emphasis being placed on age at time of migration from rural to urban areas.

Significant social class differences have been observed in the incidence of both lung cancer and bronchitis in men (Winkelstein et al. 1967, Zeiberg et al. 1967). A significant social class gradient in lung cancer mortality rates has been found in all parts of England and Wales (Wicken and Buck 1964, Dean 1966). While none of these figures are standardized for differences between social classes in smoking habits, the absence of this standardization is unlikely to affect the conclusions.

Lung cancer mortality rates have been examined in relation to environmetal factors which include air pollution levels of smoke and sulful dioxide, population density, social index, and smoking (Buck and Brown 1964, Wicken and Buck 1964, Dean 1966).

Lung cancer mortality, however, appeared to be more strongly associated with smoking habits than air pollution levels. Lung cancer mortality rates were not significantly associated with smoke or sulfur dioxide levels in residential areas. An association was found between lung cancer and population density for both sexes. This discussion is not intended to discount the role of air pollution in the etiology of lung cancer, but rather to emphasize the difference between the role of air pollution in the etiology of these two disease—bronchitis and lung cancer.

Dean (1966) has concluded that the association between smoking and lung cancer is causal in nature. It is more difficult to conclude that the association between urbanization and lung cancer is causal in nature. It is difficult to avoid the conclusion that exposures to urbanization, in addition to smoking and climate factors, contribute to the etiology of this disease. The evidence of a greater liability to lung cancer is clearly elucidated in the studies of British-born emigrants to New Zealand (Eastcott 1956, 1960), South Africa (Dean 1959, 1961), Australia (Dean 1962), and U.S.A. (Haenszel 1961). This liability is the same for men as for women and is consistent at all ages over 35 (Eastcott 1956). The conclusion is that the immigrants are affected by their former environment and the effect is related to the length of exposure in that environment. The differences cannot be explained on the basis of personal habits or consumption of tobacco.

There is accumulating evidence that air pollution and cigarette smoking add their respective insults to the respiratory system, with smoking as the dominate causative agent. There also appears a saturation point where smoking precipates cancer of the lung in all those predisposed to it and the addition of air pollution causes no appreciable increase in lung cancer mortality (Reid 1958).

Buell and Dunn (1967) have concluded that evidence of the etiologic roles of urban living and cigarette smoking in the development of lung cancer seem to be complete. However, the question of how these etiologic agents interact has yet to be resolved. Namely, do other etiologic factors interact with smoking in a competitive, additive, or multiplicative (synergestic) way? Obviously the alternatives preented would affect the design of epidemiologic studies. The literature on occupational exposures gives no clear answer, possibly because interactions are rarely studied. The literature on air pollution exposures and the urban factor contains inconsistencies, and the authors have tried to argue against the too ready acceptance of the multiplicative hypothesis.

Smoking is, without question, the dominant factor in the etiology of lung cancer. The "smokers" represent a restricted group who are intermittently exposed to high concentrations of cigarette smoke. Air pollution on the other hand may affect the whole population as a consequence of continuous exposure throughout life. Therefore "air quality criteria" should reflect consideration of the nonsmoker although smoking is the dominant factor. The difficulty arises when attempts are made to isolate quantitatively air pollution exposures contributing to lung cancer causation.

Emphysema

The relationship between emphysema and air pollution is not as clear as for bronchitis; however, an "urban factor" has been shown. In the United States, urban mortality rates from emphysema are approximately twice rural rates, and growing in importance. Differences between the United States and England rates may be explained in part by differences in the reporting of chronic bronchitis and emphysema.

Bronchial asthma

Bronchial asthma is of particular interest in studies of air pollution health effects. Resulting from an allergic response the effects are manifested in susceptible (sensitive) individuals. The increased frequency of asthmatic attacks during periods of high oxidant levels in the atmosphere are but an example (Schoettlin and Landau 1961). A long list of stimuli are capable of triggering asthmatic attacks which includes both irritant air contaminants as well as aeroallergens.

Aerollergens which are the primary causitive agents in inducing bronchial asthma, include pollen, spores, rusts, and smuts. Individual human responses vary widely with the disease having a high national incidence.

There are a number of localized area of particular note in studies of allergic responses to acute, subacute, and long-term air pollution exposures.

"Tokyo-Yokohama respiratory disease" is a case in point (Huber, et al. 1954; Smith, et al. 1964). It was observed as early as 1946 among American troops stationed in the highly industrialized Yokohama area of Japan. The condition also appeared among dependent family members living in the area. Later the condition was observed in our military personnel in the Tokyo area. Individuals treated did not respond to the usual medication for asthma, but the symptoms generally disappeared when the individuals were evacuated from the area. If allowed to remain in the area a few months to a year, many individuals develop permanent lung disability.

It is estimated that 3 to 5 percent of the exposed American personnel develop the disease; it occurs in epidemic form in late fall and early winter, when the usual allergens of plant origin are not present in the air; it is unclear whether the native population have the disease, although it is known that they do have true allergic bronchial asthma; few of the patients report a history of allergy in themselves or their families; removal from the area, especially if early, results in recovery in most cases; and there is a good correlation between frequency of attack and air pollution levels.

A second example is "New Orleans asthma" which occurs in epidemic proportions under particular local wind conditions (Weill, et al.

1964). The condition suggested an association with a particular type of pollutant sources but when one such silica particulate source was eliminated the episodes continued.

A third example, "Hawaiian asthma," may provide useful information on the effects of air pollution. There is a higher rate of asthma attacks in Hawaii than in any other States. The disease affects primarily the very young children. As yet, no cause has been defined.

While categorization of dose effect relations for these diseases may not be possible within the World Health Organization framework, these particular air pollution allergies and other may provide useful insights into respiratory disease etiology and the role of air pollution.

Acute nonspecific respiratory disease

The termed "acute nonspecific upper respiratory disease" is synonymous with the common cold and generally accepted as being due to a group of viruses. There is, however, substantial evidence that this condition is also etiologically associated with air pollution as a predisposing cause. This evidence comes from laboratory experiments on the effects of irritants on the development of deeper, more severe, acute, and subacute lung disease. The conclusion is that acute respiratory infectious diseases, both in upper and lower respiratory tract and deeper in the lungs, are etiologically associated with irritant air contaminants.

There are other indications that air pollution causes host changes which predispose the individual toward diseases due to infectious agents. Another possibility is that the organisms, themselves, are altered by pollutants into a new, more virulent characteristics infectious agent.

Studies involving children are very useful because they remove the effects of smoking and occupation. The study by Douglas and Waller (1966) was very informative in this regard. Results consistently showed that upper respiratory tract infections in children were not related to general air pollution exposure, but lower respiratory tract infections were related for each age group selected. The frequency and severity of lower respiratory tract infections was observed to increase also as air pollution exposures increase. Both boys and girls were similarly affected with no difference being found between children in middle class and working class families.

Another useful study which defines the role of air pollution in the etiology of respiratory disease was conducted by Toyama (1964). Two groups of schoolchildren, 10 to 11 years old, in Kawasaki, Japan, were selected. One group was living and going to school in a relatively unpolluted area and the other in a very heavily polluted area. A significant correlations was found between monthly dust fall and lead peroxide sulfation rates, and in respiratory flow rates. The children from the polluted area had a lower pulmonary function, as measured by the Wright peak flowmeter, which was lowest during the periods of highest air pollution exposures. Also these children had a higher frequency of nonproductive cough, mucous membrane irritation of the upper respiratory tract, and increased mucus secretion. Lung vital capacity in the two groups was similar.

Watanabe (1965) has reported differences in pulmonary function in schoolchidren, ages 10 to 11. Air pollution levels were measured as atmospheric dustfall and sulfur dioxide concentrations. A significant decrease in lung function measured as the Wright peak expiratory flow rate of the children was found in the polluted area as compared to the less polluted areas.

A study of comparative respiratory illnesses among very young schoolchildren was conducted in Sheffield (Lund, et al. 1967). Upper respiratory tract illnesses were measured in terms of micro-purulent nasal discharge and three or more colds yearly, and showed an association with pollution levels. Lower respiratory tract illnesses were measured in terms of persistent or frequent cough, colds going to the chest, and episodes of pneumonia and bronchitis, showed a similar pattern. Illness was found to be less common in the clean than in the dirty area. Measurements of forced expiratory volume and forced vital capacity were significantly reduced in the children in the most heavily polluted area, and were unaffected by socioeconomic class or area, except in the most polluted area. Respiratory illness was found to be more common in the areas of higher atmospheric pollution, while socioeconomic factors such as social class, number of children in the house, and whether or not they shared bedrooms, appeared to have little influence.

Colley and Holland (1967) were able to assess the varying influences of smoking, area of residence and place of work, overcrowding, family size, social class, and genetic factors, in the etiology of chronic respiratory disease. In the fathers, smoking and social class (as classified by occupation) showed an influence on the prevalence of cough, while the effect of area of residence was not demonstrable. In mothers the influence of social class (as identified by husband's occupation) was absent but the effect of smoking and area of residence was manifested. The effect of area residence and degree of pollution was demonstrable on the children, also.

3.5 Systemic Effects of Air Pollution

There are a number of air contaminants which must be considered on the basis of their effects on the total body system (systemic effects), the effects usually depending upon the human body burden of the contaminant. These air contaminants include arsenic, asbestos, cadmium, beryllium compounds, mercury, manganese compounds, carbon monoxide, fluorides, hydrocarbons, mercaptans, inorganic particulates, lead, radioactive isotopes, carcinogens, and insecticides.

The results of nonoccupational exposure to asbestos, discussed earlier in this report (sec. 3.4), suggest that asbestos dust may present a long-term hazard to members of the general population other than occupationally exposed groups.

Beryllium and beryllium compounds have been shown to produce beryllìosis among nonoccupationally exposed persons living in the vicinity of industrial plants even though neither they, nor any member of their households, had connection with the plant (Eisenbud, et al. 1949). The exposure is reflected as increased excretion of beryllium and in increased amounts in the tissues of the body.

Prior to the first reports of berylliosis there was a Norwegian report that pneumonia occurred with an unusually high frequency among people living near a manganese processing plant (Elstad 1939). Further study in Italy (Paneheric 1955) suggest confirmation of such a medical relationship between airborne manganese compounds and pneumonia although it is difficult to be sure of the causal relation because of the similarity of the pneumonia clinically to cases not associated with manganese as an etiological factor.

Recent studies have identified cadmium as a causative agent in the development of hypertension in humans. Very little is known about both the sources of ambient air levels of cadmium or the mechanism of human absorption and ingestion. However, zinc and cadmium are the only contaminants that show a marked correlation with heart disease. While further studies are needed to determine whether cadmium is related to the entire cardiovascular death spectrum, there is ample evidence that cadmium is related to the hypersensitive component. This evidence is augmented by higher "body burdens" in the kidneys of hypertensive patients and the experimental production of heart disease in rats exposed to cadmium. (Schroeder 1964, 1965; Schroeder and Balassa 1961; Perry and Schroeder 1955; and Carroll 1966.)

In addition to the effect of cadmium, increased vanadium in the environment contributes statistically to an increase in the heart disease mortality rates. (Hickey, et al. 1967). Although the effects of vanadium in this study were statistical only, physiological effects of vanadium on several forms of life have been reported (Underwood 1962, Schutte 1964). There is no reason to believe that other metals may not also contribute to mortality rates. A particular enzyme function may be destroyed by a number of toxic metals which replace the essential metal of the enzyme additively, but with vary degrees of efficacy.

Recent reports suggest that some chemicals may induce genetic damage in man, similar to those induced by radiation. A number of chemical mutagens—some in widespread use—have been identified; examples include ethyl methane sulfonate and methy-nitro-nitrosoguanidine. Of significance is the fact that many of the chemicals are highly mutagenic in experimental organisms at concentrations that are not toxic and that have no overt effect on fertility.

Two kinds of effects are suggested. First, a small increase in mutation rates is produced by these chemical agents. Such small increases in mutation rate probably could not be detected in a short period of time by any direct observations on human beings. Protection from such effects must depend on prior identification of mutagenicity.

Second, a chemical compound presumed to be innocuous is in fact highly mutagenic and large numbers of individuals are exposed before the danger is realized. These effects might be conceivably detected by an increased incidence of certain genetic diseases.

These are but examples of the nature of systematic effects resulting from exposure to atmospheric contaminants. Without question they require particular attention in the development of air quality criteria.

V. REFERENCES CITED

Bradley, W. H., Logan, W. P. D., and Martin, A. E. (1958). *Mon. Bull. Minist. Hlth. Lab. Serv. 17*, 156.

Brasser, L. J., Joosting, P. E., and Zuilen, D. Van, "Sulphur Dioxide—To What Level Is It Acceptable?" Research Institute for Public Health Engineering, Report G300, July 1967.

Buck, S. F. and Brown, D. A. (1964). "Mortality from Lung Cancer and Bronchitis in Relation to Smoke and Sulphur Dioxide Concentration, Population Density and Social Index." Tobacco Research Council Research Paper No. 7.

Buell, P. and Dunn, J. E. (1967). "Relative Impact of Smoking and Air Pollution on lung cancer." *Arch. Environ. Health 15:* 291–297 (Sept.).

Carroll, R. E. (1966). "The Relationship of Cadmium in the Air to Cardiovascular Disease Death Rates." *JAMA*, Vol. 198 : 267–269 (October).

Ciocco, A. and Thompson, D. J. (1961). "A Follow Up of Donora Ten Years Later: Methodology and Findings" *Am. J. Pub. Health* 51 155–164.

Clifton, M., Kerridge, D., Pemberton, J., Moulds, W. and Donogue, J. K. (1959). "Morbidity and Mortality from Bronchitis in Sheffield in Four Periods of Severe Air Pollution (Great Britain)" Proceedings of Intnl. Clean Air Conference, London, P. 189.

Clifton, Marjorie (1967), "Pollution Data for Health Studies", reprinted from the Transactions of the International Chest and Heart Conference, held at Eastbourne 4th–7th April, 1967, pp. 143–50.

Colley, J. R. T. and Holland, W. W. (1967). "Social and Environmental Factors in Respiratory Disease." *Arch. Environ Health* 14 : 157 (January).

Commins, B. T. (1962) "Chemistry of Town Air," *Research* Vol. 15, pp. 421–426.

Dean, G. (1959). *Brit. Med. J.* 2 : 852.

Dean, G. (1961). *Brit. Med. J.* 2 : 1599.

Dean, G. (1962), *Med. J. Aust.* 1 : 1003.

Dean, G. (1964). "Lung Cancer in South African and British Immigrants." *Proc. Roy, Soc. Med. 57*, 984.

Dean, G. (1966). "Lung Cancer in Northern Ireland." *Brit. Med. J.* 1 : 1506.

Deane, M. (1965). "Epidemiology of Chronic Bronchitis and Emphysema in the United States." *Med. Thorac* 22 : 24–37.

Dorn, H. F., and I. M. Moriyama (1964). "Uses and Significance of Multiple Cause Tabulations for Mortality Statistics." *Amer. J. Pub. Hlth* 54 : 3.

Douglas, J. W. B. and Waller, R. E. (1966). "Air Pollution and Respiratory Infection in Children." *Brit. J. Prev. Soc., 20*:1.

Eastcott, D. F. (1956). "Epidemiology of Lung Cancer in New Zealand," *Lancet* I, 37.

Eastcott, D. F. (1960). *Conference* 1 :4.

Edwards, F., McKeown, T., and Whitfield, A. G. W. (1959) *Lancet* 1 :196.

Eisenbud, M., Wanta, R. C., Dustan, C., Steadman, L. T., Harris, W. B. and Wolf, B. S. (1949). "Non-occupational Berylliosis." *J. Indust. Hyg. Toxicol. 31*:282.

Elstad, D. (1939). *Nord. Med. 3*:2527.

Faerber, K. P., Hoffmann, A., and Schmitz, G. (1959). *Off Gesundh.-Dienst. 12*:493.

Fairbairn, A. S. and Reid, D. D. (1958). "Air Pollution and Other Local Factors in Respiratory Disease." *Brit. J. Prev. Soc. Med. 12*:94.

Firket, J. (1936). "Fog Along the Meuse Valley." *Trans. Faraday Soc. 32*:1192.

Firket, M. (1931). "The Cause of the Symptoms Found in the Meuse Valley During the Fog of December 1930." *Bull. Royal Acad. Med., Belgium 11*:683.

Glasser, M. Greenburg, L., and Field, F. (1967). "Mortality and Morbidity During a Period of High Levels of Air Pollution" *Arch. Environ. Hlth* 15 : 684.

Goodman, N., Lane, R. E. and Rampling, S. B. (1953). "Chronic Bronchitis: An Introductory Examination of Existing Data." *Brit. Med. J. 2*:237.

Graham, J. W. "The Destructiveness of Daylight," pp. 6–24, cited by A. C. Pigou in "Economics of Welfare" 4th edition, p. 184, MacMillan Co., London (1932).

Greenburg, L., Jacobs, M. B., Drolette, B. M., Field, F. and Braverman, M. M. (1962A). "Report of an Air Pollution Incident in New York City, November 1953." *Pub. Health Reports 77:*7–16 (January).

Greenburg, L., Field, F., Reed, J. I. and Ehrhardt, C. L. (1962B). "Air Pollution and Morbidity in New York City." *JAMA 182:*161–164 (Oct.).

Greenburg, L., Field, F., Erhardt, C. L., Glasser, M. and Reid, J. I. (1967). "Air Pollution, Influenza and Mortality in New York City." *Arch. Env. Health* Vol. 15, No. 4, p. 430.

Haenszel, W. (1961). "Cancer Mortality Among the Foreign Born in the United States." *J. Nat'l. Cancer Inst. 26:*37–132.

Haenszel, W., Loveland, D. B., and Sirken, M. G. (1962). "Lung cancer mortality as related to residence and smoking histories. I white males" Int Nat'l Cancer Institute 28:4, pp. 947–1001 (April).

Haenszel, W., and Taeuber, K. E. (1964) "Lung Cancer mortality as related to residence and smoking histories. II White Females" Int Nat'l Cancer Institute 32:4, pp. 803–838 (April).

Hamilton, E. M., and Jarvis, W. D. "The Identification of Atmospheric Dust by Use of the Microscope." Monograph, Central Electricity Generating Board, March 1963.

Heggestad, H. E. (1966), "Ozone as a Tobacco Toxicant," *Journal of the Air Pollution Control Assn.*, Vol. 11, No. 12, pps. 691–694 (December).

Heggestad, H. E., and Middleton, J. T. (1959)," Ozone in High Concentrations as Cause of Tobacco Leaf Injury," *Science 129:* 208–210.

Hickey, R. J., Schoff, E. P. and Clelland, R. C. (1967). "Relationship Between Air Pollution and Certain Chronic Disease Death Rates." *Arch. Environ. Health 15:*728 (December).

Hill, Sir Austin B. (1962). Alfred Watson Memorial Lecture to the Institute of Actuaries. *Jnl. Inst. Actu. 88:*178.

Hill, Sir Austin B. (1965). "The Environment and Disease: Association or Causation." Proceedings of the Royal Scoiety of Medicine. *58:*295–300.

Hill, A. C., Pack, M. R., Treshow, M., Downs, R. F., and Transtrum, L. G. (1961), "Plant Injury Induced by Ozone" *Phytopathol 51:* 356–363.

Holland, W. W. (1966A). "The Natural History of Chronic Bronchitis." *J. Coll. Gen. Practnrs. 11,* Suppl. 2.

Holland, W. W. (1966B). "The Study of Geographic Differences in the Prevalence of Chronic Bronchitis." *Statisician 16:5.*

Holland, W. W., Spicar, C. C. and Wilson, J. M. G. (1961). "Influence of the Weather on Respiratory and Heart Disease." *Lancet 2:*338.

Holland, W. W. and Reid, D. D. (1965B). "The Urban Factor in Chronic Bronchitis." *Lancet 1,*445.

Holland, W. W., Reid, D. D. Seltser, R. and Stone, R. W. (1965B). "Respiratory Disease in England and the United States." *Archs. Envir. Hlth, 10:*338.

Holland, W. W., and Stone, R. W. (1965C). "Respiratory Disorders in United States East Coast Telephone Men." *Am. Jnl. of Epidemiology,* Vol. 82, No. 1.

Huber, T. E., Joseph, S. W., Knoblock, E., Redfearn, P. L. and Karakawa, J. A. (1954). "New Environmental Respiratory Disease (Yokohama asthma)." *Arch. Industr. Hyg. 10:*399.

Lawther, P. J. (1958). "Climate, Air Pollution and Chronic Bronchitis." *Proc. Roy. Soc. Med. 5:*262.

Lawther, P. J. (1965). "Air Pollution and the Public Health." Offprint from the Journal of the Royal Society of Arts, September, 1965.

Lawther, P. J., Martin, A. E., Wilkins, E. T. (1964), "Epidemiology of Air Pollution." Report on a Symposium, World Health Organization, Public Health Papers, No. 15.

Ledbetter, M. C., Zimmerman, P. W., and Hitchcock, A. E. (1954), "The Histopathological Effects of Ozone on Plant Foliage" *Contrib. Boyce Thompson Inst., 20:* 275–282.

Lee, J. A. H. (1957). "Chronic Otites Media Among a Sample of Young Men." *J. Laryng. 71:*398.

Logan, W. P. D. (1949). "Fog and Mortality." *Lancet 1:*78.

Logan, W. P. D. (1953). "Mortality in the London Fog Incident, 1952." *Lancet 1*:336.

Lund, J. E. et al. (1967). "Patterns of Respiratory Illness in Sheffield Infant School Children." *Brit. J. Prev. Soc. Med. 21*:7.

Macdowall, F. D. H., Mukammal, E. I., and Cole, A. F. W. (1964), "Direct Correlation of Air-polluting Ozone and Tobacco Weather Fleck", *Can. J. Plant Sci.*, *44:* 410–417.
Mahler, E. A. J. (1966), "Standards of emission under the Alkali Act," paper presented at International Clean Air Congress, October 1966, London, England.
Martin, A. E. (1961). "Epidemiological Studies of Atmospheric Pollution." Monthly Bulletin of the Ministry of Health and the Public Health Laboratory Service, *20* :42.
Martin, A. E. (1964). "Mortality and Morbidity Statistics and Pollution." Symposium Number 6, Section 1, Medical and Epidemiological Aspects of Air Pollution, Proceedings of the Royal Society of Medicine. (October, 1964) Vol. 57, No. 10, Part 2, pp. 969–975.
Martin, A. E., and Bradley, W. H. (1960). "Mortality, Fog and Atmospheric Pollution." Monthly Bulletin of the Ministry of Health and the Public Health Laboratory Service *19* :199.
Meadows, S. H. (1961). "Social Class Migration and Chronic Bronchitis." *Brit. J. Prev. Soc. Med. 15* :171.
Medical Research Council (1960) Committee on Aetiology of Chronic Bronchitis. Standardized Questionnaire on Respiratory Disease. *Brit. Med. J. 2* :1665.
Menser, H. A., and Heggestad, H. E. (1966) "Ozone and Sulfur Dioxide Synergism : Injury to Tobacco Plants," *Science, 153:* 424–425.
Middleton, J. T. and Paulus, A. O. (1956). *Arch. of Industrial Health 14* :526. *2* :526.
Middleton, J. T., Emik, L. O. and Taylor, O. C. (1965). *Jnl. Air Poll. Control Assn. 15* :476.
Min. of Health (1954). "Mortality and Morbidity During the London Fog of December 1952." Reports on Public Health and Related Subjects No. 95. Her Majesty's Stationary Office, London.
Min. of Health (1963). "On the State of the Public Health." Annual Report of the Chief Medical Officer of the Ministry of Health for the Year 1963.
Ministry of Housing and Local Government (1966A) "List of registrable works and noxious or offensive gases," London, England.
Ministry of Housing and Local Government (1966B), "One hundred and third annual report on Alkali and other Works," by the Chief Alkali Inspector, London, England.
Moriyama, I. M., (1963). "Chronic Respiratory Disease Mortality in the United States. *Pub. Hlth. Rep. 78* :9.
Paneheri, G. (1955). In : F. S. Mallette, ed., Problems and Control of Air Pollution, New York, p. 253.
Pemberton, J. and Goldberg, C. (1954). "Air Pollution and Bronchitis." *Br. Med. J. 2* :567.
Perry, H. M., Jr. and Schroeder, H. A. (1955). "Concentrations of Trace Metals in Urine Treated and Untreated Hypertensive Patients Compared with Normal Subjects." *J. Lab. Clin. Med. 46* :936 (December).
Public Health Service (1949). "Air Pollution in Donora, Pa., Epidemiology of the Unusual Smog Episode of October 1948." Preliminary Report, Public Health Service Bulletin No. 306.
Public Health Service (1964). "Smoking and Health." Report of the Advisory Committee to the Surgeon General. PHS Pub. No. 1103.
Public Health Service (1967A). "Air Quality Criteria for Oxides of Sulfur." PHS Pub. No. 1619.
Public Health Service (1967B). "Health Consequences of Smoking." PHS Pub. No. 1696.
Reid, D. D. (1956). "General Epidemiology of Chronic Bronchitis." *Proc. R. Soc. Med. 49:*767.
Reid, D. D. (1958). "Air Pollution and Respiratory Disease." Proc. of Conf. Nat'l. Soc. for Clean Air, England.
Reid, D. D. (1960). *Amer. J. Pub Hlth 50:*53.
Reid, D. D. (1964A). "Air Pollution and Respiratory Disease in Children." Second Int'l. Symposium, Croningen, The Netherlands.
Reid, D. D. (1964B). "Assessing the Comparability of Mortality Statistics." *British Med. Jnl. 2:*1437–1439 (Dec. 5, 1964).
Reid, D. D. (1964C). "Air Pollution as a Cause of Chronic Bronchitis." *Proc. Royal Soc. Med. 57:*10, pp. 965–968 (October).

Reid, D. D. (1966). "Studies of Disease Among Migrants and Native Populations in Great Britain, Norway and the United States: I. Background and Design." *Nat. Cancer Inst. Mon. 19*:287–299.

Reid, D. D. and Fairbairn, A. S. (1958). "The Natural History of Chronic Bronchitis." *Lancet 1*:1147.

Reid, D. D., Cornfield, J. Markusch, R. E., Seigel, S., Pedersen, E. and Haenszel, W. (1966). Studies of Disease Among Migrants and Native Populations in Great Britain, Norway and the United States. 3. Prevalence of Cardio-respiratory Symptoms among Migrants and Native-born in the United States. *Nat. Cancer Inst. Mon. 19*:321.

Rosenbaum, S. (1961). Home Localities of National Servicemen with Respiratory Disease. *Brit. J. Prev. Soc. Med. 15*:61.

Ryazanov, V. A. (1962), "Sensory physiology as a basis for air quality standards" *Arc Environ Health 5*: 480–494.

Scadding, J. G. "Unanswered questions about chronic bronchitis." Supplement No. 2 to Vol. XI of the Journal of the College of General Practitioners.

Schoettlin, C. E. and Landau, E (1961) "Air Pollution and Asthmatic Attacks in the Los Angeles Area." Public Health Reports 76:6, p. 545–548 (June).

Schrenk, M. H., Heimann, H. Clayton, G. D., Gafafer, W. M. and Wexler, H. (1949). Air Pollution in Donora, Penn. Epidemiology of the Unusual Smog Episode of October 1948. Pub. Health Bulletin No. 306, Federal Security Agency, Washington, D.C. 173pps.

Schroeder, H. A. (1964). Cadmium Hypertension in Rats. *Amer. J. Physiol. 207*:62–66 (July).

Schroeder, H. A. (1965). Cadmium as a Ractor in Hypertension. *J. Chronic Dis. 18*:647–656 (July).

Schroeder, H. A and Balassa, J. J. (1961). Abnormal Trace Metals in Man; Cadmium. *J. Chronic Dis. 14*:236–258 (Aug.)

Schutte, K. (1964). "The Biology of the Trace Elements." Philadelphia: J. B. Lippincott Co.

Scott, J. A. (1953). "Fogs and Deaths in London, December 1952." *Pub. Health Rep. 68*:474–479.

Scott, J. A. (1956). London County Council, Annual Report of County Medical Officer of Health for 1955.

Scott, J. A. (1958). *Med. Offr. 99*:367.

Scott, J. A. (1959). "Fog and Atmospheric Pollution in London, Winter 1958–1959." *M. Officer* (London) *102*:191–193.

Scott, J. A. (1963). "The London Fog of December 1952." *M. Officer 109*:250–252.

Smith, S. W., Kolb, E. J., Phelps, H. W., Weiss, H. A. and Hollinden, A. B. (1964). "Tokyo-Yokohama Asthma, An Area Specific Air Pollution Disease." *Archs. Envir. Hlth. 8*:805.

Stocks, P. (1959). "Cancer and Bronchitis Mortality in Relation to Atmospheric Deposit and Smoke." *Brit. M. J. 5114*:74–79.

Taylor, G. S., and Rich, S. (1961), "Tobacco Fleck Controlled with Antiozonants," *Phytopathol, 51*: 579.

Thomas, M. D. (1961). Effects of Air Pollution on Plants. In: Air Pollution; Monograph Series No. 46, World Health Organization, p. 233.

Toyama, T. (1964). "Air Pollution and its Health Effects in Japan." *Arch. Environ. Health 8*:153.

Trub, C. L. P. and Posch, J. (1959). Zbl. Bakt. Hyg. *176;* I, 3/6, 207.

Underwood, E. J. (1962). "Trace Elements in Human and Animal Nutrition," ed. 2 New York Academic Press.

Wagner, J. C. (1965). Epidemiology of Diffuse Mesothelial Tumors: Evidence of an Association from Studies in South Africa and the United Kingdom. Annals of the New York Academy of Science *132*: 575–578.

Wagner, J. C., Sleggs, C. A. and Paul, Marchand (1960). Diffuse Pleural Mesothelioma and Asbestos Exposure in the North-western Cape Province. Brit. J. Industrial Medicine *17*: 260–271.

Waller, R. E. and Lawther, P. J. (1955). Some Observations on London Fog. British M.J. *2*: 1356–1358.

Waller, R. E. and Lawther, P. J. (1957). Further Observations on London Fog. British M.J. *2*: 1473–1475.

Watanabe, H. (1965). Air Pollution and Its Health Effects in Osaka, presented at 58th Annual Meeting of Air Pollution Control Association, Toronto, Canada, June 20–24.

Weill, H., Ziskind, M. M., Derbes, V., Lewis, R., Horton, R. J. M., and McCaldin, R. O. (1964). Further Observations on New Orleans Asthma. Arch. Env. Hlth. *8*: 184.

Wicken, A. H. and Buck, S. F. (1964). Report on a Study of Environmental Factors Associated with Lung Cancer and Bronchitis Mortality in Areas of North East England. Tobacco Research Council Research Paper No. 8.

Wilkins, E. T. (1954). Air Pollution and the London Fog, December 1952. Journal of the Royal Sanitary Institute. Vol. 74, No. 1.

Wilson, E. D., Gates, I., Owen, H. R., and Dawson, W. T. (1926). Street Risk of Carbon Monoxide Poisoning. J.A.M.A. *87*: 319.

Winkelstein, W., *et al* (1967). "The Relationship of Air Pollution and Economic Status to Total Mortality and Selected Respiratory System Mortality in Men," *Arch. Environ. Health 14*: 162 (January)

World Health Organization (1958). "Air Pollution," Fifth Report of the Expert Committee on Environmental Sanitation, Technical Report Series, No. 157, Geneva.

World Health Organization (1961). "Air Pollution" Compendium, Monograph Series No. 46, 442 pps.

World Health Organization (1964). "Atmospheric Pollutants", Technical Report Series No. 271, Geneva.

Zeiberg, L. D., Horton, R. J. M., and Landau, E. (1967). "The Nashville Air Pollution Study: V. Mortality from Diseases of the Respiratory System in Relation to Air Pollution," *Arch. Environ. Health, 15*: 214 (August).

Zimmerman, P. W. (1950). Effects on Plants of Impurities Associated with Air Pollution. In: Air Pollution, Proceedings U.S. Technical Conference, McGraw-Hill, New York, p. 127.

C. Weather Effects

<u>Restoring the Quality of Our Environment</u>. Report of the Environmental Pollution Panel, President's Science Advisory Committee. The White House, November 1965. Pages 119-27.

Some scientists believe that the cumulative effect of air pollution may be to reduce the amount of sunlight reaching the earth and thus have a long term cooling effect which would cause vast dislocations in agriculture and other human activities. Others believe that the increased quantities of carbon dioxide produced by combustion will have a long term warming effect which may even melt the polar icecaps, raise the level of the sea, and cause immense flooding of coastal regions. All seem convinced however that mankind is inadvertently indulging in a vast experiment affecting its climate and perhaps its very existence.

The selection reproduced below was written by a subpanel of the Science Advisory Committee. The Subpanel was chaired by Roger Revelle, Director of the Center for Population Studies of Harvard University.

PROBABLE FUTURE CONTENT OF CARBON DIOXIDE IN THE ATMOSPHERE

We can conclude with fair assurance that at the present time, fossil fuels are the only source of CO_2 being added to the ocean-atmosphere-biosphere system. If this held true throughout the last hundred years, the quantity of CO_2 in the air at the beginning of the present decade was about 7% higher than in the middle of the last century (see Table 3).

Throughout these hundred years, the rate of fossil fuel combustion, and thus of CO_2 production, continually increased, on the average about 3.2 percent per year. The amount produced in 1962 was almost 25 times the annual production in the mid-1860's. The rate of increase may be accelerating. During the eight years from 1954 to 1962, the average rate of increase was 5%.

We can ask several questions about the future CO_2 content of the atmosphere. Two of these questions are:

(1) What will the total quantity of CO_2 injected into the atmosphere (but only partly retained there) be at different future times?

(2) What would be the total amount of CO_2 injected into the air if all recoverable reserves of fossil fuels were consumed? At present rates of expansion in fossil fuel consumption this condition could be approached within the next 150 years.

The second question is relatively easy to answer, provided we consider only the estimated recoverable reserves of fossil fuels. The data are shown in Table 5. We may conclude that the total CO_2 addition from fossil fuel combustion will be a little over 3 times the atmospheric content, and that, if present partitions between reservoirs are maintained, the CO_2 in the atmosphere could increase by nearly 170 percent.

The answer to the first question depends upon the rate of increase of fossil fuel combustion. Table 6 shows that if this combustion remains constant at the 1959 level, the total CO_2 injected into the atmosphere by the year 2000 will be about 28 percent of the atmospheric content in 1950. If the average rate of increase of combustion continues at 3.2 percent per year, the quantity injected into the atmosphere by the year 2000 will be about 42 percent; if the 5% rate of increase during the last 8 years persists, the quantity injected will be close to 60 percent. Assuming further that the proportion remaining in the atmosphere continues to be half the total quantity injected, the increase in amospheric CO_2 in the year 2000 could be somewhere between 14 percent and 30 percent.

Based on projected world energy requirements, the United Nations Department of Economic and Social Affairs (1956) has estimated an amount of fossil fuel combustion by the year 2000 that with our assumed partitions would give about a 25 percent increase in atmospheric CO_2, compared to the amount present during the 19th Century. For convenience, we shall adopt this figure in the following estimate of the effects on atmospheric radiation and temperature.

TABLE 5.—*Estimated Remaining Recoverable Reserves of Fossil Fuels*

	10^9 Metric Tons	Carbon Dioxide Equivalent, 10^{18} gms	As % of Atmospheric CO_2 in 1950
Coal and Lignite [1]	2, 320	5. 88	252
Petroleum and Natural Gas Liquids [2]	212	. 67	29
Natural Gas [3]	166	. 43	18
Tar Sands [2]	75	. 24	10
Oil Shales [2]	198	. 63	27
Total	2, 971	7. 85	336

[1] Assumed to be 20 percent lignite containing 45 percent carbon, and 80 percent bituminous coal containing 75 percent carbon.

[2] Assumed carbon content of petroleum, natural gas liquids, and hydrocarbons recoverable from tar sands and oil shales=86 percent.

[3] Assumed composition of natural gas by volume: $CH_4=80$ percent, $C_2H_6=15$ percent, $N_2=5$ percent.

Source: Computed from data given by M. King Hubbert, "Energy Resources, A Report to the Committee on Natural Resources of the National Academy of Sciences—National Research Council," NAS Publication 1000–D, 1962, pp. 1–141.

TABLE 6.—*Estimates of Carbon Dioxide From Fossil Fuel Combustion in Future Decades, Assuming Different Rates of Increase of Fuel Use*

Year	As percent of Atmospheric CO_2 in 1950		
Growth rate, percent/year	0	3. 2	5. 0
1959	13. 80	13. 80	13. 80
1969	17. 30	18. 00	18. 47
1979	20. 79	23. 79	26. 15
1989	24. 28	31. 94	37. 90
1999	27. 77	41. 96	58. 75
2009	31. 26	57. 04	93. 14

POSSIBLE EFFECTS OF INCREASED ATMOSPHERIC CARBON DIOXIDE ON CLIMATE

One of the most recent discussions of these effects is given by Möller (1963). He considers the radiation balance at the earth's surface with an average initial temperature of 15°C (59°F), a relative humidity of 75 percent, and 50% cloudiness. We may compute from his data that with a 25 percent increase in atmospheric CO_2, the average temperature near the earth's surface could increase between 0.6°C and 4°C (1.1°F to 7°F), depending on the behavior of the atmospheric water vapor content. The small increase would correspond to a constant absolute humidity, that is, a constant weight of water in the atmosphere. The larger increase would correspond to a constant relative humidity, that is, as the temperature rose, the water vapor content would also rise to maintain a constant percentage of the saturation value. A doubling of CO_2 in the air, which would happen if a little more than half the reserves of fossil fuels were consumed, would have about three times the effect of a twenty-five percent increase.

As Möller himself emphasized, he was unable to take into account the vertical transfer of latent heat by evaporation at the surface and condensation aloft, or of sensible heat by convection and advection. For this reason he was unable to consider the interactions between different atmospheric layers in a vertical column. In consequence, Möller's computations probably over-estimate the effects on atmospheric temperature of a CO_2 increase. A more comprehensive model is being developed by the U.S. Weather Bureau. This includes processes of convection and of latent heat transfer through the evaporation and condensation of water vapor. Meaningful computations should be possible with this model in the very near future. But climatic changes depend on changes in the general circulation in the atmosphere, and these will be related to the spatial distribution and time variation of carbon dioxide and water vapor. The ratio of CO_2 to water vapor is higher in the polar regions than in low latitudes, higher in winter than in summer, and much higher in the stratosphere than near the ground. For example, the volume of carbon dioxide in the atmosphere at high latitudes is about half the volume of water vapor, while near the equator it is less than a tenth of the water vapor volume. As a result, the radiation balance of the earth will be affected differently at different seasons, latitudes, and heights by changes in the atmospheric CO_2 content. Without a comprehensive model incorporating both the fluid dynamics and the radiation transfer processes of the atmosphere it is not possible to predict how these effects will perturb the general circulation. Such a model may be available within the next two years.

Models of atmospheric thermal equilibrium in which vertical convection is allowed to maintain the observed vertical temperature gradient have recently been constructed by S. Manabe of the U.S. Weather Bureau (Manabe and Strickler, 1964; Manabe, 1965). These show that the effect of infra red absorption from the present atmospheric carbon dioxide at mid latitudes is to maintain a ground temperature about 10°C (18°F) higher than would prevail if no CO_2 were present. An increase in the CO_2 content without a change in absolute humidity would, according to these models, produce a somewhat smaller surface temperature rise than that estimated by Möller. But a considerable change would occur in the stratosphere, where the CO_2 concentration by volume is perhaps 50 times that of water vapor. A 25% rise in carbon dioxide would cause stratospheric temperatures to fall by perhaps 2°C (3.6°F) at an altitude of 30 kilometers (about 100,000 feet) and by 4°C (7°F) at 40 kilometers (about 130,000 feet).

One might suppose that the increase in atmospheric CO_2 over the past 100 years should have already brought about significant climatic changes, and indeed some scientists have suggested this is so. The English meteorologist, G. S. Callendar (1938, 1940, 1949), writing in the late 1930's and the 1940's on the basis of the crude data then available, believed that the increase in atmospheric CO_2 from 1850 to 1940 was at least 10%. He thought this increase could account quantitatively for the observed warming of northern Europe and northern North America that began in the 1880's. From Table 2 and our estimate of the CO_2 partition between the atmospheric, the ocean, and the biosphere, we see that the actual CO_2 increase in the atmosphere prior to 1940 was only 4%, at least from fossil fuel combustion. This was probably insufficient to produce the observed temperature changes. [But it should be noted that up to 2.5% of the atmospheric carbon dioxide (after partition with the ocean and the biosphere) could also have been added by the oxidation of soil humus in newly cultivated lands.]

As Mitchell (1961, 1963) has shown, atmospheric warming between 1885 and 1940 was a world-wide phenomenon. Area-weighted averages for surface temperature over the entire earth show a rise in mean annual air temperature of about 0.5°C (0.9°F). World mean winter temperatures rose by 0.9°C (1.6°F). Warming occurred in both hemispheres and at all latitudes, but the largest annual rise (0.9°C or 1.6°F) was observed between 40° and 70° N latitudes. In these latitudes, the average winter temperatures rose by 1.6°C (2.8°F).

The pronounced warming of the surface air did not continue much beyond 1940. Between 1940 and 1960 additional warming occurred in northern Europe and North America, but for the world as a whole and also for the northern hemisphere, there was a slight lowering of about 0.1°C (0.2°F) in mean annual air temperature (Mitchell, 1963). Yet during this period more than 40% of the total CO_2 increase from fossil fuel combustion occurred. We must conclude that climatic "noise" from other processes has at least partially masked any effects on climate due to past increases in atmospheric CO_2 content.

OTHER POSSIBLE EFFECTS OF AN INCREASE IN ATMOSPHERIC CARBON DIOXIDE

Melting of the Antarctic ice cap.—It has sometimes been suggested that atmospheric warming due to an increase in the CO_2 content of the atmosphere may result in a catastrophically rapid melting of the Antarctic ice cap, with an accompanying rise in sea level. From our knowledge of events at the end of the Wisconsin period, 10 to 11 thousand years ago, we know that melting of continental ice caps can occur very rapidly on a geologic time scale. But such melting must occur relatively slowly on a human scale.

The Antarctic ice cap covers 14 million square kilometers and is about 3 kilometers thick. It contains roughly 4 x 10^{16} tons of ice, hence 4 x 10^{24} gram calories of heat energy would be required to melt it. At the present time, the poleward heat flow across 70° latitude is 10^{22} gram calories per year, and this heat is being radiated to space over Antarctica without much measurable effect on the ice cap. Suppose that the poleward heat flux were increased by 10% through an intensification of the meridional atmospheric circulation, and that all of this increase in the flow of energy were utilized to melt the ice. Some 4,000 years would be required.

We can arrive at a smaller melting time by supposing a change in the earth-wide radiation balance, part of which would be used to melt the ice. A 2% change could occur by the year 2000, when the atmospheric CO_2 content will have increased perhaps by 25%. Since the average radiation at the earth's surface is about 2 x 10^5 gram calories per square centimeter per year, a 2% change would amount to 2 x 10^{22} calories per year. If half this energy were concentrated in Antarctica and used to melt the ice, the process would take 400 years.

Rise of sea level.—The melting of the Antarctic ice cap would raise sea level by 400 feet. If 1,000 years were required to melt the ice cap, the sea level would rise about 4 feet every 10 years, 40 feet per century. This is a hundred times greater than present worldwide rates of sea level change.

Warming of sea water.—If the average air temperature rises, the temperature of the surface ocean waters in temperate and tropical regions could be expected to rise by an equal amount. (Water temperatures in the polar regions are roughly stabilized by the melting and freezing of ice.) An oceanic warming of 1° to 2°C (about 2°F) occurred in the North Atlantic from 1880 to 1940. It had a pronounced effect on the distribution of some fisheries, notably the cod fishery, which has greatly increased around Greenland and other far northern waters during the last few decades. The amelioration of oceanic climate also resulted in a marked retreat of sea ice around the edges of the Arctic Ocean.

Increased acidity of fresh waters.—Over the range of concentrations found in most soil and ground waters, and in lakes and rivers, the hydrogen ion concentration varies nearly linearly with the concentration of free CO_2. Thus the expected 25% increase in atmospheric CO_2 concentration by the end of this century should result in a 25% increase in the hydrogen ion concentration of natural waters or about a 0.1 drop in pH. This will have no significant effect on most plants.

Increase in photosynthesis.—In areas where water and plant nutrients are abundant, and where there there is sufficient sunlight, carbon dioxide may be the limiting factor in plant growth. The expected 25% increase by the year 2000 should significantly raise the level of photosynthesis in such areas. Although very few data are available, it is commonly believed that in regions of high plant productivity on land, such as the tropical rain forests, phosphates, nitrates and other plant nutrients limit production rather than atmospheric CO_2. This is probably also true of the oceans.

Biological processes are speeded up by a rise in temperature, and in regions where other conditions are favorable higher temperatures due to increased CO_2 might result in higher plant production.

OTHER POSSIBLE SOURCES OF CARBON DIOXIDE

We are fairly certain that fossil fuel combustion has been the only source of CO_2 coming into the atmosphere during the last few years, when accurate measurements of atmospheric carbon dioxide content have been available. Carbon dioxide may have been produced by other sources during earlier times but it is not now possible to make a quantitative estimate. However we can examine the order of magnitude of some of the possible inputs from other sources, on the basis of our knowledge of the processes that might be involved.

Oceanic warming.—The average temperature of the ocean cannot have increased by more than 0.15°C during the past century, since any greater warming would have caused a larger rise in sea level than the observed value of about 10 centimeters. A more probable upper limit is .05°C, because most of the sea level rise can be accounted for by glacial melting. An average .05°C rise would correspond to 0.5° in the top 400 meters. This would cause a nearly 3% rise in the CO_2 partial pressure.

After equilibration with the atmosphere, the partial pressure in both the air and the uppermost ocean layer would be higher by about 2.5%.

Burning of limestone.—Annual world production of carbon dioxide from the use of limestone for cement, fluxing stone, and in other ways, is about 1% of the total from fossil fuel combustion, or 4×10^{-5} of the atmospheric CO_2 content per year.

Decrease in the carbon content of soils.—Since the middle of the Nineteenth Century, the world's cultivated farmland has been enlarged by about 50%. This is an increase of close to a billion acres or 1.6 million square miles, corresponding to 2.7% of the land area of the earth, and perhaps to 5% of forests and grass lands. Most soil humus is believed to be concentrated in forests and grassy areas. Assume that the total humus is equal to twice the amount of carbon in the atmosphere and that half the carbon in the humus of the newly cultivated lands has been oxidized to carbon dioxide. The total injected into the atmosphere from this source becomes less than 5% of the atmospheric CO_2.

Change in the amount of organic matter in the ocean.—About 7% of the marine carbon reservoir consists of organic material. Since a 1% change in the carbon dioxide content of the ocean changes the CO_2 pressure by 12.5%, a decrease by 1% in the marine organic carbon (which would increase the total oceanic carbon dioxide by .07%) would raise the carbon dioxide pressure of the ocean and the atmosphere by about 1%. An increase in the temperature of water near the surface, during the past one hundred years, could have speeded up the rate of oxidation of organic matter relative to its rate of production by photosynthesis. Measurements of the content of organic matter in the ocean are neither accurate enough nor sufficiently extended over time to allow a direct estimate of this possibility. A change of several percent could have occurred without detection.

Changes in the carbon dioxide content of deep ocean water.—The deep ocean waters contain about 10% more carbon dioxide than they would if they were at equilibrium with the present atmospheric content. This is a result of the sinking of dead organic remains from the surface waters and their subsequent oxidation in the depths. The combination of biological and gravitational processes can be thought of as a pump that maintains a relatively low carbon dioxide content in the surface waters and in the atmosphere. If the pump ceased to act, the atmospheric carbon dioxide would eventually be increased five fold. Variations in the effectiveness of the pump could have occurred without detection during the past 100 years, and could have caused notable changes in the atmospheric carbon dioxide content.

Changes in the volume of sea water.—During the Ice Age the volume of sea water varied by about 5%. Changes of this magnitude would change the carbon dioxide content of the atmosphere by 10 to 15%.

But during the last several thousand years, variations in oceanic volume have been small. During the past hundred years, world average sea level has varied by less than 10 centimeters. This very small volume change would have no appreciable effect on the atmospheric carbon dioxide.

Carbon dioxide from volcanoes.—Over geologic time, volcanic gases have been the principal sources of new carbon dioxide injected into the atmosphere. On the average the influx of volcanic CO_2 must have balanced the extraction from the atmosphere by rock weathering. The present rate of influx of volcanic CO_2 is close to a hundred fold less than that from fossil fuel combustion. No data exist on the worldwide level of volcanic activity over geologic time. It is conceivable that the level has fluctuated by two orders of magnitude, and that the fluctuations persisted for millenia, or even for millions of years.

Changes due to solution and precipitation of carbonates.—Calcium and magnesium carbonate precipitation on the sea floor lower the total CO_2 content of ocean water, but increase the carbon dioxide pressure and the free CO_2 content. Conversely, chemical weathering of limestone and dolomite on land lower the atmospheric CO_2 and the free CO_2 content of the sea, but increase the total oceanic CO_2. The rates of these processes are about one order of magnitude lower than the present rate of production of carbon dioxide by fossil fuel combustion.

We conclude that the only sources of carbon dioxide comparable in magnitude to fossil fuel combustion during the last 100 years could have been a decrease in soil humus due to the increase in the area of cultivated lands, a decrease in the content of "dissolved" organic matter in the ocean, or a lowering of the carbon dioxide content of deep ocean waters. Marked changes in the oceanic regime would have been necessary for the latter two processes to have significant effects. As we have shown, none of the three processes are likely to be significant at the present time. Nor are any oceanographic data available which suggest that the required changes in the ocean occurred during the last hundred years.

CONCLUSIONS AND FINDINGS

Through his worldwide industrial civilization, Man is unwittingly conducting a vast geophysical experiment. Within a few generations he is burning the fossil fuels that slowly accumulated in the earth over the past 500 million years. The CO_2 produced by this combustion is being injected into the atmosphere; about half of it remains there. The estimated recoverable reserves of fossil fuels are sufficient to produce nearly a 200% increase in the carbon dioxide content of the atmosphere.

By the year 2000 the increase in atmospheric CO_2 will be close to 25%. This may be sufficient to produce measurable and perhaps marked changes in climate, and will almost certainly cause significant changes in the temperature and other properties of the stratosphere. At present it is impossible to predict these effects quantitatively, but recent advances in mathematical modelling of the atmosphere, using large computers, may allow useful predictions within the next 2 or 3 years.

Such predictions will need to be checked by careful measurements: a series of precise measurements of the CO_2 content in the atmosphere should continue to be made by the U.S. Weather Bureau and its collaborators, at least for the next several decades; studies of the oceanic and biological processes by which CO_2 is removed from and added to the atmosphere should be broadened and intensified; temperatures at different heights in the stratosphere should be monitored on a worldwide basis.

The climatic changes that may be produced by the increased CO_2 content could be deleterious from the point of view of human beings. The possibilities of deliberately bringing about countervailing climatic changes therefore need to be thoroughly explored. A change in the radiation balance in the opposite direction to that which might result from the increase of atmospheric CO_2 could be produced by raising the albedo, or reflectivity, of the earth. Such a change in albedo could be brought about, for example by spreading very small reflecting particles over large oceanic areas. The particles should be sufficiently buoyant so that they will remain close to the sea surface and they should have a high reflectivity, so that even a partial covering of the surface would be adequate to produce a marked change in the amount of reflected sunlight. Rough estimates indicate that enough particles partially to cover a square mile could be produced for perhaps one hundred dollars. Thus a 1% change in reflectivity might be brought about for about 500 million dollars a year, particularly if the reflecting particles were spread in low latitudes, where the incoming radiation is concentrated. Considering the extraordinary economic and human importance of climate, costs of this magnitude do not seem excessive. An early development of the needed technology might have other uses, for example in inhibiting the formation of hurricanes in tropical oceanic areas.

According to Manabe and Strickler (1964) the absorption and reradiation of infrared by high cirrus clouds (above five miles) tends to heat the atmosphere near the earth's surface. Under some circumstances, injection of condensation or freezing nuclei will cause cirrus clouds to form at high altitudes. This potential method of bringing about climatic changes needs to be investigated as a possible tool for modifying atmospheric circulation in ways which might counteract the effects of increasing atmospheric carbon dioxide.

REFERENCES

Arrhenius, Svante, 1903: Lehrbuch der kosmischen Physik 2. Leipzig: Hirzel.

Bolin, Bert, and Eriksson, Erik, 1958: Changes in the carbon dioxide content of the atmosphere and sea due to fossil fuel combustion. Rossby Memorial Volume, edited by B. Bolin, Rockefeller Institute Press, New York, 1959. pp. 130–142.

Bolin, B., and Keeling, C. D., 1963: Large-scale atmospheric mixing as deduced from the seasonal and meridional variations of carbon dioxide. *Journal of Geophysical Research,* Vol. 68, No. 13. pp. 3899–3920.

Broecker, Wallace S., 1963: C^{14}/C^{12} ratios in surface ocean water. Nuclear Geophysics: Proceedings of conference NAS/NRC Publ. 1075, Washington, D.C. pp. 138–149.

Broecker, Wallace S., 1965: Radioisotopes and oceanic mixing. Manuscript Report, Lamont Geological Observatory, Palisades, New York.

Brown, Craig W., and Keeling, C. D., 1965: The concentration of atmospheric carbon dioxide in Antarctica. Manuscript submitted for publication, Scripps Institution of Oceanography.

Callendar, G. S., 1938: The artificial production of carbon dioxide and its influence on temperature. *Quarterly Journal Royal Meteorol. Soc.,* Vol. 64. p. 223.

Callendar, G. S., 1940: Variations in the amount of carbon dioxide in different air currents. *Quarterly Journal Royal Meteorol. Soc.,* Vol. 66. p. 395.

Callendar, G. S., 1949: Can carbon dioxide influence climate? *Weather* Vol. 4. p. 310.

Callendar, G. S., 1961: Temperature fluctuations and trends over the earth. *Quarterly Journal Royal Meteorol. Soc.,* Vol. 87, No. 371. pp. 1–12.

Chamberlin, T. C., 1899: An attempt to frame a working hypothesis of the cause of glacial periods on an atmospheric basis. *Journal of Geology,* Vol. 7. pp. 575, 667, 751.

Conservation Foundation, 1963: Report of conference on rising carbon dioxide content of the atmosphere.

Craig, Harmon, 1957: The natural distribution of radiocarbon and the exchange time of carbon dioxide between atmosphere and sea. *Tellus,* Vol. 9, No. 1. pp. 1–17.

Eriksson, Erik, 1963: The role of the sea in the circulation of carbon dioxide in nature. Paper submitted to Conservation Foundation Conference, March 12, 1963.

Eriksson, Erik, 1963: Possible fluctuations in atmospheric carbon dioxide due to changes in the properties of the sea. *Journal of Geophysical Research,* Vol. 68, No. 13. pp. 3871–3876.

Hubbert, M. King, 1962: Energy resources; a report to the committee on natural resources of the National Academy of Sciences—National Research Council. Publication 1000–D, National Academy of Sciences—National Research Council, Washington, D.C.

Kaplan, Lewis D., 1959: The influence of carbon dioxide variations on the atmospheric heat balance. *Tellus,* Vol. 12 (1960), No. 2. pp. 204–208.

Kraus, E. B., 1963: Physical aspects of deduced and actual climatic change. *Annals of New York Academy of Sciences,* Vol. 95. pp. 225–234.

Lamb, H. H., and Johnson, A. I., 1959: Climatic variation and observed changes in the general circulation. *Geografiska Annaler,* Vol. 41. pp. 94–134.

Lamb, H. H., and Johnson, A. I., 1961: Climatic variation and observed changes in the general circulation. *Geografiska Annaler,* Vol. 43. pp. 363–400.

Lieth, Helmut, 1963: The role of vegetation in the carbon dioxide content of the atmosphere. *Journal of Geophysical Research,* Vol. 68, No. 13. pp. 3887–3898.

Lysgaard, L., 1963: On the climatic variation. *Changes of Climate,* Rome Symposium by UNESCO–WHO; Published in Arid Zone Research Volumes, 20, Paris. pp. 151–159.

Manabe, Syukuro, and Strickler, Robert F., 1964: Thermal equilibrium of the atmosphere with a convective adjustment. *Journal of Atmospheric Sciences,* Vol. 21, No. 4. pp. 361–385.

Manabe, Syukuro, 1965: Dependence of the climate of the earth's atmosphere on the change of the content of some atmospheric absorbers. Text of talk made at summer study session of NAS Panel on Weather and Climate Modification.

Mitchell J. Murray, Jr., 1963: Recent secular changes of global temperature. *Annals of New York Academy of Sciences,* Vol. 95. pp. 235–250.

Mitchell, J. Murray, Jr., 1963: On the world-wide pattern of secular temperature change. *Changes of Climate,* Rome Symposium by UNESCO–WHO; Published in Arid Zone Research Volumes, 20, Paris. pp. 161–181.

Möller, F., 1963: On the influence of changes in the CO_2 concentration in air on the radiation balance of the earth's surface and on the climate. *Journal of Geophysical Research,* Vol. 68, No. 13. pp. 3877–3886.

Pales, Jack C., and Keeling, C. D., 1965: The concentration of atmospheric carbon dioxide in Hawaii. Manuscript submitted for publication, Scripps Institution of Oceanography.

Plass, Gilbert N., 1955: The carbon dioxide theory of climatic change. *Tellus,* Vol. 8 (1956), No. 2. pp. 140–154.

Plass, Gilbert N., 1956: The influence of the 15μ carbon-dioxide band on the atmospheric infra-red cooling rate. *Quarterly Journal Royal Meteorol. Soc.,* Vol. 82. pp. 310–324.

Plass, Gilbert N., 1961: Letter to the editor concerning "The influence of carbon dioxide variations on the atmospheric heat balance," by L. D. Kaplan, *Tellus,* Vol. 12 (1960) pp. 204–208. *Tellus,* Vol. 13 (1961), No. 2. pp. 296–300.

Revelle, Roger, and Suess, Hans E., 1956: Carbon dioxide exchange between atmosphere and ocean and the question of an increase of atmospheric CO_2 during the past decades. *Tellus,* Vol. 9 (1957), No. 1. pp. 18–27.

Suess, H. E., 1965: Secular variation of the cosmic ray-produced carbon–14 in the atmosphere and their interpretations. Submitted for publication to *Journal of Geophysical Research,* Scripps Institution of Oceanography.

United Nations Department of Economic and Social Affairs: World energy requirements in 1975 and 2000. Proceedings of the International Conference on the Peaceful Uses of Atomic Energy, 1956. pp. 3–33.

United Nations, 1961–64: World energy supplies. Statistical Papers, Series J. United Nations, New York.

IV

Land Use Control

A. In General

Environmental Quality. The First Annual Report of the Council on Environmental Quality, Transmitted to the Congress, August 1970. Pages 165-92.

The work of the Council on Environmental Quality, and the significance of its first Annual Report has already been described, see Chapter I. The excerpt from the report, reproduced below, which deals with land use has two noteworthy features. First, it stresses the dynamic nature of land uses. Changing economic, sociological, and technological trends have changed land uses in the past dramatically and will undoubtedly do so in the future as well. An active national land use policy will inevitably be dealing with trends and forces rather than facts and conditions. An appropriate institutional framework must take this into account. Second, the report stresses the enormous influence which the federal government has on land use policy. The question, seen in this light, becomes not whether the federal government should influence land use decisions, but how it should. As the report notes, land uses today are the product of numerous state, local, and federal decisions, as well as those of private institutions, and many of those decisions are uncoordinated and sometimes even conflicting.

A National Land Use Policy has been proposed both by President Nixon and by Senator Henry M. Jackson. Unquestionably, should such a proposal ever become law it will be based at least in part on some of the premises set forth below.

IX

Land Use

The first men upon this land, the American Indians, treated it with reverence, blended with it, used it, but left hardly a trace upon it. Those who followed have been less kind. They brought with them a different creed which called on man to conquer nature and harness it for his own use and profit.

Now more than halfway into the 20th century what they have done is being turned back upon them. Misuse of the land is now one of the most serious and difficult challenges to environmental quality, because it is the most out-of-hand, and irreversible. Air and water pollution are serious, hard to manage problems, too. But they are worked at with standards, with enforcement tools, and by institutions set up for those specific antipollution purposes. Land use is still not guided by any agreed upon standards. It is instead influenced by a welter of sometimes competing, overlapping government institutions and programs, private and public attitudes and biases, and distorted economic incentives.

The 50 States comprise about 2.3 billion acres of land. Of that, 1.9 billion acres lie in the contiguous 48 States. Nearly 58 percent of the land area is used for crops and livestock. More than 22 percent is ungrazed forest land. Less than 3 percent is in urban and transportation uses, although it is increasing. Areas designated primarily for parks, recreation, wildlife refuges, and public installations and facilities account for about 5 percent. The rest—12 percent—is mainly desert, swamp, tundra, and other lands presently of limited use by man.

Almost 59 percent of the land is in private hands. Thirty-four percent is owned by the Federal Government. Some 94 percent of all Federal lands lie in the westernmost States—about half in Alaska alone. About 2 percent of the Nation's land is held by Indians. The rest—about 5 percent—is State, city, or county owned.

But land is not just acreage. Land embraces the complex biological systems of the soil and the plants and animals which are all part of a continuing life cycle. Man's understanding of these biological processes, particularly of the permanent damage that begins subtly with piecemeal alterations of the land, is still limited. Yet his dependence upon its stability is enormous.

In the Nation's early history, easy availability of land prodded millions to join the massive migrations west and along the major river basins. The wilderness was to be tamed, the trees cleared, and the soil put to crops. Much of the wilderness is now gone, and most of what is left is far distant from the three-quarters of the Nation that lives in the cities and the suburbs. The landscape visible to most Americans is cluttered with traffic, neon signs, powerlines, and sprawl. Flood plains are not just for the river but are subjected to intense development. Open space, the elbow room for urban man, continues to dwindle. Unfortunately, traditions of land use have derived from an assumption that land is a limitless commodity—not a finite biological community.

Government spurs much of this land development by where it locates and how it designs airports and highways, insures home loans, permits filling of wetlands, and lays water and sewer lines. Local governments exercise the primary authority over land use. But effective public influence is hampered by a lack of agreement on objectives, by misplaced economic incentives, and by failure of local governments to harmonize land use.

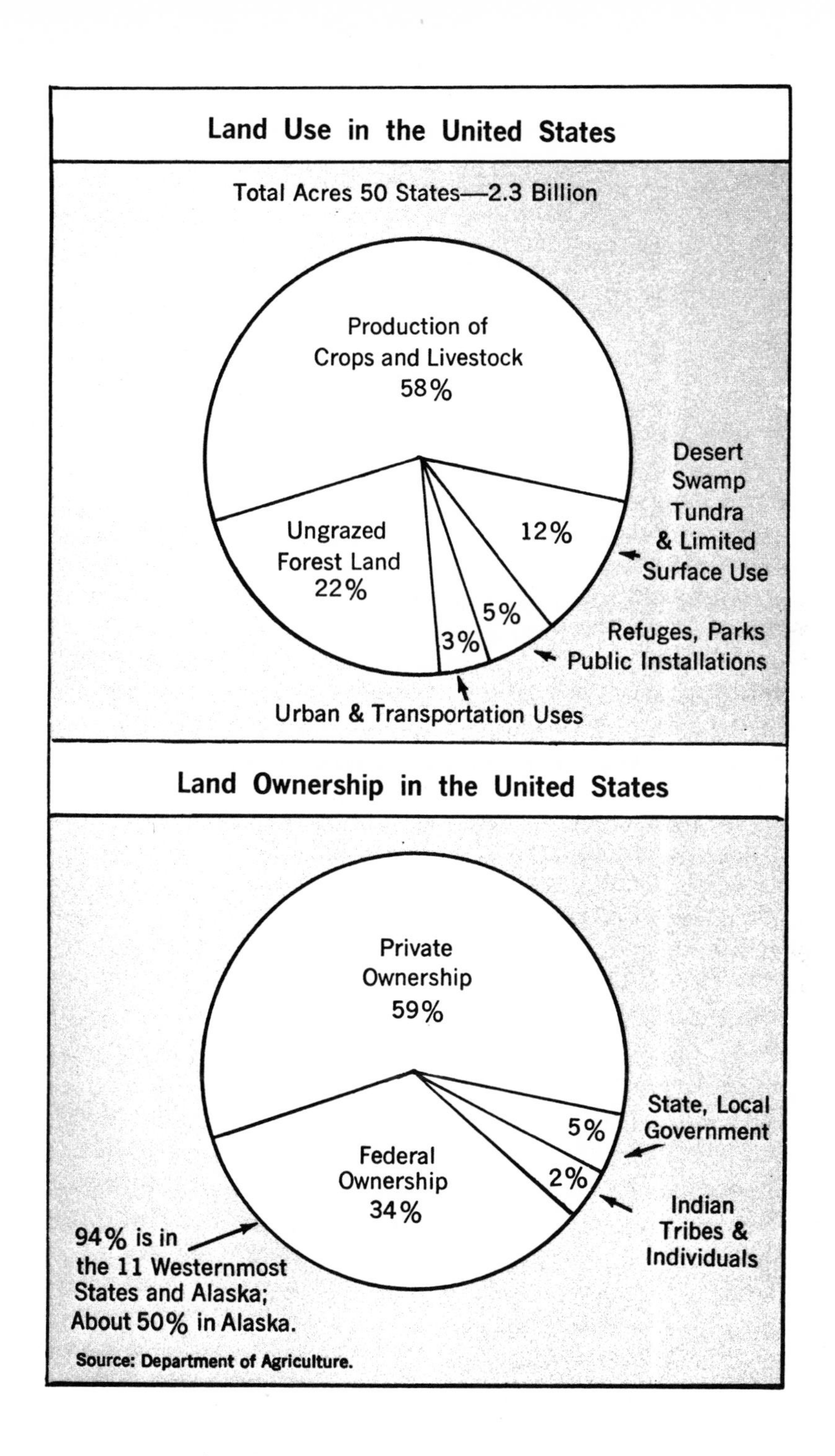

Source: Department of Agriculture.

THE STATUS OF THE LAND

The Urban Environment

Twentieth-century America has known two great population displacements—the first from farms to cities and the second from the cities to the suburbs. Three out of four Americans now live in urban environments—in incorporated settlements of at least 2,500 population and in metropolitan areas consisting of central cities and suburbs. Metropolitan areas alone are home to two of every three Americans, and the percentage is rising.

Residential patterns in many metropolitan areas resemble a series of sharply defined concentric circles. The decaying inner city houses the majority of the urban poor, usually nonwhites who ordinarily have little choice but to stay. Surrounding the inner city are neighborhoods commonly less old and less dense, usually populated by white working-class families. These families often lean toward leaving the central city because of rising property taxes, deteriorating school systems, and racial changes. But they have stayed because they could not afford the newer suburban housing or because of strong ties to their neighborhoods. The 1970 census may show some shifts from this pattern. Beyond the city is suburbia. The suburbs are a magnet for those seeking escape from the burdens of urban life while retaining some of its advantages. (In some areas the pattern is different; the poor often live on the fringes, and the affluent in the city itself, and mixed patterns occur in other metropolitan areas.)

The financial plight of the cities is well known. The influx of the poor and the exodus of the middle class and the wealthy, among other factors, have drained the cities of many of their revenue sources. Public services, such as police protection, park maintenance and sanitation, have deteriorated.

The most visible effects of these economic pressures are the rapid pace and nature of physical change. This is clearly evident in downtown areas where the constant cycle of construction and demolition is often considered a city's badge of growth. Pressed for revenues, many cities bow to the demands of developers to replace historic buildings and distinctive architecture with almost uniform steel and glass box office buildings. Unfortunately, this construction may simply put more people on the sidewalks and more cars on the streets, more monotonous skyscrapers towering above, and more noise and congestion below. Much downtown rebuilding has furthered the trend toward daytime cities with facilities such as offices and banks, which have no nighttime uses. Cities lose their uniqueness as their historic buildings and neighborhoods are replaced by the dullest in modern architecture. The result is often a dreary sameness in the appearance and character of downtown areas.

Physical deterioration is overtaking the housing in many areas of central cities. The oldest housing traditionally filters down to the poorest families as previous occupants better their lot and move to new apartments or suburban homes. Of the 1.7 million substandard housing units in urban areas, nearly all lie in central cities. In slum neighborhoods of some large cities, the amount of abandoned housing is as high as 15 percent.

With the well-known reliance of the suburbanite on the automobile, often to commute to the center city, many downtown and other urban centers have gradually become auto dominated. Much of the change and reconstruction in downtown areas is for freeways, parking garages, and lots—lavish users of space. However, the alternative, public transportation, is constantly frustrated by rising costs, the sprawling nature of the suburbs, and dependence on the automobile.

Many cities have lost the spirit to attract people downtown. With the growing number of shops and other services locating in the suburbs and with crime threatening many local businesses and frightening people from the streets, much of the vitality is missing from the central cities. Many city officials, however, are trying to bring people and excitement back. Apartments in downtown areas, sidewalk cafes, outdoor concerts, bicycle routes, even saving the cable cars in San Francisco—all are small but key efforts to recapture this lost spirit. Rehabilitation, successfully carried out in Philadelphia, and a few other cities signals that the character of the urban environment can be revitalized.

Few cities have kept pace with parkland needs. Trees have not been planted to shade busy avenues at the same rate that they have been felled for street widening. Too often only a few species are planted, and they are often blighted by disease and insect invasions so common to unstable ecosytems. In the Midwest, Dutch elm disease has decimated row on row of beautiful elms.

The immediate economic pressure on a city to permit a parking lot or building on what might remain open space, or to use parklands as part of a freeway route, is often insurmountable. Nevertheless, the accelerating cost of land acquisition and the growing need to preserve open space in a crowded urban environment make the purchase of open areas a sound long-range economic practice, which continues to pay immeasurable dividends. The Boston Common, New York's Central Park, Washington's Rock Creek Park, San Francisco's Golden Gate Park, and many other city parks are tributes to the foresight of early planners in saving large ópen areas. The protection of streams, ponds, and marshes within cities permits the survival there of numerous species of wildlife, including small animals, birds, and waterfowl which adapt surprisingly well to the urban environment.

Man requires a feeling of permanence to attain a sense of place, importance, and identity. For many persons in the city, the presence of nature is the harmonizing thread in an environment otherwise of man's own making.

The Suburban Environment

Although rural to urban migration was primarily economically motivated, the suburban impulse is largely a matter of social preference. Because of economic and social obstacles, these population shifts have affected the races unequally. Eighty percent of the blacks in metropolitan areas now live in central cities, while 60 percent of the metropolitan whites live in the suburbs. Zoning practices, subdivision controls, and the higher costs of suburban living have made it difficult for lower income minority groups to move from the city to the suburbs.

From 1950 to 1969, while the population of central cities increased only 12 percent, the population of suburban areas soared 91 percent. By 1969, there were 71 million suburbanites and 59 million central city residents.

The suburban tide—Since the late 1800's a blend of town and country has stood for the optimum residential environment. The suburb is thought to offer the best of both worlds. Besides enjoying ready access to the large city with its concentrated economic and cultural facilities, the suburban resident seeks a crime-free neighborhood amid clean air, open lawns, and quiet and uncrowded living. Streetcar and occasional subway lines started what at first was but a trickle of people from the central cities to the outskirts of town. The Federal Government quickened the outward flow after World War II by providing mortgage assistance, which enabled many central city residents to become suburban homeowners. Later the vast urban freeway systems turned the flow into a flood. The automobile now controls suburban life.

To the couple living in an apartment downtown, the birth of a second child is a common signal to abandon the city. And often families with school-age children leave for the better educational systems of the suburbs. As industries convert to modern, single-flow assembly production processes, manufacturers also forego the city for rambling suburban plants. Since World War II, space needs per industrial worker have quadrupled, and three of every four new manufacturing jobs have been created in suburbs.

Each year, expanding urban areas consume an estimated 420,000 acres of land in an undiscriminating outward push. Development moves out from the city along transportation corridors, branching out from the highways and expressway interchanges. Extension of water and sewer services generates whole new developments on quickly divided farms. After outlying areas are built up at moderate densities, developers often return to land which was passed over as undesirable or too costly in the first wave.

Many suburban communities zone to assure that house lots are large and apartment houses few, a practice that assumes that land is abundant. This zoning practice, in seeking to attract moderate to high income families, tends to exclude those in greatest need of the jobs opening in the suburbs. Excluding them deepens the concentration of poverty and unemployment in the central city ghettos.

Highways and freeways become congested as the tide of suburban commuters to the city grows. This congestion, together with the lengthening distances from suburban homes to downtown offices produces tension and robs the typical suburbanite of time with his family.

The impact of growth—Although the impact of the rural-to-suburban shift of land use varies greatly throughout the country, certain effects tend to be common to this change. Open space is continuously eaten up by housing, which, with most present subdivision practices, provides few parks but instead only offers each family its individual front and back yard. Space is likewise diminished by other facilities required by suburban development. Shopping centers and highway interchanges, made necessary by dependence on the automobile and truck, consume large portions of land. Airports, commonly constructed in suburban or exurban areas and constantly growing in size and number, pose similar problems on an even larger scale, attracting a vast conglomeration of light industry and housing. Consequently, the growing suburban population finds less and less public open space.

Building and construction practices, together with the quickened pace of development and complementary zoning, often end in severe abuse of the land and are ultimately costly to the public. The popular practice of stripping subdivisions of all cover before commencing construction destroys tree and plant cover and can trigger heavy soil runoff. Sedimentation from this runoff in urbanizing areas loads nearby streambeds and ultimately river channels. This can cause costly downstream dredging, upstream flood control and destruction of the esthetic quality of lakes and rivers.

Public pressure for flood control projects is often spurred by suburban development along flood plains, which usually contain fertile soil supporting an abundant variety of native plant and animal life. Construction over aquifer recharge areas, where the groundwater is normally replenished, accelerates rapid runoff, increases flooding, and contributes to water shortages.

Suburban development often spreads across ridges and slopes which should be left alone because of their beauty and because their trees and plant cover absorb rain and inhibit flooding. Trees are not only important for their esthetic qualities and as habitat for birds and wildlife, but they affect temperature and air pollution as well. Building on steep slopes can affect soil stability, causing severe erosion which then undermines foundations. Nevertheless, few cities or counties adequately control development of flood plains, steep slopes, or land above aquifer recharge areas. Important data concerning aquifers, subsoil composition, cover, wetlands, and wildlife are not considered by many planning and most zoning boards.

Esthetically, this current pattern of growth triggers at least three adverse consequences. First, much commercial development along roads and highways through suburbs is of cheap and unimaginative construction. Gaudy neon signs, billboards, powerlines, and clutter characterize this development. Second, many residential subdivisions are visually boring—block after block of treeless lawns, uniform setbacks, and repetitious housing designs and street layouts. Finally, wooded streambeds, slopes, and ridges, which could help break the monotony of uniform housing developments, are often destroyed.

The future—Many local governments and developers are learning that careful site planning can forestall abuse of the natural environment and help meet both housing and environmental goals.

The combination of open space with cluster zoning or planned unit development lowers the initial community service costs because of smaller networks of roads and utilities, and it makes a more livable environment for the long term. These developments can vary from new satellite communities like Columbia, Md., to small, sensitively designed clusters of townhouses around a common green. The challenge is to change traditional ways of building and development and to break down the economic and institutional barriers that obstruct widespread use of these innovations.

The Rural Environment

Although polls have consistently uncovered a strong American preference for life in small towns and rural areas, the last few decades have seen a huge exodus to urban centers. About half of the Nation's 3,000 counties, most of them rural, lost population between 1960 and 1970. Of the 53 million Americans living in rural areas, the number living on farms has dropped below 10 million. The remaining 40-odd million rural residents live in villages and towns of less than 2,500 people or on small plots of country land.

Changing rural land use—Statistics on rural land use show conflicting trends. The Department of Agriculture estimates that 2 million acres of land each year, excluding surface mining, are converted to nonagricultural use. Half of this is shifted to such uses as wildlife refuges, recreation areas, and parks. The other 1 million acres are converted to more intensive uses. Of that, 160,000 are covered by highways and airports. An estimated 420,000 acres become reservoirs and flood control projects. The remaining 420,000 acres are developed for urban uses. About half the national acreage converted every year to urban uses is cropland and grassland pasture. The rest is forest and other land.

Food and fiber production by U.S. farmers in 1967 exceeded by 38 percent the 1950 volume. Yet during the same period the number of farms diminished by 2.5 million, and the number of acres harvested declined by a net of 34 million. In 1969, about 334 million acres of land produced all the country's crops for domestic consumption and export. Another 98 million acres of cropland lay idle, in soil improvement crops or in pasture.

From 1944 to 1964, roughly 22.5 million acres were brought into production through private and Government-assisted drainage and irrigation programs. At the same time, in parts of the East not yet urbanized, much of abandoned cropland changed to grassland. It was managed as such for pasture or simply left idle in the first stage

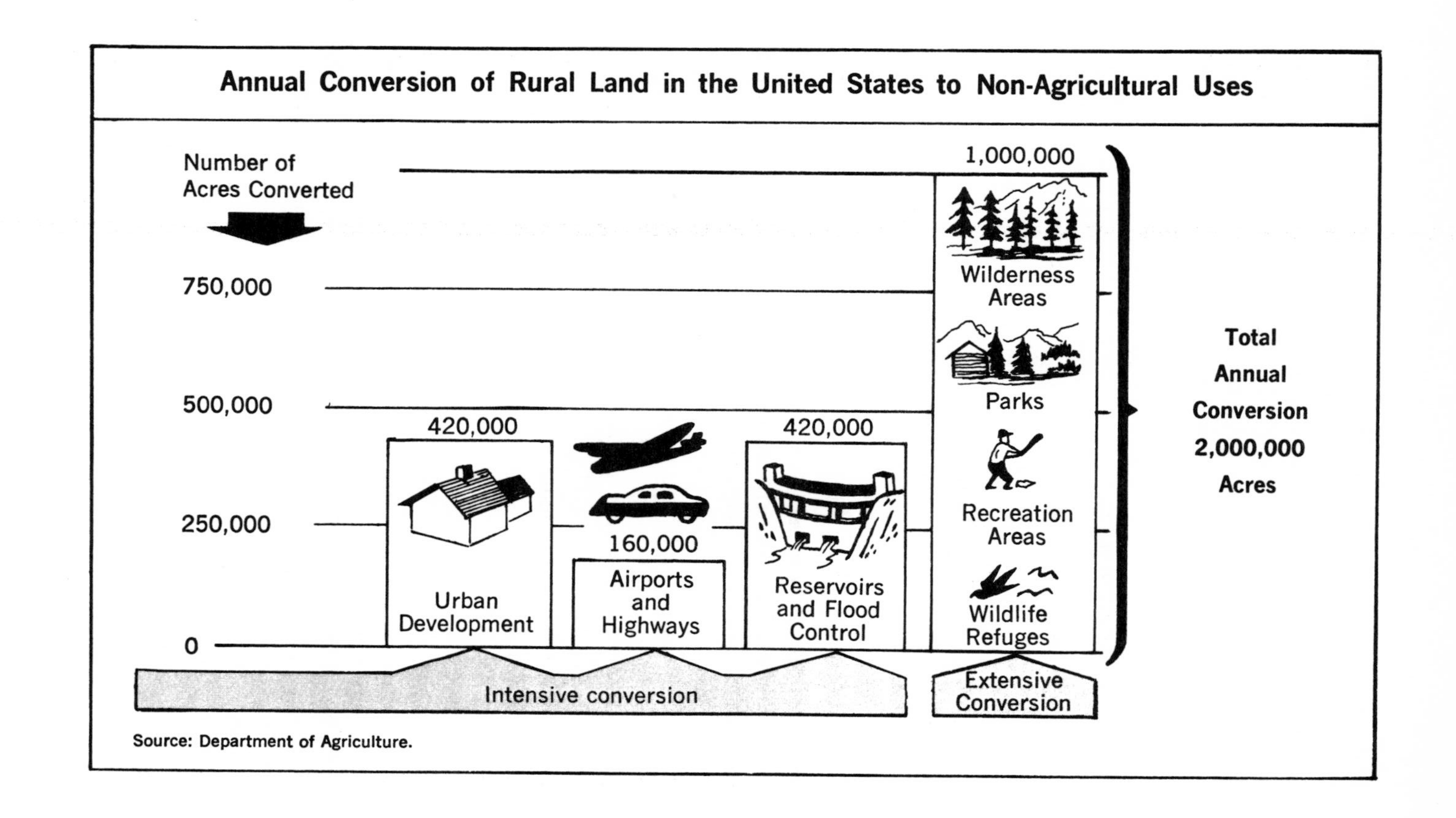

Source: Department of Agriculture.

of natural revegetation. Some of this land now is undergoing or has passed through grass and brush stages to a final forest state.

Effects—The agricultural land use pattern today reflects a highly technical, mechanized, and product-specialized food and fiber industry. It has used the most productive land in the Nation, and it has altered other lands for crop production through vast Government expenditures—often at the expense of marshes, swamps, and bottom lands needed as waterfowl and wildlife habitat. The impact that this highly intensive use of land has on the preservation and diversity of our wildlife and on the integrity of our soils is just beginning to be understood.

Much of the character of farming today is dominated by monoculture, the production of single cash crops. Monoculture has increased production efficiency but has reduced the plant and wildlife diversity essential to a stable ecosystem. These less complex ecosystems are highly susceptible to attack by insects and diseases which can devastate a standing crop or single species regionwide. Moreover, monoculture has forced a heavy dependence on pesticides and fertilizers.

Farmland near cities is increasingly disappearing into suburban development. The scenery of diversified rural landscapes—with their fields and hedgerows and woodlots—is an important esthetic resource, particularly to urban populations. Although not entirely comparable, much of the charm and beauty of European countrysides derives from such landscapes. Unfortunately, insufficient effort has been made to keep the most attractive rural lands near cities from being consumed in the massive conversion to urban life.

The Coastal Environment

In the land and water areas of the American coasts lie some of the most fertile parts of the environment. The coastal zones, which include beaches, estuaries, tidal flats, bays, marshlands, lagoons, and sounds, with their adjacent lands, comprise areas of great biological diversity and productivity. The coastal zone includes urban, suburban, rural, and natural areas and faces all the problems of each. It is, however, a unique system which has important national significance and is subject to intense and conflicting manmade pressures.

Its surface is relatively small—only 15 percent of the U.S. land area. But 33 percent of the Nation's people are concentrated on the coasts, four-fifths of them in urban areas. While the national population increased 46 percent from 1930 to 1960, the population in coastal counties increased by 78 percent.

Life at the edge—Because of the natural mixing of fresh and salt waters, the estuarine environment produces a wide variety of living organisms, from microscopic species to large numbers of fish and shellfish, birds, and mammals. Many species, such as clams and oysters, spend their entire life cycles in the estuaries. Others, particularly shrimp, migrate from the sea to estuarine nursery areas. In these rich waters, they grow to sub-adult size before returning to the sea to complete their life cycles. The anadromous species, such as salmon and striped bass, pass through the estuaries to their spawning grounds farther upstream, and the young return through the estuaries to the ocean. At least two-thirds of the animal populations in the oceans spend an essential portion of their life cycle in estuarine waters or are dependent on species that do. Innumerable waterfowl and shorebirds depend on the plant and animal organisms of the coastal zone for their food. Many winter and nest in these waters.

The base for all animal life in estuaries is the abundant variety of plant growth, from mangroves to eelgrass and algae. They are supported by the mixing and flushing action of the tides and the organic nutrients which collect to produce the rich bottoms and wetlands. While estuarine zones are physically varied, they all share the slow mixing action of the seaward flow of fresh water with the landward tides of the sea. Because of the concentration of people within the coastal zone, the estuaries receive large volumes of all kinds of waste, which are thereby trapped and concentrated. When estuarine waters are polluted, vast numbers of important fish and shellfish are affected as well as the numerous birds, reptiles, and other wildlife which are part of this food chain.

The conflicts—Competition for the use of the limited coastal zone is intense. Shipping activities are increasing, with larger vessels needing deeper channels. Mining and oil drilling in coastal waters grows daily. Urban areas expanding throughout the coastal zone continue to enlarge their influence over these waters. Industrial and residential developments compete to fill wetlands for building sites. Airport and highway construction follows and further directs growth patterns in the coastal zone. Recreation—from enjoyment of the surf and beaches to fishing, hunting, and pleasure boating—becomes more congested as available areas diminish. Since over 90 percent of U.S. fishery yields come from coastal waters, the dependence of the commercial fisheries industry on a stable estuarine system is obvious.

Although some uses of coastal areas are undoubtedly necessary, many are not. Much industry, housing, and transportation could be sited elsewhere.

Dredging and filling—Besides water pollution, the major adverse effect on the coastal lands and waters stems from physical alteration of submerged and adjacent land and habitat—particularly the shallow marshes and wetlands. The major alterations of wet and submerged coastal lands comes from draining, dredging, and filling. Cumulatively these actions can entail the disappearance of the essential food base for practically all organisms in these waters.

The consequences of dredging and filling, because they often represent a series of incremental activities, do not usually become apparent until much of the permanent damage is done. Some 2 million to 3 million waterfowl used to nest and feed in San Francisco Bay before a large part of it was gradually filled. Their numbers are now down to less than 600,000. The State of California, with support from citizens and communities in the Bay area, has now formed the Bay Conservation and Development Commission to provide regional control over that irreplaceable estuary.

To the developer with little appreciation of the biological importance of estuaries, wetlands represent attractive waterfront acreage in particular demand by industrial and commercial concerns and home buyers. Relatively inexpensive to dredge, fill, and bulkhead for building sites, shallow wetlands attract many industries which are not dependent on waterfront sites but which find an economic advantage in developing these low-priced lands. Too often local governments acquiesce, anticipating the increased tax revenues. Consequently, natural coastal areas are being nibbled away. The long-range economic and ecological costs of this process are borne not just by the particular local community but by the people of the State and the region, and no less by the rest of the Nation.

The vacation industry—A growing part of the development pattern spreading throughout the coastal zone is the growth of vacation homes. The Department of the Interior has estimated that over 68 percent of the total recreational property values along the coasts and Great Lakes are accounted for by shorefront homes. They occupy over 90 percent of the recreational lands on developed coasts.

Only 6 percent of the land that can be classed as recreation shoreline is in public ownership, and not all of that is accessible to the public, particularly the many miles of Department of Defense holdings.

Management—Ownership of the wetlands in many States is a confusing tangle of State, local, and private claims, and in some coastal States valuable State-owned wetlands have been transferred to private interests for the specific purpose of development. Likewise, restrictions on development of the contiguous lands and wetlands are, with some exceptions, inadequate. Only a few States, notably Massachusetts, Connecticut, and North Carolina, have wetlands protection laws which require permits or other controls before alterations can be made to private coastal lands. Even fewer States have exercised any statewide powers over the contiguous dry lands. Hawaii, Wisconsin, and to some extent, Oregon, are among the exceptions.

The most important Federal permit authority affecting coastal zones is held by the Army Corps of Engineers. Dredging, filling, and all other structural changes in navigable waters require Corps permits under the Rivers and Harbors Act of 1899. The Fish and Wildlife Coordination Act and an interagency agreement also provide for review of all Corps permit applications by the Department of the Interior and State fish and game agencies. Their role is to review fish and wildlife, water pollution, recreation, and other environmental considerations. The National Environmental Policy Act of 1969 bolstered the authority of the Corps to consider all environmental aspects of a particular permit application. And the Water Quality Improvement Act of 1970, as a condition to granting permits, now requires State certification that the particular activity will not violate State water quality standards.

The Administration has recommended legislation to Congress to assist the Great Lakes and coastal State governments to manage and protect their coastal resources. The proposed legislation seeks to encourage these States, in cooperation with local units of government, to prepare and implement comprehensive land use plans for their coastal zones. To be eligible for Federal grants, the States would be required to assume regulatory and eminent domain powers. At a minimum, this would assure that local zoning actions would comply with statewide plans. Further, public notice, public hearings, and review of all Federal and federally assisted State and local projects would be conditions of grant eligibility.

The Natural Environment

Natural plant and animal communities—besides satisfying man's yearnings for beauty and recreation—perform indispensable roles in his welfare. Wild areas are part of man's past and essential to his future. The biological rhythms of modern man, which shape his life, are often tied to the natural forces beyond his urban surroundings. And yet wild areas are not fully free from man's influence. Some—for instance, where man is present for short periods and leaves hardly a trace—are little influenced. Others—lands actively mined, logged, or grazed—are heavily influenced.

For many years, the American people have realized that the most unique and spectacular of these natural areas should be protected and saved by Government.

The Federal lands—Management of most Federal natural lands falls under four agencies:

	Million acres
Department of the Interior:	
Bureau of Land Management	470.4
Fish and Wildlife Service	30.5
National Park Service	27.1
Department of Agriculture: Forest Service	186.9

Many other forests, parks, and recreation and conservation areas have also been set aside by State and local governments and by citizens' groups with an interest in the natural environment.

Their management varies from protective preservation to multiple use. Most of the Federal lands are managed to combine outdoor recreation, range, timber, watershed, and fish and wildlife uses. Many of the management programs permit the extraction of oil, gas, and other minerals; others aim primarily at esthetic values. Although Federal land management is meant to maximize public benefit and not to reflect only economic gain, land management agencies have been criticized for policies that overemphasize developmental values.

As other natural areas are staked out for intensive resource development, they are bound to come into even more serious conflict with preservation and recreational interests. Statutory resolution of many of these potential conflicts may be necessary, as may administrative reforms by the Executive Branch. The report of the Public Land Law Review Commission will serve as a focus for the debates that will shape such changes. The report has just been published at the time of this writing; consequently, a review of its findings and recommendations is not attempted here.

Federal lands managed primarily for the protection of their natural characteristics and the preservation of their fish and wildlife can take only so much use by man. Allowing people to enter these areas in large numbers must be weighed against their impact on the region's ecology. Protected areas must be managed so that constraints are placed on certain uses, including recreational uses requiring motorized transport.

The wilderness system—Some large areas under Federal control are expressly set aside as wilderness. The Wilderness Act of 1964 already protects approximately 10 million acres, although less than one percent of this is east of the Mississippi River. The Wilderness Act requires the Secretaries of Agriculture and the Interior to submit plans to Congress for including in the National Wilderness Preservation System additional natural areas having substantial wilderness qualities and presentlymanaged by the Forest Service, the National Park Service, and the Fish and Wildlife Service. These areas are to be of sufficient size, permanently roadless, and without commercial development.

Delays in setting aside wilderness areas, especially those within national parks, have stirred a public controversy. Sixty Forest Service areas have already been designated by Congress in accordance with a 10-year plan to have all wilderness areas set aside by 1974. But not a single National Park area and only one small area in a National Wildlife Refuge have been officially designated. The process of designating wilderness areas requires surveys and master planning by the lead agency, followed by public hearings, before the proposals go to Congress. Conservation groups have complained that the Federal agencies are not adequately staffed to complete the planning and hearing procedures. Conservation leaders also have criticized Forest Service and Park Service proposals for not including some additional areas with outstanding wilderness qualities in their recommendations for Wilderness Act protection.

Scattered throughout the Federal lands under various agencies are 336 "research natural areas"—a specific classification for lands left undisturbed for purposes of research and education. A number of ecosystems, however, such as the tall grass prairie of the Central Lowlands, are sparsely represented or are not included at all. Plans are now under consideration to add approximately 300 areas so that all major ecosystems will be represented.

Acquisition of natural areas—The major source of funds to purchase most Federal natural areas is the Land and Water Conservation Fund. Revenues for it come primarily from a tax on motorboat fuel, sale of surplus Federal property, Treasury appropriations, user fees from Federal recreation areas, and—to bring the total to $200 million each year—Outer Continental Shelf mineral receipts. For the first time, the full authorization of $200 million is being requested by the President for appropriation in fiscal year 1971. The President has also asked that the unappropriated balance from past years be tapped to increase funding for 1971 to $327 million, compared to $124 million in the 1970 budget. Finally, the Administration has recommended raising the $200 million annual authorization of the Fund to $300 million. This would take effect in fiscal year 1972.

Over half the allotment from the Fund this year will go to the States and through the States to local governments. These funds are matched dollar-for-dollar by the States for the planning, acquisition, and development of outdoor recreation areas. The remaining funds are budgeted for land acquisition by the Federal Government to the National Park Service, the Forest Service, and the Bureau of Sport Fisheries and Wildlife (for rare and endangered species' habitat and for recreation lands adjacent to wildlife refuges). Major recent Federal purchases have been made from the Fund, notably the Redwood National Park in California and land within the national seashores. The new funding requested by the Administration will permit the Government to consolidate and protect many other important natural areas, including Point Reyes National Seashore in California.

The Migratory Bird Conservation Act of 1929 authorizes the Government to purchase additions to the 30-million acre National Wildlife Refuge System. Funds for this come mostly from the Migratory Bird Hunting Stamp Act of 1934, which requires all waterfowl hunters to purchase a Federal hunting stamp. Although about 25 million acres in the present 328-unit system are Federal lands which were withdrawn for refuge purposes, most of the national wildlife refuges acquired in the past 15 years have been bought with funds from sale of these "duckstamps."

Crowding—Natural areas under the direction of the National Park Service have become immensely popular with the millions who visit them each year. Yet the enjoyment diminishes as the number of tourists rises. Some of the more popular natural areas in our parklands have become clogged with traffic, noise, litter, smog, and most of the other elements of our technological society from which the visitors are trying to escape.

Identifying the best mix of uses for natural areas is difficult because scales for measuring esthetic value and natural phenomena are elusive. In the future, limits on use may have to be set at crowded parks or in ecologically fragile areas. California has attempted a system of advance reservations in some of its crowded camping areas in State parks. Some national parks may soon follow suit. Control of auto traffic will be forced upon many areas to preserve the values that the park or forest seeks to maintain. Alternative modes of transportation may be necessary. Yosemite National Park, long threatened by overcrowding, has excluded cars from the eastern part of the valley and converted other roads to one-way traffic. Where traffic is prohibited, access is provided by small buses. As Americans are more frequently attracted by their natural heritage to places where they can know nature firsthand, controls of these kinds will become more common.

Exploitation—Timber is cut and sold from the national forests, and to a lesser extent, from lands administered by the Bureau of Land Management. This use is a growing one. Cattle and sheep are permitted to graze on approximately 273 million acres of Federal land. Mineral leasing, mostly oil and gas, now uses over 64 million acres. Most of the public lands are open for recreation in one form or another. Across all these lands are spread diverse habitats which support varied species of wildlife.

Overgrazing, widely practiced during the latter part of the 19th and early parts of the 20th century—and still a problem today—has dramatically affected these lands. The semiarid and arid climate of the West has added to the destruction. Dry years have usually coincided with falling market prices. And when that has happened, cattle and sheep ranchers short of cash have often overstocked already depleted ranges. Much of this land, particularly the vast public domain, remains today in desperate condition, as wind, rain, and drought have swept over them and eroded their exposed soils. Although the effects of overgrazing in rich pastures or prairie farmland can be quickly corrected, the process is often irreversible on the limited soils and arid climate of much of the public lands.

What overgrazing does to the soils and habitat of the vast range lands, poor lumbering practices do to the forests. Mismanagement of timber cutting, in addition to being an eyesore, can also damage natural systems far beyond the forests. Debris and erosion of the limited soils of the forested West often choke streambeds far downstream. The tendency toward monoculture for the benefit of lumber and pulp has its environmental results. Not the least of these is the loss of many native species. The greatest loss of timber is not from fire and disease but from insects, which can devastate large forests of single species. Widespread application of insecticides often follows.

Wildlife—Ultimately these and other uses of the Federal lands fall most harshly upon things wild and free, for protecting and improving habitat is the key to saving wild species.

Wildlife protection in the United States today paints a complex picture. The populations of most game species are high, and their status constantly improves, particularly deer, elk, and various game birds. However, until recently, nongame species of wildlife have been largely ignored and their condition is less well known.

Wildlife in the United States has long been equated with "game" and most public attention, research and management efforts have been directed towards species of sport value. However, trends indicate that persons wanting only to observe wildlife are likely in the future to outnumber hunters. Unfortunately, this change in public attitude has not yet been reflected in increased funds for nongame species. In 1969, total funding from all sources—Federal, State, and private—aimed at wildlife research, management, and habitat protection was about $142 million. Only $6 million of that was clearly related to nongame species.

Nevertheless, sportsmen in this country have contributed impressively to wildlife preservation through the Government use of duck stamp revenues, and tax revenues on sporting arms and ammunition to heavily support State fish and game agencies. Numerous outdoor organizations have also helped conserve land and teach wildlife conservation to their members and the public.

As a group, the large predators stand in greatest danger of extinction. The belief that most predators should be exterminated was central to the early days of ranching and wildlife management. In some areas this unfortunate belief endures. There are still strong pressures on State and Federal wildlife agencies to continue predator control programs for certain bird and mammal species, despite their diminishing numbers and their importance in making room for young, vigorous animals by killing off the old, the weak, and the sick. Nonetheless, most States are making progress in providing legislative protection for the large predators. Some notable exceptions are Arizona, where mountain lion hunting is not controlled, and Alaska, where bounties are still paid for wolves.

Threatened species are getting more attention at Federal and State levels, but limited funds still restrict effective management and research. The new Endangered Species Act should strengthen control over marketing illegally obtained species. It also will place the United States in a better position to cooperate with other governments in the protection of their threatened wildlife. Its effect is already visible in the diminished poaching of alligators in the Everglades. Their numbers have increased because of it.

The emphasis here is on wildlife. Fish are dealt with in the water pollution chapter and in the foregoing section on the coastal zone. Unfortunately, the status of fish and shellfish species is less known than that of terrestrial wildlife.

Epilog—These pages have tended to emphasize the worst effects of grazing, lumbering, recreation, and other uses. The point is that many uses can be compatible and often are. Many modern timber practices can improve forest conditions, and increasing numbers of stockmen are sensitive to range conditions. It is the excess, promoted by overemphasis of the short-term economic gain, which needs curbing. The lesson of man, of land, and of wild things is hard to learn. Aldo Leopold put it well: "A land ethic of course cannot prevent the alteration, management, and use of these resources, but it does affirm their right to continued existence, and at least in spots, their continued existence in a natural state."

THE INFLUENCE OF GOVERNMENT

As the Nation has grown, so have the number of local agencies empowered to decide how land is used. But because of their limited geographic scope, they cannot provide anything resembling a land use system. The narrow authority of each permits it to ignore what the decisions of all will do to the natural and human systems regionwide.

Land Use Controls

Public officials can now call upon a growing inventory of sophisticated land use tools at local, regional, and State levels. But often these devices are little used, and seldom in a way that rationally shapes or directs development.

A wise mix of land use techniques and powers can often balance private economic interest and public benefit. However, the relative importance of particular devices depends on the degree and nature of the control desired at a specific time in a given community.

Conflicts among the rights of an individual landowner, his neighbors, and the community should be resolved by the unit of government best able to take into account broad regional interests. Urban growth has outstripped the ability of small government units to handle environmental decisions that have metropolitan, regional, or even statewide impact. To offset this, many State governments have begun taking back some of the land use powers that they had delegated to municipalities. Although it has been long in coming, Americans are recognizing the need to examine carefully what government can do to assure that land is treated as a resource to be managed and not merely as a commodity to be marketed.

The police power—This term covers a wide array of land use devices designed to protect the welfare of the general public. They limit the free use of property rights in return for protecting the general community. All exercises of the police power have a common characteristic—they are subject to easy change. This is a disadvantage of the power as a land use tool—although flexible, its exercise may be only temporary.

Zoning—Zoning is the major police power employed to control land use. It classifies and segregates the land according to the permitted uses. It can curb those uses up to the point of taking private property without compensation. Within each classification, zoning can set limits on the nature, extent, and improvement of land. Although it has worked in well-established communities where there is little land speculation or pressure for new commercial facilities, it has had less success preserving open space or channeling growth in developing areas. While ideally zoning should implement sound land use plans, it does not necessarily do so. It is usually honored in the breach by the granting of variances or amendments once the pressure of development is on.

The close relationship between the land developers and members of zoning boards in many jurisdictions often jades public confidence in zoning as a tool for sensitive land planning.

In many suburbs zoning has become a device to exclude less desirable residential, commercial, and industrial newcomers. In such communities, it is used primarily to discourage the kind of development that will cost more in municipal spending than it will pay in property taxes. This often means that apartment houses which accommodate large families, public housing, and even one-family homes on small lots are outlawed because they raise the tab for schools, parks, and other public facilities without bringing in commensurate revenues. Zoning accomplishes this by limiting construction to single family homes on large lots with large side yards and setbacks.

Subdivision controls—Subdivision controls embody some of the most promising tools for regulating urban growth. They comprise all the local ordinances and regulations that tell the landowner what he can and cannot do in dividing his land into lots and selling them or developing them. Where controls are used solely to benefit the developer economically, they can lead to tedious and unattractive developments with uniform setback and lot size, unimaginative street patterns, and little provision for open space or for commercial facilities within walking distance of homes. On the other hand, when subdivision controls permit cluster development, open space preservation, planned unit development, and other imaginative innovations, they can bring a sense of community and vitality. The flexibility of subdivision controls allows increased sophistication in the development of urban and suburban areas. For example, by the proper use of timed development, a community places all of its expenditures for new public utilities and services in one geographic area at a time, allowing for ordered growth at lowest public cost.

Sewer and water permits—Before water and sewer service may be extended to new housing and other facilities, most communities require certification that public capacities are adequate to supply the services and that equipment and other facilities on the property meet local specifications. When combined with adequate control over the use of wells and septic tanks, these permits can dictate the pace and direction of urban growth. Nevertheless, sewer and water lines today more often are installed in response to already uncoordinated and poorly thought-out development.

Suburban and regional shortsightedness has left some towns with inadequate water supplies and sewage treatment facilities for new residential, commercial, and industrial growth. Because of stricter enforcement of water quality standards, it has been necessary for regional and State authorities to prohibit further sewer hookups in large parts of several metropolitan areas, including Washington, D.C., Cleveland, and the northern New Jersey urban region. The disruptive conditions which led to these recent decisions dramatize the need for proper land use planning techniques to assure a predictable rate and direction of growth and development and to prevent economic hardship for builders, businessmen, industrial concerns, and homebuyers.

Eminent domain—Eminent domain gives to government the right to acquire lands. It is an outgrowth of the old English common law principle that the Crown held ultimate power to buy any lands in the realm and, the landholder unwilling, to seize them and pay compensation. The right of eminent domain in this country likewise may be invoked in the case of a reluctant seller, provided the property is used for the public benefit and there is just compensation. Urban renewal and highway construction projects have used eminent domain to purchase vast numbers of properties. But eminent domain is also available for smaller public benefits—such as purchasing an easement for access to a public beach or preserving an historic building.

Land purchase or acquisition of limited land rights by a public body is, in theory, an effective way to control urban sprawl. But so far, little of this has been done, either to assemble large parcels for proper development or to preserve open spaces. For financial and political reasons, local governments in rural areas have not acted while prices are still low and unaffected by urban pressure. Once an area starts urbanizing, local governments are unwilling and often unable to pay the inflated land prices. By that time, it is also too late to purchase limited rights, such as scenic easements to preserve the land, because they often cost nearly as much as full purchase.

The planning function—Comprehensive urban planning is a commitment by government to help foresee and direct growth and change within its borders. With it development decisions can be made more knowledgeably. Planning is a natural extension of the traditional goals of police power and eminent domain power to promote public welfare. When planning determinations affect private property, they require the same standards of reasonableness and fairness to the owner.

Planning has often been misunderstood. The public often fails to understand that a comprehensive plan is a living, changing mechanism and not a static set of maps. Many citizens believe that the planning process has ended when the municipality approves a master plan. The long period of time required to prepare land use, public facilities, and transportation plans for large metropolitan regions has caused many citizens to regard planning as an academic exercise. Often this feeling is only strengthened by those planners who ignore the opinions of local residents. However, growing numbers of cities, metropolitan areas, and regions adhere to and implement their plans. As planners begin to take greater cognizance of public attitudes, the value of planning as a predictive and directive tool may be better understood.

Planning has been criticized because it has often been unable to influence important land use decisions. This is not necessarily the fault of the planning process, although in many communities the planning organization is often pressured to conform its plan to unwise developments. It may be due instead to inadequate institutional arrangements that have been made to implement the plan. In many localities, the planning bodies are distinct from the action agencies. Consequently, the planning function becomes little more than advisory, and the planning office spends inordinate time and resources inside city hall lobbying the departments possessing the large budgets and the authority.

Finally, planning efforts to direct all development forces in a rational manner are often inadequate. What should be "comprehensive" sometimes deteriorates into a complementary set of specific functional planning goals for transportation, open space, schools, or some other single goal. In the process perspective is lost. Economic growth and efficiency become design criteria. Environmental aspects are often neglected for short-term economic gain. Planning that is not comprehensive and long range cannot hope to evaluate ecological factors properly.

The role of private agreements—Covenants, conditions, easements, and like restrictions on property use comprise a special set of land use devices which traditionally have been employed by neighboring property owners. Recently they have also been invoked by government and private groups to protect land. For example, citizens of several towns in Massachusetts have executed agreements with local conservation trusts, funded in part by the towns, to protect their wetlands from further development. Covenants, conditions, and easements can be tailored to carry out specific purposes. They are well established in the law and continue in their effect even if the property is sold.

Permits—The Federal and sometimes State Governments require permits for a range of activities on land and submerged lands. Although the control of individual land use decisions regarding private land is primarily local and varies widely, these permits may influence major growth patterns and economic development.

Federal permits are issued when public lands are involved. Leases, use permits, and licenses are granted to extract oil, gas, coal, and certain other minerals, and for grazing, recreation, and timbering. These permits affect not only the public lands, but also may influence the use of nearby privately owned land.

The permit authority of the Army Corps of Engineers is a significant tool in land use decisions affecting navigable waters in the United States. These permits are discussed in this chapter in the section on the coastal environment. Some States have enacted legislation requiring permits for alteration of wetlands, bottomlands, and in some cases adjacent land areas. The trend is toward stronger State controls on dredging, draining, and filling of coastal areas.

Tax Policy

Tax policy is a vital cog in deciding income and profit for land investors, suburban developers, urban developers, and landlords—the big influencers of land use. Taxation therefore is an essential tool in shaping the manmade environment and preserving the natural environment.

Local property tax—A major cause of the decline of services in some core cities is the loss of actual and potential tax revenue when residential, commercial, and industrial development spreads to the suburbs. With strict zoning and subdivision controls, many suburban communities exclude all but those who will add more to the tax rolls than they will require in municipal expenditures. Rather than finance government activity for the benefit of all people in an interrelated urban complex, the property tax has abetted local isolation from regional land use problems.

Assessment procedures have improved in many municipalities, but more reform is necessary for a fair spread of the tax burden. At present, land in urban areas tends to be undervalued and the improvements on land overvalued for tax purposes. Consequently landowners in urban areas are discourged from restoring structurally sound buildings or replacing deteriorated ones with new structures, since such improvements will raise the taxes disproportionately. On the developing fringe of urban areas, low taxes on raw land have encouraged speculative purchase and leapfrog development.

As mentioned earlier, an increasingly serious problem in city cores is property abandonment. There are delinquent taxes on some of these vacant lots and buildings, but little is done to collect them. Even less often are such properties confiscated by the city and turned to public use or sold to redevelopers.

The threatened loss of the tax base in urban areas has forced some cities to offer concessions to keep industries from moving to the suburbs, often including failure to enforce air and water pollution standards and other controls of industrial activities damaging to the environment.

Finally, few communities have constructively exploited the interrelationships of zoning, subdivision controls, and other land use devices with property tax policy. The local property tax could be a useful device, particularly in encouraging cluster development and open space preservation where no such options exist. Special tax treatment for commonly owned open space and community facilities also encourages protection or enhancement of the environment by subdivision developers.

The Federal income tax—Some aspects of the Federal income tax deliberately encourage specific types of land development. Other aspects cause inadvertent effects. The Council, with the aid of its task force on tax policy, has undertaken to examine many of the leveled criticisms. Some are outlined below.

An important land use impact of the Federal income tax stems from the deductibility of local property taxes. The intent of this provision is to relieve the homeowner from paying a tax on a tax. Often, however, major benefactors of the deduction are slum landlords and land speculators. The latter are said to be relieved by this provision from the burden of local taxation as pressures of urbanization increase the value of their holdings.

Although tightened by the Tax Reform Act of 1969, depreciation provisions of Federal income tax laws are still in need of study to determine whether they adversely affect land use decisions. Accelerated depreciation and redepreciation of buildings may foster high numbers of substandard rental housing units in core cities. Such provisions have encouraged ownership of rental property for a limited number of years, then a selloff rather than improvement of the property. These so-called fast writeoffs encourage poor quality construction in new rental units since the first owner seldom plans to own the property more than a few years.

Tax laws benefit the homeowner by allowing him to deduct mortgage interest. But they do not permit rent payments to be deducted. Much has been done recently to determine to what extent present living and land ownership patterns of the slums, the central city, and the suburbs are traceable to these tax policies.

There is growing interest by private citizens in preserving forest and woodland, protecting additional animal and fish habitat, and preventing further deterioration of their own and nearby landscapes. Some landholders have sold or donated development and other rights on their land to public bodies or private conservation groups for permanent preservation. Recent Internal Revenue Service rulings have attempted to establish the value of these gifts in light of general tax rules by basing it on the fair market value of the land as restricted compared to its value previously. Some have argued that greater certainty and more liberal treatment should be provided in establishing the value of what was given in order to encourage conservation easements and similar devices for environmental protection.

The Impact of Federal Activities

A myriad of Federal loans, grants, projects, and other programs enacted for specific public purposes often has direct impact on the use of land. Because of the interrelationships of these programs with those of other levels of government and the private sector, it is difficult to assess the degree of Federal influence. The most significant Federal activities include the highway, airport, and mass transit programs, the sewer and water grant programs, home mortgage assistance, open space funding, agriculture subsidies, planning assistance, the location of major Federal facilities, and water resource projects. A brief discussion of these programs and their land use consequences is included in the Addendum following the section "What Needs to be Done."

WHAT NEEDS TO BE DONE

Land use influences, trends, practices, and controls in the United States are complex. While by no means an exhaustive survey of all the environmental problems that are land related, the above summary does indicate the need for developing standards to evaluate what is happening to this basic resource and to develop policies that will guarantee its continued integrity. In short, there is a need to begin shaping a national land use policy.

The reforms in Government activity needed to institute a national land use policy are undeveloped at this time. It will be necessary first to determine which levels of government must assume which specific responsibilities and to identify the appropriate mechanism at each level to achieve such a policy. However, this chapter has identified certain aspects of a strategy which the Council feels merit special consideration:

- The Federal Government should encourage, through project approval under existing programs, widespread use of devices such as cluster zoning and timed development.
- Although a number of Federal programs exist to preserve buildings and neighborhoods of architectural quality and historical significance, there is a need for increased attention and a comprehensive strategy if significant progress is to be achieved.
- Comprehensive metropolitan planning should identify flood plains, wetlands, aquifer recharge areas, unstable surface and subsurface characteristics, and areas of value for scenic, wildlife, and recreational purposes. Development in these areas should be controlled.
- Federal grants for sewer and water projects and open space acquisition should be directed toward communities or project areas which will use them to control development rather than to those which merely respond to uncontrolled growth.
- Home mortgage and interest subsidy programs should be used to encourage the proper siting and environmental compatibility of the subdivisions in which new housing is constructed.
- Congress should act to reverse the degradation of the coastal zones by enacting the Administration's recommended legislation, S. 3183 and H.R. 14845.
- National wilderness areas should be designated as quickly as possible; the appropriate Federal agencies should comply with the intent of the Wilderness Act even if substantial temporary reassignment of personnel is required.
- The National Park Service should accelerate its control systems experiments to prevent overcrowding and traffic congestion in the National Parks.
- Greater emphasis should be given through existing programs to acquire small parks and natural areas near cities.
- Additional areas of special ecological significance should be protected, and the Federal Government should identify and establish a national registry for research natural areas.

B. The Federal Lands

<u>One Third of the Nation's Land</u>. A Report to the President and to the Congress by the Public Land Law Review Commission. Washington, D.C., June 1970. Pages 67-88.

The Public Land Law Review Commission was established by Act of Congress in 1964, approved on September 19, 1964, and found in 78 United States Statutes at Large 982 and 43 United States Code Annotated sections 1391-1400. The Act provides that the Commission is to be composed as follows:

(a) Three majority and three minority members of the Senate Committee on Interior and Insular Affairs to be appointed by the President of the Senate,

(b) Three majority and three minority members of the House Committee on Interior and Insular Affairs to be appointed by the Speaker of the House of Representatives,

(c) Six persons to be appointed by the President of the United States, none of whom are to be employees of the United States, and

(d) One person, elected by majority vote of the other eighteen, who shall be Chairman of the Commission.

The manner of selection of the Commission members accurately reflects the intense political nature of the issues which the Commission is to explore. The Commission members elected Representative Wayne N. Aspinall of Colorado as their chairman. Although the Commission was originally appropriated $4,000,000 to carry out its work, this was subsequently raised to $7,390,000. See 81 United States Statutes at Large, approved December 18, 1967, found also in 43 United States Code Annotated section 1399, for the increased appropriation measure. Thus the Commission's report, excerpts of which appear below, can accurately be said to embody a multi-million dollar study. Literally dozens of outside consultants and contractors were employed in conducting the various research studies and programs, as well as a large in-house staff.

The final report consists of 342 pages plus thirteen pages of introductory material. One hundred thirty seven specific recommendations are made, supported by comments and text. Although almost every part of the report is related in some way with the environment and with the related pollution problem, the excerpt reproduced below is that part of the report which focuses most directly on these issues.

Of the 2.2 billion acres of land within the United States, the federal government owns 755.3 million acres, of which 724.4 million acres are specifically within the definition of "public lands" as that term is used in the Commission report. The land is administered by the agencies listed below in the percentages indicated:

Agency	Percent
Bureau of Land Management	62
Forest Service	25
Defense Department	4
Fish and Wildlife Service	4
National Park Service	3
Various Other Agencies	2

Although roughly one third of the nation's land is federally owned, the federally owned land is unevenly distributed. For example, 95 percent of all of the land in Alaska is federally owned while less than one percent of the land in Massachusetts is federally owned. Federal lands constitute more than 28 percent of the land in each of the following western states:

Location	Percent of Total Land Federally Owned
Alaska	95
Arizona	45
California	44
Colorado	36
Idaho	64
Montana	30
Nevada	86
New Mexico	34
Oregon	52
Utah	66
Washington	29
Wyoming	48

All of the remaining states have less than 13 percent of the federally owned lands. About one half of the public lands are in Alaska. Over 90 percent of the federal lands outside of Alaska are in the eleven western states listed above.

The vast areas included within the public lands offer an unexcelled opportunity for wise land use and, as the report below illustrates, meaningful guidelines are available if only we respond to the opportunity and demonstrate the will to preserve what is left of the American landscape.

Public Land Policy and the Environment

FROM THE START of our review, we have examined, in connection with each topic or subject, the impact of particular public land uses on the environment. This Commission shares today's increasing national concern for the quality of our environment. The survival of human civilization, if not of man himself, may well depend on the measures the nations of the world are willing to take in order to preserve and enhance the quality of the environment.

These problems, which are related to the public lands in varying degrees, stem from many causes, most of them resulting from the growing population and the rapid rate of technological progress. As our national living standards improve and our numbers increase, we have come to demand, among other things, more food, more fiber, more minerals, more energy, more wood products, and more outdoor recreation. The painful experience of crowding, so common now, comes not alone from population density, but from the greater impact on the environment by modern man with his automobiles, his gadgets of all descriptions, and his insatiable demand for more and more of everything. At the same time, our technology has developed artificial products of all kinds which do not disintegrate through natural processes. These solid wastes, the junk of modern life, may bury us if the technology that created them does not find a suitable way to reuse or dispose of them. Persistent insecticides, herbicides, and detergents also constitute threats derived from our rapid industrial development.

We, however, express a cautious optimism, arising from our confidence that America's growing awareness of the danger, and the taking of appropriate steps to protect and enhance our environment, will combine to bring about the necessary corrective processes.

The environmental hazards have had impacts in many ways on our public lands. The vast extent of those lands establishes that they are at the heart of maintaining environmental quality in large areas of the United States.

The variety of characteristics of our public lands requires flexibility in the methods used to achieve quality objectives. Environmental conditions differ greatly among regions, areas, and localities. The problems of environmental management are as complex as the differences in the factors of topography, geology, soil, hydrology, vegetation, wildlife, climate, and visual-spatial form.

As the owner of the public lands, the Federal Government has many laws on the books indicating an interest in the environmental impacts of the use of those lands. Most of these laws provide little statutory guidance and leave the development of standards and procedures to the individual Federal land management agencies. The obvious exceptions are the preservation-oriented statutes relating to such areas as national parks, wilderness areas, and wild and scenic rivers.

Under general constitutional authority there are Federal laws concerning air and water pollution,[1] as well as environmental impacts of highways constructed with Federal financial assistance.[2] These are across-the-board laws, i.e., not limited to Federal lands.

The National Environmental Policy Act of 1969 [3] and the Water Quality Improvement Act of 1970 [4] apply to all Federal agencies in the performance of any of their responsibilities which may have an impact

[1] 42 U.S.C. § 1857 et. seq. and 33 U.S.C. §§ 466 et. seq. (Supp. IV, 1969).

[2] 23 U.S.C. § 131 (Supp. IV, 1969).

[3] P.L. 91–190, 42 U.S.C.A. § 4331 (1970 supp.)

[4] Act of April 3, 1970, 84 stat. 91.

"on man's environment." Thus, they provide a statutory basis to bring environmental quality into planning and decisionmaking wherever gaps exist in previous laws, even though an agency may have to obtain additional legislative authority before taking final action.

As studies prepared for the Commission have revealed, land management agencies have little, if any, statutory guidance, but have developed administratively a plethora of objectives and directives to promote consideration of esthetics, wildlife, and related values. Even so, definitions, criteria, and standards for environmental quality lack operational meaning. Air and water quality standards, where applicable, appear to be the only standards that have been defined specifically enough to be reviewed and monitored. Others often must be identified and defined at the lowest level of management and applied on an *ad hoc* basis.

Our recommendations are based on a comprehensive review of existing Federal laws and administrative practices affecting environmental quality in the management of public lands. The President has required that a review and report from public land agencies on the environmental aspects of their programs be completed and submitted to the Council of Environmental Quality by September 1, 1970. Together with our report, this action should provide a fully adequate basis for early implementation of needed changes.

Within the general framework of the broad policy goals and guidelines of the recent environmental policy acts, we recommend specific environmental goals for the public lands and, in addition to authority necessary to implement them, improved planning directives and mechanisms and stricter control techniques over various land uses.

In this chapter we treat generally with the broader principles underlying our recommendations on environmental matters. More specific implementing recommendations are contained in subsequent chapters that deal with individual subjects and commodities, and that provide a more meaningful context for their understanding.

Environmental Goals

Recommendation 16: Environmental quality should be recognized by law as an important objective of public land management, and public land policy should be designed to enhance and maintain a high quality environment both on and off the public lands.

In one sense, broad administrative discretion for environmental management recognizes the great variation from place to place in environmental conditions, the variation in regional desires concerning environmental quality, and the realities of management programs. Many of the effects of good, or bad, public land management are quite localized, although some environmental effects occur far from their origin. In another sense, however, the public lands are great national assets that deserve protection from degradation, regardless of the specific local conditions. It is in this latter sense that the need for national goals and standards becomes apparent. We believe that the existing uncertainty as to the long-term effects of land use on the ability of the ecosystem to meet future demands is of national importance.

The National Environmental Policy Act of 1969, cited above, establishes highly desirable national goals for environmental quality. It establishes a national policy to "encourage productive and enjoyable harmony between man and his environment; to promote efforts which will prevent or eliminate damage to the environment and biosphere and stimulate the health and welfare of man to enrich the understanding of the ecological systems and natural resources important to the Nation . . ." In addition, it makes it the responsibility of the Federal Government to take certain actions so as to meet a set of six general goals.[5]

But this Act does not provide goals that are sufficiently specific as guides for action on public lands. The Federal Government, after all, does have direct control over the public lands and their use. The people of the country should be given a clear idea of the kind of environment to be maintained on these lands, and the Federal actions proposed to assure that environment.

The Federal policy structure for maintaining and enhancing environmental quality on the public lands is uneven and contains broad gaps. We have found that the clearest expressions of policy concern the national parks and wilderness areas, which are set aside to protect an existing environment. For other kinds of lands, where various uses of the land and its resources are permitted, we have generally found a lack of clear policy direction.[6]

We have also noted that much of the concern expressed in the existing environmental policies for public lands deals with scenery and the protection of certain kinds of ecosystems. The recent laws provid-

[5] In addition we note that the Federal Water Pollution Control Act declares its purpose to be "to enhance the quality and value of our water resources and to establish a national policy for the prevention, control, and abatement of water pollution." 33 U.S.C. § 466 (a) (Supp. IV, 1969). The Air Quality Act of 1967 states as its objectives, among others, "to protect and enhance the quality of the Nation's air resources so as to promote the public health and welfare and the productive capacity of its population." 42 U.S.C. § 1857(b(1)) (Supp. IV, 1969).

[6] Ira M. Heyman and Robert H. Twiss, *Legal and Administrative Framework for Environmental Management of the Public Lands, chs. III and IV*. PLLRC Study Report, 1970.

Improper logging practices destroy stream values and spoil the esthetic environment of a forest. Slash burning is a contributor to air pollution.

ing for water and air quality programs, and the concern over the use of pesticides and herbicides, have not been expressed in statutory public land policy and generally have not been translated into specific administrative guides.

The multiple use acts, which provide the broadest expressions of policy for the lands managed by both the Forest Service [7] and the Bureau of Land Management,[8] require that these lands be managed and uses be permitted "without impairment of the productivity of the land." The act applying to BLM also requires that "consideration be given to all pertinent factors, including, but not limited to, ecology, . . ." We believe these are necessary and important expressions of concern for some aspects of environmental quality. But we also believe that public land laws should require the consideration of all such aspects and that environmental quality on public lands be enhanced or maintained to the maximum feasible extent.

We believe that such physical and biological effects as air and water pollution, esthetic and scenic effects, and all impacts on the ecosystem, whether immediate or secondary, short-term or long-term, including those resulting from the use of pesticides, herbicides, dispersants, and other chemicals, must all be considered as significant environmental effects. We are concerned that the current aroused interest in environmental matters not be dissipated by "fads" for one or another aspect of the environment. All of them are important, and all should be considered in public land decisions. Nor should any lessening of public popularity for the issue be permitted to relegate such consideration to minor significance.

To assure that environmental quality be given the attention it deserves on the public lands, *we propose that the enhancement and maintenance of the environment, with rehabilitation where necessary, be defined as objectives for all classes of public lands.* This proposal goes beyond the existing statutes by giving environmental quality a status equivalent to those uses of the public lands which now have explicit recognition, and by indicating that through design and management, environmental quality can be improved as well as preserved.

Environmental Standards

> Recommendation 17: Federal standards for environmental quality should be established for public lands to the extent possible, except that, where state standards have been adopted under Federal law, state standards should be utilized.

A pattern of Federal-state cooperation has emerged in some of the recent legislation dealing with environmental quality. Under the air and water pollution control laws, matching funds are provided for programs that can be initiated once a state plan is approved by the Federal Government. In this way, the local interest in air and water pollution effects is recognized, while the Federal interest in these programs is also recognized by requiring that standards suggested by the states be subject to Federal approval. With respect to other environmental quality standards, we believe the states should have a reasonable time in which to develop statewide measures.

We also believe that programs on public lands should be subject to federally approved state standards as long as these standards reflect reasonable objectives for regional and local areas. It would be highly inappropriate for the Federal Government to adopt, for example, standards not consistent with state standards approved by the Federal Government for waters flowing across public lands. The lack of Federal programs encouraging the establishment of state standards for environmental quality, and the failure of the state to act on its own, should not stand in the way of the establishment of Federal standards for the public lands wherever possible. *We recommend the enactment of Federal legislation for that purpose. In the interim, where states have adopted standards, we recommend that Federal administrators require adherence to those standards.*[9]

It will be quite difficult to establish standards for some aspects of environmental quality, such as scenic beauty, which is valued in subjective terms and is not susceptible to measurement. But it is important to make an effort to establish at least relative goals and standards, to the extent possible, for all aspects of environmental quality on public lands. The Federal Government should not allow itself to be placed in a position where it can be said that it is asking others to do what it is not willing to do itself.

Federal land and resources should be retained and managed or disposed of so as to support Federal, state, and local programs for the maintenance and enhancement of environmental quality. Actions on retained lands should generally be coordinated with other levels of government so that public land programs do not conflict with those of other governmental levels. Similarly, *when public lands and resources are sold or otherwise transferred into non-Federal ownership, the Federal Government has an opportunity to aid its efforts and those of state and local governments to improve environmental quality. Such transfer can be conditioned on the recipient complying with established standards for pollution*

[7] 16 U.S.C. §§ 528–531 (1964).
[8] 43 U.S.C. §§ 1411–1418 (1964).

[9] An example is the enactment of state laws governing strip mine reclamation.

Mining operations on the public lands, as all other activities, should cause the minimum disruption possible to environmental values. Statutes should provide for reclamation provisions.

Though an accepted forestry practice for the regeneration of forests, patch cutting presents an eyesore to the passing motorist.

control or other aspects of environmental quality, both on and off the public lands.

Planning Guidelines and Mechanisms

The recommendations we make in the preceding chapter concerning land use planning by the public land management agencies are broadly applicable to the environmental considerations which must be incorporated as an important aspect of the planning process. Thus, implementation of our proposals requiring: (1) The development of meaningful land use plans; (2) specification of the factors to be considered in developing such plans and how they are taken into account; (3) better coordination among the Federal agencies and broader intergovernmental coordination; (4) the development of regional planning mechanisms; and (5) greater public participation, will promote better consideration of environmental factors in public land use planning.

Several points require particular emphasis, however, since it is evident that the public land agencies have not responded in all cases to the needs of

Litter and vandalism on the public lands cause expensive maintenance problems and call for more progressive management and enforcement efforts (top and above). The entrance to a small-tract development, carved from the public lands, warns the visitor of worse things to come (right).

environmental quality in their planning procedures. We believe this is due in part to a lack of statutory guidance, and in part to a failure by the agencies to classify their lands in advance for environmental quality management.

Classification for Environmental Quality

Recommendation 18: Congress should require classification of the public lands for environmental quality enhancement and maintenance.

In our recommendations on land use planning, we would require environmental factors as an element to be fully considered in land use plans. In this portion of the report, we detail the manner in which attention should be given to these factors.

Environmental conditions differ greatly, not only between regions, but often because of minor differences in elevation or location. Each environmental factor—topography, geology, soil, hydrology, vegeta-

tion, wildlife, climate, and visual and spatial form—has various responses to, or capacity for, a particular use or development. Thus, the ability to predict or control the impact of a particular use on the environment will require detailed information on the composition of the environment with respect to those factors. The development of knowledge about the tolerance of particular environments to various uses at an early stage is essential, both to meaningful planning for land uses in a particular area and to the development of appropriate operating rules and controls for permitted uses. Although such an approach is being followed by some of the agencies in a rudimentary fashion, studies prepared for us show that much useful knowledge about the basic environment and the effects of various uses is lacking.

Classification of the public lands to provide for different degrees of environmental quality would provide guidance for controlling the location of activities, so as to minimize their impacts. This approach—a systematic classification and inventory of important environmental considerations on each area of public lands as part of the agencies' land use decisionmaking—will give assurance that environmental effects will be taken into account in public land decisions.

We propose that the system of environmental quality classification be based on desirable levels of quality to be maintained in each area for the major components of the environment, such as water, air, esthetics or scenery, and composition of the ecosystem. This should be done in close cooperation with the states, and where the states or local governments have developed satisfactory classifications, as, for

Disruption of the permafrost in Alaska causes serious erosion problems. This is a tractor trail near Canning River, Alaska (left). Above is a view of the Santa Barbara oil spill.

example, in connection with water quality standards, these would be incorporated in the public land classifications.

The management zones identified in multiple-use planning by the Forest Service evidence a sensitivity to environmental factors, particularly those related to scenery and vegetative cover.[10] However, this does not go far enough.

We recommend that a standard system of environmental quality classification should be developed and, after congressional approval, employed by the Federal land administering agencies in classifying the public lands for environmental management. As indicated above, there is an urgent need for workable guidelines for administering the public lands for environmental quality control. We recognize that no single standard can be promulgated and applied to the diverse conditions found throughout the public land regions. Yet, we believe the many unclear standards and guides to environmental management in the various manuals and regulations of the administering agencies are incomplete as to their coverage of major components of the environment and so general and vague as to be of little value in program operations.

We believe it is possible to devise and apply a framework of standards for use in environmental management of public lands that is clear and practical, and also flexible enough to be applied in diverse circumstances and localities. A possible approach is to establish a hierarchy of classes for categorizing each major component of the environment.

[10] n. 6, supra.

Land and water quality control should be among the primary goals of public land management.

In this approach the entire environment can be viewed as having four major components or elements: water, air, quality experience and the biosystem. Water and air, as fundamental elements of the natural environment need no definition. We suggest a separate category of "quality of experience" that embraces all those intangible visual and aural attributes of our surroundings. This category includes the often overlooked need to reduce noise pollution. Included also is the qualitative effect on the psyche of litter, refuse, overcrowding, and the form and location of constructed works, such as roads, dams, buildings, and powerlines.

The fourth category of biosystem is concerned primarily with the living elements of the environment, the vegetation and animal life including their different associations and interrelationships in various locales.

It is possible to specify two or more levels of quality for which each of these major components of the environment can and should be maintained or managed. Each quality level could be defined in terms of a purpose or an end to be served by maintaining the particular quality level.

To insure continuing quality levels so defined, the desired condition for the four basic categories, i.e., water, biosystem, quality of experience and air must be specified and maintained. The technical conditions suggested as possible guidelines to maintain each quality level might provide the basis for completing the description of each zoning or subcategory. The constraints to be imposed on each type of land use that occurs or is contemplated in each zone would be specified.

In sum, the zoning analogy is to be applied. Each major component of the environment would offer variable levels of quality to be maintained for each important environmental element. This in turn would lead to the specification of a set of different degrees of land use constraint for all types of land and resource use for each category.

An example of how environmental quality zoning classes could be used in public land administration is set forth in the accompanying table and illustration.[11] Any viable system must remain flexible and subject to change and refinement lest it become, like some city zoning measures, a procrustean bed.

The utility of this approach lies in the classification of public lands for environmental management using a verifiable method for determining what uses can be advocated, or what constraints must be placed on uses, in order to achieve a desired level of environmental quality. The desired level of environmental quality and the specific use constraints that are necessary for each area of public lands will be determined by topography, soils, vegetative cover, climate, and the whole calculus of variables peculiar to different public land locations. The area managed need not be designated by size but may be zoned to assure a given level of quality maintenance within each major component of the environment.[12]

Land Use Planning Includes Environmental Factors

Recommendation 19: Congress should specify the kinds of environmental factors to be considered in land use planning and decision-making, and require the agencies to indicate clearly how they were taken into account.

The National Environmental Policy Act[13] does not define the term "environment," nor is it defined in any other Federal statute, although there are many of them that are addressed to environmental matters. We think that clarification of the term would be desirable as a general principle, and would be particularly appropriate in setting forth the environmental factors to be considered in Federal land use planning. Thus, in such planning, the public land agencies should consider the impact of possible uses of land on the land itself, as well as on air, water, climate, vegetation, wildlife, and man, the latter from the viewpoint of his health and safety, his economic well-being, and his esthetic sense.

The National Environmental Policy Act requires a "detailed statement" on the environmental impact of, and possible alternatives to, proposed actions "in every report on proposals for legislation and other *major* Federal actions significantly affecting the quality of the human environment." (Emphasis supplied.) We endorse that principle. However, we would also apply it to *all* public land use plans and decisions, not only to those deemed by the land manager to be "major." There should be a record available with all such plans and decisions, from which can be determined the extent to which environmental factors were considered. This should be accepted as a normal process in land management.

(Text continued on page 80.)

[11] The table shows an example of one possible approach to a classification system for environmental management of public lands. The illustration portrays graphically how the classification system might appear if applied to an area of public lands. The Commission is not recommending that this table be adopted without consideration being given to possible alternative approaches.

[12] Two studies conducted under contract for the Commission described alternative approaches to systematic classification of environmental quality and related factors. Landscapes, Inc., *Environmental Quality and the Public Lands.* PLLRC Study Report, 1970. Steinitz Rogers Associates, Inc., A General System for Environmental Resource Analysis. PLLRC Study Report, 1970.

[13] n. 3, supra.

EXAMPLE

POSSIBLE CLASSIFICATION SYSTEM FOR ENVIRONMENTAL MANAGEMENT

Environmental Category	Quality Related to Purpose	Environmental Attributes to be Monitored and Managed	Management Actions
		WATER	
W–1	Fishery, and other components of the biotic system.	High level of dissolved oxygen. Exacting tolerances for temperature, trace minerals, pH, toxic chemicals, nutrients, . . . Low silt and organic matter.	Prohibit land grading, landfills, vegetative clearing, burning, mining, except where environmental review and impact studies prove that stringent measures can keep changes to environmental attributes within tolerances.
W–2	Domestic water supply, swimming, industrial uses requiring high quality water.	High to moderate levels of dissolved oxygen. Moderate temperature fluctuation. Limits on trace minerals, pH, toxic chemicals and nutrients over a range of tolerances related to resource uses.	Permit moderate disturbances (prohibited above), but only upon determination of each developments' disturbance factors and contribution to the stream or lake's budget for sediment, etc.
W–3	Irrigated agriculture, industrial cooling water.	Control of dissolved salts and toxic materials.	Strict controls on activities that disturb soils and lead to leaching of salts or flow of acidic or otherwise toxic materials from public lands.
		BIOSYSTEM MAINTENANCE	
B–1	Perpetuation of full natural biosystem—for recreation, education, scientific study.	Minimum of man-induced changes in species composition, biomass, food chains, habitat conditions, predator-prey relationships, and population dynamics.	Perpetuate natural ecosystem processes or manage to compensate. Logging, mining, and construction, etc. normally excluded.
B–2	Limited modification of biosystem to produce specific goods or services (native range management, selective cutting in mixed hardwoods).	Minor changes in plant and animal species composition. Minor changes in habitat for preferred species. Some alteration of wildlife populations.	Alter natural system only when environmental review and impact studies allow full prediction and control over specific changes. Mitigative and corrective measures to be specified in resource management plans for timber, recreation, etc.
B–3	Major modification to maximize output of a particular product or use (single species management for commercial timber production; primary management for elk).	Large scale vegetative type conversions. Major change of habitat for preferred species.	Intensive uses or developments normally permitted if environmental review and impact studies indicate biosystem losses are offset by value of goods and services.

QUALITY OF EXPERIENCE

E–1	Visual and esthetic environments as related to recreational, residential, and travel purposes.	High capacity for direct and detailed sensory involvement. Natural dominance of form, scale, and proportion. High constraint, vividness, image creation, and unity.	Avoid disturbance of natural pattern. Prohibit intrusions of logging, mining, intensive recreation, roads, power lines, etc., except insofar as environmental design studies indicate that intentional display of resource management is consistent with scenic management objectives.
E–2	Cultural, historical, and informational values for recreational and educational purposes.	Unique, archetypal, rare, or transitory artifacts or locations relative to the environmental context.	Preservation or restoration. Prohibit competing land uses. Protect from overuse by recreationists and collectors.
E–3	Personal and social experiences free from crowding, development, and noise.	High capacity for isolation and interaction with national environment. Minimum intrusion of man-made structures and facilities and man-induced changes. Low artificial noise levels (vehicles, aircraft, radios).	Limit number of recreation visitors through rationing of physical design. Prohibit or minimize noise producing intrusions. Prohibit development of structures except where design studies show minimum disturbance.
E–4	Natural biological and physical features.	Unique or dramatic landforms or features (not necessarily of biological importance). High capacity for orientation (as with landmarks). Rare or especially archetypal geologic formations.	Modify resource management practices to enhance such features. Prohibit or restrict extractive or product-oriented uses except as they may be shown to complement feature-oriented uses.

AIR QUALITY

A–1	Human health protection (respiration; sight; skin).	Hold levels and combinations of oxides of sulfur and carbon, hydrocarbons, photochemical oxidents, and particle (solid) matter to tolerances required to support each purpose of air quality maintenance on both a 24 hour and annual basis. Maintain natural background levels of particle matter in ambient air in rural areas to the extent possible. Specific conditions to be maintained depend on different meteorological conditions, climate (wet, dry) topography, and latitude-longitude.	Control use of internal combustion engines on public land areas to hold hydrocarbon and particle matter below necessary levels. Control dust generated by mining and logging, and by recreational vehicles and logging and ore trucks traveling on unpaved land. Control stack emissions from on and offsite pulp and paper mills, concentrate mills, organic fueled power generating plants, and other industrial plants to hold particle matter and gaseous pollutants to necessary levels. Burning of logging waste and controlled burning of forest and rangelands for management purposes to be regulated daily and seasonally to meet necessary air quality requirements.
A–2	Natural biosystem protection (carbon dioxide-oxygen exchange balance; foliage burn).		
A–3	Materials protection (corrosion; etching; stain).		
A–4	Esthetics protection (haze; odors).		

Studies of Environmental Impacts

Recommendation 20: Congress should provide for greater use of studies of environmental impacts as a precondition to certain kinds of uses.

Beyond the consideration given to the environment in general land use planning, as well as the likely effect of certain kinds of uses, some uses, entailing severe, often irreversible, impacts, should be permitted only if a decision is based on a detailed study of their potential impact on the environment. The kinds of uses that should require impact studies because of the severity of their effect, include transmission lines, roads, dams, open-pit mining operations, timber harvesting, extensive chemical control programs, mineral operations on the Outer Continental Shelf, and high density recreational developments. The need for and depth of such studies would vary directly with the nature of the proposed use and the sensitivity of the environment upon which it would operate.

The agencies are now doing this administratively, particularly with respect to use of national forest lands for transmission lines, dams, and roads. However, the principal problem in many cases appears to be one of timing, in that the public land agencies are brought into the picture at so late a date that when impact studies are made, they are often done concurrently with the implementation of the project.[14] Unless the agencies are brought in at an early stage, these studies can at best serve a limited function, i.e., mitigation of adverse impacts. They cannot be used, as may be appropriate in some cases, to provide the basis for a decision to select alternative sites, routes, etc., or even not to proceed with the project at all.

Expanded Research

Recommendation 21: Existing research programs related to the public lands should be expanded for greater emphasis on environmental quality.

Such an expanded research effort is required in order to provide the information and expertise necessary to give proper attention to the environmental aspects of public land management.

This would not necessitate a new program, but simply an extension of existing programs under several statutes,[15] which form the basis of Forest Service and independent grant research programs. The Commission's recommendation to merge the Forest Service with the Department of the Interior, made elsewhere in this report, would make the Forest

[14] n. 6, supra.

[15] See 16 U.S.C. § 581–581j and 16 U.S.C. §§ 582a–582a–7 (1964) as to the Forest Service and 42 U.S.C. §§ 1961b (Supp. IV, 1969) as to water resources.

Service research program more responsive to research needs on national parks, refuges, and Bureau of Land Management lands. Greater emphasis on environmental quality research should include efforts to provide better measurements, to the extent possible, of esthetic factors and other nonquantifiable amenities.

Mandatory Public Hearings

> Recommendation 22: Public hearings with respect to environmental considerations should be mandatory on proposed public land projcts or decisions when requested by the states or by the Council on Environmental Quality.

An Executive order [16] implementing the National Environmental Policy Act directs all Federal agencies to "develop procedures to ensure the fullest practicable provision of timely public information and understanding of Federal plans and programs with environmental impact in order to obtain the views of interested parties," including *"whenever appropriate,* provision for public hearings" (emphasis added). We believe that this does not go far enough.

In our general land use planning recommendations, we suggest, among other things, mandatory public hearings at an appropriate stage in the planning process. This will permit public participation in developing information on all relevant subjects, including environmental factors.

While we have generally favored leaving the use of public hearings to agency discretion in specific land actions, *in situations where significant environmental considerations are involved, we recommend mandatory public hearings.* As the best indication of the "significance" of particular environmental situations, we think a request by either a state or the Council on Environmental Quality [17] is of appropriate dignity to require a hearing. Individuals or groups that may have particular concerns would not be precluded from urging the agencies to hold a discretionary hearing, but when a state or the Council on Environmental Quality are convinced of the importance of their cause, a hearing would then become mandatory.

Adequacy of Existing Control Authority

With certain exceptions, our review of the statutory authority of the land administering agencies shows that it is satisfactory—even though no adequate guidelines exist—to permit the agencies to employ a wide range of control techniques to prevent or minimize the adverse environmental impacts of various lands.

Under the contractual and licensing authority which governs most uses of the public lands, there is ample authority to include protective provisions in such control instruments as timber sale contracts, mineral leases, grazing permits, and recreational and other special use permits. The major exceptions to this general situation under the existing system, which would be rectified under recommendations we make in this report, concern recreation activities by the general public on multiple use areas, mining activity under the Mining Law of 1872,[18] and certain occupancy uses, particularly road construction and utilization. The failure of the agencies, particularly the Bureau of Land Management and the Forest Service, to make greater use of such authority as they have, emphasizes the need for explicit statutory guidelines. Such guidelines for protection of the public lands are recommended elsewhere in this report.

Control of Offsite Impacts

> Recommendation 23: Congress should authorize and require the public land agencies to condition the granting of rights or privileges to the public lands or their resources on compliance with applicable environmental control measures governing operations off public lands which are closely related to the right or privilege granted.

Because there is often a direct connection between public land resource rights and privileges granted to various industrial users and later environmental impacts caused by the utilization of the resource off the public lands,[19] the agencies should be authorized and directed to control the adverse environmental impacts of activities off the public lands as well as on them caused by those using public land resources.

For example, public land timber may supply a woodpulp mill causing air and water pollution and the degradation of landscape esthetics. Smelters processing public land minerals may cause similar adverse environmental impacts.

This recommendation is premised on the conviction that the granting of public land rights and privileges can and should be used, under clear congressional guidelines, as leverage to accomplish broader environmental goals off the public lands.

However, we recognize that considerable restraint

---

[16] Executive Order No. 11514, March 7, 1970, 35 Fed. Reg. 4247.

[17] Created by National Environmental Quality Act of 1969, n. 3, supra.

[19] See, for example, case study 5, Rocky Mountain Center on Environment, *Environmental Problems on the Public Land.* PLLRC Study Report, 1970.

Fires in the forest, whether wild or employed as a silvicultural tool, are another source of air pollution.

must be used in implementing this recommendation. *We recommend that the activities against which such indirect leverages should be employed ought generally to be limited to those that bear a close relationship to the use of the public lands and that would have an adverse effect on the environment off the public lands.*

Where Federal, state, or local environmental quality standards have been established, firms that are violating these standards would be identified by the applicable level of government. Such firms should not be eligible for obtaining public land resources for use in the plant where violations occur. Federal privileges granted should be conditioned on continued compliance, and should be subject to termination for violation, in addition to other applicable penalties under Federal, state, or local law.

We believe that this policy would not be unduly intrusive as long as it is restricted to the stages of processing that involve the use of resources in essentially the same form as they leave the public lands, and to violations of clearly established environmental standards by the particular plant processing the resources. In other words, we do not propose that a Federal public land lease be denied a company in Utah or Alaska because that company's unrelated activity in a manufacturing plant is accused of polluting the Hudson River in New York.

In the preceding discussion, it is demonstrated how the recommendation we make permits the United States to use its licensing power to protect adverse environmental impacts off the public lands. Similarly, the Federal Government should at all times manage its public lands so that its own actions will not degrade the surrounding environment. *To support this conclusion, we recommend that the land management agencies should be required by statute to control fire, insect, and disease outbreaks on public lands, including wilderness areas, to assure that there is no adverse impact on any adjacent area.*

Covenants and Easements

> Recommendation 24: Federal land administering agencies should be authorized to protect the public land environment by (1) imposing protective covenants in disposals of public lands, and (2) acquiring easements on non-Federal lands adjacent to public lands.

Activities carried out on non-Federal lands in proximity to public lands can and do adversely affect the environment of the public lands. In addition to degrading the scenic values of the public lands, adjacent or nearby land uses can cause air and water pollution with attendant impacts upon the natural biosystems and the health of public land users.

We have confidence that, because of their mutual concern, such activities in the vicinity of the public lands will be appropriately regulated by state and local authorities in close cooperation with the Federal agencies. But we must not risk failure, and, *therefore, recommend that if cooperation is not prompt and successful, the agencies should be empowered to take direct action in furtherance of the preservation of the public land environment.*

Although some of our contract studies suggest that direct Federal regulation or zoning, in the limited situations with which we are concerned, would be appropriate and constitutionally permissible, we do not favor such an approach. Rather, *we recommend*

[18] A major deficiency is contained in 43 U.S.C. § 932, an 1866 act granting rights of way for the construction of highways over the unreserved public lands which may be initiated and constructed without federal approval. This precludes meaningful federal control over the location and design of such highways to protect environmental values.

action through the use of traditional public land acquisition and disposal techniques: The agencies should be authorized and directed to (1) include in patents and leases of public lands covenants to preserve environmental values on adjacent or nearby federally owned lands; and (2) acquire easements over lands in non-Federal ownership when necessary to protect environmental values on the public lands they manage.

Improved Control Techniques

We have found in our review that, although there are provisions in regulations and other administrative directives to prevent or minimize environmental abuses of the public lands, there are important gaps in authority and practice.[20]

The public land agencies must be in a position, and have controls available, to respond to adverse environmental impacts as their nature becomes known. For example, the use of pesticides and herbicides grew rapidly after World War II, but knowledge of the possible adverse consequences of such chemicals lagged until recently.

Not only must the Federal agencies have statutory authority for controlling uses of the public lands in the interests of environmental quality, but they must have programs for monitoring activities on the public lands. Recent examples of failure to maintain proper and authorized controls over oil drilling on the Outer Continental Shelf have, for example, resulted in major adverse environmental impacts. To some extent, these resulted from a lack of personnel and an occasional laxity on the part of the public land agencies as well as avoidance of controls by users.

In this context we note, as we do in Chapter Eleven, that public notice should not only be given of "operational orders," but waivers of such orders or regulations should also be publicized. Sometimes the waiver of an order is more significant than the regulation, and the public should be informed.

We believe it is important that public land agencies develop regular procedures for monitoring all activities and adherence to regulations where ignorance, negligence, or violation could result in adverse environmental impacts. We recognize that there is a need for an environmental monitoring system to observe generally and evaluate modifications in the environment. However, where environmental effects are generally widespread, as with air and water quality, we do not believe there should be an extensive monitoring system established just for the public lands. *We do recommend that when and if a nationwide monitoring system is established, the public land management agencies should participate in it and make certain that the specific requirements for knowledge concerning the public lands are met.*

[20] See, e.g., n. 19 supra; case studies 4, 9, 11, and 16.

Industrial pollution should be controlled by public land laws if the industry's raw material originates on public lands.

Responsibilities of Users

Recommendation 25: Those who use the public lands and resources should, in each instance, be required by statute to conduct their activities in a manner that avoids or minimizes adverse environmental impacts, and should be responsible for restoring areas to an acceptable standard where their use has an adverse impact on the environment.

Natural areas should be given recognition as a proner use of the public lands in the statutes immediately, so they can be protected from other uses.

Many uses of the public lands are not controlled by permit or contract. And even if the recommendations of this Commission are adopted to require permits for additional uses, some, such as permits for general recreation use or for hunting and fishing, will not create a relationship between the United States and the user that will permit the establishment of specific control measures to protect the environment.

Where public lands and resources are used or obtained under a contract or permit issued for a specific purpose, the situation is quite different.

In such cases, the Federal Government is able to, and to some extent does, establish environment conditions that must be maintained in connection with the use. Forest Service and Bureau of Land Management timber sale contracts require that the contractor build roads in a specified manner, remove obstacles to the normal flow of water, remove slash from some areas, and so on. These and similar requirements imposed on operators holding mineral leases place the burden and cost of meeting the requirements directly on the operators. They must take the estimated costs of all contractual obligations into consideration when obtaining the contract.

A major difficulty is that the requirements are uneven and will remain so in the absence of a statutory foundation. Where the casual user has caused damage, or where there has been a failure to have a proper requirement in a lease, the Federal Government must bear the cost of restoration, rehabilitation, or the minimum cleanup of the area. *We recommend that there be a statutory requirement that all users be made responsible for maintaining or restoring en-*

vironmental quality to an acceptable level at their own expense.

Flexibility must be given to the administrators to include specific reasonable conditions in permits and contracts. In other chapters of this report we recommend means of implementing this recommendation with respect to uses that are known to have potential impact on the environment. Furthermore, we emphasize that the measures required of the user, and the type of rehabilitation required, be made known before the user enters into a contract with the government, and that they then be made part of the agreement so that the user has a clear understanding of what is expected of him before he initiates his use of the public lands.

The cost of maintaining a quality environment thereby becomes an element in determining the economic feasibility of an enterprise. In some instances, where the production of a commodity or the furnishing of a service is desirable to meet a national need, it may not be possible for private enterprise to undertake the activity if the full cost of avoiding adverse impact or of subsequent rehabilitation is charged to the user. *We, therefore, recommend that on a pilot basis, Federal departments and agencies be authorized to share in those costs after a formal finding that there is an urgent requirement for the proposed use, and that the level of rehabilitation should be higher than could reasonably be expected from private enterprise alone as in the case of oil shale development* (see Chapter Seven).

In the situations not controlled by contractual relationships, we recognize that there will be greater difficulty of enforcement. Nonetheless, we believe statutory liability of the user must be established and some efforts be made to shift from the Federal Government at least some of the cost of restoring damage caused by noneconomic users. Excluding the cost of cleaning up litter and garbage, the Intermountain Region [21] of the National Forest Service alone spent over $220,000 in fiscal year 1969—almost 8 percent of the total allocated for the maintenance of recreation areas—to undo what is termed vandalism damage. Nationwide, throughout the Forest Service, over $2 million of such damage was caused in fiscal year 1969.

As more and more people utilize the open multiple-use lands under the management of the Bureau of Land Management for picniking, camping, hunting,

[21] The Intermountain Region encompasses all of Utah and Nevada, a portion of western Wyoming and southern and central Idaho.

Off-road vehicular use must be controlled in public land areas that are easily erodible.

Spraying for insect control could have serious consequences, from the standpoint of ecological balance and adverse effect on animals and the human population.

fishing, and other leisure time pursuits, there will be increased threats to the environment unless we take strong steps now to avoid them. Elsewhere in this report, we have pointed out the fact that trespass control has been difficult, and *we recommend that statutory authority for policing Federal lands be granted to those agencies, such as the Forest Service and Bureau of Land Management, not now having such authority.* We believe that the knowledge that the law makes the user liable to restore damaged areas, and that the agency having responsibility has policing authority, will in itself act to curb abuse of the environment. In any event, the new policing authority will provide the United States with a tool that it does not now have in the apprehension of vandals and others who cause environmental disturbance.

Environmental Rehabilitation

> Recommendation 26: Public land areas in need of environmental rehabilitation should be inventoried and the Federal Government should undertake such rehabilitation. Funds should be appropriated as soon as practical for environmental management and rehabilitation research.

Past activities on the public lands have resulted in lowered environmental quality in many places. As indicated above, there have been many causes for the degradation. It is impracticable, except where contract provisions have been violated, to try now to seek out those responsible and ask them to effect rehabilitation. *Nonetheless, it is essential that damage to the environment be corrected, and we recommend that actions be taken to restore or rehabilitate such areas.* The first step in this direction is an inventory of all instances of lowered environmental quality generated by past uses of the public lands.

Concurrently with the inventory, *we recommend an immediate accelerated program of research into the procedures and methods of maintaining and restoring environmental quality on the public lands.* We found that such efforts have been virtually nonexistent in the past. Because some adverse impacts have occurred, and more will occur if management practices are not improved, research is essential without delay.

In considering legislation for this purpose, Congress should keep in mind the considerable receipts generated from the sale and use of the public lands and their resources.

We see no alternative to making the Federal Government responsible for rehabilitating areas that were abused in past years. Where those whose actions resulted in lowered environmental conditions can be identified, and the terms under which they were using the public lands made them responsible for maintaining high quality environmental conditions, they should be required to fulfill their obligation. Generally, however, there were no such conditions, and to impose this responsibility on them now would, in our opinion, be unfair.

Natural Areas

> Recommendation 27: Congress should provide for the creation and preservation of a natural area system for scientific and educational purposes.

By 1968 Federal agencies had designated nearly 900,000 acres of public lands as natural areas, the individual units ranging in size from a few acres to 134,000. Similar preservation efforts have been undertaken on private and state owned land by states, educational institutions, and private organizations.

Natural areas are protected to permit natural biological and physical processes to take place with a minimum of interference. The preservation of such areas is for the primary purposes of research and education. As the need to understand the ecological consequences of man's activities has become more evident, the preservation of examples of all significant types of ecosystems has become important to provide a basis for comparisons in the study of the natural environment. It appears that these requirements can be met with a relatively small amount of land. We approve preservation measures of this kind.

The Federal land-managing agencies have proceeded quite independently in establishing natural areas, with no uniform guidelines for agency designations. We believe Congress should give formal status to the natural area program and provide for coordination to assure that all essential scientific and educational needs are met. The coordination we urge, perhaps by the Office of Science and Technology in the Executive Office of the President, would provide an inventory of sites valuable for ecological study, a plan to assure representation of all important natural situations, and the avoidance of duplication of effort.

We also propose that educational institutions be encouraged to assume administrative responsibility for federally-owned natural areas under permit or lease arrangements with the Federal land agencies. Such arrangements offer assurance that other uses, such as recreation, would not be allowed to interfere

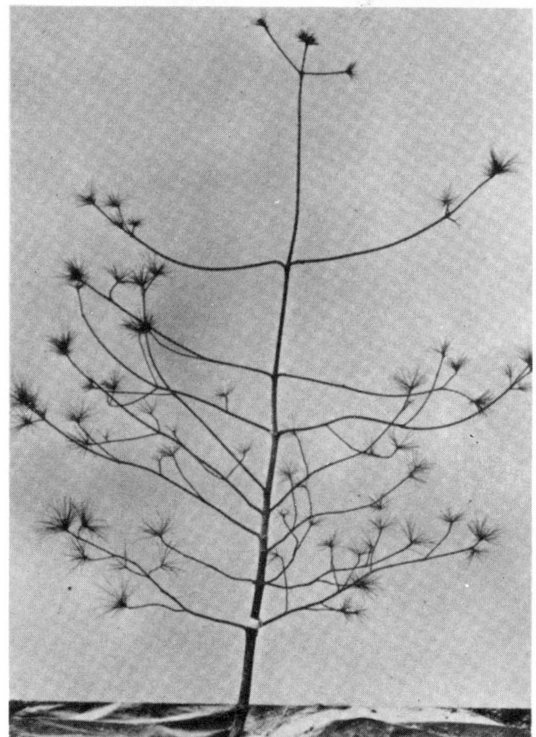

Tests by the U.S. Department of Agriculture show the effect of urban pollution on white pines. Tree above was grown free of polluted air. The culprit is either sulfur dioxide, ozone, or an interacting mixture of both, which are primary ingredients in the urban pollution mix.

with the educational and scientific purposes of these natural areas, will place administrative responsibility with those who will be conducting research and, at the same time, will lessen the cost borne by the Federal Government.

Summary

The sum total of the recommendations in this chapter is to make the public lands of the United States examples for the rest of the country in how to manage and use lands and resources with due regard for the environment. It is essential that this be done if we are to hope that citizens will engage in the practices that government urges.

By expressing our concern for what happens in lands adjacent to the public lands, as well as the environment on the public lands, we give recognition to the central factors of ecology which has been repeated many times, but of which we must not lose sight: Everything is connected to everything else. It is this fact that may make effective environmental quality goals and controls on the remote public lands meaningful in fighting the environmental degradation that has already occured in the highly industrialized and urbanized areas of the country. The immediate, more direct benefit, of course, will be that we protect the public land areas from being subjected to pollution and other forms of blight that plague so much of the Nation.

We submit that the recommendations we make in this report will accomplish the objectives we believe to be so essential.

C. Highway Policy

Highway Environment Reference Book: Prepared by: Environmental Developmental Division, Office of Environmental Policy, with Special Assistance from the Office of Public Affairs. November 1970. U.S. Department of Transportation, Federal Highway Administration. Pages 1-46.

Air and water pollution may fade away but a highway is forever. Perhaps for this reason controversies surrounding highway location and design tend to arouse more citizen interest and often litigation than most other kinds of environmental problems. Too often in the past highway engineers seemed to deserve their reputation of considering the cheapest or shortest route as the only route between two points--irrespective of environmental consideration. It was not until 1968 that the Environmental Development Division was established, and to many, its effect is not yet apparent. Part of the problem, of course is traceable to the relatively late promulgation of adequate legislative guidelines by the Congress. It was not until Federal-Aid Highway Act of 1968 that Congress insisted that park and similar land not be used for highway purposes unless there is no feasible alternative.

In protesting federal highway location decisions citizens' groups have often been handicapped by a lack of knowledge concerning the Federal Highway Administration's own policies and procedures. It is believed that the selection reproduced below will help to remedy that situation.

APTER 1: INTRODUCTION

VA Organization and Function

The Federal Highway Administration (FHWA) of the Department of Transportation is organized outside its central office into ten Regional Administrations. Nine of these regional offices administer the Federal-aid highway program through 52 division offices (one in each State, the District of Columbia, and Puerto Rico). The division offices administer the program through State highway departments, which are responsible for the planning, location, design, and construction of highway projects. The division offices are responsible for continuous monitoring of Federal-aid highway projects to insure that they are consistent with FHWA policy and with the policies and guides of the American Association of State Highway Officials.

The three western regional offices are also responsible for direct Federal highway projects, including design, contract letting, and supervision of construction. The tenth regional office handles all direct Federal highway projects in the eastern national forests and parks.

It is through these offices that programs to protect and enhance the environment are carried out. FHWA policies apply to all highway projects involving Federal funds. Outside direct Federal construction, Federal funds are granted the States on a formula basis and are generally matched 50-50 by the States. However, in those States which have large areas of public domain or Indian reservation lands, the Federal Government pays a proportionately greater share. For highways on the Interstate system, the funds are generally 90% Federal and 10% State.

The policies which the States must follow in order to qualify for Federal aid are based on legislated authority contained in Title 23, U.S.C., and are embodied in Policy and Procedure Memorandums (PPM) and, to a lesser extent, in Instructional Memorandums (IM). Related information is transmitted by Circular Memorandums (CM).

ghway Contribution and Evolution

The Federal Highway Administration and the State highway departments are fully cognizant of the tremendous economic and social import of highways. The highway system in the United States has been one of the most important instruments for achieving the high standard of living that the great majority of our citizens enjoy. The highway and personal motor vehicle have made possible unprecedented mobility for most of us, permitting us to enjoy the varied wonders of our Nation and contributing to social and commercial interaction, binding the States and regions together.

Realizing the immense impact that highways have on our lives and communities the Federal Highway Administration has long been concerned with the compatibility of highways with their environment, increasingly so in recent years. In the early stages of developing roads, the sole concern was to enable people to move themselves and their goods from one place to another. But now that our technology has assured this objective we have reached a stage where the highway must be considered in the context of its overall impact upon the social-economic-cultural fabric, of which it is a part. Other criteria, in addition to efficiency, have risen in importance in judging the value of a highway: safety, aesthetics, and its effect on the social and physical environment, both urban and rural. These aspects have become important and widespread public concerns.

FHWA's current efforts in the area of environmental concern run through the entire spectrum of the highway program, from overall systems planning through route location planning, right-of-way acquisition, and design, to construction, operations and maintenance. In fact, in today's highway program as much importance is given to land use compatibility, various amenities, ecological factors, noise reduction, air and water quality, and other environmental considerations as to engineering elements. This compendium is an attempt to provide an overall view of measures being taken by FHWA in cooperation with the State highway departments to protect and improve the environment surrounding Federal-aid highways. It is recognized that any attempt to separate environmental aspects of highway construction and operation from the other elements that are involved in undertakings of such importance, scope, and complexity must be arbitrary, since planning, designing, building, and maintaining a modern highway is an integral process consisting of many diverse but interrelated elements, many of them affecting in various ways and various degrees the impact of a highway on its total environment.

Environmental Development Division

Although every office of the Federal Highway Administration is concerned with environment elements, it is the primary concern of the new Environmental Development Division of the Office of Right-of-Way and Location.

The Environmental Development Division was created in 1968 as an interdisciplinary team of architects, urban and regional planners, landscape architects, sociologists, economists, appraisers, and engineers. Its general purpose is development and promotion of the construction of highways that are compatible with their environments and that help to further the goals of the communities which they serve. This purpose is accomplished by providing professional assistance and guidance to FHWA central and field offices,

State highway departments, State planning commissions, and other State and local agencies in the planning and design of highways. Primary emphasis is given by this division to the preliminary engineering phases of highway projects, beginning with route location studies. The principal objective of this division is to insure in the planning and design of highways:

1. Comprehensive consideration of social, economic, and environmental factors, especially those most affecting or affected by a highway decision.

2. Optimum utilization of the potential for joint development of other facilities with the construction of highways and for multiple use of highway rights-of-way.

3. Use of multi-disciplinary design teams as staff advisors to agencies and jurisdictions responsible for highway and related programs in order to achieve a coordinated highway and environmental program.

4. Use of intergovernmental policy groups in a comprehensive highway project planning process to develop integrated and coordinated highway and environmental plans and programs.

5. Use of citizen and unofficial community groups as advisors to agencies and jurisdictions responsible for highway and related programs.

Environmental Expenditures

Just as it is difficult to separate environmental aspects from the entire highway program, so it is difficult to assign a dollar figure for expenditures to preserve and enhance the environment. Many measures taken in the process of planning, location,and design may have as their justification the preservation of some environmental element; other measures may have benefits to the environment as fortunate side effects.

Nevertheless, it is possible to compute those expenditures for distinct measures which can be associated with environmental protection or improvement. It is estimated that about 12% of the total cost of Federal-aid highway projects are devoted to environmental considerations. With the immense size of the highway program this amounted to approximately $580 million for fiscal year 1969. Of this total about $50 million were obligated for direct measures taken for improved environmental quality and nearly $85 million for indirect measures, i.e. actions which have another primary purpose, but also beneficial environmental aspects.

Other financial obligations included in this total were for positive programs such as roadside improvement for motorists and residents alike, research on the general environmental impact of highways, relocation assistance for those displaced by highway construction, and the highway beautification program. These programs amounted to $445 million in FY 1969.

AASHO Committees

In addition to these programs and internal organizational arrangements and concerns, FHWA works with the States to improve measures for environmental protection and enhancement through its participation in the American Association of State Highway Officials and its various committees. Environmental policies and standards are principally the concern of the Committee on Right-of-Way and the Subcommittee on Urban Transportation Planning, Socio-Economic and Environmental Factors. In addition, AASHO has created the Special AASHO Liaison Committee on Environmental Quality, consisting of selected highway department and FHWA personnel together with representatives of influential conservation organizations, governmental agencies, educational institutions and private businesses.

The following chapters have been organized on a functional basis and contain discussions of the numerous highway program elements which are responsive and applicable to the nationwide concern for improved environmental quality.

The appendices, which correspond in number and sequence to each chapter, cite the legal bases for the various program elements and list Policy and Procedure Memorandums, Instructional Memorandums and Circular Memorandums as appropriate. Other pertinent papers, publications and articles have also been cited in the appendices.

APTER 2: PLANNING

avel Demands

By the year 2000 the population of the United States will total 300 million, an increase of 100 million in 30 years. Two-thirds of this growth is expected to occur in about 50 large metropolitan areas of more than one million inhabitants. To accommodate this increase it will be necessary to build as many dwelling units as have been built since the landing of the Pilgrims 350 years ago. Transportation demands will be even greater. Highway travel will double by 1985 and double again by 2000. Despite the continually growing transportation demands, proportionally less urban land is now devoted to highways than in the past. A major consultant's study of land use trends in 48 large American cities compared recent land use survey results with older land use data and found consistent declines in the proportions of developed land devoted to highways. It is also noteworthy that while there are now 3.7 million miles of roads and streets in the United States, over 80 percent of this mileage was in existence when the Federal-aid highway program was instituted in 1916.

Reasons for the decline in developed land devoted to highways include the greater usage of freeways and the emphasis on upgrading existing roads and highways. Freeways provide an efficient means of moving vehicles and, compared to city streets, carry three times more traffic per square foot of pavement. Upgrading existing streets has also allowed capacity increases as well as safety improvements.

Although future travel patterns and growth may be modified by concentration of population and employment centers, some substitution of communications for travel, and the development of new modes of travel, more leisure time and more disposable income will continue to offset any consequent reduction in travel demands.

This growth highlights the central role of highway planning. The nation and our cities cannot thrive without transportation, and the American people depend greatly on the automobile to provide this transportation. Within the country's 233 urbanized areas (urban areas which contain at least one city with a population of 50,000 or more), 99 percent of all person-trips and 98 percent of all person-miles of travel are by highway vehicle. Although there is a very strong reliance on highways, public acceptance of new facilities is directly related to the measures taken to ensure their compatibility with the environment, urban as well as rural, cultural as well as natural.

These concepts broaden the scope of highway development from that of simply providing a means of moving people and goods efficiently and safely to that of significantly influencing urban, regional, and national development, and in many ways the quality of life.

Urban Transportation Planning

Of all the provisions broadening the range of concern in the highway program none is more far reaching than the urban transportation planning provisions of the 1962 Federal-Aid Highway Act. This law prescribes that "the Secretary shall not approve ... any program for projects in any urban area of more than 50,000 population unless he finds that such projects are based on a continuing comprehensive transportation planning process carried on cooperatively by States and local communities ... ."

Today transportation planning is underway in all of the 233 urbanized areas as required by the 1962 Act, and in a number of smaller areas. Because of the process there is now a much better understanding of the problems both in the planning and the development of transportation improvements by officials at the various levels of government. Better administrative arrangements for more efficient utilization of available resources have been devised (as evidence of this, the transporation planning programs in 172 of the urbanized areas are being supported by both Federal-aid highway planning funds and 701 planning funds administered by the Department of Housing and Urban Development.) Communications between the professionals of the various disciplines and between the professionals and the decision makers have improved. And, in addition to the significant advances made in technical procedures, the process has contributed significantly to the establishment of more meaningful cooperative planning programs.

These cooperative planning efforts include the joint land use-transportation planning programs established in many metropolitan areas. The Joint Land Use-Transportation Planning Program for the Twin Cities Metropolitan Area (Minneapolis-St. Paul) is a good illustration of what can be done in such combined urban area studies. Through the use of an extensive attitude survey of 4,600 area residents, public attitudes about housing, employment, transportation, government, recreation, and other aspects of their environment were measured. The result was a body of information on public values and attitudes that formed the foundation for the definition of alternate area development plans. Alternate plans then underwent public and official scrutiny and debate through a system of committee and public meetings to arrive at a recommended development plan. Some states are extending such planning to embrace smaller communities and even rural areas on a statewide basis.

mmunity Values

The study of social and community values as related to land use development and to transportation is an integral part of all urban transportation studies. In the development of transportation plans full consideration is given to these factors in order to raise the standards of the urban area. Open space, parks, and recreational facilities are important environmental factors. More and more in transportation planning additional attention is being given not only to the preservation and enhancement of existing open space, but also to the provision of additional open space in anticipation of future development.

Increasingly greater concern is being given in the planning process to the avoidance of historic sites and park and conservation lands or, where avoidance is not feasible, special care in design to minimize harm; coordination with appropriate state agencies to control erosion and to protect fish and wildlife; flood control measures; scenic enhancement; and relocation assistance.

Care is also exercised to select locations for new transportation facilities so that neighborhoods are not disrupted. To the maximum extent possible cutting through school districts, fire station districts, etc., is avoided. The appearance of the facility not only from the viewpoint of the motorist, but from that of the pedestrian and the nearby residents is considered.

An example of how esthetic, social and economic effects are being brought into the decision-making process at this stage is the stress being put on the development of several alternative transportation system proposals for each metropolitan area. Each of the plans is an outgrowth of a somewhat different mix of priorities of the esthetic, social and economic effects. The several alternates, with the merits and demerits of each, are available for review and debate by elected officials and the general public. This procedure allows wide participation by the public and by elected officials in the formulation of community objectives and in the selection of transportation plans to meet these objectives.

There is no way to combine all the good and bad effects of highways into some all-inclusive net figure. Such an attempt would involve formidable problems of trying to measure intangible social and esthetic aspects. But to fully exploit a highway's potential for making physical and social surroundings better there are being developed by the National Highway Planning Division in the Office of Planning methods for appraising and combining not only the economic costs and benefits of alternate highway programs, but also the less tangible but equally valid environmental and social costs and benefits. Evaluation of possible highway programs can no longer be considered valid unless there are included estimates of their environmental impact. Accordingly, research is underway to develop procedures for estimating the likely impact of alternate highway plans on nine variables:

- economic activity and population distribution
- recreational and cultural opportunities
- displacement of families and businesses
- structure of neighborhoods
- land use patterns and land values
- air pollution
- noise
- national defense
- visual quality of the environment

Planning Criteria

Pending development of these procedures, planning criteria are spelled out in Policy and Procedure Memorandum 50-9, which sets forth the requirements and guidelines for the urban transportation planning process established pursuant to the planning requirements of the Federal-Aid Highway Act of 1962 (Sec.134, 23 U.S.C.). This PPM spells out the state-local cooperative nature of the continuous planning process, which is meant to determine whether specific highways are needed and, if so, the character of a highway and its general corridor location. Written memoranda of understanding with local communities are required to ascertain that planning decisions reflect the needs and desires of communities.

PPM 50-9 is now being revised to place even greater stress on the development and articulation of goals and objectives, transit planning, public involvement, the evaluation of alternative plans, and in general the need for open channels of communication between the communities and the planning staffs during all stages of the planning process. It will also include further definitions of the social and community value factors to broaden this element of the process.

Recent Studies and Programs

Specific programs that have been undertaken recently in the area of planning directed at enhancement of the environment include the following:

. Sign and Junkyard Study

As an aid to implementation of the Highway Beautification Act of 1965, advertising signs, displays, and devices, and junkyards within 660 feet of the rights-of-way of Interstate and Federal-aid primary routes were inventoried.

. Rest Area Study

Beginning in 1968 all rest areas adjacent to Federal-aid and State highways were inventoried and usage data for selected rest areas were collected. Rest area facilities vary considerably, and the data collected will aid in designing better rest areas that will provide restful, esthetic and useful facilities for the motorist.

. Statewide Transportation Studies

States undertaking statewide transportation studies generally incorporate environmental considerations when selecting the highway system plan. In Wisconsin, for example, one function of the highway system is to serve the State recreation plan and to provide for recreational travel. The State highway arterial system provides access to scenic areas. Along many sections a choice exists between an arterial highway and a parallel scenic highway. In Connecticut, a State goal is to provide adequate recreational sites within 25 miles of the urban centers, and the highway system provides access to the recreational areas. In Rhode Island, activities related to the statewide transportation study were a statewide inventory of historical sites, designation of land for parks to be purchased with funds under the open space program, and submittal of projects to the legislature for water reservoirs and recreation. Many other States have developed recreation plans to preserve sites of unusual environmental value.

Under guidelines developed by the Current Planning Division for statewide transportation studies, it is suggested that the State highway system plan serve the general goals of the State, including the preservation and enhancement of the environment. Also, inclusion of ecological and esthetic considerations in the selection of highway corridors is promoted, such as the location of a highway corridor along the sides of hills to preserve prime agricultural land in the valleys and to open up scenic vistas from the highway.

The purpose of this guide is to provide the best possible living environment by insuring residents an adequate supply, choice, and quality of development in a balance consistent with their stated and implied values and goals.

- "Urban Highways in Perspective" published by the Automobile Safety Foundation (1968). The report discusses the community impact of urban highways. It delineates areas of agreement, problems remaining to be solved and steps towards solution of these problems.

- "A Strategy for Evaluating a Regional Highway Transit Network," a report from the Baltimore Planning Council (1968). One section of the report delineates four classes of criteria for evaluation of urban transportation plans: public investment requirements, movement considerations, environmental considerations, and implementation considerations.

- "A Framework for Action" (1969), a report from the Cleveland Seven County Transportation-Land Use Study. This describes the evaluation and selection of an urban transportation system using a comprehensive set of goals, objectives and evaluation criteria.

- Technical reports

  Two technical reports are currently in preparation; one on noise and one on air pollution. These reports will describe methods for measuring the effects of alternative transportation systems on noise and air pollution.

- The Office of Planning is providing technical assistance to HUD on their extension of a study "Methods for Evaluation of Effects of Transportation Systems on Community Values." This study seeks to develop improved methods of identifying, measuring, and valuing five types of transportation effects:

  1. Accessibility to educational, cultural, social, and employment opportunities.

  2. Real estate, property development, and property tax base.

  3. Disruption and relocation.

  4. Air and noise pollution.

  5. Esthetic and open-space values.

- Training

The Urban Planning Division conducts a 2-week Urban Transportation Planning Course which covers the technical aspects of urban transportation planning. Pertinent lectures which are given include land use forecasting, community values, goals and objectives, and evaluation of alternative transportation systems-goal achievement. This course has been given 22 times to 757 planners, engineers, economists, and other interested persons.

A 3-day Urban Transportation Planning Short Course is also conducted by the division. It is directed at administrative level personnel requiring a general knowledge of urban transportation planning. Lectures are given on community values, goals and objectives, and evaluation of alternative transportation systems. This course has been given 10 times to 650 persons.

- Urban Studies

The trend in urban transportation studies is to give greater consideration and weight to environmental factors in the evaluation of transportation plans. One of the planning groups to give early recognition to the importance of environmental factors in urban transportation planning is the Southeastern Wisconsin Regional Planning Commission (SEWRPC). In 1962, SEWRPC conducted a natural resource inventory which covered mineral resources, forests, soils, air, surface and ground water, flooding, fish and wildlife, water quality, and recreational and open space. This information lead to the identification of primary and secondary environmental corridors. The protection of these corridors was a major factor in the evaluation of alternative land use and transportation study. Many of the technical reports from SEWRPC have been distributed nationwide, and the Bureau has participated in SEWPRC activities since 1962.

Other urban areas are conducting natural resources studies and developing more sophisticated methodology for evaluating the effects of environmental factors on transportation systems.

CHAPTER 3: LOCATION, DESIGN AND CONSTRUCTION

Location

Early highway route selection procedures were generally of a limited nature, since prior to the inauguration of the Interstate System highways were to a great extent upgradings of roads which had more or less evolved over varying periods of time. In this process highway construction decisions became largely economic in nature, reflecting considerations of traffic patterns, service benefits, and construction costs.

But the Federal-Aid Highway Act of 1956 mandated a 41,000 mile system built principally on new rights-of-way, bringing to the foreground the concept of locating and constructing highways that are compatible with and take full advantage of both the natural and cultural environment. Making the most of this opportunity, highway designers now base their location decisions not simply on economic and safety considerations but on a full range of human values.

It is during the location stage that special concern for protection of the environment is particularly important. Beginning with preliminary engineering for corridor and route selection, measures are taken to assure preservation of scenic and environmental resources and the application of geometric alinement principles for complete integration of the highway into existing land forms and urban developments. Highway department engineers consider all the various feasible alternatives for a highway corridor or route location and then survey - among other things - all scenic features and environmental factors affected by these alternatives. Taking advantage of the hearings and coordination required by law and FHWA directives, the alternate which will provide the highway which is the most functional and compatible with the environment is determined.

The decade of the 1960's saw the establishment of some very significant milestones toward implementation of this concept of environmental considerations in the location and design of highways. The Federal-Aid Highway Act of 1962, which established comprehensive cooperative urban transportation planning requirements, has been discussed in Chapter 2.

Early in 1963 a special effort was instituted by the Bureau of Public Roads to emphasize aesthetic considerations in location and design decisions (Circular Memorandum "Aesthetics in Highway Location and Design," March 1, 1963). Also in 1963 there was instituted a requirement that Federal-aid Highway projects be coordinated with theprojects and concerns of State agencies responsible for protecting fish and wildlife resources (Instructional Memorandum 21-5-63 "Coordination of Public Interests of Highway Improvements with Those of Fish and Wildlife Resources").

This requirement was expanded in 1964 to include coordination with the agencies responsible for the protection and improvement of park and other outdoor recreational and historic resources (Circular Memorandum- "Consideration of the overall interests of the public in the Federal-aid highway programs and programs for the protection or improvement of parks and other outdoor recreational and historical resources", May 25, 1964.)

Legislation enacted in 1966 and amended by the Federal-Aid Highway Act of 1968 (as Sec. 138, 123 U.S.C.) provides further protection for parks, recreation and conservation areas, and historic sites in the process of determining the location of highways. As specified in this legislation, it is now national policy that in the building of highways special effort is to be made to preserve public park and recreation lands, wildlife and waterfowl refuges, and historic sites if these lands are considered, by the officials having jurisdiction over them, to be of national, State or local significance. Such lands are not to be used for highway purposes unless there is no feasible alternative. In those instances where such lands cannot be entirely avoided in locating a highway, all possible measures must be taken to minimize harm to their beauty and principal use.

In addition, a national historic preservation law (P.L. 89-665) was enacted in 1966 providing for the establishment of a National Register of Historic Places and an Advisory Council on Historic Preservation. Section 106 of this law stipulates that the head of any Federal agency having direct or indirect jurisdiction over a proposed Federal or Federally-aided project must, prior to any expenditure of Federal funds, take into account the effect of the project on any area, site, structure, or object included in the National Register, and must afford the Advisory Council a reasonable opportunity to comment on the project.

As noted in the Circular Memorandum, "Preservation of Historical Sites" Dec. 29, 1966) FHWA has interpreted this requirement to include not only those highway projects having direct physical involvement with historic sites, but also those projects which may have an adverse effect on the setting or surrounding area.

The States are provided up-to-date lists of historic sites included in the National Register, the latest having been printed in the Federal Register of February 3, 1970, and distributed to State highway departments by Circular Memorandum, "Preservation of Historic Sites" (March 24, 1970).

blic Hearings

Greater public participation in decisions regarding highway location and design was ensured by the issuance of Policy and Procecure Memorandum 20-8, "Public Hearings and Location Approval", in January, 1969, requiring at least two public hearings for major highway construction projects, coordination with appropriate public and private agencies, and consideration of a wide range of factors in determining location and design. These requirements and the encouragement of additional hearings and meetings early in highway department deliberations provide opportunity for the public and relevant agencies to participate in highway location and design by presenting their views and by providing information relative to social, economic, and environmental effects of possible highway locations.

The public hearings required by PPM 20-8 are intended to provide a formal means of ascertaining and documenting all possible community, transportation, and environmental factors affecting or affected by highway developments. Environmental factors to be considered in highway location and design, as directed by PPM 20-8, include: effects of recreation and park facilities; aesthetics; public health and safety; residential and neighborhood character; conservation, including erosion, sedimentation, wildlife, and general ecology of the ara; natural and historic landmarks; noise; air and water pollution; property values; and replacement housing.

At least two public hearings must be held for major highway construction projects, such as a new highway; improvements which change an existing highway to such an extent that it greatly alters its social, economic, or environmental effect on the area; or improvements which significantly change the operation of connecting roads.

The first of these, a corridor public hearing, is held after the highway department has determined that there is a need for highway construction or improvement in a certain area and after it has studied the alternative possible locations for the highway. The need for the project and the alternative locations are presented at the hearing for the information of the affected public. Anyone is free to comment on the proposal at the hearing and for at least 10 days after the hearing.

A second hearing, a highway design public hearing, is held after a location has been approved and after the highway department has reviewed its design alternatives and the comments of the public. At this hearing it presents alternative possible designs, such as elevated or depressed roadway, alternative interchanges, etc. Again the public has the opportunity to comment and suggest changes.

Hearings are generally not held for minor improvements, such as resurfacing or installing traffic lights or turn lanes. However, a public hearing will be held if any such improvement requires the purchase of right-of-way or if it changes the relationship of the highway to connecting roads or abutting property.

After the hearing or hearings are held, the final plans are drawn up and sent to the local office of the Federal Highway Administration for approval. If they are approved, the highway department publishes a notice so that the public may know what the final decision is.

ght-of-Way

Once a highway location has been determined, further provisions are available to help protect the environment, provisions particularly applicable in urban areas. A new federal revolving fund has been set up by the Federal-aid Highway Act of 1968 (amending Sec. 108, 23 U.S.C.) so that property needed for a highway may be acquired up to seven years prior to construction This advance acquisition fund is important since it allows the States to (1) acquire needed right-of-way before development occurs, thus reducing acquisition costs, (2) have right-of-way available when construction is necessary and (3) minimize undue hardship to those who need to dispose of their property after location becomes known and ready purchasers are not available.

To preserve environmental amenities, States are authorized by Par. 5, Policy and Procedure Memorandum 80-1, "Right-of-Way Procedures (General Principles and Coordination With Other Government Agencies)" to use Federal funds to aid in obtaining and improving various kinds of replacement sites for property taken when this is considered to be in the public interest.

Section q of this same paragraph authorizes the States to acquire whole parcels or portions of remainders to a street line or some other logical boundary in order to provide a highway more in conformity with the neighborhood through which it passes. If consistent with the highway design, any portion of the right-of-way can be used for green strips, small parks, play areas, parking or other highway-related public use, or for any other public or quasi-public purpose which might assist in integrating the highway into the local environment and enhancing other publicly supported programs.

Sometimes excess right-of-way must be purchased when a highway is being constructed but later is not required for the operation of the highway. Policy and Procedure Memorandum 21-4.1, "Right-of-Way Procedures (State Acquisitions Under Federal-Aid Procedures)" permits relinquishment of such land if the land will not be needed for highway purposes in the forseeable future, if the right-of-way being retained is adequate under current standards, and if the release will not adversely affect the highway or traffic thereon. Policy and Procedure Memorandum 80-5, "Right-of-Way Procedures," added to these qualifications a determination that the lands to be relinquished are not suitable for retention in order to restore, preserve, or improve the scenic beauty adjacent to the highway.

In a measure related to this, Instructional Memorandum 80-2-68, "Disposition or Relinquishment of Excess Right-of-Way", stipulated that park and conservation agencies should be afforded specific opportunities to acquire, insofar as State laws permit, excess right-of-way of the Federal-Aid highway systems when such lands have potential use for park, conservation, recreation, or other similar purposes. State highway departments are to notify appropriate agencies of their intention to relinquish such potentially useful excess right-of-way.

Design

With regard to highway design the general aim is to provide a system of highways that are efficient, safe, pleasing, and compatible with both natural and man-made environments. This objective can be achieved only if environmental and aesthetic considerations form an inherent part of the entire engineering process in highway development, from the preliminary steps of corridor and route selection through design and construction to roadside development and maintenance.

Some of the basic aesthetic principles of current highway engineering have become familiar to American drivers through their use of freeways: limited access, separation of roadways, and grade separation of highway crossings. That these are also safety principles is not simple coincidence, but rather a reflection of the basic aesthetic principle that form follows function. The better a highway is designed to serve its purpose of effectively moving traffic, the safer it will be, and the better it will look. For example, good design of the roadway and judicious development of the roadside can provide visual cues to aid drivers naturally and without strain.

When, in addition to the beauty of this functional form, the highway is located and designed so as to blend with its urban or rural setting and to provide vistas of natural wonder and cultural landmarks, not only are motorists afforded maximum beauty as they drive, but the highway itself becomes a positive aesthetic contribution to the areas through which it passes, an integral part of its environment. This blending is accomplished through a variety of measures: fitting the highway to the existing landscape, with subtle transitions between constructed slopes and existing ground; taking advantage of the topography whenever economically feasible to provide separate alinements and profiles for divided roadways; joining of slopes and plantings with native roadside growth, to serve in some instances as a buffer for noise or light; designing structures based on principles of architectural excellence; and installing miscellaneous necessary structures, such as walls, fencing, guardrails, and signs, in ways that do not detract from the highway environment.

A number of specific measures recommended or required by FHWA are used by the States to make highways positive aesthetic contributions to their environment and to minimize any possible ecological disturbance. Among these practices are the following:

Design: By the use of independent roadway design, flat and transition curves, contour grading, and other measures the roadway is made to complement and to harmonize unobtrusively with its environment.

Techniques and equipment: The use of aerial surveys, the computer, and other new techniques and equipment enable engineers to provide more refined blending of the highway with the terrain.

Right-of-Way: Greater than minimum right-of-way widths needed for a highway may be acquired to allow for gentle slopes, to preserve existing plant growth along a highway, and to provide a buffer between the highway and surrounding land uses.

Cross-section: The use of liberal cross-section elements, wide lanes, and, where possible, wide medians improves not only the appearance of a highway, but, by reducing congestion, noise, and air pollution, its overall effect on the environment.

Grade: Emphasis on flat grades reduces the level of noise and air pollution.

Slopes: Flattening of slopes and rounding the intersections of slope planes presents a more natural appearance of roadsides and also reduces erosion.

View: By the use of such practices as curvilinear alignment, careful grading and selective clearing, a pleasing view from the road is provided motorists.

Depression: By the use of depressed roadways where feasible in urban areas, open spaces are increased and noise diminished.

Uniformity: The conscious effort to attain uniformity of design details throughout the highway systems, especially the Interstate system, makes highway facilities a more integrated whole and, thereby, a more aesthetic endeavor.

Structures: By appropriate use of structures, such as walls, tunnels, and bridges, unattractive deep cuts and fills are minimized. These structures can be designed to be positive contributions to the area. For example:

> New high strength steels and new designs have made it possible to build attractive deck bridges rather than bulky through truss bridges.
>
> In multi-level interchanges or other elevated structures girders are fabricated so as to fit the curving alinement of ramps and roadways. The use of box girder design enhances the appearance of the underside of the structure.
>
> Single column piers are used, where the width of the structure permits, to improve the graceful appearance of elevated highways.
>
> The use of greater lateral clearances for piers and abutments presents a graceful open appearance of grade separation structures.

Cuts: Presplitting and serrating rock cut slopes improves the appearance of cuts that are necessary and reduces erosion.

Retaining walls: Where walls are necessary because of restricted right-of-way or other reasons, various geometries and surface textures are used to harmonize with the surroundings. Arch configurations incorporating planting boxes and stone masonry have been used for this purpose.

Water Impoundment: Where appropriate highway departments, working in conjunction with State fish, wildlife, and conservation agencies, construct highways so that embankments serve as dams for the impounding of water to form ponds and small lakes. (An example of this practice along Interstate 94 in North Dakota is noted in a news release in the appendix to this Chapter.)

Hydraulics: Hydraulic engineering is an essential element of highway design in order to keep the roadway usable and to prevent erosion or flooding. Measures considered include provision for dispersion or concentration of water, dissipation of energy, ponding, and silting basins. Drainage structures are installed flush with the median grade so that they do not present a hazard or detract from the appearance of the roadside. Several cities have participated in the construction of joint-use storm sewers built in conjunction with a highway project. Other cooperative projects include the construction of drainage and flood control measures as planned by other Federal, State and local agencies.

Lighting: The use of "high level", "high mast" or "tower" lighting is becoming increasingly common with some installations already in place and a number of additional projects in the design stage. This system utilizes the area lighting principles and is accomplished by the use of a relatively small number of tall poles or towers, each supporting several luminaires. Instead of concentrating the light on roadways and adjacent surfaces, the entire interchange area is illuminated. A wider panoramic view of all portions of the interchange, rather typical of daytime conditions, is afforded the motorist. This is considered an aid in discerning the location and arrangement of all roadways, structures and other visual cues which guide the motorist. In addition to safety and economic advantages, this lighting technique should overcome the unsightly "forest of poles" which has frequently been criticized as presenting a cluttered interchange appearance during daylight hours.

Utilities: The use of highway rights-of-way by utilities is limited for aesthetic and safety reasons. Restrictions are particularly strict in scenic, park, and conservation areas. In some cases highway structures provide a way for utilities to cross a stream or some other natural feature.

These practices are encouraged and disseminated by FHWA memoranda and by standards and policy publications of the American Association of State Highway Officials. In addition, training courses are conducted throughout the year by FHWA for field personnel, trainees, and State highway engineers.

lti-discipline Teams

In recent years increasing emphasis has been given to a potent and valuable force in highway location and design, namely multi-discipline teams, composed of various mixtures of architects, engineers, planners, sociologists, economists, landscape architects, ecologists, and other relevant professions, to work with highway engineers toward the creation of a highway facility that will be a positive preserving and developing force in a particular area. These teams seek to identify and to preserve community and environmental values and to maximize the benefits which can be realized through coordinated planning and development of highway and non-highway facilities, providing opportunity for appropriate use of space adjacent to, above, or below highways.

This is not an entirely new concept. Early attempts successfully carried out by the Bureau of Public Roads include the George Washington Memorial Parkway (Mount Vernon Parkway), built during the early 1930's, the Blue Ridge Parkway, and the Baltimore-Washington Parkway. On these highways engineers worked together with architects as a team, from the location and design stages to completion of construction, giving full consideration to their effects on the scenic and cultural environment.

Since the beginning of Urban Federal-aid highway programs in 1944 and especially since the beginning of the Interstate program, the States have been encouraged to make full use of relevant outside disciplines in the planning, location, and design of specific urban freeway projects. Several meetings have been held by FHWA with outside organizations and State highway officials for the purpose of exploring the role of multi-discipline teams. As a result of these meetings and of experience in a number of difficult urban situations in which solutions for freeway locations and designs had been found through broad social and economic considerations, multi-discipline teams have been formed in a number of cities, notably Baltimore, Chicago, and Phoenix. In Baltimore the team pointed out that the Interstate routing through the city could serve as a buffer against land use development incompatible with residential neighborhoods. The Chicago team devised a unique comprehensive 4-block wide corridor development plan for a crosstown expressway, with industrial and commercial establishments to be developed between the separated roadways, and with a potential tree-lined buffer through residential areas. The Papago Freeway in Phoenix has been planned to take full advantage of the opportunity it presents for joint development of highway and other urban facilities.

In addition to these teams of outside experts several highway departments, as well as FHWA, have established internal multi-disciplinary staff capability to permit an environmental design approach. In whatever form they take, their purpose is to improve the design of urban highways, to promote joint development and multiple use, and to exploit the potential of highways to serve as a positive influence on urban forms and functions, thereby promoting the quality of the urban environment.

Construction

For direct Federal construction projects, specifications issued by FHWA and supervison by its field personnel provide protection of scenic features and the environment surrounding a highway corridor. On Federal-aid projects, a project agreement is executed which stipulates that the State will comply with the terms and conditions of Title 23, U.S.C., regulations issued by the Secretary of Transportation, and policies and procedures promulgated by FHWA, including those regulations pertaining to protection of the environment.

Generally, these regulations pertain to preservation of air and water quality and the prevention of soil erosion and are discussed in a separate chapter. Other regulations prescribe that pits from which material is taken for highway construction shall be located and graded to create a minimum disruption to the landscape, that rodent control measures be taken on all federal-aid highway projects, that roadsides shall be cleaned up following construction, and that disposal of construction debris should be done in a way that is satisfactory to the public and in accord with the aims of the Beautification Act of 1965.

Continuous surveillance of measures taken for environmental protection is maintained by FHWA field personnel on Federal-aid projects during all stages in order to provide professional guidance and technical assistance as well as to insure that the project is being carried out in accordance with appropriate laws and regulations and any special provisions of the project agreement. Field review teams may also travel from the Washington office to evaluate particular aspects of a project.

In addition to these measures taken to protect the environment, other actions are taken when the opportunity is available to improve the scenic and recreational aspects of the area. An outstanding example of this practice is the conversion of borrow pits, from which roadbed material had been taken, to form a chain of 50 lakes bordering Interstate 80 in Nebraska. Boating, fishing, and swimming are now available where no lakes existed before the construction of the highway. Similar lakes and ponds have been created along many other Interstate routes for recreational purposes. One other protective program warrants mention here. The Federal-Aid Highway Act of 1956 (Sec. 305, 23 U.S.C.) authorized the use of federal-aid funds for the salvage of archeological and paleontological remains uncovered during highway construction.

Taking advantage of this authority FHWA, as spelled out in Policy and Procedure Memorandum 20-7, "Archeological and Paleontological Salvage," and subsequent Instructional and Circular Memorandums, encourages the States to work closely with appropriate authorities to avoid, if possible, or to preserve threatened remains that appear to be of paleontological or archeological value. Federal funds may be used for any archeological surveys that may be judged necessary, for excavation of discovered remains, and for measures necessary to preserve the remains.

Since 1956 a total of 160 projects have been undertaken in 26 States at a cost of more than $1.8 million, of which almost $1.6 million were federal funds. A description of projects conducted during 1969 is included with Circular Memorandums in the appendix to this chapter.

Taken together, these practices are convincing evidence of the committment of FHWA and the State highway departments to protect and improve the environmental aspects of our cities and rural areas as affected by highways. To exemplify some of these practices the following list has been compiled of some specific actions that have been taken by highway departments with FHWA participation to protect or enhance a highway-affected facility or area.

ARKS

1. A small area of Brandywine Park in Wilmington, Delaware, was crossed by I-95 on a high, graceful bridge, which an artist described as a contemporary masterpiece. The area under the bridge will continue to be used as park land.

2. A city park and recreation area in Albuquerque, New Mexico. The highway department acquired an 80-foot wide by two blocks long strip of excess land during right-of-way acquisition for I-25. The land, which is adjacent to a city park, was deeded to the city and subsequently developed as a part of their park.

3. Interstate 40 in North Carolina passes through the Pisgah National Forest and parallels the Great Smoky Mountains National Park. This corridor was virtually inaccessible by motor vehicles. Close cooperation with the U.S. Forest Service and National Park Service greatly improved the access to the proposed camping areas in one of the more scenic areas of the Nation.

ISTORIC SITES

1. Overland Trail Park and Museum near Sterling, Colorado. Prior to I-80S, this little museum was neglected and seldom used. The city then made a $15,000 addition. Subsequent to the improvements, the museum had 8,000 visitors in 1966; 9,000 in 1967; and 14,500 in 1968.

2. The Dorsey Mansion and property, reportedly the oldest mansion in Howard County, Maryland, was acquired for I-95 right-of-way. At the request of the Historical Society, this mansion is being constructed to reduce needed right-of-way. The building is vacated and it will require considerable expense to restore the historical qualities of the site.

3. Beverly Mill Historic Landmark in Virginia dates back to 1749. It is located on Route 55 in Thorofare Gap. The I-66 freeway alignment avoided the mill at the request of the Garden Club of Virginia. A natural barrier was also preserved in the nearby Broad Run stream

FISH AND WILDLIFE RESOURCES

1. About 40 miles east of Little Rock, I-40 passes through the Wattensaw Game Management Area. There are 15,010 acres of land covered by the management area. Of this area, 264 acres were donated for freeway right-of-way with the provision that borrow areas needed would be located by the Game and Fish Commission, that a rest area would be built west of the White River West Relief structure and that a road would be located under this structure for turning movements. Agreement on this location required coordination between the Bureau of Public Roads, Arkansas State Highway Department and Game and Fish Commission, Federal Wildlife Game and Fish Department and the U.S. Forest Service.

2. Relocated Route 72 was to bisect the Round Meadow Swamp, a breeding ground for migratory wild fowl, in Connecticut. At the request of various conservation organizations, the State shifted the alignment and built a box culvert under Route 72 to allow access to the game refuge. The access structure cost was $100,000 and the additional right-of-way cost was $140,000.

3. The effect of haying operations upon game bird nesting success is being studied on I-94 right-of-way in North Dakota by the State University The research objective is to determine the need for preserving wetlands and nesting cover for game birds. The right-of-way will be mowed in one-mile sections on alternate sides of the highway to permit a study of the differential in nesting from mowed to unmowed areas. The goal of this project is to establish a workable wildlife right-of-way management plan to provide both an acceptable highway appearance and good nesting cover for game birds.

SCENIC AND RECREATIONAL AREAS

1. A safety rest area on I-70 in Kansas has some unusual rock formations, some of which have inscribed dates going back to the turn of the century. These rock formations resemble giant-sized mushrooms and are representative of the type of formations found in Mushroom Rocks State Park in Kansas. Prior to construction of the freeway, these rock formations were not easily accessible. The rest area is a means of attracting the attention of the travelling public to the interesting geological features of the area.

2. The Minnesota Department of Highways acquired right-of-way along the west bank of the Mississippi River for the river crossing of I-90. The low area, that originally flooded during high water, is being built up above flood plain and developed as a safety rest area and welcome center. The view at this location is excellent. The adjacent area, planned as a recreational boating facility, is to be included in the State park. Joint public use will provide acess by land and water for full utilization of the area.

3. The Route 63 highway embankment and dam was constructed by the South Dakota Department of Highways and the Department of Game, Fish and Parks in cooperation with Public Roads, U.S. Bureau of Indian Affairs, and the Sioux Tribe. The dam forms Eagle Feather Lake with an area of 60 acres providing fishing, boating and swimming as well as water storage and conservation. Clearing of the lake bed and placement of concrete block riprap was done by the Sioux Tribe. The extra cost of the dam was partially defrayed by the construction work performed by the Tribe.

LLEGES, HOSPITALS, CEMETERIES, CHURCHES, ETC.

1. Trinity Episcopal Church is near the fringe of the business district in Wilmington, Delaware. The construction of I-95 necessitated the realignment of Adams Street away from the side of the church and the widening of Delaware Avenue in front. The widening required a narrow strip of frontage property. To compensate the church for this taking, the highway department transferred the west half of the old Adams Street right-of-way to the church. The eastern half reverted to the church, as the abutting property owner, automatically upon its abandonment. The State also plans to lease a remaining triangular parcel to the church for a parking area.

2. The acquisition of right-of-way for I-35 through the Emporia State Teachers College was amicably accomplished by involving the college in the planning. The Kansas State Highway Commission exchanged 19.25 acres of land for 21.1 acres needed for right-of-way. A concrete box culvert provides access to divided land and serves as access to the city utilities. Borrow material was removed to form a lake for campus use.

3. The Resaca Public School in Brownsville, Texas, was compensated $16,487 for a playground acquired for U.S. 83 right-of-way in 1964. The land acquired was low ground. The replacement lots which were acquired by the school were on higher ground and more desirable for a playground. No improvements were taken.

HER PUBLIC-USE FACILITIES

1. The Greater Cincinnati Airport is located in Kentucky approximately 13 miles from downtown Cincinnati. Existing I-75 is located approximately three miles east of the airport with the main airport connection on Kentucky 236. Interstate 275 and I-471, presently under design, will improve airport access from the tri-state area. Interstate 275 will pass to the north of the airport and interchange with a four-lane connector to the airport. The airport plans for expansion includes the closing of the existing access from Kentucky 236; therefore, the airport expansion is very closely tied to the completion of I-275.

2. The Minnesota Department of Highways and the city of Albert Lea, Minnesota, made joint acquisition of right-of-way for I-90 and for the Municipal Airport runway extension. This eliminated severance damage

payments and reduced the expenditure of public funds in acquisition of whole parcels of land. The location and design of the freeway was made with consideration of the airport facility.

IAPTER 4: JOINT DEVELOPMENT AND MULTIPLE USE

In trying to preserve the environment and to improve our cities, planners and builders are restricted by the limits of available time, space and money:

> Time, because our urban population will double its current size during the next 35 years, meaning that in that same time dwellings, transportation capacity, commerce, and industry will also have to be doubled. In fact, they will have to be much more than doubled, since growth of demand is much more a consequence of a rise in our standard of living than of an increase in our numbers.
>
> Space, because active competition for the same urban space often means that one community need can be met only at the expense of other needs.
>
> Money, because the limits of financial resources to provide for these burgeoning demands are real. The demands are already enormous and in order to prepare today for tomorrow's demands, as we obviously must, our limited dollars must be put to the most effective use possible; they will have to be made to do double or triple duty.

These limitations all militate against our efforts to preserve and to improve the overall quality of our cities; and if we are to provide any significant improvement, it is apparent that new techniques will have to be employed. Principal among these are programs of joint development; programs through which cities can meet some of their increasing needs for better housing, commercial facilities, recreation areas, and open spaces by combining their development with urban highway construction. Rather than simply building a new highway they have the opportunity to develop entire corridors with multiple, complementary uses, fitting new or reconstructed highways into their environment as harmonious, integral, essential elements. As the 1968 Highway Needs Report noted, "The improvement of our cities is a National goal of high priority, the achievement of which requires, among many things, the planned integration of transportation facilities with all other elements of the urban environment."

egislative Authority

Sec. 111, 23 U.S.C., and Sec 1.23,23 C.F.R., provide for multiple use of highway rights-of-way, authorizing States to use or to permit the use of airspace above and below highways for such purposes as will not impair the full use and safety of the highway. Two pieces of legislation passed in 1968 took further steps to permit implementation of the joint development concept. The Federal-Aid Highway Act of 1968, amending Sec. 128(a), 23 U.S.C., requires that public hearings for a Federal-aid Highway location include consideration of "social effects of such a location, its impact upon the environment, and the consistency with the goals and objectives of such urban planning as has been promulgated by the community."

The Intergovernmental Cooperation Act of 1968 stresses that "The economic and social development of the Nation and the achievement of satisfactory levels of living depend upon the sound and orderly development of all areas," and calls for a Presidential review of "Federal programs and projects having a significant impact on area and community development, including programs providing Federal assistance to the States and localities to the end that they shall most effectively serve these basic objectives." It directs that "to the maximum extent possible, consistent with National objectives, all Federal aid for development purposes shall be consistent with and further the objectives of State, regional, and local comprehensive planning. Consideration shall be given to all developmental aspects of our Total National community, including but not limited to housing, transportation, economic development, natural and human resources development, community facilities, and the general improvement of living conditions."

In addition to these Federal government measures over half the States have at least some authority to acquire and/or control land near highways. Some States' laws permit acquisition of land, within certain limits, in addition to that needed for right-of-way, which land can later be reconveyed with restrictions on development that will be permitted.

Thus, some legal authority currently exists to permit joint development. In order to aid the program further, FHWA has prepared model State legislation authorizing joint development and multiple use in the hope that legislatures will expand highway department authority to meet their states' emerging needs.

Benefits

Some specific benefits that can be realized through joint development illustrate how the program can accomplish its general purpose. Among these benefits are:

Provision of an environmental catalyst, melding the highway facility into a larger whole with a minimum of disruptive effect.

Assurance of compatible land uses adjacent to the highway.

Provision of a transistion zone between different and inconsistent kinds of land uses, where it is desirable to separate such uses.

Conservation of valuable urban space for appropriate public or private uses at minimum cost.

Reduction of costs to local government of public services and structures on or near the highway right-of-way.

Addition of ratables to the tax base.

Improvement of the amenities of a neighborhood or area by maintenance of continuity and by making possible the provision of facilities for recreation, for education and social contact, and other kinds of public and community services that otherwise would be far more difficult and costly to provide.

Provision of employment opportunities for neighborhood and community residents.

The objectives of joint development thus include not only satisfying highway transportation needs, but also attaining compatibility of a highway with its urban environment, and helping to achieve community planning goals. To best meet these sometimes disparate objectives with optimum solutions requires continuous comprehensive highway project planning by a partnership of highway and other public and private agencies preparing specific plans, designs, and programs for their mutual and community benefit prior to completion of a highway project. Through coordination and mutual adjustment between the agencies it is possible to reduce the negative effects anticipated between highway and environment, to utilize most of the positive effects of each to the advantage of the other, and to have each allow for improvements by the others.

Efforts have been made for several years by FHWA to foster joint development, particularly since the recommendations of the Hershey Conference in 1962 that solutions to urban problems be achieved through combinations of highway, urban renewal, and other public and private resources to integrate freeway construction with companion urban renewal projects.

Interim PPM 21-19

Realizing both the legal limitations of State and Federal authority and the comprehensive nature of joint development, FHWA issued preliminary policy procedures to be followed in joint development planning activities related to new highways in urban areas. As pointed out in Interim Policy and Procedure Memorandum 21-19, "Joint Development of Highway Corridors and Multiple Use of Roadway Properties" (Jan. 17, 1969), the role of highway departments in joint development is two fold: First, joint development reconnaissance for highway locations. This is undertaken after transportation studies have shown the need for a new highway. Joint development reconnaissance analyzes the needs and characteristics of an area and compares the effects each alternative location would have on community values and goals. For each alternative it will indicate the relationship of the highway to the local land use policies and the opportunities for joint action among participating organizations.

Second, highway corridor joint development planning. This is part of the detailed design of a highway after a particular location has been selected. This planning will be primarily the responsibility of the local jurisdiction in which the highway improvement is to be located, and is to be carried out in cooperation with the highway department. The purpose is to draw up a detailed schedule of actions to be undertaken jointly to carry out highway corridor development as a coordinated effort. These actions may include urban renewal, zoning, private development, etc.

Federal financial participation is available for the planning costs of joint development in the highway corridor and, where necessary, in the costs of feasibility studies of proposed uses in the right-of-way.

Use of Airspace and Acquisition in Limited Vertical Dimension

Non-highway use of the space above and/or below highways is permitted under PPM's 80-10.1 and 80-10.2, provided that such use will not adversely affect the safe and efficient operation of the highway and that the highway use will not be a danger to the other uses permitted. Properly planned utilization of airspace can make a positive contribution to enhancement of the environment through which the highways pass, and can also benefit those displaced by highway construction. The community and/or local government may also receive monetary benefit from the use of airspace since disposition of the income received is the State's responsibility.

Acquisition in Limited Vertical Dimension, covered in PPM 80-5, is another right-of-way tool that leads to highway facilities jointly sharing the corridor with non-highway facilities. In "ALVD", a three-dimensional "air tunnel" is acquired to accommodate the highway proper, together with permanent R/W interests for construction and maintenance of the support members and temporary easements to permit entry for construction of the highway facilities. ALVD is likely to result in substantial monetary savings in congested urban areas, as well as effect improved relations within the community by minimizing the disruption of neighborhood activities.

In recognition of the interdisciplinary nature of joint development, project planning and design teams, staffed with land use and urban planners, engineers, architects, sociologists and other related specialists, are being utilized by highway departments where projects warrant, and some departments have been increasing their in-house capabilities. As noted in the introductory chapter, FHWA has established its own interdisciplinary team, the Environmental Development Division, including experts in architecture, landscape architecture, sociology, economics, regional and urban planning, and land acquisition and appraisal, as well as engineering.

ojects Underway

A number of major joint development projects are underway. The Chicago Crosstown Expressway is an extensive, comprehensive project in the Cicero Avenue area, involving a separate roadway design encompassing a 4-block-wide corridor, with special provisions for commercial and industrial development between the roadways, and plans to make expressway serve as a buffer between this comercial-industrial area and bordering residential areas. Expected relocation of existing businesses into the corridor may help open up new areas for park, recreation, and open space uses.

The Papago Freeway in Phoenix, the product of an interdisciplinary team of consultants working with community groups, was laid out only after extensive environmental, land use, and transportation studies were completed. The result is a design for a highway that is economical to build, will serve traffic demands effectively, and will be a dramatic addition to the city.

The Center Leg of the Inner Loop Freeway in Washington, D. C., includes plans for joint use of scarce inner city land by acquiring whole blocks, decking over the highway, and using the resulting area for a park and for low and moderate price housing.

Recent proposals for freeway development in Pittsburgh involve a renewal project granting an easement for the East Street Corridor and selling residential land to private developers.

On the Massachusetts Turnpike an interchange at Newton will include the use of both highway and rail airspace for a motel, parking garage, office building, restaurant, bank, various stores, and plazas.

An extensive pictorial listing of joint development projects undertaken through the United States is included in the FHWA publication Highway Joint Development and Multiple Use.

Experience with these and other joint development projects show clearly that although the primary function of a highway remains the safe, efficient, and economical movement of people and goods, coordinated planning of highway and non-highway facilities in the highway corridor can produce new development desired by the local community and increase the total benefits to the public.

While most of the emphasis on joint development has been in urban areas, it is recognized that, as one example, coordinated highway-recreation planning and development in rural areas, particularly near cities, will enhance the appearance of the roadside while providing facilities for recreational activities. This is of increasing importance in view of the decreasing recreational facilities for our growing population. Preliminary steps toward such joint development have been taken in the form of three studies in New York, California, and Illinois. Brief summaries of these studies are included in Circular Memorandum, "Three Reports on Integrating Highways and Recreational Facilities" (April 30, 1968).

Multiple Use

An additional aspect of joint development is multiple use, which permits the use of highway rights-of-way, where feasible, for other than highway purposes. As noted eariler, such use is authorized by Sec. 111, 23 U.S.C. Policy and Procedure Memorandum 80-1, "Right-of-Way Procedures (General Principles and Coordination With Other Government Agencies," March 20, 1969), authorizes the States, in purchasing right-of-way for highway construction, to acquire whole parcels or portions of the remainders to a logical barrier or boundary line and to devote the land not specifically required for the safe operation of the highway to other public or quasi-public uses, such as small parks, green strips, or play areas. Instructional Memorandum 21-2-69, "Federal Participation in the Development of Multiple Use Facilities on the Highway Right-of-Way" (Jan. 17, 1969), encourages the States to take full advantage of this authorization with the cooperation of local governments in order to make the highway conform to its environment and to obtain the greatest benefit for the community

This program has been particularly successful. In the past 4 years more than 500 requests have been processed from almost every State in the Nation for the use of highway land for nonhighway purposes. Examples of multiple use in various States are found in Highway Joint Development and Multiple Use.

Fringe Parking

A specific legal provision in the area of joint development and multiple use is Sec. 11 of the Federal-Aid Highway Act of 1968, which authorized a 2-year demonstration program of providing fringe parking facilities on or adjacent to Federal-aid highway rights-of-way. The program is limited to urban areas of 50,000 or more population, and any parking facility constructed under the program must be connected to existing or proposed mass transit facilities. The purpose of the program, aided with 50% Federal financing, is to provide convenient transfer between private vehicles and public transit facilities, encouraging the use of public transit facilities and thereby lessening congestion and pollution and the need for extensive highway construction and improvements. Guidelines for the use of the States in implementing this program were issued with Instructional Memorandum 21-1-69, "Guidelines for Administering Demonstration Fringe Parking Facility Projects" (Jan. 16, 1969).

A further practice, followed where situations permit, is joint use of a transportation corridor by highway and rail. Notable examples of this type joint development of a corridor are the Kennedy, Eisenhower and Dan Ryan Expressways in the Chicago metropolitan area where rapid rail transit facilities have been built in the medians of these expressways. More common is the use of existing rail right-of-way for new highway construction. Where a railroad traversing an urban area has a wider right-of-way than is needed, its use for highway purposes reduces the need for displacing families and businesses and consequent disruption of the surrounding area.

CHAPTER 5: AIR AND WATER QUALITY AND NOISE ABATEMENT

If pollution is a many-sided thing, any measure to combat the evil is even more so. Essentially pollution is a threat to our ecologic system and personal health as well as an affront to our esthetic sensibilities. But measures or programs that militate against such an affront or threat cannot always be easily separated out and identified as such.

So it is that any digest of programs which help to maintain or improve air and water quality and reduce noise will be incomplete, since other helpful measures or considerations will be hidden among programs or activities that have other primary objectives. For example, the effect of traffic noise is considered in highway location and design; or discussion with recreation and conservation agencies may influence the location of a certain highway to preserve recreation or conservation lands.

Nonetheless, there are a number of measures that are directly aimed at the achievement of air and water quality and noise abatement.

Air Quality

The Air Quality Act of 1967 authorized the Department of Health, Education and Welfare to regulate pollution emissions from new motor vehicles. In that year motor vehicles were responsible for causing 72% of the carbon monoxide and 49% of hydrocarbons in the Nation's atmosphere. As a result of HEW-imposed standards there has been a downtrend in these two pollutants since 1967 However, the engine modifications used to control carbon monoxide and hydrocarbons have increased the emission of nitrogen oxides, the only other significant pollutants from internal combustion engines. And despite controls and modifications carbon monoxide and hydrocarbon levels will continue to rise as more motor vehicles are added to urban highways.

Actions taken by FHWA to relieve air pollution brought by highway traffic have been indirect but effective. Pollution is reduced whenever traffic is permitted to flow more smoothly, i.e. when congestion and stop-and-go type driving are reduced. This objective is accomplished primarily through the provision of freeways and expressways in urban areas. As an example of the effectiveness of expressways in relieving air pollution it has been found that traffic on central business district streets adds .42 pounds of carbon monoxide per vehicle mile, whereas traffic on expressways contributes about one-fourth that amount, .11 pound per vehicle mile.

The principal reason for this difference in carbon monoxide emissions is that internal combustion engines burn fuel most efficiently at relatively constant speeds, but the amount of unburned fuel increases sharply when a motor vehicle is accelerated or decelerated. For example, when an automobile is cruising, less than 2% of the fuel supplied to the engine is emitted unburned, whereas during deceleration this jumps to 18%. Average speed also plays an important role: An increase in speed from 20 mph to 30 mph reduces polluting emissions by one-third.

Freeways help reduce air pollution levels in urban areas in another indirect way, namely by aiding in the dispersal of industry, thereby permitting the atmosphere to provide natural clearing of industrial pollutants.

Less dramatic but still helpful results are obtained whenever traffic flow is improved. An important program in this regard is the Traffic Operation Program to Increase Capacity and Safety (TOPICS). This program takes cognizance of the fact that it is neither desirable nor feasible to accommodate all of our rapidly growing urban transportation needs by building new highways. Such an attempt would be too expensive and too disruptive and would always lag behind demand. Under TOPICS traffic engineering improvements are made on existing streets to increase traffic flow without resorting to major construction.

Although the program has been in effect for some years, separate Federal-aid funds were first authorized specifically for this purpose by the Federal-Aid Highway Act of 1968, which amended Sec. 135, 23 U.S.C. Policy is spelled out in Policy and Procedure Memorandum 21-18, "Urban Traffic Operations Program to Increase Capacity and Safety." Ordinarily improvements carried out under this program will be of a relatively minor nature in accordance with an areawide plan, as for example installing left turn lanes at intersections, providing bus turn-out lanes to facilitate passenger boarding without delaying traffic, providing reversible-flow traffic lanes, and building pedestrian overpasses. TOPICS improvements are also designed to contribute, as far as possible, to the effectiveness of public transit through such measures as reserving special lanes for buses during rush hour traffic.

Air pollution resulting from highway construction activity is also a possibility. Specific measures recently undertaken by FHWA to combat air pollution during construction include recommendations that State highway departments require, where applicable, nonburning techniques for the disposal of brush and timber removed by highway projects in urban areas, and recommendations that dust collection systems be used on hot-mix asphalt plants and other types of plants used primarily for highway construction. All but six States now require such dust collection systems.

Further evidence of the growing concern within FHWA over air pollution is found in the creation in March, 1969, of a Task Force on Noise and Air Pollution, which included representation from each major office in FHWA. The purpose of the group was to consider the involvement of air pollution, including noise, in connection with highway improvement programs. It functions included:

Collecting data, studies, research reports, and other information available on noise and air pollution resulting from the construction or operation of highways.

Reviewing and evaluating such information and developing recommendations for further research.

Developing a course of action for implementation of the Task Force's findings by establishing policy statements and standards and by dissemination of information to field offices and State highway departments.

Reflecting increasing concern for environmental protection and enhancement, this task force was replaced in March 1970, by a permanent Task Force on Environmental Considerations with representation from every major office to assure that every phase of the highway program is adequately involved in this concern.

Water Quality

Recent specific actions by FHWA toward the maintenance of water quality and reduction of erosion followed the issuance of Executive Order 11258 in 1965 (revised by Executive Orders 11288, 11507 and 11514), which required that specific actions be taken by each department of the Federal Government to provide leadership in the nationwide effort to improve water quality through prevention, control, and abatement of water pollution. Subsequent to the issuance of the Order, division engineers of FHWA discussed the matter with highway department representatives to determine whether the States' practice and procedures would achieve the objectives of the Executive Order. After consultation with the Federal Water Pollution Control Administration (now the Federal Water Quality Administration) and the Department of Interior, Instructional Memorandums were issued with guidelines outlining procedures for maintaining water quality and reduction of possible soil erosion occurring during and following highway construction, and for the draining of storm water from Federal-aid and direct Federal highway construction projects. Since January 1, 1967, Federal aid plans, specifications, and estimates must contain provisions to keep pollution of all waters by highway construction to a reasonable minimum. Similar requirements were imposed for direct Federal projects.

Major emphasis was placed on reduction of erosion and on the control of siltation, although pollution by chemicals, raw sewage, lubrications, fuels, and the like was not ignored. Sample specification language was furnished for consideration by those States whose existing provisions needed improvement. Measures that were suggested included limiting the area of raw, erodable earth exposed at a given time; construction of silt basins; timely planting of erosion control grasses and plants; limitations on fording streams; and reasonable restrictions for bridge and culvert construction. Cooperation with other public agencies having an interest in this matter was emphasized, as well as consistency with their laws and regulations.

To assure compliance with the issued guidelines, all States were requested to review their existing construction specifications and incorporate any necessary changes to accomplish the objectives and intent of the guidelines and the President's Executive Order. Some of the major practices and procedures included in the guidelines to aid in promoting the abatement of water pollution and soil erosion on Federal-aid highway construction projects are:

> Highway locations are to be selected with due consideration of the problems associated with the basic elements that will greatly reduce erosion during and after construction.
>
> During the construction of a project, the contractor must exercise every reasonable precaution throughout the life of the project to prevent silting of rivers, streams, and impoundments.
>
> Prior to the suspension of construction operations for any appreciable length of time, the contractor shall shape the top of earthwork in such a manner that will permit the runoff of rain water with a minimum of erosion.
>
> Temporary erosion and sediment control measures, such as berms, dikes, etc., deemed necessary by the engineers shall be provided and maintained during construction until permanent drainage facilities and erosion control features are completed and operative.
>
> Frequent fording of live streams with construction equipment will be held to a minimum and, where necessary, temporary bridges or other structures shall be used.

Contractors shall provide adequate sanitation facilities on all construction projects meeting the standards established by State health authorities.

The final condition of borrow or waste pits shall be finished or covered with vegetation in such a manner that it will not be a contributing factor to water pollution.

Excavation and embankment operations will be closely correlated with the seeding and mulching treatment of cut and fill slopes so that their surfaces will not be exposed for an extensive period of time and thereby contribute to soil erosion.

These controls, required on all Federal-aid and direct Federal highway construction projects, are now being included also as directives or special provisions in the majority of state highway specifications. Other steps to control erosion and to prevent water pollution during highway construction may be directed by special provision on some projects, e.g. seeding slopes as soon as possible as work progresses; step-cutting of slopes to help seeding get started; and constructing brush filter zones at the toes of fill slopes.

A highway built to current standards has few erosion problems after its completion, particularly if good maintenance practices are followed. All highway agencies recognize the potential detrimental effects of erosion within the highway right-of-way and accordingly, give special attention in design to preventive measures where needed. The success of these measures is evidenced by the many miles of highways now serving the traveling public without serious erosion scars.

Highways not properly located, designed, constructed, or maintained are at times subject to erosion and may contribute to stream pollution. Serious erosion not only results in unsightly conditions and increased maintenance costs, but sometimes causes safety hazards.

Problems encountered in finding feasible ways to minimize erosion are varied and complex. Several disciplines of science and engineering are required to reach an acceptable solution to most erosion problems. Highway designers, project engineers, and maintenance personnel use the advice of hydrologists, hydraulic engineers, soil engineers, soil scientists, agronomists, landscape architects, and other specialists to minimize erosion problems.

Erosion control guidelines encompass all phases of highway engineering to realize economical and effective control of erosion that might occur. National guidelines for the control of erosion must necessarily be of a general nature because of the wide variation in climate, topography, geology, and soils encountered in different parts of the country. For example, erosion control must be given careful attention in the design of a highway traversing an area of rough topography, erodible soils, high and constant wind velocities, and heavy precipitation. A high degree of erosion control is required in a watershed that is the collecting area for a public water supply or a recreational facility.

Erosion is controlled to a considerable degree by geometric design as well as by drainage and landscape development. It is minimized by the use of flat side slopes rounded and blended with the natural terrain; drainage channels designed with due regard for depth, width, slopes, alinement, and protective treatment; facilities for ground water interception; protective devices, such as dikes and berms; and protective ground covers and plantings.

Complete explanation of measures taken to minimize erosion related to highways is found in Guidelines for Minimizing Possible Soil Erosion from Highway Construction, a report presented to Congress in 1967 and distributed to State highway departments by Instructional Memorandum 20-6-67.

With regard to drainage, it is recognized that for more effective and efficient control of water pollution communities should attempt to construct separate storm and sanitary sewer systems. Drainage provisions for Federal and Federa-aid highways are to reflect this trend. The State highway departments are advised that the drainage courses or to existing storm sewers if the latter have sufficient capacity and can coveniently serve the highway's needs.

When it is not practicable to discharge highway drainage into an adequate natural drainage course or an immediately accessible separate storm sewer of adequate capacity, one of the following alternate solutions is to be adopted:

> Construction of a separate storm sewer to handle drainage resulting from the highway construction, such sewer to empty into an adequate natural drainage course or be connected to another existing storm sewer of ample capacity to handle its present flow plus that contributed by the highway drainage.
>
> Connection to an existing combined sewer provided that the sewer has adequate capacity to handle present flows plus the additional flow, if any, resulting from the highway drainage.

The selection of an alternate solution is to be made in the earliest stage of project planning based on the following considerations:

> Impact on appropriate water quality standards.
>
> Cost versus benefits to be derived for each alternate.

Pollution control (reduction of pollution impact on receiving waters must be optimized).

Sewer separation plans of the state or local jurisdiction.

Plans for control and/or treatment of combined sewage in excess of collection system or treatment plant capacity.

Compliance with state statutes and local ordinaces and regulations.

Many highway projects in urban areas involve widening of existing streets in areas currently being drained into combined sewers. Although the continued drainage of surface water from these highway improvements into a combined sewer is not considered desirable, alternate solutions for disposal of the surface water will frequently not be feasible because the local government has not provided a plan and an implementing program for separation of combined sewers. In these instances drainage of the highway to the combined sewers may be continued if the project does not result in significant changes in the hydrology of the area drained. If the highway construction results in significant additional surface water discharges to combined sewers, separate storm drainage is to be provided, or other alternative solutions are to be determined by consultation between the Federal Water Quality Administration and other agencies involved.

Noise Abatement

Highway noise considerations pertain to construction activities as well as traffic operations. The average noise level in the United States, particularly in urban areas, is constantly rising and with it, increasing concern for its detrimental effects. Medical research indicates that loss of hearing is by no means the only ill effect of noise. Loud sounds cause blood vessels to constrict, skin to pale, muscles to tense, and adrenal hormone to be injected into the blood stream. In general noise interferes with communication, sleep, and privacy, leading to personal annoyance.

Accordingly, the Federal Highway Administration is currently considering new measures to abate noise from highway construction and operation. A study recently completed by a consultant for the Highway Research Board and funded by Federal Highway Planning and Research money and titled Highway Noise - A Design Guide for Highway Engineers will provide basic techniques for engineers to predict noise levels of proposed highway projects, suggest acceptable noise levels, and offer various ameliorative measures which can be taken to control noise, e.g. installing accoustical barriers, elevating or depressing the roadway, and providing different road surface conditions.

Preliminary work has been completed on development of a lecture on highway noise for presentation to highway officials and engineers.

The intent of this presentation will be to convey a greater awareness of the need for the consideration of noise in the highway program and to indicate the direction that this consideration should take. Similar lectures will be prepared regarding other kinds of pollution.

In addition to the measures undertaken by FHWA, the Department of Transportation has set up an Office of Noise Abatement, which is concerned with transportation-created noise of all kinds, including highway and motor vehicle noise.

CHAPTER 6: RELOCATION ASSISTANCE

It is inevitable that construction of new highways will necessitate the displacement of some families and businesses, and it is recognized that adequate provisions should be made for those who must move, since their moving is involuntary and is done for the benefit of the general public.

Awareness of this responsibility was reflected in the Federal-Aid Highway Act of 1962, which required all States to provide relocation advisory assistance for residential moves and authorized relocation payments of up to $200 for residential moves and $3,000 for business moves. Following this the Federal-Aid Highway Act of 1966 directed that a study be undertaken of the highway relocation assistance problem to ascertain the compensation currently available and to suggest additional assistance that might be recommended. A report on this study by FHWA was published in 1967, followed by provisions in the Federal-Aid Highway Act of 1968, adding Chapter 5 to 23 U.S.C. and authorizing unprecedented compensation and assistance for those displaced by highway construction. The Act also required that the States make these provisions available by July 1, 1970.

In March of 1970 the Secretary of Transportation strengthed the program, making it mandatory that relocation housing be available for those to be displaced-even if it had to be built - before Federal approval would be given for construction of any transportation facility.

In order to provide guidance for the States to implement the program, FHWA issued Instructional Memorandum 80-1-68, "Relocation Assistance and Payments - Interim Operating Procedures" (Sept. 5 , 1968) detailing the specific measures that must be implemented to aid displacees in compliance with the 1968 Act.

The Act authorizes payments to displaced owners and renters that will enable them to buy or rent dwellings that are comparable to those from which they were displaced and that meet at least certain minimum standards of decency, safety and sanitation. Home owners may receive up to $5,000 above the amount which they received for the property acquired for highway purposes and renters may receive up to $1,500 over a two-year period as a rent supplement.

Decent, safe, and sanitary standards for replacement housing are spelled out in IM 80-1-68 and include minimum standards for water, kitchen and bath facilities, heat, lighting, and fire safety.

This program has a noteworthy impact on the environment since it includes as one of its fundamental provisions that no highway construction will be undertaken unless and until adequate relocation housing is found to be available in the general area for those who must be moved. Coupled with the requirement that the relocation housing available be in accord with stipulated standards of decency, safety, and sanitation, communities are provided assistance in preventing deterioration of their neighborhoods and help in improving their general environment.

From October 1, 1968, to December 31, 1969, a total of 27,516 dwellings were required for Federal-aid highway projects. The relocation assistance program provided financial and advisory assistance for 79,457 individuals, at a cost of over $18 million, helping them to obtain equal or better housing. Of this amount almost $5 million was for replacement housing costs required for owner-occupants and renters to secure comparable and decent, safe, and sanitary housing.

CHAPTER 7: BEAUTIFICATION

The Highway Beautification Act of 1965 was drawn up against a background of more than 30 years active concern for the environment by those charged with responsibility for the Nation's highway program. Landscaping and roadside development as a part of the normal costs of construction were first authorized by the Federal-Aid Highway Act of 1938. Authorization of land acquisition for the preservation of natural beauty in the highway corridor was added in 1940, and the first Federal provisions for outdoor advertising control were enacted in 1958.

These enactments reflected the growing awareness of the need to protect the highway corridor and to blend the highway into the existing environment, an awareness leading to the Highway Beautification Act of 1965 and its ongoing implementation.

Control of Outdoor Advertising

Title I of the Highway Beautification Act, revising Sec. 131, 23 U.S.C., provides for the control of outdoor advertising along the Nation's Interstate and Federal-aid primary highways, a total of 265,000 of the Nation's most important roads.

The law enacted in 1958 had attempted to control outdoor advertising by offering the States bonus payments for providing and enforcing billboard limitations along Interstate highways. However, success of this program was limited. By the time the offer expired in 1965 only 25 States had entered in to agreements to provide advertising controls. At best, only 18,000 miles of Interstate highways would have been protected under this legislation.

Measures brought by the 1965 Act are considerably broader and stronger than the 1958 Federal legislation. Instead of just the 42,500 miles of Interstate highways, the Federal-aid Primary system, totaling more than 200,000 miles, is covered as well. And, instead of a bonus program it provides a penalty system and permits the Secretary, under certain conditions to withhold 10% of a State's Federal-aid funds. Although there is nothing to prevent the States from enacting more stringent controls, the Act provides minimum standards for the regulation of outdoor advertising by the States.

Basically two distinct actions are required of the States by Title I. As spelled out in Policy and Procedure Memorandum 80-9, "Acquisition Procedures for the Control of Outdoor Advertising Signs and Junkyards and for Landscaping and Scenic Enhancement (Highway Beautification Act of 1965)," each State must, first, provide for control of future signs so that they will be permitted only in areas allowed by the Highway Beautification Act and only under the conditions prescribed. Specifically, the States, under current standards, are to control outdoor advertising within 660 feet of the right-of-way of Federal-aid primary and Interstate highways. Only the following signs, certain of which are subject to size, spacing, and other standards, are to be permitted:

1. Directional and other official signs, including motorist service signs (specific gas, food, and lodging information) on Interstate highways, and signs pertaining to natural wonders and scenic and historic attractions:

2. Signs advertising sale or lease of the land on which they are located:

3. Signs advertising activities on the property where they are located:

4. Signs in commercial or industrial zones and areas, to comply with size, lighting and spacing criteria, to be agreed upon by the State and the Secretary.

Second, the States must effect the buying and removal of nonconforming signs. Federal-aid for this purpose may be used, when available, to pay 75% of the cost of such removal. When Federal-aid funds are not available, removal need not be enforced by the States.

Control of Junkyards

Title II of the Act adds Section 136 to Chapter 1, 23 U.S.C., providing for the control of junkyards, scrap metal processing facilities, trash dumps, and the like. All such facilities alsong Interstate and Federal-aid primary highways, except those in industrial zones and areas, are to be screened, such as by shrubbery or fencing, if they are within 1,000 feet of the right-of-way and are visible from the roadway. If they cannot be effectively screened, they are to be removed.

As in the removal of outdoor advertising devices, Federal aid, when available, may pay 75% of the cost of removing junkyards.

In addition to providing guidance to the States for the screening or removal of junkyards, the FHWA has also been cooperating with other governmental agencies, such as the Business and Defense

Services Administration in the Department of Commerce and the Bureau of Mines in the Department of Interior on the problems of permanent disposal of junked automobiles.

Roadside Enhancement

Title III of the Beautification Act, revising Section 319, 23 U.S.C., concerns additional aspects of protecting and enhancing roadside environment. This Title of the Act provides authorization for landscaping and roadside development, building rest and recreation areas within the highway right-of-way, and the purchase of scenic easements on and improvement of strips of land in order to preserve, restore, and enhance scenic beauty adjacent to highways. This provision applies to all primary and secondary highways on the Federal-Aid system as well as the Interstate highways, a total of more than 900,000 miles.

As revised by the Beautification Act, Section 319 authorizes the use of both general funds and funds from the Highway Trust Fund for landscaping; the purchasing of scenic easements; and roadside development, including the building of rest areas. General treasury funds authorized by the Beautification Act for these purposes need not be matched by the States, but expenditures from the Highway Trust Fund must be matched by the States in the same ratio as construction funds, i.e. 90% Federal - 10% State for projects on Interstate highways, 50-50 for other highways on the Federal-aid system.

Federal policy regarding landscaping and roadside development is spelled out primarily in Policy and Procedure Memorandum 21-17, "Landscaping and Scenic Enhancement." As noted in this PPM landscaping carried out in conjunction with highway construction or major reconstruction includes a combination of engineering and landscape activities in all highway phases t are essentially expansions of on-going programs, and have enjoyed enthusias response. However, controversy over the other portions of the act, primari control of outdoor advertising, has hampered its full implementation. Congress fialed to authorize any program funds for Fiscal Years 1968 and 1969, leading to confusion and indecision, particularly among the State legislatures. In an attempt to revitalize the program FHWA has been engaged in a restudy of it during the past years to see whether it can be made more workable. A report on this study should be completed in the near future.

In the meantime, 32 States, the District of Columbia, and Puerto Rico have legislation relating to outdoor advertising control. Forty States, the District of Columbia, and Puerto Rico have laws controlling junkyards. And the number of State highway department personnel employed in landscape development work has more than doubled since passage of the Highway Beautification Act.

nancial Aspects

Since passage of the Act, most of the Federal funds expended for outdoor advertising control have been used for an inventory of signs adjacent to Federal-aid primary and Interstate highways. However, of the $2 million authorized for appropriation for outdoor advertising control during Fiscal Year 1970, $1.2 million are being distributed in the form of bonus payments in accordance with the 1958 Bonus Act. A total of $3.7 million in bonuses has been paid out since the start of the program.

Also, close to $9 million of Federal funds have been set aside thus far for the screening of 1,421 junkyards and the removal of 125 for which screening was not considered feasible. This represents about 10% of the junkyard screening and removal expected to be accomplished under the Act.

More than $123 million of beautification funds have been authorized for expenditure on projects involving landscaping and scenic enhancement. Of this amount $63 million have been expended or authorized for the construction of 511 rest areas; $28 million have been scheduled for the purchase of 5,257 scenic easements; and $32 million have been used or set aside for use in landscaping projects.

In addition to these beautification expenditures of general treasury funds, $75 million in Federal-aid from the Highway Trust Fund have been expended since March, 1965, on the construction of 613 rest areas, most of which are on the Interstate System. (A recent inventory has disclosed that there are now more than 7,000 rest areas, with parking spaces for three or more vehicles, along the Nation's Federal-aid highway system, nearly 1,000 of which are on the Interstate System. About 550 are in urban areas.) Eighty-four million dollars have been spent for landscaping, of which $73 million has been spent on the Interstate System, and well over half, $49 million, has gone toward projects in urban areas.

The largest single amount of Federal funds expended for protection of the environment, however, has been for measures carried out as a part of highway construction for erosion control. Since March 1965, $220 million have been spent expressly for this purpose, including $160 million for Interstate projects. Erosion control practices are discussed in the chapter on "Air and Water Pollution Abatement."

Highway Beauty Awards Competition

In addition to these programs the Federal Highway Administration conducts an annual Highway Beauty Awards Competition. Under this program, begun in 1967, FHWA presents awards to State and local governments, authorities, organizations, and businesses for achievements in the national program of highway beauty. By thus drawing national attention to outstanding examples of highway-roadside beauty coordination, FHWA fosters aesthetic highway design and treatment.

Utilities

One further measure providing for protection of roadside areas is that controlling the use of highway right-of-way by utilities. Use of highway right-of-way for location of utilities is permitted to the extent that this does not impair the flow of traffic or the scenic appearance of the highway. Since revision of Policy and Procedure Memorandum 30-4.1, "Accommodation of Utilities," in November, 1968, new utility installations are not permitted in scenic strips and overlooks, rest areas, or recreation areas, or within the right-of-way of highways which are adjacent to any of these lands or which pass through public parks, recreation areas, wildlife and waterfowl refuges, and historic sites. Underground utility installations may be permitted where this will not impair the appearance of such areas, and high voltage power lines may be permitted if other locations or undergrounding are not feasible and if they do not impair the appearance of these areas.

V

Solid Waste

<u>Solid Waste Management</u>. A Comprehensive Assessment of Solid Waste Problems, Practices, and Needs. Prepared by Ad Hoc Group for Office of Science and Technology, Executive Office of the President, Washington, D.C., May 1969. Pages 1-39, 108.

Less than half of all municipalities have an adequate solid waste disposal system. With average per capita production of solid wastes running between two and five pounds per day the sheer quantity of domestic solid wastes which are inadequately disposed of becomes staggering. When, to this amount, commercial, industrial, agricultural, and other solid wastes are added it can be seen that solid waste management is an important facet of the overall pollution picture. Congress recognized this when it passed the Solid Waste Disposal Act of 1965, approved October 20, 1965, and found in 79 United States Statutes at Large 997 and 42 United States Code Annotated section 3251-3259. The act provides both for research and development and for financial assistance to appropriate agencies.

On March 8, 1968 the President directed the Director of the Office of Science and Technology to undertake a comprehensive review of solid waste technology. The resulting report, written under the direction of Rolf Eliassen, a consultant for the Office of Science and Technology, ran 111 pages plus nine pages of introductory material. The excerpt reproduced below is the most authoritative discussion of the problem. In the portion not reprinted, proposed programs to help deal with the program are explored. In general they suggest additional research and development and the use of a "systems approach" to solid waste management.

RESEARCH, DEVELOPMENT, AND DEMONSTRATION: THE KEY TO SOLID WASTE MANAGEMENT

I. STATEMENT OF THE SOLID WASTES PROBLEM.

The constant increase in per capita production of refuse, coupled with population growth and concentration in the urban areas, is a major consideration in management of solid wastes. The size of the problem, as well as the practices or lack of practices used in its solution, is responsible for the nation's crisis in solid waste management.

The various public statements in the past, whether expressed in tons of solid wastes[1/] or in dollars of expenditure, have usually been limited to the quantities of wastes for which municipalities have been responsible or represent an estimate of gross municipal costs. It is estimated that almost one and one-half billion pounds of urban solid wastes are generated in the nation each day and that costs of community collection and disposal approximate $4 billion per year.

The traditional view of solid wastes, including urban garbage and rubbish, augmented by quantities of abandoned vehicles, demolition-construction wastes and various other items, requires some reexamination. There is more to the problem than just handling and disposing of community or urban solid wastes, despite the fact that even this one aspect of the problem is far from solution. Congress directed attention to the overall solid waste problem in the Solid Waste Disposal Act of 1965 which defined solid wastes as "garbage, refuse, and other discarded materials, including solid-waste materials resulting from <u>industrial</u>, <u>commercial</u>, and <u>agricultural operations</u>, and from community activities." (underlining supplied) Disposal was also broadly defined to include "the collection, storage, treatment, utilization, processing, or final disposal of solid waste."

1/ The term solid waste describes that material which is normally solid, and which arises from animal or human life and activities and is discarded as useless or unwanted. Solid waste also includes deposited waste particulates, even when temporarily suspended in air or water. It refers to the heterogeneous mass of throwaways from the urban community as well as the more homogeneous accumulations of agricultural, industrial, and mineral wastes.

Public Health and Ecological Aspects of Solid Waste Management

The relationship between public health and improper disposal of solid wastes has long been recognized. Rats, flies, and other disease vectors breed in open dumps and in residential areas or other places where food and harborage are available. A recent literature search by the Public Health Service indicated association between solid wastes and 22 human diseases. Another study investigated the occupational hazards of workers who provide solid waste collection and disposal services. The statistical data illustrated that the illness-accident rate for sanitation workers was several times higher than that for industrial employees.

Implications for public health and other problems associated with water and air pollution have been linked to mismanagement of solid wastes. Leachate from open refuse dumps and poorly engineered landfills has contaminated surface and groundwaters. Contamination of water from mineral tailings may be especially hazardous if the leachate contains such toxic elements as copper, arsenic, and radium. Open burning of solid wastes or incineration in inadequate facilities frequently results in gross air pollution. Many residues resulting from mismanagement of solid wastes are not readily eliminated or degraded. Some are hazardous to human health; others adversely affect desirable plants and animals. Although attempts to set standards for air and water quality have been made and are continuing, in many cases harmful levels of many individual contaminants and of their combined or synergistic effects are not yet known.

Nature has demonstrated its capacity to disperse, degrade, absorb, or otherwise dispose of unwanted residues in the natural sinks of the atmosphere, inland waterways, oceans, and the soil. But the major concern are those residues that may poison, damage, or otherwise affect one or more species in the biosphere, with a resultant ecological shift.

Practical problems in solid waste operations have often been treated without full use of available information. Short-range remedies have held sway over long-range solutions. Local views have prevailed over state or regional planning. Standards set may have met single objectives but have failed to meet other objectives. In some instances, decisions may be made on the basis of dollar cost-benefit analysis alone.

Yet, in most situations overriding benefits to health, conservation, recreation, and aesthetics must enter and dictate solutions.

It is impractical to think of impeding urban affluence, progress of industry, and advances of agriculture by strict control, but the trend for the future must include the development of solid waste management systems that permit progress without harmful side effects. The preservation of public health and aesthetics of the environment should be of paramount consideration in management decisions involving solid wastes.

Natural Resources Aspects of Solid Waste Management

The earth's mineral and other resources are not unlimited. Man extracts metals from ores and transforms other materials from a natural state to man-made products, but when these products have fulfilled their usefulness, and are classified as solid wastes, the valuable and nonrenewable materials are often lost. Iron, for example, is a nonrenewable resource concentrated in ores over periods involving millions of years of geological processes. This element is extracted, processed, and widely dispersed over the earth to serve many useful purposes, from cans to automobiles. When discarded, these objects rust away and that amount of iron may never be available again for use by future generations. Shortages in resources for some of the less common elements have been created already by many indiscriminate waste disposal practices.

A prime solution to the resource depletion problem created by profligate use of metals and other elements would be to improve methods for recycling and reuse. To some extent, the salvage industry has traditionally served this purpose, but the concept might well be more widely applied. It is not inconceivable that present-day tailings dumps, landfills, and automobile graveyards may be looked upon in the future as "mines" for minerals whose natural ores have been depleted, or that remain only in deposits which can be mined only at greater cost than required for recycling of wastes.

Economic Aspects of Solid Waste Management

Implicit in public attitudes toward waste materials is the problem of economics. National concern must transcend both the concept of what the public can "afford" to pay and the question of why the expenditure of about $4 billion each year for solid waste collection and disposal has not staved off the present mounting problems with solid wastes. The truth is that the national standard of living depends to a significant degree upon processes leading to waste generation. It is reasonable to presume that a departure from traditional wastefulness of resources might reduce the volume of wastes to be managed. Thus, there is a need for finding ways in which wastes themselves may be salvaged, reworked, and recycled back as a part of resources that are being processed.

A system for managing solid wastes must be economically as well as technologically feasible. In waste management, as in many other fields, a particular facility or item of equipment may be technically suitable, but prohibitively expensive. Many communities are unable or unwilling to pay the price necessary to take advantage of presently available solid waste disposal devices. In general, only the larger cities find incineration economically feasible and incineration may become even more costly as legislation requires increased investment in equipment to meet air quality standards. Several compost plants have been constructed for processing municipal solid wastes; some have failed, largely because there had been no clear market potential for the finished product. As competition grows for the remaining open space in cities and suburban areas, land for sanitary landfills is becoming increasingly scarce, and this disposal method for municipal solid wastes, therefore, is becoming increasingly expensive.

Location and quality of operation of solid wastes disposal facilities may have an important impact upon the economics of an area, affecting land use patterns and the value of surrounding property. Although incinerators, sanitary landfills, and compost plants generally do not enhance the value of surrounding property, there are isolated examples in Southern California and other areas in which expensive residential neighborhoods are found adjacent to sanitary landfill sites, planning for ultimate conversion to parks, golf links, and other recreational uses. Three-quarters of the $4 billion yearly cost of U.S. solid waste disposal is attributed to the collection and transport of solid wastes. Yet this expenditure has purchased only hit-or-miss collection systems with little satisfaction for the majority of American communities. The one-quarter of expenditures currently allocated for processing and disposal of wastes has, despite obvious public objections, resulted in the use of the open dump as the predominant disposal method.

It is apparent that even more money must be devoted to improving management of solid wastes, or improvements must be introduced to the present system while achieving economic efficiencies, or both. Clearly, comprehensive solid waste management is a legitimate and necessary expense of living, a fact not yet appreciated by the public.

Technological Aspects of Solid Waste Management

The problems of waste management are greatly influenced by industrial technology and market considerations. Planned obsolescence, for example, is a way of life in the production and sales of some consumer goods industries. The consumer is urged to buy the new and trade in or throw away the old. The rise of the nonreturnable container, in spite of the fact that it costs the user more than the old type, is another illustration of the close relationship between market strategy and the generation of solid waste. In spite of the fact that industrial technology and market incentives result in more goods and a higher rate of solid waste production, little current thought is being devoted to the true disposability of the product. With the exception of the new biodegradable detergents introduced since 1965 and current efforts in the field of pesticides, there is virtually no instance in which product design consideration has included that of ultimate disposability. "Disposability" has been interpreted only in terms of user convenience--aconvenience characterized by single-service discardability. Markedly absent have been objectives of reusability and ease of ultimate disposal.

Political--Administrative Aspects of Solid Waste Management

Cities throughout the world have traditionally dealt with wastes by transporting them beyond their own immediate confines and discarding them in the least expensive way the public might tolerate. No serious problems arose from this practice as long as cities or jurisdictions were separated by space occupied only sparsely by human beings.

But today, urban sprawl is erasing space between cities. A major obstacle to organizing efficient high-quality metropolitan refuse collection and disposal systems is the multitude of local governmental units, many of which are remnants from another era. These local political subdivisions may be separated by natural boundaries such as lakes, rivers, mountains, by the political boundaries of satellite communities that surround an inner city, or by state boundaries.

Improved solid waste management is impeded by the failure by many communities, particularly the smaller ones, to combine their efforts and financial resources for collection and disposal. Certain efficiencies -- economies of scale -- can be realized with optimum unit size. Without cooperation, communities may not be able to afford the equipment and qualified personnel necessary for adequate solid waste management, or they may not be able to use what is, from a technical viewpoint, the most desirable disposal site.

Area-wide refuse disposal service is being provided in a few metropolitan areas by special purpose districts, by counties, or by cooperative agreements between cities and other local political subdivisions. The economy-of-scale advantage of such area-wide disposal service not only reduces unit costs, but also makes it possible to operate on a more sanitary and acceptable basis and to avoid needless duplication of investment.

This regional approach to solid waste management usually provides the answer to the problem even when a region includes interstate areas. Failure to take the regional approach may be a matter of local pride or inertia; there may be legal barriers to interjurisdictional cooperation. In either case there is need for more realistic legislation or administrative response.

An important factor in effective solid waste management is operational planning. Even though collection and disposal of solid wastes remains an undeniable community need, in some cities and communities the need is not considered in basic community facility planning. Like other public services, solid waste management is related to population growth, density, and industrial and commercial zoning. Land is required. Planning calls for a long-range look at the total area involved. Communities are inclined to conceive of solid waste collection and disposal activities in terms of how to deal with community refuse for the next few years. Communities might best consider and act in terms of what must be done to meet the total area needs for 25 to 50 or even 100 years.

Perhaps the basic political fact of solid waste management is citizen response to factors and solutions regarding solid waste problems. The citizen, together with municipal administrators and industrial executives who are deeply involved in finding solutions to the solid waste problems, must be made aware of the need for a new concept of solid waste management. The Solid Waste Disposal Act of 1965 has accelerated the pace of solid waste technology and research; past and present technological information related to solid wastes is being accumulated on a more consistent and thorough basis; people with varied skills and from many disciplines are seeking to solve the problems caused by the ever-increasing quantities of solid wastes from many sources.

SOURCES OF SOLID WASTES

Solid wastes fall into five major source categories: urban (including domestic, commercial and municipal), industrial, agricultural, mineral and Federal establishments. Table 1 presents estimates of the quantities of solid wastes generated from all these sources in 1967. The total amounted to 3.65 billion tons, equivalent to 100 lbs. per capita per day for a national population of 200 million. By 1980 this figure is expected to increase to 5 billion tons.

TABLE 1

GENERATION OF SOLID WASTES FROM FIVE MAJOR SOURCES IN 1967

Source	Solid Wastes Generated	
	lbs/cap./day	million tons/yr.
Urban		
domestic	3.5	128
municipal	1.2	44
commercial	2.3	84
sub-total	7.0	256
Industrial	3.0	110
Agricultural		
vegetation	15.0	552
animal	43.0	1563
sub-total	58.0	2115
Mineral	30.8	1126
Federal	1.2	43
TOTALS	100.0	3650

Urban Waste Sources

Traditionally, urban refuse disposal has been concerned only with wastes collected from households and then principally with the garbage component. This limited view of community responsibility has been altered by the rapid urban expansion that followed World War II. Customary concepts of solid waste management are being revolutionized by the need for area-wide plans and for solutions beyond the capabilities of any of the myriad political jurisdictions of the modern consolidated community. Even when one city is located at some distance from another, particularly where explosive population increases have had maximum impact, urbanization has progressed beyond the limits of cities and satellite suburbs to become a characteristic of rural areas. With high land values and heavy investment in fixed installations of such specialty enterprises as dairies, poultry and egg production, and animal feeding facilities, the rural area of the community now surrounds the urban-industrial-suburban sector as a shell that has lost the elasticity traditionally associated with rural areas. The rural area, moreover, exists in sufficient depth to have a profound effect on both its own and the city's problem of waste management.

Solid waste management practices have also been influenced by the development of new materials, planned obsolescence, and a never-ending variety of new products from industry. These include single-use containers and nonreturnable bottles, as well as all the other gadgetry and conveniences that swell the amount as well as the types of wastes to be managed within a community. The increases in plastics use has affected the degradability of refuse; nonreturnable containers have influenced combustibility of refuse and the appearance of the countryside; and industry creates some 2,000 new products each year with a resultant increase in the variety and amount of wastes.

What exactly are urban solid wastes? They include food wastes (garbage); paper, paper products, wood, bedding metals, tin cans, glass, crockery, dirt (rubbish), and ashes; dead cats, dogs, sweepings and leaves, and abandoned cars and trucks; such residue from demolition and new construction projects as lumber, masonry, metals, paints, and concrete; certain radioactive materials explosives, pathologic wastes, and a myriad of similar materials from hotels, institutions, stores, and industries (Table 2). Lack of a common terminology leads to uncertainty and even error in interpreting published information. Consequently in this report all solid wastes in urban areas have been grouped by the three major sources: domestic or residential, commercial-institutional and municipal.

Residential Waste Sources. The estimated quantities of these wastes are shown in Table 1. Their composition is indicated in Table 2. Terminology differs all over the country. A city may use one of the many synonyms such as "garbage," "rubbish," or "refuse" to describe all the materials it collects and disposes of. For example, some cities collect table scraps, waste paper, tin cans, bottles, and other miscellaneous household wastes and call them "garbage," while other cities use the term "rubbish" to mean the same material.

TABLE 2. COMPOSITION OF SOLID WASTES FROM URBAN SOURCES

Urban Sources	Waste	Composition
Domestic, household	Garbage	Wastes from preparation, cooking and serving of food; market wastes from handling, storage, and sale of food.
	Rubbish, trash	Paper, cartons, boxes, barrels, wood, excelsior, tree branches, yard trimmings, metals, tin cans, dirt, glass, crockery, minerals.
	Ashes	Residue from fuel and combustion of solid wastes.
	Bulky wastes	Wood furniture, bedding, dunnage, metal furniture, refrigerators, ranges, rubber tires.
Commercial, institutional, hospital, hotel, restaurant, stores, offices, markets	Garbage	Same as domestic.
	Rubbish, trash	Same as domestic.
	Ashes	Same as domestic.
	Demolition wastes, urban renewal, expressways	Lumber, pipes, brick masonry, asphaltic material and other construction materials from razed buildings and structures.
	Construction wastes, remodeling	Scrap lumber, pipe, concrete, other construction materials.
	Special wastes	Hazardous solids and semiliquids, explosives, pathologic wastes, radioactive wastes
Municipal, streets, sidewalks, alleys, vacant lots, incinerators, power plants, sewage treatment plants, lagoons, septic tanks	Street refuse	Sweepings, dirt, leaves, catch basin dirt, contents of litter receptacles, etc.
	Dead animals	Cats, dogs, horses, cows, marine animals, etc.
	Abandoned vehicles	Unwanted cars and trucks left on public property.
	Fly ash, incinerator residue, boiler slag	Boiler house cinders, metal scraps, shavings, minerals, organic materials, charcoal, plastic residues
	Sewage treatment residue	Solids from coarse screening and grit chambers, and sludge from settling tanks

Commercial Waste Sources. Commercial waste sources are wholesale and retail trade, hotel, restaurant, hospital and institutional, finance, insurance, corporate, and general offices, and services excluding private households. The commercial sector of the American economy primarily manages its solid wastes on an individual basis. In many of the downtown areas municipal collection services are provided to these commercial enterprises. However, the bourgeoning growth of American suburbs has placed the focus of many commercial enterprises beyond downtown business areas and therefore beyond the normal purview of municipal collection services. These suburban commercial enterprises, therefore, must frequently engage private collection and disposal agencies. Current storage equipment or practice usually consists of the common 20- to 30-gallon containers; the containerized system; open wooden bins; or, promiscuous dumping onto the ground at the backside of the commercial enterprise.

At the current time, no accurate information is available on a national basis on the amounts and characteristics of solid wastes generated and discharged by these commercial enterprises. Rough estimates are presented in Table 1. It may be surmised, for example, that the assortment of solid wastes to be stored, collected, and disposed of in the operation of commercial, health-related institutions, such as hospitals, nursing homes, clinics and sanitariums, require special management because of safety problems, but there is little published data on this problem.

Municipal Waste Sources. Municipal sources of urban solid wastes include street litter, discarded auto bodies, power plant and incinerator ashes and residue, and sewage sludge (Table 2). Old cars are being abandoned on city streets, highways, and vacant lots in ever-increasing numbers as owners find this cheaper than to have them towed to a collection point. These junked cars add considerable volume to municipal wastes. In 1964 in New York City, 13, 600 cars were abandoned on streets and parking lots, five times the number abandoned in 1960. By 1966, this figure had almost doubled again -- to 25, 000 vehicles. Nationwide, over half a million cars were abandoned in 1967. The growing accumulation of junked automobiles along roads that encircle cities, and in fields and vacant lots are a valuable source of metallic and nonmetallic materials, including heavy and light-gauge steel, cast iron, aluminum, zinc, copper, lead, stainless steel, rubber and plastics. Based on current scrap prices, a junked car is worth about $65 if separated into its different scrap components. This separation is time consuming and expensive, however, and does not give municipal waste collection systems a method of disposal at present other than numerous auto graveyards dotting the landscape. It has been estimated that 180 million passenger tires are currently being produced each year. New car registration for 1968 has been estimated at 10 million. With the increasing number of automobiles, there is also an increasing number of discarded tires, which flow into the municipal waste source.

In 1965, electric power generating plants consumed over 240 million tons of bituminous coal and lignite, generating about 24.5 million tons of residue, approximately 80 percent fly ash and 20 percent bottom ash, or cinder. By 1970, the annual output of the pulverized-coal-fired generating plants will exceed 30 million tons.

When collected, wastes from urban sources are handled by private contractors and municipal collection agencies and are transported to private and municipal disposal facilities. However, substantial amounts of urban wastes do not go through channels of collection. They may be incinerated in apartment house or home incinerators or burned in backyards of residences and commercial establishments or on demolition sites. Some portions of urban wastes are disposed of in garbage grinders connected to sewers. Consequently quantities of solid wastes collected in municipalities, even when known, do not account for all such wastes which are generated. A similar situation exists in the industrial sector. Data on the quantities of urban wastes actually collected are shown in Tables 3 and 4. Average values of solid wastes generated in urban centers across the country were shown in Table 1.

Reported estimates on the size of the garbage fraction of domestic-commercial refuse range from 5 to 15 percent. Reported ratios of combustible and noncombustible rubbish also vary, the range in these classifications extending from 15 to 25 percent for the noncombustibles and 64 to 75 percent for the combustibles. A range of reported values also exists for breakdowns within these totals. Examples of the more common constituents are: paper, 42 to 57 percent; metals, 1.5 to 8 percent; glass, 2 to 15 percent; rags, 0.6 to 2 percent; garden debris, 10 to 12 percent; ashes, 5 to 19 percent. Domestic waste components,

TABLE 3. COLLECTION OF RESIDENTIAL SOLID WASTES IN NINE U.S. CITIES (1966)

City	Residential solid wastes (lbs/cap/day)
Niagara Falls, New York	2.70
Los Angeles, California	2.59
Philadelphia, Pennsylvania	2.10
New York City, New York	2.42
St Petersburg, Florida	2.14
San Francisco, California	3.16
Miami, Florida	1.75
Wilton, Connecticut	2.40
Weston, Connecticut	2.10
AVERAGE	2.37

TABLE 4. COLLECTION OF RESIDENTIAL AND COMMERCIAL SOLID WASTES IN NINE U.S. CITIES (1966)[a,b]

City	Lbs per capita per day		
	Municipal collection	Private contractor [c]	Tota
Glendale, California	3.38	.83	4.21
Los Angeles, California	3.36	.06	3.42
San Francisco, California	2.54	3.00	5.54
Miami, Florida	3.11	.64	3.75
Baltimore, Maryland	4.18	.09	4.27
Cleveland, Ohio	1.73	.34	2.07
Philadelphia, Pennsylvania	2.41	.08	2.49
Woonsocket, Rhode Island	2.58	.56	3.14
Norfolk, Virginia	4.36	.65	5.01
AVERAGES	3.07	.69	3.81

[a] Measurements made primarily at disposal sites.
[b] These cities are shown because they were reported on by the Refuse Removal Journal and because the industrial segment could be broken out of the data.
[c] Designates solid wastes collected by private contractors and disposed of in private facilities.

daily per capita generation, and the total annual generation was compiled for Santa Clara County in 1967 and may perhaps be considered typical. (Table 5). A more detailed sample of metropolitan refuse as-received at a municipal incinerator in the East is shown in Table 6.

Lack of uniform and reliable information on generation, collection and disposal of solid wastes has been a major drawback to needed perspective in the field of solid wastes.

The first nationally comprehensive view of solid waste data and practices throughout the country is just now becoming available through the National Survey of Community Solid Waste Practices.[1] This National Survey was designed largely by the Solid Wastes Program of the Public Health Service and is being carried out primarily by the states under the state planning grant program. Data are not complete because communities do not, for the most part, measure the solid wastes which are handled either in collection or disposal. Tonnage estimates of wastes collected have been made from the National Survey. These values are compared in Table 7 with estimates of solid wastes generated previously shown in Table 1.

The collection categories are those used in the standard Federal questionnaire-interview form which was designed to allow for variations in community collection practice. Even the ten categories of urban and industrial wastes inadequately portray all the permutations in common use. The category of "combined household and commercial" refuse, for example, probably includes some industrial wastes and may also cover wastes defined in this report as municipal.

It is estimated that urban and industrial wastes generated in the United States average 10 lbs. per capita per day, whereas collected wastes average only 5.12 lbs. per capita per day, only 51% of the total generation. The uncollected balance is generally an unknown factor in urban life. It is accounted for in part by apartment house, institutional or home incineration (with only the ash or residue appearing in collection data), and also by home and commercial garbage grinding. In the present disarray in the management of solid wastes, considerable responsibility is left to the individual household or entrepreneur to find some place for getting rid of his wastes. Many send their residual solids to unauthorized open dumps.

---

[1] National Survey of Community Solid Waste Practices, 1968, Solid Wastes Program, October 1968

TABLE 5. ESTIMATE OF COMPONENTS OF DOMESTIC REFUSE IN SANTA CLARA COUNTY, CALIFORNIA (1967)[a]

Classification	Percent of total	Lbs/cap/day	Tons/yr
Garbage	12	0.42	75,240
Rubbish			
paper	50	1.75	313,500
wood	2	0.07	12,540
cloth	2	0.07	12,540
rubber	1	0.04	6,270
leather	1	0.04	6,270
garden wastes	9	0.31	56,430
metals	8	0.28	50,160
plastics	1	0.04	6,270
ceramics & glass	7	0.24	43,890
nonclassified	7	0.24	43,890
TOTALS	100	3.50	627,000

[a]A per capita production of 8 lbs per day, of which 44 percent is domestic re′ ‑e, was assumed in estimating values.

TABLE 6. COMPOSITION AND ANALYSIS OF COMPOSITE MUNICIPAL REFUSE (1966)[a]

	Components	Percent by weight
1	Corrugated paper boxes	23.38
2	Newspaper	9.40
3	Magazine paper	6.80
4	Brown paper	5.57
5	Mail	2.75
6	Paper food cartons	2.06
7	Tissue paper	1.98
8	Wax cartons	0.76
9	Plastic coated paper	0.76
10	Vegetable food wastes	2.29
11	Citric rinds and seeds	1.53
12	Meat scraps, cooked	2.29
13	Fried fats	2.29
14	Wood	2.29
15	Ripe tree leaves	2.29
16	Flower garden plants	1.53
17	Lawn grass, green	1.53
18	Evergreens	1.53
19	Plastics	0.76
20	Rags	0.76
21	Leather goods	0.38
22	Rubber composition	0.38
23	Paint and oils	0.76
24	Vacuum cleaner catch	0.76
25	Dirt	1.53
26	Metals	6.85
27	Glass, ceramics, ash	7.73
28	Adjusted moisture	9.05
	TOTAL	100.00

[a]Source: Kaiser, E. R. Chemical analyses of refuse components, 1966.

TABLE 7. COLLECTION COMPARED WITH GENERATION OF SOLID WASTES IN 1967

Solid Waste Generation		Solid Waste Collection [(a)]	
Source	lbs. /cap. /day	Source	lbs. /cap. day
Urban		Urban	
domestic	3.50	household	1.17
commercial	2.30	combined house. & commercial	2.45
		commercial	0.39
		institutional	0.04
		demolition-construction	0.18
sub-total	5.80	sub-total	4.23
municipal	1.20	street & alley	0.09
		tree & landscaping	0.06
		catch basin	0.01
		sewage solids	0.04
sub-total	1.20	sub-total	0.20
total urban	7.00	total urban	4.43
Industrial	3.00	Industrial	0.69

[(a)] Categories and data reproduced from National Survey run of September 7, 1968. Data covers collection by public agencies, private collectors, individual householders and commercial and industrial establishments.

Industrial Waste Sources

Municipal incinerators, refuse dumps and sanitary landfills, and acres of old automobile hulks are common manifestations of solid wastes. Less obvious, in warehouses and back lots of industrial properties, are industrial wastes,[1] the piles and acres of sludges, slags, waste plastics, scrap metal, bales of rags and paper, drums of off-grade products. Solid wastes of industrial origin comprise a category of major magnitude (Table 1). The magnitude of the problem of their disposal is compounded by the fact that industrial wastes are as diversified as industry itself. Unlike residential solid wastes, with characteristics generally similar throughout the United States (Tables 6 and 7), the quality and quantity of industrial solid waste varies from industry to industry. As a result, solid waste management techniques cannot be generalized for industry as a whole.

Industrial scrap metal is generated at a rate in excess of 15 million tons annually. The paper industry generates over 30 million tons of paper and paper-product waste. A recent study of industry indicated that in the overall approximately 115 million tons of industrial solid wastes are being generated each year. At present there is little data available on amounts of industrial wastes collected, transported, or disposed of, nor on amounts stored on company property. Two factors influence the size of these stored "inventories": adaptability to reprocessing or reuse, and cost of storage.

In general, waste materials are reintroduced into the industrial process generating them whenever possible, although the percentage of such incorporation is often so nominal as to result only in a net gain of the waste. In other instances it is not feasible to incorporate the waste in the original or any other process; then, the cost of storing the waste on company property must be evaluated against any other form of disposal. For some industrial wastes no other form of disposal now is possible. This approach to waste management leads to an obvious conclusion; even without the motivation provided by national interest, many manufacturing concerns are impelled toward more scientific and practical methods of solid waste management.

---

[1] By definition, industrial solid wastes are any discarded solid materials resulting from an industrial operation or deriving from an industrial establishment. They include processing, general plant, packaging and shipping, office, and cafeteria wastes. Specifically excluded are dissolved solids in domestic sewage and any dissolved or suspended solids in industrial waste waters.

A fortunate few industries produce wastes for which some use exists. Others have faced up to the problem earlier and have found satisfactory solutions to their solid waste disposal problems; in the main, however, industry has ignored solid waste problems insofar as possible, so that localized solutions are as varied as the sources generating the waste.

It may be expected that two types of solid wastes will be generated by the 22 major industrial groups (Standard Industrial Classification Manual) (Table 8): (1) wastes that are common to all types of industry; (2) wastes that are peculiar to the individual industry by reason of the nature of the raw material processed and the products.

The common grouping includes wood, fiber, paper, metal, and plastic containers in which raw materials and supplies for maintenance and operation are received by the industry; scraps of metal, wood, paper, and plastics used by the industry in painting, crating, or otherwise preparing its product for shipment; and the normal debris from general housekeeping and operating and maintaining an industrial plant. In this category are such items as wastepaper, floor sweepings, broken wooden pads, rubber tires, broken glass, and the junk of obsolescence or breakage of equipment. Residues of paint, industrial chemicals, dyes, adhesives, and solvents remaining in discarded containers are also common to all industries, although the amount and kind varies from industry to industry and from plant to plant (Table 8).

In the list of solid wastes generated by specific industries, residues from plating, pickling, painting, and similar operations were included. For the most part these processes produce liquid residues that are managed separately from the solid wastes. Nevertheless, oils, paints, dyes, plating and pickling liquors, solvents, and adhesives, do appear in such solid wastes as metal cuttings, sawdust, paper, sludges, wood, cloth, and junk. Without adding significantly to the volume of wastes to be managed, they may have a public health or nuisance significance in certain methods of disposal; hence their presence is noted in the list of industrial waste materials.

The waste disposed of by the major industries in 1965 totaled approximately 182 billion lbs. The sawmill industry was the largest contributor with 65,000 million lbs. However, because of new utilization processes there, it seems likely that waste for disposal in the western sawmill industry will be virtually eliminated by 1975.

TABLE 8. SOURCES AND TYPES OF INDUSTRIAL WASTES[a]

Code	S. I. C. Group Classification	Waste generating processes	Expected specific wastes
17	Plumbing, heating, air conditioning Special Trade Contractors	Manufacturing and installation in homes, buildings, and factories	Scrap metal from piping and duct work; rubber, paper, and insulating materials, misc. construction and demolition debris
19	Ordnance and accessories	Manufacturing and assembling	Metals, plastic, rubber, paper, wood, cloth, and chemical residues
20	Food and kindred products	Processing, packaging, and shipping	Meats, fats, oils, bones, offal vegetables, fruits, nuts and shells, and cereals
22	Textile mill products	Weaving, processing, dyeing, and shipping	Cloth and fiber residues
23	Apparel and other finished products	Cutting, sewing, sizing, and pressing	Cloth and fibers, metals, plastics, and rubber
24	Lumber and wood products	Sawmills, mill work plants, wooden container, misc. wood products, manufacturing	Scrap wood, shavings, sawdust; in some instances metals, plastics, fibers, glues, sealers, paints, and solvents
25	Furniture, wood	Manufacture of household and office furniture, partitions, office and store fixtures, and mattreses	Those listed under Code 24, and in addition cloth and padding residues

[a] Source: Standard industrial classification manual, 1967

TABLE 8. SOURCES AND TYPES OF INDUSTRIAL WASTES (cont'd)

Code	S. I. C. Group Classification	Waste generating processes	Expected specific wastes
25	Furniture, metal	Manufacture of household and office furniture, lockers, bedsprings, and frames	Metals, plastics, resins, glass, wood, rubber, adhesives, cloth, and paper
26	Paper and allied products	Paper manufacture, conversion of paper and paperboard, manufacture of paperboard boxes and containers	Paper and fiber residues, chemicals, paper coatings and fillers, inks, glues, and fasteners
27	Printing and publishing	Newspaper publishing, printing, lithography, engraving, and bookbinding	Paper, newsprint, cardboard, metals, chemicals, cloth, inks, and glues
28	Chemicals and related products	Manufacture and preparation of inorganic chemicals (ranges from drugs and soups to paints and varnishes, and explosives)	Organic and inorganic chemicals, metals, plastics, rubber, glass, oils, paints, solvents and pigments
29	Petroleum refining and related industries	Manufacture of paving and roofing materials	Asphalt and tars, felts, asbestos, paper, cloth, and fiber
30	Rubber and miscellaneous plastic products	Manufacture of fabricated rubber and plastic products	Scrap rubber and plastics, lampblack, curing compounds, and dyes
31	Leather and leather products	Leather tanning and finishing; manufacture of leather belting and packing	Scrap leather, thread, dyes, oils, processing and curing compounds

TABLE 8. SOURCES AND TYPES OF INDUSTRIAL WASTES (cont'd)

Code	S. I. C. Group Classification	Waste generating processes	Expected specific wastes
32	Stone, clay, and glass products	Manufacture of flat glass, fabrication or forming of glass; manufacture of concrete, gypsum, and plaster products; forming and processing of stone and stone products, abrasives, asbestos, and misc. nonmineral products	Glass, cement, clay, ceramics, gypsum, asbestos, stone, paper, and abrasives
33	Primary metal industries	Melting, casting, forging, drawing, rolling, forming, and extruding operations	Ferrous and nonferrous metals scrap, slag, sand, cores, patterns, bonding agents
34	Fabricated metal products	Manufacture of metal cans, hand tools, general hardware, nonelectric heating apparatus, plumbing fixtures, fabricated structural products, wire, farm machinery and equipment, coating and engraving of metal	Metals, ceramics, sand, slag, scale, coatings, solvents, lubricants, pickling liquors
35	Machinery (except electrical)	Manufacture of equipment for construction, mining, elevators, moving stairways, conveyors, industrial trucks, trailers, stackers, machine tools, etc.	Slag, sand, cores, metal scrap, wood, plastics, resins, rubber, cloth, paints, solvents, petroleum products

TABLE 8. SOURCES AND TYPES OF INDUSTRIAL WASTES (cont'd)

Code	S. I. C. Group Classification	Waste generating processes	Expected specific wastes
36	Electrical	Manufacture of electric equipment, appliances, and communication apparatus, machining, drawing, forming, welding, stamping, winding, painting, plating, baking, and firing operations	Metal scrap, carbon, glass, exotic metals, rubber, plastics, resins, fibers, cloth residues
37	Transportation Equipment	Manufacture of motor vehicles, truck and bus bodies, motor vehicle parts and accessories, aircraft and parts, ship and boat building and repairing motorcycles and bicycles and parts, etc.	Metal scrap, glass, fiber, wood, rubber, plastics, cloth, paints, solvents, petroleum products
38	Professional, scientific controlling instruments	Manufacture of engineering, laboratory, and research instruments and associated equipment	Metals, plastics, resins, glass, wood, rubber, fibers, and abrasives
39	Miscellaneous manufacturing	Manufacture of jewelry, silverware, plated ware, toys, amusement, sporting and athletic goods, costume novelties, buttons, brooms, brushes, signs, and advertising displays	Metals, glass, plastics, resins, leather, rubber, composition, bone, cloth, straw, adhesives, paints, solvents

The quantities of wastes generated by a sample of 21 of the standard industrial code groups was the subject of a recent survey (Figure 1). The groups were selected for comparison because waste quantities were relatively easy to estimate and because of the likelihood that these quantities would be large. (Included in the survey were two nonmanufacturing groups--"demolition" and "supermarkets"--which are categorized in Table 2 of the report as commercial rather than industrial). The median industry in the 21-group sample was estimated to generate over 2300 million lbs. of waste per year. Sawmills, the highest, produced 28 times this much and the demolition industry over 16 times as much. On a per capita employee basis, the industrial worker in the median group was found to contribute some 8 times as much solid waste as the average community resident. In general it was found for each manufacturing industry surveyed that the average employee generates as plant trash, (office papers, lunch scraps, coffee cups and other containers, paper towels and other sanitary items and newspapers) about as much solid waste during his working day as his per capita generation while at home.

The six industries discussed below are significantly varied in their solid waste problems and approaches to offer a good general introduction to industrial waste types and quantities.

Food Processing Industry Wastes. The growing, harvesting, processing, and packaging of fruits, vegetables and other food crops, generate large tonnages of solid wastes. It has been grossly estimated, for example, that of the total weight of corn crop grown for canning about 50 percent is field waste, about 30 percent is process waste, and less than 20 percent is actual corn in the can. The fast developing technology in mechanical harvesting is causing a shift in the solid waste handling location from the rural to the urban setting. Former hand harvesting methods were more selective in sorting out the usable crop portion. Culls, plant parts, and other wastes were left in the field where they were simply disposed of by plowing in or burning. With more sophisticated mechanical harvesting, a greater percentage of the harvested crop may be waste and is transported with the usable crop portion to a central location, often in the urban complex. Here, previous waste handling techniques often cannot be used. Such wastes then add to urban waste management problems and complexity of solutions.

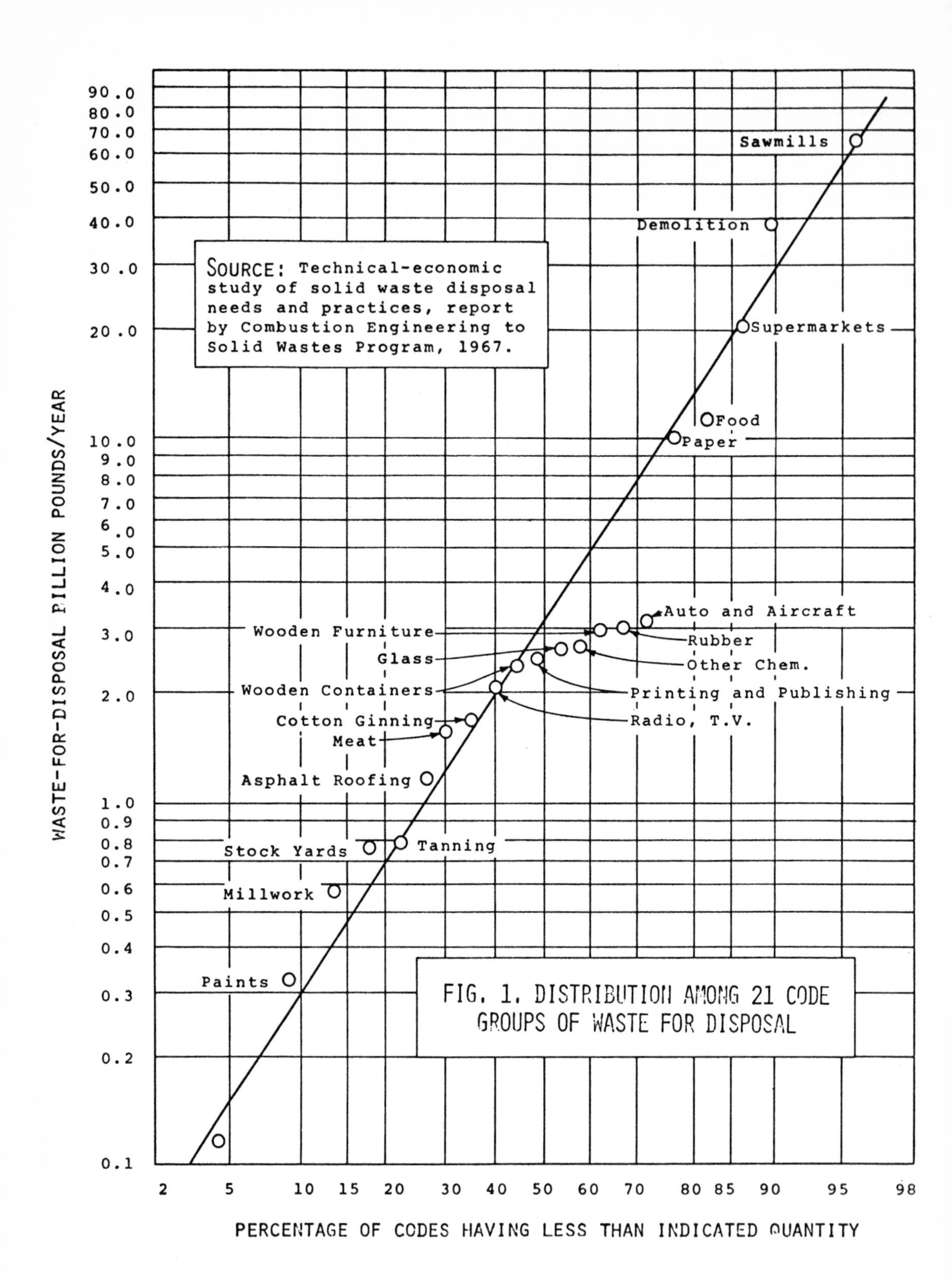

FIG. 1. DISTRIBUTION AMONG 21 CODE GROUPS OF WASTE FOR DISPOSAL

Production of processed foods has more than doubled in the past 25 years. In the face of continued population growth and consequent increased food requirement, this production and preservation of foods must continue to accelerate and will result in concomitant increases in the generation of solid waste. Solid wastes originating from forest and fiber crop production and processing are also considerable. These organic materials are often more refractory and difficult to handle and manage than those from animal product and food crop processing.

Chemical Industry Wastes. The chemical industry, one of the key industries in the United States, embraces a complex of subindustries, the borders of which are often indistinct. The industry is characterized by its large size, complexity, and rapid rate of change. Since 1950, industry sales have risen from $16 billion to over $35 billion dollars. The number of products has increased to over 7,000 commercial chemicals. There is also an increasing tendency to provide many different grades of a single basic product in order to meet varied engineering requirements for highly specific end uses. For example, the number of major plastics groups covered in Chemical Economics Handbook has increased from 9 in 1950 to 22 currently, and the available forms and grades of each have multiplied.

The manufacture of industrial chemicals generates unusable or unsaleable by-products, many of which are in solid form. The quality and quantity of these process solid wastes can vary greatly among industries, thus complicating assessment of the disposal problem. There are 1,905 establishments engaged in producing industrial chemicals, employing 250,000 people and located mainly in the Middle and South Atlantic and West South Central States. The chemical industry produces a highly heterogeneous waste because of specific manufacturing processes. By the very nature of the industry, much of the waste is liquid, containing dissolved and suspended solids. It is disposed of by various means, such as normal sewage treatment processes, pit incineration, deep-well injection, dumping at sea, etc. With increased disposal regulations it may be necessary to dewater, filter, or treat the liquid waste, thus generating a solid waste for disposal.

A recent survey estimates that about 113 million lbs. of solid wastes are generated each year by this industry -- 20 percent being process wastes and 80 percent plant trash and shipping wastes. It also indicated that the majority of these wastes are disposed of by private disposal operations.

Drug Industry Wastes. This industry has been growing steadily for many years, but its growth was greatly accelerated during the second world war by the enormous demands of the armed forces. Acceleration of this industry continued during the following years, so that in 1963 a total of 1,272 establishments were involved in the manufacturing of drug products. Each year new products are being discovered and manufactured. The complexity of the drug waste problems is increased by the variation in product composition -- dry powdered extracts, liquid galenicals, volatile oils, dried herbs, etc. Many plants treat their own waste waters, extracting solids for ultimate disposal.

Rubber Industry Wastes. A wide variety of chemicals is employed in the production of intermediate and final products of the synthetic rubber industry. The chemical characteristics of these compounds are well known when referred to the reactions that are essential in producing the quality of materials needed for production of superior grades of synthetic rubber. This knowledge, however, does not extend to the behavior of many of these compounds in the environment and their effects once they become wastes.

Rubber waste includes those products which are not completely used up in the process of manufacturing, and those products discarded by the consumer after use. In 1965 over 2.3 million long tons of rubber were consumed in the United States. Of this total, about 22 percent was natural rubber, 66 percent was synthetic rubber, and less than 12 percent was reclaimed rubber.[1] It has been reported that 2.9 billion lbs of waste are generated within the rubber industry: 50 percent as plant trash, 40 percent rubber and rubber trimmings, and 10 percent solvents and pigments.

Automobile Industry Wastes. The automobile industry generates 4 to 5 percent of the gross national product. During the first three quarters of 1967, personal consumption expenditures for automobiles and automobile parts were running close to $29 billion out of a total of more than $71.6 billion expenditures for all durable goods.

Assembly operations for automobiles are carried on in 51 plants located in 38 cities, with production ranging between 7 and 9 million cars a year. The quantity and quality of solid wastes generated from the automotive industry is virtually unknown, as are problems of storage, handling, and collection and disposal. Because of the size

[1]Reclaimed rubber is scrap natural or synthetic rubber that has been sorted, ground, difibered, and devulcanized.

of the industry and its economic influence on the community, the quantity and character of solid wastes and the problems of collection and disposal should be identified.

Solid wastes from direct production operations in the automotive industry include metal scraps and chips, sanding grit, oily sludges from stamping and assembly operations, scrap plastic, rubber, and sludges from parts and accessory manufacture. Continuing changes in production technology and materials of construction, as well as the development of new or improved air and water pollution abatement techniques that often produce solid wastes are factors of importance in assessing the solid waste problem in the automobile industry.

Household Appliance Industry Wastes. In 1967, household appliances totaled $5,067 million for manufacturers' shipments. The industry as a whole constitutes approximately 700 establishments employing over 144,000 people.

Cold-rolled steel is the major material used by the household appliance industry, but the demand for plastics is increasing steadily. The solid waste from the industry is varied and includes office waste, packing and shipping wastes, process wastes of metal scrap, rubber, plastic, and wire. While the character of these wastes may be similar to commercial wastes, there is little information on the real character and quantity of solid wastes from this industry and the impact on the environment.

An appliance, such as an automatic washer, is generally an assembly of a number of basic components (electric motor, cabinet, rotating basket, electric timer to control the cycle, gear case, etc.). Different manufacturers, however, utilize different materials and configurations for each of the basic components. There are, therefore, considerable differences in the processes used in fabricating each of the basic components, with resultant differences in the quantity and character of the solid wastes generated.

Agricultural Waste Sources

The principal agricultural and forestry wastes generated in the United States are animal manures, vineyard and orchard prunings, crop harvesting residues, animal carcasses, greenhouse wastes, and pesticide containers.[1] The composition of these wastes is indicated in Table 9.

A gross estimate of the total animal and vegetation solid waste generated by the agricultural segment of the country was obtained from the 1967 United States Statistical Abstract. The waste generation factors for individual crops and animals were obtained from a number of reference sources. Major agricultural crop wastes amounted to 550 million tons per year and animal wastes, 1,560 million tons per year (Tables 1, 10, and 11). Not included in the gross estimate of agricultural crop wastes was residue and trash in forestry operations; over 25 million tons of logging debris are left in forests during the average year.

Domestic animals produce over 1 billion tons of fecal wastes a year and over 400 million tons of liquid wastes. Used bedding, paunch manure from abattoirs, and dead carcasses make the total annual production of animal wastes close to 2 billion tons.

As much as 50 percent of animal wastes is generated in concentrated growing operations close to urban areas. Large-scale animal raising operations have developed rapidly in the last 20 years. One operation with 10,000 head of cattle on a feedlot can produce 260 tons of manure per day. An operation involving 400 milk cows produces about 14 tons of solid waste daily. A poultry enterprise with 100,000 birds produces 5 tons of waste daily. Production of farm animals has become a big business in the United States resulting in a yearly manure waste load of 1,562,721,000 tons (Table 11).

Mineral Waste Sources

During the past 30 years, well over 20 billion tons of mineral solid wastes have been generated in the United States by the mineral and fossil fuel mining, milling, and processing industries. It is estimated that the present generation of mineral solid wastes is 1.1 billion tons per year (Table 1). The task of contending with the mineral solid wastes generated in the past or being generated now is serious, but the future promises even greater difficulty. By 1980 the nation's mineral industries are estimated to be generating a minimum of 2 billion

---

[1] The pesticide containers are listed as an agricultural waste because they are used and discarded in agricultural operations; in other words, their use has become an essential feature of agricultural practice.

TABLE 9. COMPOSITION OF SOLID WASTES FROM AGRICULTURAL SOURCES

Agricultural Source	Waste	Composition
Farms, ranches, greenhouses, livestock feeders, and growers	Crop residue	Cornstalks, tree prunings, pea vines, sugarcane stalks (bagasse), green drop, cull fruit, cull vegetables, rice, barley, wheat and oats stubble, rice hulls, fertilizer residue
	Forest slash	Trees, stumps, limbs, debris
	Animal manure (Paunch manure)	Ligneous and fibrous organic matter, nitrogen, phosphorus, potassium, volatile acids, proteins, fats, carbohydrates
	Poultry manure	Same as animal manure
	Animal carcasses, flesh, blood, fat particles, hair, bones, oil, grease	Ammonia, urea, amines, nitrates, inorganic salts, various organic and nitrogen-containing compounds
	Pesticides, insecticides, herbicides, fungicides, vermicide and microbicide residues and containers	Chlorinated hydrocarbons, organo-phosphorus compounds, other organic and inorganic substances, e.g., strychnine and lead arsenate

TABLE 10. SOLID WASTE GENERATION FROM MAJOR AGRICULTURAL CROPS (1966)[a]

Crop	Acres harvested (thousands)	Field waste load (tons/cap/yr)	Field waste load (mil. tons/yr)	Nature of waste
Corn, for grain	56,888	4.5	255.996	Leaves, stalks
Wheat " "	49,843	1.3	64.796	Stubble
Oats " "	17,848	1.8	32.126	"
Barley " "	10,226	1.8	18.407	"
Rye " "	1,283	1.3	1.668	"
Mixed grains	1,000	1.3	1.300	"
Rice	1,967	3.0	5.901	"
Flaxseed	2,627	0.8	2.102	Leaves, stalks
Alfalfa-clover seed	1,312	0.8	1.050	"
Sorghum, for grain	12,837	3.0	38.511	"
Cotton	9,595	2.0	19.190	"
Beans, dry	1,519	2.0	3.038	"
Peas, dry	344	2.0	.688	"
Soybeans, for beans	36,644	2.0	73.288	"
Peanuts, for nuts	1,436	3.0	4.308	"
Potatoes	1,479	3.0	4.437	Leaves, vines
Sweet potatoes	187	3.0	.561	"
Tobacco	967	0.5	.488	"
Sugar cane, for sugar	630	0.5	.315	Leaves, stalks
Sugar beets	1,161	3.0	3.483	Leaves
Vegetables	3,636	3.0	10.908	Leaves, stalks, culls
Fruits, nuts	4,699	2.0	9.398	Prunings, leaves, culls
TOTALS	218,137		551.959	
			2.760	tons per capita per year
			5,520.	lbs per capita per year
			15.1	lbs per capita per day[b]

[a]Does not include non-waste producing crops such as hay, silage, etc.

TABLE 11

SOLID WASTE GENERATION BY MAJOR FARM ANIMALS (1966)

Animal	Number on Farms (thousands)	Waste load (manure) (tons/unit/yr)	Waste load (manure) (thousand tons/yr)	
Cattle	108, 862	10	1, 088, 620	
Hogs	47, 414	8	379, 312	
Sheep	21, 456	3	64, 368	
Horses, mules	No estimate since 1960	-------	---------	
Poultry				
Broilers	2, 568, 338	.0045	11, 557	
Turkeys	115, 507	.025	2, 888	
Layers	339, 921	.047	15, 976	
Ducks, etc.	No estimate	-------	---------	
TOTAL			1, 562, 721	
			7, 814	tons/cap/yr.
			15, 627. 2	lbs/cap/yr.
			42. 8	lbs/cap/day

tons solid waste annually; if, as expected, ocean and oil shale mining become major commercial enterprises, approximately 4 billion tons of waste will be generated. Increased waste generation can be expected in nearly every commodity area, including coal, phosphate rock, clay, and mica among others, not only because of increased production but also because of the need for treating lower-grade ores.

Although some 80 mineral industries generate wastes, 8 industries alone are responsible for 80 percent of the total. Of these, the copper industry contributes the largest tonnage, followed by the iron and steel, bituminous coal, phosphate rock, lead, zinc, alumina, and anthracite industries. Smelting, nonmetallic mineral mining including sand and gravel, gold dredging, stone, and clay, and the chemical processing of ores and products account for most of the remaining 20 percent of the mineral solid waste generated (Table 12).

Mineral mine wastes are, for the most part, barren overburden or submarginal grade ore from open pit or surface mining of copper and iron ores, phosphate pebble rock, a multitude of nonmetallic minerals such as clay, mica, and kyanite, and from the underground mining of lead and zinc ores. These mountains of waste, often hundreds of feet high and covering extensive land areas, accumulate over the landscape adjacent to the mining operations. Large tonnages of solid waste from copper and iron ore mining have been disposed of in remote or sparsely populated areas. Equally large tonnages, however, of the wastes from these mining operations as well as from mining of lead and zinc ores, and nonmetallic minerals have been deposited in or near populated or frequently visited areas.

Culm banks and coal waste piles are largely residual wastes from processing coal, and contain a high percentage of carbonaceous matter. A recent survey has identified waste piles from 470 bituminous coal cleaning plants, and 800 culm banks each containing 10,000 tons or more of material from the cleaning of anthracite coal. Many of these waste piles and culm banks are located in or near heavily populated urban areas. The possibility of another rock and mud slide from such towering piles, such as that which occured in Aberfan, Wales, is an ever present threat to life and property in the vicinity of these banks. Wind erosion of the banks contaminates the nearby areas with dust. When the banks are ignited, as frequently occurs, they pollute the air with noxious fumes which destroy the surrounding vegetation, corrode

TABLE 12. GENERATION BY TYPE OF SOLID WASTES FROM THE MINERAL AND FOSSIL FUEL INDUSTRIES (1965)

Industry	Mine waste	Mill tailings	Washing plant rejects	Slag	Processing Plant wastes	Total (thousands of tons)
Copper	286,600	170,500	---	5,200	---	466,700
Iron and Steel	117,599	100,589	---	14,689	1,000	233,877
Bituminous coal	12,800	---	86,800	---	---	99,600
Phosphate rock	72	---	54,823	4,030	9,383	68,308
Lead-zinc	2,500	17,811	970	---	---	20,311
Aluminum	-----	---	---	---	5,350	5,350
Anthracite coal	-----	---	2,000	---	---	2,000
Coal ash	-----	---	---	---	24,500	24,500
Other[a]	-----	---	---	---	---	229,284
Totals	419,571	288,900	144,593	23,919	40,233	1,146,500

[a] Estimated waste generated by remaining mineral mining and processing industries.

domestic and commercial structures and cause accident provoking smog. In 1964 there were 495 burning refuse deposits located in 15 states. The banks were estimated to contain 500 million cubic yards of material.

Vast accumulations of finely ground tailings from the beneficiation of low-grade copper, lead, zinc, phosphate rock, mica, talc, and a variety of other ores and nonmetallic minerals, or from chemical treatment of copper, uranium, gold, and other ores are found throughout the nation. These wastes cover mountainsides or are impounded behind huge dikes on level terrain, or in canyons. The piles are not only eyesores, but create dust on dry, windy days, because they lack vegetation. During rainy periods, silt and chemicals washed from the piles pollute the streams and rivers.

A variety of slags and related smelter wastes is generated each year by the metallurgical processing industries. The slags are generated from iron and steel, copper, lead, and zinc smelters, dusts from foundry operations, mill scales, and a myriad of other processing plants. Blast furnace and open hearth slag is being utilized currently as concrete aggregate, road material, mineral wool, and soil conditioner. Slag from past iron and steel smelting operations, however, remains unused and other smelter slags and manufacturing plant slag wastes are accumulating throughout the nation. For example, about 100 million tons of blast furnace and open hearth slags have accumulated in the Pittsburgh, Pennsylvania, area alone. Ferrophosphorus slags from the electric furnace production of phosphorus from phosphate rock and basic oxygen furnace dust are typical examples of the smelting wastes which are disposed by simply dumping into unsightly piles. In 1966, 4.3 million tons of basic oxygen furnace dust were generated and it is estimated that by 1969 the tonnage generated will be 6.5 million tons.

The metallurgical and chemical processing of mineral concentrates also generates solid wastes of numerous types. Typical examples are asbestos processing wastes, metal plating plant waste solutions and sludges, steel pickling plant wastes, gypsum wastes from chemical conversion of phosphate rock to fertilizer materials, and the red muds generated in preparing alumina from bauxite. Some of the wastes are toxic and present serious local air, water and land pollution problems. For example, each year about 1.7 million tons of this slimy, caustic red mud waste are discharged directly into the Mississippi River by two plants in Louisiana, and an estimated 3.6 million more tons are stored in waste ponds in other parts of the nation.

The sand and gravel, rock, gold dredging, and stone and clay industries produce a relatively small volume of waste but because of the nature of the operations they are responsible for the devastation and defacement of extensive areas of land, estimated to be about 1.3 million acres, or roughly one-half the land despoiled to date by strip and surface mining. The sand and gravel rubble piles and churned up areas, and to some extent stone quarrying activities, are only too common in or near heavily populated centers. The type of waste generated is similar to that caused by highway and urban construction.

Coal strip mine overburden wastes present problems differing from those caused by mineral mine wastes. Although these overburden wastes occur nation-wide, they are of particular concern in the extensive coal regions of the Eastern United States where the accumulation of mixtures of carbonaceous slate, shale, and earth, commonly termed spoil banks, may be thousands of feet long and over a hundred feet high. These spoil banks produced by stripping barren overburden from coal seams are a source of acid water contributing to stream pollution. They frequently will not support vegetation, and because of their rough topography, are unsuitable for recreational or industrial purposes and constitute an aesthetic insult to the land environment.

The breadth of the environmental effects of waste accumulations is evident from an inventory of the land disturbed by the mineral and fossil fuel mining and processing industries. Prior to 1965, an estimated 5 million acres--equivalent to about 7,000 square miles-- were either covered with unsightly mineral and solid fossil fuel mine and processing waste, or the land was so devasted by current and past operations as to be not only useless and ugly, but in some cases a hazard to human life and property. Currently, about 20,000 active surface and strip mining operations are carving into the landscape at an estimated rate of about 153,000 acres annually. By 1980, it is expected that more than 5 million acres will have been defaced by these operations.

Federal Waste Sources

There are approximately 6,000 major defense installations located in the United States and throughout the world, with approximately 3.5 million persons in the uniformed military services. Civilian employees of government number approximately 2.6 million, and the varied nature of their work results in solid wastes of every conceivable nature.

Some solid wastes produced at Federal installations, and particularly those associated with the military, require special considerations in disposal, or certain aspects of the solid waste system must satisfy special requirements. For example, disposal of defective bombs presents problems seldom, if ever, encountered outside of government.

Besides the sheer volume of wastes, another important reason for focusing attention on Federal waste sources is the obligation of government to set an example in the field of environmental pollution control. Unless the Federal Government appears willing to take whatever steps are necessary to provide adequate disposal for its own wastes, persuasion of local governments and industry to improve their solid waste management systems will be considerably less effective.

The character of solid wastes produced by civil departments and agencies of the Federal Government may vary with the agency's mission and the nature of its activities. For example, wastes from research installations may contain glass, chemicals and other materials not found among the wastes from agencies which serve predominantly administrative functions.

Military Installations. Military installations vary in size from small isolated bases to large and extensive complexes similar to a modern city with its industrial area. At these installations there are all of the varied activities characteristic of a civilian community, along with those specialized operations associated with defense activities. The types of wastes associated with normal community and industrial activity are generated, as well as certain special problems, such as the residues from munitions and defense material. Food wastes, packaging materials, and all of the miscellaneous debris and residues characteristic of an industrialized urban society are included. From the information available there does not appear to be any significant difference in the type or amounts of such wastes generated in these military communities as compared with civilian communities of comparable size.

The large-sized industrial complexes of the military departments, as in the civilian industry, generate considerable amounts of solid wastes, including residues of metal machining, packing materials and military material, which has either outlived its usefulness or does not, for one reason or another, meet quality standards. The outfitting and provisioning of military units and naval vessels involve large quantities of packing materials and containers, which present a major need for efficient waste management.

Another source of solid waste generated by the military results from the construction and demolition of buildings and facilities. This creates large amounts of construction waste material to be processed and disposed of. In some ways this is comparable to the problems presented by urban renewal or large-scale community development programs. On large military installations, there are natural resources conservation programs, dealing with the problems associated with the maintenance of forests and related agricultural operations. In the case of the U. S. Army Corps of Engineers, civil works activities, reservoir clearing, the accumulation of solid wastes associated with river and harbor operations are special problems. There are also unique problems relating to the manufacturing, testing, and disposal of specialized military material and munitions.

VI. SUMMARY: GENERAL OBSERVATIONS AND RECOMMENDATIONS

Many citizen and conservation groups have begun to recognize solid wastes as posing a costly and pressing national problem that can seriously menace health and welfare in this country. There are only three repositories for all waste materials: the earth, its waters, and its atmosphere. One result of unsuccessful waste management is possible pollution of one or more of these. Open dumping or the improper location or operation of refuse landfills, for example, may cause gross pollution of land and the groundwater beneath the land. Open burning of solid wastes and faulty incineration cause air pollution in many American cities. The solid waste problem has been compounded by the lack of technological and manpower development. New technologies are required for the disposal of new and nearly indestructible materials reaching the solid waste stream, and effective methods must be developed by which such materials can be salvaged and recycled. Collection of solid wastes, the most costly phase of solid waste systems, could be made more economical and efficient if talent from industry, the universities, and research institutes were made available to apply new ideas and principles.

There is, at present, an obvious shortage of qualified and well trained personnel to design and administer the processing, collection, and disposal of the great volumes of solid waste. The shortage of professional personnel adds to the problem.

This report makes a comprehensive assessment of solid waste management problems and practices. It establishes the need for research and development on these problems taking account of waste from various sources: urban, industrial, agricultural, mineral and Federal establishments. The present national solid waste practices are immensely costly when all environmental factors are considered, including the waste of valuable resources. Reduction of these costs is a major goal of solid waste management. Accomplishment of rational solid waste management depends in large part on research and development and on demonstration in actual practice of the results of the research. The role of the Federal Government is to conduct research on solid wastes with its own staff and facilities, to direct research in this area by means of contracts and to financially support solid waste research by grants to research organizations.

VI

Pollution and Agriculture

A. In General

Wastes in Relation to Agriculture and Forestry. U.S. Department of Agriculture, Miscellaneous Publication No. 1065, March 1968. Pages 1-16.

Agriculture is both injured by pollution, and causes it. Agriculture is a source of pollution when agricultural chemicals such as insecticides, herbicieds, fungicides, and fertilizers are used excessively and become problems to others. Other agriculturally produced pollutants include soil carried through air and water, as dust and sediment respectively, and organic wastes such as plant residues and animal waste. It is not widely known that animal excrement far exceeds human excrement in volume. Thus although our human population is only slightly over 200 million, waste production by our domestic animals is equivalent to that of a human population of two billion people. One cow produces, for example, fecal matter equivalent to that produced by fifteen humans. Because of the emergence of large feed-lot, dairy, and poultry operations in concentrated surroundings, this problem has recently assumed far more significant proportions than previously.

Agriculture is also adversely affected by pollution. Air pollution such as is common near our major cities will stunt or kill many beneficial plants. Water pollution is a hazard to livestock. And plants or animals exposed to radioactivity or other polluting substances may become unfit for human consumption.

The study from which the following material was excerpted was written by Cecil H. Wadleigh, Director of the Soil and Water Conservation Research Division of the Agricultural Research Service of the Department of Agriculture. The entire report consists of 112 pages plus seven pages of introductory matter.

WASTES IN RELATION TO AGRICULTURE AND FORESTRY

INTRODUCTION

An enduring cliché proclaims that the only certainties in man's existence are death and taxes. Massive evidence is being accrued in our affluent society that a third inevitable should be added—wastes.

Wastes are ubiquitous. They occur on the farm and in the factory, in the high-rise apartment and on the rural estate, on the city streets and the open highway, in murky mines and verdant forests, in sedate offices and carefree parks, in the waters of lakes and streams, and in the air around us.

One can take the optimistic view that these wastes are merely resources for which we have not found a use. This view offers hope that with understanding, adequate technology, and the proper economic and legal constraints, there is a possibility that the problems inherent in the production, management, and disposition of wastes may be resolved. This is a good objective. But the probability of attainment is another matter.

We set great store by the operations of the marketplace. As is proper, net profit is a key guideline to decision. In the process, that which was previously a resource may become a waste. Waste disposal problems proliferate.

Within the agricultural economy, animal manures provide a case in point. Only a few decades ago, manure was esteemed as a prime source of soil fertility. Much technical information was available on proper handling to attain maximum benefits on the land. Even today, the "Plain People" of the Pennsylvania Dutch region carefully husband manure supplies toward maintaining highly productive lands. But these people have virtually a religious dedication to conserving and improving their land and their farms. They are, first of all, devoted stewards of the land, and only secondarily participants in the marketplace. They have no problems on how to dispose of animal wastes.

But these conscientious Amish are exceptions to the conventional thinking in commercial and agricultural enterprises.

With the abundant availability of synthetic nitrogen fertilizers at low cost following World War II, it became more economical for the farmer to supply plant nutrients to his fields from the fertilizer bag rather than meet the expense of hauling manure from barnyard or feedlot. Further, the general trend in mechanizing and streamlining production was given impetus by labor shortages during and after the war. Poultry enterprises, cattle and hog feedlots, and dairies became enormous by previous standards. Manure accumulated in vast quantities. Disposal of the stuff became a problem in terms of both engineering efficiency and esthetics.

Whether wastes are looked upon as resources, or resources as wastes, is, in the words of the Canadian Council of Resource Ministers, "a matter of attitudes."

Drive along a rural road on a balmy Sunday afternoon. Cast an eye along the roadside. All to be seen is not natural beauty. Many a traveler has obviously contributed to the mess of cans, bottles, and other assorted refuse along the way. The reasoning, or lack of reasoning, that compels humans to contribute to this desecration of the landscape would make a most interesting study.

As mentioned previously, disposition of wastes is a "matter of attitudes." There is a venturous, covetous streak not far below the surface of our people; a streak that erupts now and then into wholesale exploitation of resources. We read of the appalling slaughter of buffalo merely for sport or for their hides during the latter part of the 19th century. We see in the cutover areas of the Lake States the dismal sequel to the rapacious slashing of virgin forests in that region around the turn of the century. We see ugly scars upon the landscape in the form of barren and hideous spoil banks from irresponsible strip mining of years ago—spoil banks that even now are contributing silt and acid to streams. There is in all of these examples an utter disregard toward wastefulness.

Fortunately, there arises now and then from the multitude a man of great stature and wisdom who shows us the errors of our ways in exploiting resources and creating great wastes.

Professor E. W. Hilgard, the distinguished soil scientist and former geologist for the State of Mississippi, gave a hard-hitting lecture on poor farming practices, erosion, and sedimentation before the Mississippi Agricultural and Mechanical Fair Association at Jackson on November 14, 1872. He was gravely concerned over the poor husbandry practices of the farmers on the brown loam soils of the State—practices that were ruining the land, eroding and gullying the landscape, and silting the streams. He told his audience frankly: "If we do not use the heritage more rationally, well might the Chickasaws and the Choctaws question the moral right of the act by which their beautiful parklike hunting grounds were turned over to another race, on the plea that *they* did not put them to the uses for which the Creator intended them." It was another 75 years before effective countermeasures to land deterioration and stream siltation began to be taken in Mississippi.

One of the real giants in the history of the U.S. Department of Agriculture was Dr. Harvey W. Wiley. During the latter part of the 19th century, Wiley was Chief Chemist in the Department and undertook his classic studies on food adulteration. He found endless cases in which food processors were not careful to distinguish between that which was pure food and that which included an adulterant or even wastes. Fortunately, Wiley was also an excellent publicist and skilled agitator. Under his skillful touch, Congress passed the first Food and Drug Act in 1906.

One may not mention forest conservation—the elimination of wastefulness in forestry—without reference to the work of Gifford Pinchot in the early 1900's. Pinchot launched a crusade for sound forest management and harvesting practices, watershed conservation, disease control, and fire protection. It is regrettable that so much exploitation and waste in lumbering had to take place before the need for a Pinchot was recognized.

When the drought years of the thirties depleted the soils of cover and made vast expanses susceptible to the ravages of wind and water, another dynamic leader was at hand—Hugh Hammond Bennett, Chief of the newly formed Soil Conservation Service. He ably dramatized the tremendous duststorms of 1935 as not only depleting the soil on the Great Plains, but also creating enormous pollution of the atmosphere. He dramatized the tremendous soil losses from high runoff and the consequent contamination of streams with silt. The people were with him.

We have touched upon a few instances to emphasize the point that coping with problems in the production, management, and control of wastes related to agriculture and forestry is not a simple matter. It goes beyond the development of new and improved technology; beyond the sophisticated analyses of benefits and costs in the economic arena; beyond the police power of legislation.

One must take into account the general attitude of people towards various wastes: their appreciation for and interest in esthetic values, their concern over health hazards implicit in wastes, their willingness to accept restrictive regulations in lieu of private privileges, and their willingness to pay the added costs of maintaining a better quality of their environment.

If public agencies develop active programs to alleviate environmental contamination, the taxpayer usually pays the bill. If industry modifies its processing procedures to eliminate or abate production of wastes, any added cost will eventually be passed on to the consumer. If agriculture and forestry increase costs of production to reduce wastes, or the adverse effects of wastes, then in due course the consumer will find an adjustment in the price tag.

In addition to the question of how much control of wastes we want, one must ask how much control do we want to pay for? Beyond doing that which is economically justifiable, how far do we go in meeting health standards and social desirability? We must differentiate between the things we would like to do, those we should do, those we must do, and those we can do.

Compromises will need to be made among various objectives. For example, people express concern over pesticide residues. They want no part of the possible ill effects from accidental or inadvertent ingestion of these chemicals. They are just as adamant against biting into an apple and finding half a worm. Both extremes are easily avoidable.

Economic and social goals may need to be abetted by legal constraints.

There must be a mutuality of interplay between the course of technology pertaining to wastes and development of the conflux of legal-social-economic objectives and restraints. Urban people sometimes build homes out among dairy farms. There are cases in which the suburbanites who did not like the odors of dairy farming succeeded in imposing legal constraints on the farmers, thereby incurring the farmers' economic and social distress.

Consideration must be given to zoning laws that protect agricultural areas from untoward encroachment. California has shown the way with the passage of A.B. 2117—the California Land Conservation Act of 1965.

There must be enlightened leadership. Those graced with the good fortune to stand on mountains and discern the bright horizons of a pristine environment have the right—even the duty—to help turn unawareness and indifference into understanding, concern, and action.

Nothing gets done without someone being motivated.

APPROACH

This report provides terse consideration to 10 major categories of entities that contaminate the air, water, and soil of our environment in relation to agricultural and forestry endeavor (see app. I). These categories are radioactive substances, chemical air pollutants, airborne dusts, sediments, plant nutrients, inorganic salts and minerals, organic wastes, infectious agents and allergens, agricultural and industrial chemicals, and heat. A brief discussion is also presented on economic evaluation.

A few comments are offered on each category about the importance of the problems, the extent to which agriculture and forestry are involved, contributions that have been made to ameliorate the problems by research in agriculture and forestry, and an indication of the need for new or better information and technology towards meeting pressing problems. More complete discussion is presented in the four appendices.

It is imperative to recognize that different kinds of wastes may not be completely allocated to distinct pigeonholes, such as appears to be the procedure in this report. There are many interrelated effects.

The consequences of animal wastes moving into a stream must be considered with reference to the burden of oxygen-demanding organic wastes that the stream may be carrying from sewage, municipal garbage, processing plants, and dead or dying algae and other aquatic life. Discussion of animal wastes cannot be wholly discrete from that of infectious agents of animal origin. Both may arise from the same barnyards, feedlots, and poultry houses. The effects of animal wastes may not be discussed as something quite apart from plant nutrients. Obviously, animal wastes are a good source of nitrogen, and there is evidence that elevated levels of nitrate found in some ground waters are associated with accumulations of these wastes. Barnyard runoff may carry as much as 1,000 p.p.m. (parts per million) of phosphorus.

In discussing organic wastes as contaminants of water, one must recognize the importance of heat as a water pollutant. The elevated temperature of a stream that has been diverted to provide cooling for industry can have just as adverse an effect on poor assimilation of wastes as an increased load of biochemical oxygen demand.

A sediment burden in the water may lower oxygen content and thereby accentuate the adverse effects of organic wastes bringing in biochemical oxygen demand. Plant nutrients moving into waters may not be fully evaluated without considering the amounts adsorbed on the sediment being carried. Sediment from surface soils carries phosphorus into streams. Nutrient-poor sediment from the subsoils of construction areas may deactivate the soluble phosphate of a stream by tenacious adsorption. Sediment may be the prime carrier of chlorinated hydrocarbon insecticides that are occasionally detected in streams.

The interplay of these various categories of wastes in their impairment of our environment related to agricultural and forestry endeavor must be evaluated. The full perspective of waste problems cannot be gained by wearing blinders.

Radioactive Substances

Radioactive substances, which are possible contaminants of air, water, and soil, may be produced by several sources. Radioactive wastes resulting from mining and refining radioactive minerals such as uranium and thorium may wash into streams. There may be accidental or unauthorized releases of radioactive isotopes from power reactors or from research laboratories. There is atmospheric fallout from testing nuclear weapons. This has diminished with the existence of the test ban treaty involving the United States, the U.S.S.R., and Great Britain. Unfortunately, all nations having nuclear capability are not members of this treaty and therefore do not adhere to the ban. Potential for radioactive contamination under nuclear war staggers the imagination.

Agriculture and forestry contribute but very little to either actual or potential contamination from radioactivity—some phosphate rocks used for fertilizer production contain very low levels of radioactive uranium and thorium. Agriculture has experienced minor adverse effects from this source of contamination. During the spring of 1966, milk was confiscated in Nevada because of an unduly high level of iodine-131 in the area as a result of accidental venting during and following an underground weapons test. Potential damage to agriculture from radioactive fallout could be serious under unrestricted testing of nuclear weapons. Under nuclear warfare, the situation would be catastrophic.

Much information has been attained through research on the behavior of radionuclides in soils, plants, and animals. Attention was especially focused on the behavior of strontium-90. This isotope justifies serious concern because of its relatively long half-life and its chemical behavior so analogous to that of calcium in soils, plants, bones, and milk.

Engineering research has designed equipment to decontaminate soil that has been surface-coated with radioactive fallout. One procedure involves skimming off the contaminated surface, and another involves burying the surface by plowing deeply—2 feet—and adding a chemical root inhibitor to the buried layer.

There is need of much better information on the duration and effects of retained fallout particles on plants to improve validity of prediction of radiation doses to plants and animals. There must be much better techniques in terms of effectiveness, cost, and rapidity of operation for decontaminating forage, food plants, and soils. Research should ascertain the possibilities for specific genetic lines of feed and food plants that have the physiological capability to exclude radioactive elements.

Agriculture should support a relatively modest program of research on measures counteractive to radioactive wastes. If ever the unfortunate time arrives when such information is urgently needed, it will be too late to start.

Chemical Air Pollutants

Airborne chemical contaminants include such substances as sulfur dioxide, fluorine, carbon monoxide, ozone, nitrogen oxides, peroxyacetylnitrate, sundry hydrocarbons, and just plain smoke. Thirty million tons of sulfur dioxide are spewed into the atmosphere over the United States every year from smelters, use of fossil fuels for heat and power, and burning refuse. Seventy-five million tons of the other chemical contaminants are emitted into our atmosphere by exhausts from internal combustion engines, industrial mills, refuse burning, home heating, forest fires, and agricultural burning.

Crop plants, ornamentals, and trees are subjected to chronic injury in and near every metropolitan area. Livestock have been afflicted with fluorosis near certain industries, such as steel, aluminum, and phosphate mills. In southern California acute injury is widespread. About 14,000 square miles in that State are now afflicted with airborne toxicants. Losses to agriculture and forestry in the United States because of the flood of these noxious chemicals into the atmosphere are estimated as exceeding $500 million annually.

Agricultural endeavor makes a small local contribution to chemical air pollution in the form of hydrocarbons and smoke emitted from burning crop residues in fields. Major forest fires produce tremendous quantities of smoke. Fires in our forests produce upwards of 500,000 tons of hydrocarbons annually. In nearby areas, these hydrocarbons can become more serious by being converted to photochemical toxicants.

During the past 20 years, considerable research has been underway on (a) evaluating the relative sensitivity of different kinds of economic plants, trees, and livestock to various levels of specific air pollutants; (b) determining the actual components in polluted air that cause damage; (c) measuring the acute and chronic effects of given levels of specified pollutants in different plants and animals; (d) developing management practices that would minimize or ameliorate the adverse effects of given air pollutants; and (e) developing and selecting varieties of crops and trees with higher tolerance to air pollutants. Very promising research is underway towards prediction and abatement of forest fires due to electrical storms.

In meeting the problems inherent in the damages of airborne toxicants to agriculture and forestry, it would be most advantageous if these chemical effluvia could be eliminated at the source. Some accomplishment has been attained on this objective—for example, virtual discontinuance of sulfur dioxide emanation from the smelters in Trail, British Columbia, Canada—and more certainly will be accomplished. But there is little hope that this approach will adequately cope with the growing problems of air pollutants in agriculture and forestry.

There is an urgent need to improve the validity of the techniques available for assessing damages from air toxicants. Information is particularly weak for assessing chronic damages.

Little is known of the actual physiological mechanisms by which the harmful effects of air toxicants take place. Ignorance on these fundamental problems is a roadblock to valid understanding of damages and possible countermeasures.

One of the most promising avenues for attaining countermeasures to air toxicants is that of developing and selecting species, varieties, and genetic lines of crops, ornamentals, and trees that are tolerant, if not immune, to the major effects of chemical air contaminants.

There is evidence that expansive areas of vegetation, particularly woods, have a cleansing action on toxic wastes in the air. Much more information is needed on this effect.

The possibility of developing chemical protectants against air pollution damage needs much further exploration and study.

Even though total contribution to air contamination from forest fires seems relatively small, the total damages from these fires go far beyond their contribution to air pollution. All possible effort to develop better technology and better information to curb such fires should be pursued.

In view of the magnitude of the growing problems, and the paucity of research effort on air toxicants by agriculture and forestry in the past, it is urgent that research effort on the problems in this general area be expanded and pursued without delay.

Airborne Dusts

During the average year, 30 million tons of natural dusts enter the atmosphere. Most of this arises as soil blowing from inadequately protected fields under cultivation, deteriorated rangelands, and sand dune areas. A small amount arises from highway and industrial construction sites. Susceptibility to wind erosion is the dominant problem on 55 million acres of cropland in the United States.

Industries such as cotton ginning, alfalfa mills, lime kilns, cement plants, smelters, and mining operations release 17 million tons of dust into the atmosphere annually. This dust often coats the foliage of nearby crops, ornamentals, and trees. Growth and quality of product may be impaired.

Windblown soil not only robs the land of good topsoil, it also becomes a serious air pollutant. Respiratory ailments of man and animals are accentuated. Eyes are afflicted. Highway and air vision is impaired. Machinery is scourged by paint removal and dirt in bearings. Offices, homes, schoolrooms, and factories are completely permeated by dust. Pesticides may be carried long distances. Farmsteads, fence rows, ditches, and roads may be partially buried by drifts of dust.

Wind erosion research and its application has contributed importantly to the effectiveness of today's conservation and other advanced farming methods in reducing the threat of future "Dust Bowls" in the Great Plains. Meteorological records show that the drought in Kansas during the mid-1950's was just as bad as that during the 1930's. Yet records show that soil blowing was not nearly as serious during the fifties as it was 20 or 25 years earlier. At Dodge City, Kans., in 1936–37, there were 120 days of blowing dust. At the same location in 1956–57, under comparable meteorological conditions, there were only 40 days of blowing dust.

Research has developed a wind erosion prediction equation used by action agencies to provide technical guidance in farm and ranch planning.

In expanding the research on the problems of soil blowing in the Great Plains consideration must be given to various concepts for describing and delineating the influence of atmospheric wind and turbulence on soil detachment and transport; effect of wind and related atmospheric factors in soil drying and creation of soil surface conditions susceptible to high rates of detachment; modification of tillage practices and machines to provide effective and lasting crop residues and soil cloddiness conditions to resist soil detachment; the tolerance of crop plants to wind abrasion; the development and selection of better strains and varieties of grasses and crops to tie down soils susceptible to blowing; design of better windbreaks; and selection of better shrubs and trees to serve as windbreaks.

Much better information is needed on techniques to curb dust production on large feedlots.

Research has developed greatly improved screens to permit continuous operation of cotton gins while reducing lint fly. But much-improved engineering technology is needed to effectively filter the air from cotton gin operations as well as to handle trash without contaminating the air with dust or smoke or carrying diseases and insects back to the fields.

The current emphasis by the Federal Government to do whatever is feasible to attain clean air over the United States must include research to control or abate dusts from agricultural sources.

Sediment

Sediments are primarily soils and mineral particles washed into streams by storms. Although most of the sediment comes from land, a relatively small amount is contributed from the spoil banks of mining operations and smelters.

Some 4 billion tons of sediment are washed into tributary streams in the United States each year. About one-fourth of this sediment is transported to the sea. Water erosion is the dominant problem on 179 million acres of cropland and a secondary problem on an additional 50 million acres. At least half of the sediment is coming from agricultural lands. Some 30 percent of total sediment delivery arises as geologic erosion such as that found on the tributaries of the Missouri, Colorado, Rio Grande, Red, and Arkansas Rivers. For example, the Badlands of South Dakota are a tremendous source of geologically eroded sediment. Five to ten percent of the silt delivered comes from forests and associated rangelands. Streambanks erode and streambeds degrade, but aggradation may fully compensate for sediment losses. Urban, industrial, and highway construction sites and roadbanks also contribute eroded materials. Although such sources make but a small contribution to the national sediment delivery, they may contribute over half of the silt to the streams in a local watershed undergoing intensive construction activity.

Research conducted on many different soils varying in characteristics, slope, cover, and prevailing climatic conditions provided a mass of empirical data on soil loss as induced by rainstorms. The data were subjected to complex mathematical analysis by a digital computer which enabled the formulation of a universal erosion equation. This equation aids action agencies to predict what soil losses will be on a given soil of specified slope, under varying cropping conditions, with a given rainfall energy input.

Watershed research has shown that land cover is the major deterrent to sediment delivery into tributary streams. Extensive empirical information has been accrued on the entrainment, transportation, and deposition of sediments in the upstream watersheds towards making valid predictions of sediment accumulation behind flood-detention structures.

Improved technology in logging and construction of forest roads has reduced sediment delivery from forestry operations.

Forestry research on the abatement of wild fires (fires not started by man) has made a major contribution in diminishing sediment delivery from forested lands. Soil loss from a burned area is frequently tremendous.

Most all research in this area has been empirical. There is an urgent need for fundamental studies on the energetics involved in the detachment and movement of soil particles by raindrop splash and flowing water. Surface sealing and related phenomena that result in decreased infiltration need to be understood in order to be corrected. Water intake and movement through soil during freezing and thawing is especially important, but the interplay of forces involved is inadequately understood. Control practices need to be developed and integrated into systems that will reduce runoff a maximum amount and provide for the removal of surface drainage without appreciable erosion.

There is an urgent need for new concepts and procedures for identifying critical sediment sources and predicting sediment delivery from areas affected by climatic factors, soils, geology, topography, stream channel characteristics, and watershed protection measures. Special emphasis must be given to developing better technology for stabilizing and revegetating gullies, spoil banks, roadbanks, strip mines, overgrazed rangelands, forest burns, and badly disturbed construction sites.

Research towards diminution of forest fires and overgrazing of rangelands would greatly contribute to reducing sediment delivery.

Criteria for engineering design of sediment traps and debris basins could stand much improvement.

There needs to be a much better understanding of sediment transport in tortuous upstream tributaries, with emphasis on better understanding of the hydraulic forces necessary for design and maintenance of stable stream channel systems.

Limited information is available on the role of sediment as a transporting agent for pesticides and other chemicals.

Since comprehensive river basin planning is moving forward rapidly under the leadership of the Water Resources Council, expansion of effort to gain new or better information and technology on sediment problems in upstream watersheds should keep pace.

Plant Nutrients

Surface Water

Plant nutrients, as here used, are inorganic chemicals essential for the mineral nutrition of plants. When present in surface waters they may become contaminants in that they provide for unwanted growth of aquatic plants. Nitrogen and phosphorus are the two elements principally involved, but potassium is sometimes involved.

The algae "blooms" that frequently develop in nutrient-laden waters cause an off-taste and an unpleasant odor to the water. When streams and lakes reach the "green soup" stage of algae growth, the odor of decaying plants becomes offensive, fish are killed because of reduced oxygen content of the water, and there is interference with boating, swimming, and water skiing.

These nutrients enter surface water by discharges of raw or treated sewage, some industrial wastes, and runoff and seepage from land. Barnyards and feedlots yield nutrients. Use of chemical fertilizers on lands is sometimes suspected as being a significant source of plant nutrients found in streams and lakes. Such growths can ruin farm ponds. They harm the usefulness of the permanent pools of upstream watershed structures. Along with other water weeds, they can seriously impair the flow in irrigation and drainage ditches.

Detergents used in households and industry provide an abundance of soluble phosphate to sewage effluent. Evidence indicates that sewage delivery of phosphate amounts to 2 pounds per person per year. If the sewage effluent from 1 million people enters a stream with an average annual flow of 5,000 cubic feet per second, the average phosphorus content will be 0.2 p.p.m. This level is more than ample to enable excellent growth of algae without the phosphorus from other sources.

Each 1,000 tons of suspended sediment can be expected to carry about 1,000 pounds of phosphorus, most of which is in the adsorbed or fixed state unavailable for plant growth. Depending on the source, usually not more than 10 percent of the phosphorus on sediment is available for plant nutrition. Sediment derived from subsoils low in phosphate may actually deactivate much of the soluble phosphorus in a stream.

Phosphorus is moving into streams from agricultural lands. Many years of research by soil scientists using lysimeters and other techniques show that only infinitesimal amounts of phosphorus move through the soil in the soluble state. Some of this may be carried as complex organic molecules. Surface runoff may well carry phosphate as part of the suspended particles. Barnyard wastes may carry 1,000 p.p.m. of phosphorus. Water in the drains from fertilized fields in irrigated areas in rare instances have been found to carry as much as 1 p.p.m. of phosphate.

Very little modern information is available on the phosphorus content of runoff from soils varying in geochemical characteristics, treatments with chemical amendments and fertilizers, cultural practices, and hydrologic conditions.

Far more study is needed on methods of determining phosphorus content of water in terms of biological significance. Total phosphorus values are meaningless. We ought to be able to distinguish four different states of phosphorus in water samples: (*a*) That which is in true solution as orthophosphate; (*b*) that which is in solution in a nonpolar form as an organic polyelectrolyte; (*c*) that which is adsorbed on suspensoids; and (*d*) that which is a component part of the mineral of suspensoids.

In view of the widespread concern over eutrophication of surface waters, a major effort should be launched as rapidly as possible to attain evidence as to the cause and effect sequence. Even on large bodies of water such as Lake Erie, there is frequent accusation that use of fertilizer on farmers' fields is the causative source. The dearth of accurately accrued quantitative and qualitative information on phosphorus in runoff water from the land does little to allay argument.

Nitrate in Ground Water

Nitrate in drinking water can cause methemoglobinemia in babies—"blue babies." It is also toxic to livestock. The biochemical status of a baby's stomach readily reduces nitrate to nitrite, as does that of a ruminant. Thus, one hears of nitrate in well waters causing "blue babies" or nitrate poisoning in ruminants. Cases of "blue babies" associated with nitrate in well waters have been reported in the Middle West.

When nitrate is found in ground water, the sources may be several—sewage or septic tank effluent, feedlots or barnyards, field fertilization, or the natural accumulations such as found in the caliche of semiarid regions. The Public Health Service Standards specify that the nitrate-nitrogen content of drinking water should not exceed 10 p.p.m.

Research evidence in Missouri indicates that virtually none of the nitrate in ground water comes from field fertilization. Evidence from a secluded valley under concentrated study in California indicates that sewage effluent and nitrification processes in the semiarid soils may be the main sources of the rather high levels—100 p.p.m.—of nitrate found in the ground water. Studies in Hawaii on high nitrate levels in the aquifer indicated that nitrogen fertilization of irrigated sugarcane fields may be the source.

The diversity of observations and conclusions with respect to nitrate from various sources moving into ground water indicates the need for better information that may be gained by expanded field studies. The degree to which nitrite may be associated with nitrate during downward percolation needs careful study. The very modest level of research in this area needs to be increased appreciably to gain reliable answers to pressing questions.

Inorganic Salts and Minerals

This category includes neutral inorganic salts, mineral acids, and fine suspended metal or metal compounds. These substances enter into streams or onto soil from the effluent of various smelting, metallurgical, and chemical industries; from drainage from mines; and from natural sources. Industrial sources of these chemicals are comparatively minor in relation to total dissolved solids carried by the Nation's streams.

The Public Health Service estimated in the early 1930's that 2.7 million tons of sulfuric acid were produced annually by mines and delivered into tributary streams. Although the amount of acid delivery from mines has decreased, awareness of the problem has increased. Acid mine drainage kills fish and spoils water for domestic, livestock, irrigation, and recreation uses. Agriculture and forestry have an interest in the elimination of this problem.

Salts normally present in the soils and geologic materials of arid regions move into streams and onto irrigated farms. Colorado River water at Yuma, Ariz., carries about 1.2 tons of salt per acre-foot. Average annual flow of the Colorado at Hoover Dam is about 15 million acre-feet annually. At this location, the river transports more than

15 million tons of salt. Three thousand freight trains, 100 cars long, would be needed to carry this much salt. Much of it comes from geological deposits or accumulations in arid lands. It means that a farmer in the Imperial Valley of California who applies 5 acre-feet of irrigation water to his crops, also applies 6 tons of salt per acre. Hence, he must remove at least 6 tons of salt per acre in the drainage water from his farm.

Salinity is a hazard on about half the irrigated acreage in the Western States. Crop production on one-quarter of this acreage is already impaired by salt-affected soils. Production is threatened in irrigated projects the world over.

Over the past couple of decades, research has enhanced the capability of coping with salt problems affecting agriculture.

Improved procedures for evaluating water quality for irrigation have been developed, using electrical conductivity as the primary criterion and sodium-absorption-ratio as the secondary criterion.

Far better standards have been developed for leaching procedures in the reclamation of soils and for leaching requirements during continued management.

Much advancement has been attained in understanding the physico-chemical behavior of salt-affected soils.

Since water is the vehicle by which salt moves in soil, an understanding of the physical principles of water movement is absolutely essential. Excellent progress has been made in measuring the energetics of water retention and movement in soils.

Good progress has been made in characterizing the salt tolerance of important crop plants.

There is a real need to develop better means of assaying salty-soil problems as related to characteristics of irrigation waters. The relationships are very complex, and oversimplification can lead to poor technical guidance.

Recent advancements in making key measurements on the physical forces involved in soil water must be exploited towards far better appraisal of field conditions related to water.

There is an increasing need to attain an understanding of the biochemical and physiological mechanism in plants that determine their tolerance to various salt-affected soils. The role of climatic conditions in modifying these mechanisms also needs clarification. There is an urgent need to develop crop varieties higher in salt tolerance. An understanding of basic mechanisms in each important species would obviate the present superficial approach of cut-and-try testing.

The management practices followed in an irrigated field that is subject to salinization need far better quantitative characterization.

Hard experience in the Western States has shown that successful irrigation requires effective drainage. Accumulating salts must be continually leached out. Excessive leaching wastes water and nutrients. There is an urgent need to develop and test better mathematical expressions for the calculation of leaching requirements, drainage requirements, and the proper salt balance that ought to be maintained.

One of the pressing needs for information in irrigation agriculture is the development and evaluation of alternative procedures for the disposal or reclamation of return flow—procedures that seek to maximize beneficial use of total water available, while minimizing contamination for downstream users.

The heavy salting of highways in Northern States to facilitate traffic movement following snowstorms has caused serious damage to right-of-way vegetation and has increased silt damage to streams when erosion followed killing of the vegetation. The consequences of highway salting need far better evaluation.

The Federal Council of Science and Technology has recommended that agricultural research involved in this general area should increase fourfold during the current 5-year period.

Organic Wastes

This category includes material such as sewage, animal wastes, crop residues, forest trash, and food and fiber processing wastes. When these substances are carried in water, they incur a high biochemical oxygen demand. When dry upon the land, some are combustile, some produce odors, and some attract flies and vermin.

Sewage

About 60,000 acre-feet of water flows from municipal sewage facilities on the average day. Some of this could be used for irrigation even

297-533 O - 68 - 2

though that which received no treatment or only primary treatment could not be used in the production of food crops. Such water may be used for irrigating feed and fiber crops. Further, effecting movement of such contaminated water through soil is the best means of attaining purification.

The State agricultural experiment stations and the U.S. Department of Agriculture have already developed useful information on irrigation with sewage effluent both in terms of crop production per se, and in terms of overirrigation to provide ground water recharge with reclaimed water.

The good prospects of this research in sustaining agricultural water supplies in the future, in addition to providing an effective system of water purification, require that this line of endeavor be supported to the fullest extent feasible.

Animal Wastes

Domestic animals produce over 1 billion tons of fecal wastes a year. Their liquid wastes come to over 400 million tons. Used bedding, paunch manure from abattoirs, and dead carcasses make the total annual production of animal wastes close to 2 billion tons. In fact, waste production by domestic animals in the United States is equivalent to that of a human population of 1.9 billion.

As much as 50 percent of this waste production may be produced in concentrated supply. Big operations have developed rapidly in the last 20 years. An outfit with 10,000 head of cattle on a feedlot produces 260 tons of manure a day. Economic research reveals that it is cheaper for the farmer to supply fertility to his fields from the fertilizer bag than to meet the cost of hauling manure to the field. What is to be done with this manure?

If it accumulates, it effuses offensive odors into the surrounding area; it provides a spawning ground for vermin; on drying, it is a source of unsavory dusts; in rainstorms it produces runoff high in biochemical oxygen demand; and it may be the source of certain infectious agents found in streams.

Stockmen have been subjected to lawsuits on the grounds that such wastes were a public nuisance. Dairy farmers have had to make expensive moves to remote areas. Poultry enterprises have offended and been placed under restrictive curbs.

There is a pressing need to develop basic design criteria that are amenable to some adjustment to meet the widely varying constraints associated with different enterprises in different parts of the country. Elements of the problem include characteristics of manures; removal of manure from livestock quarters; storage; transport; feasibility of use on land; and disposal by burning, using lagoons or similar facilities, or burying. Other disposal problems include handling carcasses, milkroom wastes, and silage effluents.

Much effort should be allocated to identify the odor-producing organisms prevalent in manures, and to develop techniques to destroy such organisms.

Treatments of manure that would lower its attractiveness as a breeding ground for flies and vermin are especially needed.

In applying manure to cropland, more acceptable procedures are needed for storage and distribution without emission of offending odors and possibility of contamination of runoff water.

Use of lagoons for disposal has not been fully satisfactory. They tend to be underdesigned, overloaded, and misused. Anaerobic fermentation with the accompanying odors sometimes takes place. Some study is underway and must be continued to provide artificial stirring and better oxidation of water systems used for manure disposal.

The looming urgency of disposal problems on animal wastes, associated with a serious dearth of needed technology on efficient and economic methods, adds up to an immediate need for major expansion of research effort to attain sound answers to these problems.

Plant Residues

Plant residues from crops and orchards impair the environment in two ways: (1) They act as reservoirs of plant diseases and pests, and (2) they emit smoke and hydrocarbons into the surrounding area when burned. Rice straw as well as that of other grains is burned. Burning of residue from grass seed production is used extensively as a sanitation measure. Alternate methods of handling such residues are needed.

If residues are left on the soil, they can aid in preventing wind and water erosion. However, the soil microflora may be altered in nature and activity, as a result of a mulch of crop residue, and can adversely affect production of a succeeding crop.

Although problems pertaining to the handling of crop residues may not be regarded as top priority, new and better information is needed to economically handle these wastes without detrimental side effects.

Trash in Forests and Forestry Operations

Twenty-five million tons of logging debris are left in woods during the average year. This is a reservoir for tree diseases and insects. It is also an exceedingly serious fire hazard. The average size of forest fires originating in logging waste is more than seven times that of fires originating in uncut areas where trash has not accumulated. Average annual losses from forest fires are $600 million. In the average year, wild forest fires produce 160 cubic miles of smoke, 34 million tons of particulates, and 338,000 tons of hydrocarbons. Forest burns have excessive runoff and very high sediment delivery. They contribute to flood damages.

The trash and deadwood from elms killed by Dutch elm disease and oaks killed by oak wilt must be destroyed or treated to prevent vector transmission of the disease to healthy trees.

At the present time, research has produced no economically feasible technique to dispose of forest trash en masse other than by controlled burning. Chipping is useful in isolated instances.

Research needs to be expedited towards the improvement of techniques for controlled burning that reduce atmospheric contamination. All possible effort must be allocated to research that would aid in reducing the incidence of wild fires originating in forest trash.

Processing Wastes

Oxygen-demanding wastes from processing of agricultural and forestry products include runoff or effluent from sawmilling; pulp, paper, and fiberboard manufacturing; fruit and vegetable canning; cleaning dairies; slaughtering and processing of meat animals; tanning; manufacturing cornstarch and soy protein; sugar refining; malting, fermenting, and distilling; scouring wool; and wet processing in textile mills. These wastes, on entering a stream, greatly increase the demand for oxygen. They may make the water unsightly, unpalatable, and malodorous.

The oxidative requirements of the effluent from the woodpulp, paper, and paperboard industries exceed those of the raw sewage from all of the people in the United States.

In a year's time, the canning industry produces effluent with oxidative demands that are double those of the raw sewage from Metropolitan Detroit; the meatpacking industry, double those of Metropolitan Chicago; and the dairy industry, four times those of Metropolitan Boston.

Research has contributed to the abatement of processing wastes by (*a*) developing a commercially useful product out of that which had been a waste—the manufacture of insulating board out of sugarcane bagasse; (*b*) improving the procedure so that less wastes are produced—a new polysulfide modification of the kraft process of pulping results in greater pulp yields, less waste, and reduced air and water pollution; and (*c*) developing methods of waste treatment before disposal in stream—development of oxidative lagoons for potato-processing wastes.

Because of the tremendous contribution of processing wastes to the oxidative demands on our streams, it is obvious that research on farm and forest products should be enhanced by every means possible to give much greater emphasis to (1) developing useful products out of materials that are now considered wastes; (2) developing and improving processing procedures that lessen waste production; and (3) improving techniques for processing waste treatment.

Dilution of Organic Wastes in Streams

Depletion of dissolved oxygen in a stream is conditional upon the load of oxygen-demanding wastes added to a stream and the amount of streamflow available to waste assimiliation. Low flows in late summer and fall in many rivers and streams may be only one-fifth or one-tenth the average annual. Capacity of the stream to assimilate organic wastes is concomitantly affected. Hence, river basin planners must give major emphasis to operational plans that would minimize low flows. This will involve both structural and land treatment measures. Forest management and land treatment can affect water yield and abate low flows, but structural measures will undoubtedly have the major influence.

In view of the major emphasis being given to comprehensive river basin planning at this time, it is urgent that adequate research on forested and agricultural watersheds be directed towards gaining new technology and better information for determination of optimal combination of land treatment and structural measures on upstream watersheds to minimize the adverse effects of low flows.

Infectious Agents and Allergens

Through history, water contaminated with infectious agents has caused human disasters. It can still happen. In late May and early June of 1965, 18,000 people in Riverside, Calif., were infected with *Salmonella typhimurium* that by some unexplained means had entered the city water supply.

Soil, water, and air may be transmittal mediums for numerous organisms that afflict man as well as other animals and plants.

Those unfortunates suffering from allergies will not be consoled by evidence that 1.7 million tons of pollen move into the atmosphere over the United States every year.

Animal Disease Agents

Agricultural losses caused by infectious agents of livestock and poultry carried by air, water, and soil have been heavy. Some of the diseases so transmitted are leptospirosis, salmonellosis, hog cholera, mastitis, foot-and-mouth disease, tuberculosis, brucellosis, histoplasmosis, ornithosis, infectious bronchitis, Newcastle disease, anthrax, blackleg, footrot, coccidiosis, blackhead of turkeys, erysipelas, and transmissible gastroenteritis. A number of these may afflict humans.

The presence of coliform bacteria in water has been used as an indicator of bacteria for about 75 years. Although the coliform bacteria are not pathological, their presence has been taken as an indication that infectious bacteria might also be present.

Recent evidence suggests that more definitive information on bacterial contamination of water may be gained by making counts of both fecal coliform and fecal streptococcal bacteria. Empirical observations indicate that if the bacterial contamination of water is coming from animals, the f. coliform/f. streptococci ratio is less than 0.5. If the bacterial contamination is from human sources, this ratio usually exceeds 4. On this basis, a water bacteriologist may use the following as a general guide: If this ratio is less than 1.0, the pollution is coming from nonhuman sources; if the ratio exceeds 2.5, the bacterial pollution is most probably coming from human sources. Ratios between 1.0 and 2.5 suggest that the pollution is a mixture of human and nonhuman sources.

On the basis of the foregoing, bacterial assays on water samples from the upper Potomac Basin indicate that most of the bacterial pollution of this river system is coming from nonhuman sources.

Much needs to be learned about the complete reliability of the f. coliform/f. streptococci ratio as an indicator of the source of infectious agents in river water. Comparable studies need to be made on other river systems.

Excellent progress has been made in veterinary research in the United States over the past 85 years. Bovine tuberculosis has been reduced to a remarkably low point. Counteractive measures involving environmental sanitation have been highly successful in curbing brucellosis and hog cholera.

Even though much has been accomplished, much needs to be done.

Research underway on parasitic diseases will contribute to reduction of environmental contamination in two distinct ways: First, by the development of methods to reduce parasitism in our livestock population which will, in turn, reduce environmental contamination by parasite-infested animal wastes; second, by reducing the opportunity for environmental contamination by pesticides through development of methods of reducing or eliminating parasitism based on biological control, immunization, improved management, or more efficient use of better chemicals.

Expansion of research on developing counteractive measures to animal diseases is warranted fully on the basis of the huge annual losses these diseases impose on agriculture.

In view of the key objectives for clean air around us and clean water in our river basins, it is essential to have (*a*) far better information on the extent to which infectious agents in our surface waters come from agricultural sources; (*b*) possible procedures to curb movement of infectious agents from the farm or feedlot to the water; and (*c*) continued progress in eliminating animal diseases that contaminate air, water, and soil. Some animal diseases—for example, encephalitis—are spread by mosquitoes that are often spawned under irrigation agriculture. Needed research should be supported to the fullest extent feasible.

Plant Disease Agents

Diseases of crops, ornamentals, and trees have caused losses in billions of dollars. Many of these diseases are transmitted via air, water, or soil—that is, by contamination of the environment.

Black shank of tobacco has been a very serious disease in several tobacco-growing areas. It is spread by contaminated water, soil, or plants. The disease spores can contaminate ponds or streams into which infested fields drain. The disease persists for many years in some soils but not in others. Research on this problem has developed resistant plant varieties and a system of field sanitation.

Red stele disease of strawberry is an example of a plant disease fungus that persists in contaminated soil for many years even if strawberries are not present.

Stem rust of wheat is an example of a plant disease that is often spread, with disastrous results, by contaminated air. This disease has cost agriculture billions of dollars in losses. Countering the broad transmission of strains of this disease by tainted breezes has gained through development of disease-resistant varieties of grains. Since the disease organism mutates and attacks previously resistant varieties, a continuous breeding program is essential. In fact, one of the most laudable contributions of research in the United States is that of developing control measures for stem rust of wheat—a disease that can ravage millions of acres of a food crop by contaminated winds.

White stringy root rot of conifers is a disease caused by the fungus *Fomes annosus* Fr. It is present throughout our softwood forests. Airborne spores inoculate freshly cut stumps. Once tree roots are infected, the fungus may survive below ground for 50 years or more.

Although the potato blight epidemic that caused such widespread starvation and misery in Ireland occurred over a hundred years ago, we still know very little about the spread of such plant disease epidemics by a contaminated environment.

Research on the control of plant pathogens that contaminate the soil needs particular attention, since few are subject to control by either chemicals or plant breeding. High priority must be given to studies on changes in microbial populations in soils on incorporation of specific crop residues, with especial emphasis on the extent to which populations of plant pathogens on the soil are suppressed.

Plant disease agents carried by air, water, and soil can be just as disastrous to man's welfare as an infectious agent such as *Salmonella typhosa*—the cause of typhoid fever. Research towards minimizing the prevalence of these agents is tantamount to improving environmental quality for man's welfare. This is a high priority objective.

Allergens

The Public Health Service reports that in the average year, there are 12,646,000 sufferers from asthma or hay fever, or both. Of these, about 5 million are asthma sufferers, and about 75 percent of the asthma cases are due to pollen. The remaining 7,646,000 suffer from pollen allergies. Costs in terms of medical treatment and workdays lost are enormous.

There is very little research underway designed specifically for the control of plant species that produce allergenic pollen. In fact, there is no complete catalog of allergenic pollen.

Chemical control by herbicides having high physiological specificity offers a real possibility for the weeds that are troublesome. Pollen is carried long distances by wind, and any control measure must significantly reduce pollen counts.

The evidence suggests that very rewarding research could and should be undertaken in this area of serious aerial contamination.

Agricultural and Industrial Chemicals

The use of synthetic organic chemicals has been beneficial to man and his environment. But the discharge of some of these chemicals into the environment has induced problems. These chemicals include such substances as household detergents and the more recent insecticides, herbicides, fungicides, and nematocides.

Detergents

Agriculture and forestry have interest in problems caused by detergents. The excellence of these chemicals as cleansing agents is related to their high capability to disperse colloidal particles. They also disperse soil colloids. A dispersed soil has relatively low hydraulic conductivity in either the saturated or unsaturated state. An installation such as a septic tank with its distributing field at a farmhouse, in a rural community, or on a forest recreation area depends on reasonably good hydraulic conductivity of the soil around the distributing tiles. Reports of malfunctioning septic systems are common. Design of septic systems could be improved; they could be better adjusted to soil characteristics and to degree and nature of probable detergent load.

No hazard to humans or animals is apparent from the presence of detergents in surface waters.

Detergents carry phosphate as an active component. When sewage effluents enter rivers, lakes, and estuaries, this phosphate becomes a key plant nutrient promoting the development of obnoxious algal blooms. This role of phosphate is discussed in the section on plant nutrients. Much more factual information is needed on the respective contributions of phosphate from detergents and sewage in general compared to that from land runoff.

Insecticides

The potential contamination of the environment by pesticides, particularly the chlorinated hydrocarbons, has been a matter of public and private discussions. Within a few years after DDT began to be used extensively as a field and forest insecticide, DDT and its metabolites were found in the fatty tissues of fish and wildlife, both living and dead. Newspapers carried features on fish and wildlife losses with an implicit indictment of agriculture and forestry for having used insecticides.

The use of these chemical tools has made a tremendous contribution to man's health and welfare over the past 25 years. Unfortunately, in some instances, they have also been abused and misused without due consideration to their impact on the nontarget organisms.

In 1964, 470 million pounds of insecticides were used on 83 million acres of land. In a few instances, adverse effects have occurred in agriculture and forestry. Insecticides that have broad spectrum activity are just as effective on many nontarget insects as on the target ones. For example, use of certain insecticides has resulted in losses of honeybees and other beneficial insects.

The application of certain insecticides to cotton, corn, or other crops may lead to insecticide residues in soybeans or peanuts grown in the soil a year or more later. Certain chlorinated hydrocarbons can persist in soil for many years. In one experimental study 50 percent of the original high-level application of DDT was found in the soil 8 years after the material had been applied.

The pesticide monitoring program maintained by the U.S. Department of Agriculture on widespread soils and waters is, therefore, exceedingly important. Evidence revealed from these monitoring activities does not show cause for concern on pesticide buildup. But this does not warrant complacency over the diverse problems stemming from insecticide use.

Over the long pull, research efforts towards avoiding the adverse effects of insecticide residues in the environment must continue to receive major attention.

Far better information is needed on insect population trends with the objective that emergency use of chemicals may be avoided.

Research on developing insecticides with higher biodegradability and lower persistence in the environment must move forward apace.

There needs to be far better information on the chemical behavior of insecticides in soils varying widely in physico-chemical attributes. The extent and level of pesticides in waters and in soil-water systems must be better evaluated and the significance of the findings better understood.

Certain residue problems arise as a result of spray or dust drift beyond the target area. There are good possibilities of developing better methods of application or lower application rates. These must be pursued.

The use of physical attractants to aid in nonchemical control of economic insects needs to be more fully explored.

All possible support should be allocated to research for completely selective methods of controlling major insect pests. These techniques include the use of predators and dissemination of specific insect diseases; development of specific insect attractants; breeding and selection of insect diseases; breeding and selection of insect-resistant crop varieties; the use of self-destruction mechanisms such as release of sterile males; and the development of chemical insecticides that act on selective physiological systems peculiar to the target insect.

Herbicides

Weed control by use of selective herbicides has increased substantially during the past 20 years. In 1964, 184 million pounds of herbicides were sold in the United States and 97 million acres of agricultural land were treated with these unique chemicals.

Although good progress has been made in the technology of herbicide use, much research is needed to refine this technology. Herbicides applied to control weeds in one crop may leave residues in the soil which prevent the growth of certain other crops planted immediately following harvest of the treated crop. Diuron used for weed control in irrigated cotton in the Southwest may leave residues in the soil which could injure vegetable crops such as lettuce, carrots, cabbage, and cucumbers when planted in the winter following cotton harvest.

Spray drift of 2,4,5-T and 2,4-D from applications on nearby forests, cropland, roadsides, and rights-of-way has damaged chemically sensitive trees such as dogwood, paper birch, box elder, chestnut, black locust, and other shade trees, shrubs, and herbaceous ornamentals. In some situations injury has occurred as much as 10 miles downwind from the target area.

The aromatic solvents and other herbicides used to control aquatic weeds in irrigation systems are known to be toxic to fish at levels generally higher than the amounts needed for weed control. However, judicious use of certain solvents or herbicides will provide weed control with little or no fish toxicity.

Limited information is available on the behavior of some herbicides in soils. It is a complex picture. Adsorption of a herbicide by a soil is determined by the specific surface of the soil, organic matter content, nature of the clay mineral, moisture status, temperature, nature and degree of base saturation, and structure and polarity of the herbicide molecule.

Soil micro-organisms are important in detoxifying herbicides in soils. It is largely because of these soil micro-organisms and other dissipation mechanisms that most herbicide residues in soils do not show progressive buildup.

The fact still remains that the toxicology of residual herbicides on succeeding crops is often unpredictable. Consequently, there is a pressing need for far better information on the fate of herbicides in soils, including the cultural practices and climatic conditions that incur a modifying influence.

There is a continuing urgency to develop superior herbicides with greater specificity and fewer adverse residual side effects.

Concentrated research effort is needed on engineering principles to develop better techniques to reduce drift.

There would appear to be tremendous economic advantages in developing weed control chemicals that are more specific; for example, a chemical that would only inhibit pollen production on ragweed.

The extensive use of herbicides and the problems arising from residues in soil and water or drifts in the atmosphere focus on the need to emphasize investigative studies on the pertinent array of problems.

Fungicides

Fungicides do not appear to be of major concern in environmental contamination. The organic mercury compounds are generally the most hazardous to man, but these are used mostly for seed treatment. Accidental feeding of such treated seeds to livestock and poultry sometimes causes poisoning.

Apple orchards and vineyards that have had a long record of spraying with copper fungicides may have very high levels of copper in the surface soils. Zinc and manganese can accumulate to toxic level in soils, but evidence indicates that application of fungicides is not at fault.

In terms of residues contaminating the encironment, there are many other substances that cause more concern than fungicides.

Heat

Heat acts as a water pollutant because the amount of oxygen that water can hold in solution diminishes with increasing temperature. The introduction of heat in any substantial quantities into surface water has the net effect of introducing additional oxygen-demanding wastes.

Heat also has a detrimental environmental effect upon fish and other aquatic life. Some fish can stand only a very few degrees of increase in temperature, and a substantial increase in temperature can result in the elimination of some forms of aquatic life.

There is little in agricultural and forestry endeavor that contributes heat to streams and lakes.

One example comes to mind. Water seeping from or flowing through a recent forest burn may have a higher temperature than if the area had been in deep forest.

Industry may contribute serious heat pollution of streams. The Mahoning River in Ohio provides a good example; it is used extensively for industrial and power cooling in the Youngstown area. The temperature of the river in the summer is raised to such an extent that fish life is completely eliminated, and the river is rendered unfit for further use in either waste assimilation or additional cooling.

Agricultural and forestry interests wishing to use surface waters for recreation or fishing would be seriously affected by water unduly heated by industrial cooling.

Socioeconomic Evaluation

A whole array of decisions must be made among alternative approaches posed by wastes in the environment. How do we define quality of the environment? Where do we draw the line between contamination and noncontamination? How much are we willing to pay for specific levels of quality improvement? What level of waste treatment do we want? Do some segments of society want different levels of environmental quality than others? How do we set up standards that provide flexibility? How do we attach monetary significance to esthetic values? How do we relate these values to quantitative aspects readily evaluated in the marketplace? Do we want to maximize economic benefits, or should we seek to maximize social benefits, or should we attain an optimal combination of the two?

Answers must be sought to these and similar questions if optimal economic and social solutions to waste problems are to be attained.

Economic research on wastes should emphasize the evaluation of costs and benefits associated with waste production and waste disposal. Such evaluations are needed for each of the agents that do or may affect this quality of our environment. Such studies would identify which areas of waste production are most in need of correction in terms of dollar costs and benefits.

To this end, economic studies are underway on data from the Pesticide and General Farm Survey for 1966. Evaluation of these data in relation to those previously obtained will indicate trends.

Studies are underway on the economic significance of water quality, erosion and sedimentation processes, and the salinity of irrigation waters.

Economists maintain a continuing research program with respect to fertilizer use.

Organic wastes present a major area for economic studies toward arriving at sound decisions on alternate ways of handling animal and poultry manures and food and fiber processing wastes.

There is an immediate need for information on the status of rural waste problems. Surveys should be designed to provide for better knowledge on abatement costs, economic effects, existing institutional control arrangements, and other information as specified by survey objectives.

These surveys should highlight critical problems requiring studies in depth. Such studies should trace the economic implications of selected problems, specify feasible alternatives for solution, and provide information for the rational compromises between production efficiency and waste-free environments.

There are presently serious problems of adjustment within agriculture to meet changing concepts of environmental quality. Local ordinances and court actions have caused abrupt cessation of agricultural operations in some areas. Studies need to be made on the role that local leadership in county and district organizations can play in guiding the amelioration of rural waste problems.

There is a critical need for improved techniques for measuring secondary economic effects, and nonmonetary benefits and costs, of waste control programs.

The need for information thus is great in the complex of decision-making processes.

B. Pesticides

<u>Report of the Secretary's Commission on Pesticides and Their Relationship to Environmental Health</u>. Parts I and II. U.S. Department of Health, Education, and Welfare, December 1969. Pages 7-37.

In April 1969, Robert H. Finch, Secretary of Health, Education, and Welfare, appointed a Commission on Pesticides and Their Relationship to Environmental Health. Dr. Emil Mrak, retired Chancellor of the University of California, Davis and a distinguished authority on food science and technology, was appointed as Chairman. There were thirteen other members, mainly scientists connected with public or private university research institutes, industries manufacturing pesticides, and governmental regulatory or research bodies. The Commission employed fifteen professional people plus a number of non-professionals for its staff, and also had the assistance of many scientists and consultants through its "Advisory Panels" on Carcinogenicity of Pesticides, on Interactions, and on Teratogenicity of Pesticides.

Parts I and II of the Report were released in one volume in December 1969 and consist of 670 pages plus introductory material. The Report has already been recognized in the judicial process. See, for example, the references to it as the "Mrak Commission Report" in Environmental Defense Fund, Inc. v. United States Department of Health, Education, and Welfare, 428 F.2d 1083 (D.C. Cir. 1970). Undoubtedly, as a significant scientific study, the Report will be consulted widely by those dealing with problems relating to pesticides and will be regarded as the standard work in this area.

The excerpts reproduced below summarize the conclusions of the Report. In its introduction to the Report, the Commission made the following comments:

> "After carefully reviewing all available information, the Commission has concluded that there is adequate evidence concerning potential hazards to our environment and to man's health to require corrective action. Our Nation cannot afford to wait until the last piece of evidence has been submitted on the many issues related to pesticide usage. We must consider our present course of action in terms of future generations of Americans and the environment that they will live in."

COMMISSION RECOMMENDATIONS

Recommendation 1:

Initiate closer cooperation among the Departments of Health, Education, and Welfare, Agriculture, and Interior on pesticide problems through establishment of a new interagency agreement.

The registration of pesticides is now vested only in the Secretary of Agriculture under the Federal Insecticide, Fungicide, and Rodenticide Act (FIFRA). The regulations implementing FIFRA state that the purpose of the act is "to protect the public health before injury occurs rather than to subject the public to dangers of experimentation and take action after injury."[1] However, the present interagency agreement requires the Secretaries of DHEW and USDI to produce scientific evidence clearly demonstrating a present hazard to health or to the environment in order to remove from registered use or prevent the registration of any specific pesticide. In regard to health protection, the burden of proof should rest upon the manufacturer to demonstrate to the Secretary of HEW that appropriate tests do not produce untoward effects upon two or more species of mammals which might indicate a hazard to health. Such a procedure is entirely in keeping with the purpose of the act as stated above.

A new interagency agreement is needed to strengthen cooperative action among the Departments of HEW, USDA, and USDI to protect public health and the quality of the environment from pesticide hazards. Approval by the Secretaries of DHEW and Interior as well as Agriculture should be required for all pesticide registrations. Pesticide uses deemed by any of the three Secretaries to be hazardous should be restricted or eliminated.

The agreement should further require a continuous review of new scientific information on pesticides now in use, with a formal review made 2 years after initial registration and subsequent formal reviews by the three agencies at 5-year intervals.

---

[1] C.F.R., Title 7, Cr. 11, Sec. 362.106(d)(1).

Such an agreement and the closer interagency cooperation it would produce would have other distinct advantages:

- It would call attention to evidence suggesting concern and expedite appropriate action;
- It would encourage cooperative approaches to public education, applicator training, research on the biological effects of a pesticide, and promote the development and use of improved methods of pest control; and
- It would provide a mechanism for focusing the concerns and skills of each agency and to coordinate action on pesticide problems.

If the objective of providing to the Secretary of DHEW the authority to meet his responsibility for control of health hazards of pesticides cannot be attained by a new interagency agreement, it will be necessary to amend the Federal Insecticide, Fungicide, and Rodenticide Act (FIFRA).

Recommendation 2:

Improve cooperation among the various elements of the Department of Health, Education, and Welfare which are concerned with the effects of pest control and pesticides.

The diversified and significant responsibilities associated with the Department of Health, Education, and Welfare pesticide and pest-control activities lack sufficient coordination and direction. Several segments of DHEW have direct and implied responsibilities for proteciton of public health in relation to the use of pesticides. The problem of achieving cooperation appears to be intensified by recent reorganizations. Mechanisms should be developed to assure exchange of information between all pertinent segments of DHEW. There is a need for reappraisal of the vector control activities, educational programs, research responsibilities, monitoring, State aid programs, and other activities involving pesticides.

Recommendation 3:

Eliminate within two years all uses of DDT and DDD in the United States excepting those uses essential to the preservation of human health or welfare and approved unanimously by the Secretaries of the Departments of Health, Education, and Welfare, Agriculture, and Interior.

The uses of DDT and DDD as pesticides should be limited to the prevention or control of human disease and other essential uses for which no alternative is available. Such uses should be clearly identified and individually evaluated in relation to human hazard from exposure, movement in the natural environment concentration in the food chains of the world, and other environmental considerations. Unanimous approval by the Secretaries of DHEW, USDA, and USDI (who in turn are expected to call on Federal, State and private experts for advice) would provide for Identification of essential uses and assure that such approval will be based on sound judgment.

Abundant evidence proves the widespread distribution of DDT and its metabolites (principally DDE) in man, birds, fish, other aquatic organisms, wildlife, soil, water, sewage, rivers, lakes, oceans, and air. Evidence also demonstrates that these materials are highly injurious to some nontarget species and threaten other species and biological systems. Elimination of all nonessential uses should be achieved and the period of 2 years is recommended to assure achievement without excessive economic disruption.

Unavoidable residues of these persistent pesticides will continue to occur in the soil, water, air, and food supplies for a period of years despite restriction of usage in the United States. Reasonable methods must be established for the use of as much of the food supply as possible without hazard to human health.

Despite diminution of DDT usage, the Commission urges that research be intensified to gain further understanding of the ecological dynamics and public health implications of this example of a persistent chemical widely distributed in the environment.

It should be recognized that DDT is used in developing nations in the prevention and control of malaria, typhus, and other insect-borne diseases, and in the production of food and fiber. The control of such uses is the responsibility of those nations. They should, however, receive from the United States the full benefit of all available information and assistance on hazards, safe and effective uses of pesticides, and alternative methods of pest control.

Recommendation 4:

Restrict the usage of certain persistent pesticides in the United States to specific essential uses which create no known hazard to human health or to the quality of the environment and which are unanimously approved by the Secretaries of the Departments of Health, Education, and Welfare, Agriculture, and Interior.

Several pesticides other than DDT are persistent and cause or can cause contamination of the environment and damage to various life forms within it. These include aldrin, dieldrin, endrin, heptachlor, chlordane, benzene hexachloride, lindane, and compounds containing arsenic, lead, or mercury. We may anticipate that decreased use of DDT and the above chemicals will result in an increased use of other chemicals such as toxaphene. While there is no evidence that toxaphene undergoes biological magnification, its chemical properties suggest that it should receive close surveillance. Furthermore, the use of organometallic compounds and salts of heavy metals other than arsenic, lead, or mercury should be periodically reviewed.

The uses of persistent compounds should be fully reviewed in light of recent advances in understanding the undesired effects of some pesticides. The acceptable uses should be selected and approved unanimously by the appropriate departments. Such usage should be retricted to essential purposes, limited to the lowest effective dosage required for the production and protection of essential foods and fibers, and replaced by safer alternatives wherever possible.

It is, however, impractical to attempt to eliminate the residues of such pesticides from foods by the application of zero tolerance limits. Modern techniques have greatly increased the sensitivity of the analytical methods available when the zero tolerance concept was advanced. This fact must be recognized in judging the possibilities of hazards and establishing tolerance limits with a sufficient margin of safety to protect human health and welfare.

Recommendation 5:

Minimize human exposure to those pesticides considered to present a potential health hazard to man.

Decisions on restriction of human exposure to pesticides should be made by the Secretary of the Department of Health, Education, and Welfare. In reaching such decisions, consideration must be given to both the adequacy of the evidence of hazard to human health and possible consequences to human welfare that flow from the imposition of restrictions on human exposure to pesticides.

Accordingly, it is of utmost importance that the results of screening tests be scientifically and rationally considered. The correct interpretation of hazards to human health is sometimes extraordinarily difficult. It must involve the transfer of the results of animal experiments to prediction of human effects. In addition, the screening process frequently involves preliminary examination of the effects of massive dosages, possible contamination of test samples, and other factors which affect proper interpretation.

The health and welfare of the public must be effectively protected. However, it is not in the best interest of the public to permit unduly precipitate or excessively restrictive action based only on anxiety.

In recent screening studies in animals employing high dosage levels, several compounds have been judged to be positive for tumor induction. In similar screening studies other pesticides have been judged to be teratogenic. The evidence does not prove that these are injurious to man, but does indicate: (1) A need to reexamine the registered uses of the materials and other relevant data in order to institute prudent steps to minimize human exposure to these chemicals; and (2) to undertake additional appropriate evaluatory research on representative samples of these substances in order to guide future decisions. It is further important to have detailed knowledge of sample composition and purity. These materials are: aldrin; amitrol; aramite; avadex; bis (2-chloroethyl) ether; chlorobenzilate; p,p'-DDT; dieldrin; heptachlor (epoxide); mirex; n-(2-hydroxyethyl)-hydrazine; strobane; captan; carbaryl; the butyl, isopropyl, and isooctyl esters of 2,4-D; folpet; mercurials; PCNB; and 2,4,5-T.

The imposition of restrictions on exposure, particularly from pesticide residues in food and water, should be accompanied by periodic review and adjustment of pesticide residue tolerances. Indiscriminate imposition of zero tolerances may well have disastrous consequences upon the supply of essential food and threaten the welfare of the entire Nation. Stepwise lowering of pesticide tolerance may in some cases be an effective and flexible instrument with which to execute policy.

Currently our national resources of funds, manpower, and facilities will not permit the concurrent testing of all pesticidal compounds. Priorities for testing must be established. Effective national implementation of this policy will require continuing development and evaluaton of scientific information concerning the hazards of pesticides to human health. Additional chemicals are being or should be investigated and evaluated for potential hazards to human health, as resources permit.

Recommendation 6:

Create a pesticide advisory committee in the Department of Health, Education, and Welfare to evaluate information on the hazards of pesticides to human health and environmental quality and to advise the Secretary on related matters.

The Secretary of the Department of Health, Education, and Welfare is obligated to protect and enhance human health and welfare. In relation to pesticides, this requires that he draw upon a wide range of expert opinion and guidance. Excellent competence in some areas exists within the staff of the Department, but the advisory services of a group drawn from the professional, industrial and academic specialists in related fields can provide unique and essential services. A Pesticide Advisory Committee should be created and should include experts on human health and welfare, on environmental and agricultural sciences, and from appropriate economic and industrial areas of knowledge and experience.

In assisting the Secretary, the Pesticide Advisory Committee would:

- Intrepret new information from scientific sources and from increased national experience with pesticides.

- Assess the potential hazards of specific pesticides, based on consideration of persistence of residues, possible distribution and magnification in biological systems, and potential hazards to human health and welfare.
- Recommend improvements in administrative procedures relating to pesticides; areas requiring intensified research attention; adequate programs for monitoring and interpreting pesticide distribution; and, in conjunction with the U.S. Department of Agriculture and the U.S. Department of the Interior, recommend educational and training programs designed to improve usage of pesticides and reduce deleterious effects.
- Evaluate the complex risks and benefit considerations necessary for making responsible judgments on the uses of pesticides, and suggest means for maximizing benefits and minimizing risks.
- Provide advice on suitable standards and tolerances for pesticide content in food, water, and air to protect the public health and the quality of the environment.
- Identify gaps in knowledge and advise on needed research.
- Review and recommend test procedures and protocols to be employed by manufacturers in establishing the safety of pesticides.

The committee should receive the full benefit of the information and professional competence present in the Department of Health, Education, and Welfare, maintain strong liaison with any comparable advisory groups in other Federal agencies, and have free access to specialists and experts throughout the nation.

Recommendation 7:

Develop suitable standards for pesticide content in food, water, and air and other aspects of environmental quality, that: (1) protect the public from undue hazards, and (2) recognize the need for optimal human nutrition and food supply.

In setting tolerances for pesticide residues in or on foods, the Department of Health, Education, and Welfare should be cognizant of the need for optimal human nutrition and food supply. Because widespread environmental contamination by DDT and other persistent pesticides can cause unavoidable residues in many raw foods, the new Pesticide Advisory Committee should examine the problem of tolerances carefully. Total human exposure, actual daily intake, and total body burden of pesticide residues should be minimized whenever feasible, but unavoidable residues should be realistically considered. (See next recommendation).

There is need to abate widespread contamination of the environment in order to reduce unavoidable residues of pesticides in food, water, and air. Of equal importance is the need to take anticipatory regulatory action to prevent future problems caused by other pesticides.

The fact that DDT residues are widespread throughout the environment has lead to unusual difficulties for certain food industries. For example, the fact that DDT is concentrated from contaminated waters into the fat of coho salmon and other fish is not the fault of the fishing industry. DDT contamination of lakes, rivers, and oceans is not susceptible to immediate correction and reduction will require concerted action to prevent DDT entry from various sources. Tolerances for DDT residues in fish should be subjected to immediate review and reflect the relative importance of the food in the diet. Concurrent efforts should be made to apply processing methods capable of reducing the DDT content of fish.

The Pesticide Advisory Committee may wish to consider a graded series of regulatory actions developed in proportion to the extent of environmental contamination or risk thereof, in relation to total human exposure, actual daily intake, and total body burden of pesticide residues. Such a series might be as follows:

Grade I: No significant environmental contamination or risk thereof is judged to exist. Only periodic surveillance and confirmatory evaluation is required.

Grade II: Some environmental contamination or risk thereof is judged to exist. Active surveillance is required, including the assessment of total human body burden of pesticide residues; investigation of their relationships to various sources; evaluation of reductions by variations in food harvesting, processing, and distribution techniques; and routine regulatory controls.

Grade III: Substantial environmental contamination or risk thereof is judged to exist. Restrict total usage with active program of replacing present usage with alternative pesticides, and approve use by permit only.

Grade IV: Widespread or severe environmental contamination or general risk thereof exists. Ban all nonessential uses and remove from the general market. Approve use by permit only.

If indicated by the best available evidence, the above-graded series of regulatory actions would permit immediate classification of a pesticide into grade III or IV, thus requiring approved use by permit only. Anticipatory action can prevent harmful environmental contamination by pesticides and their movement into food, water, and air.

Recommendation 8:

Seek modification of the Delaney clause to permit the Secretary of the Department of Health, Education, and Welfare to determine when evidence of carcinogenesis justifies restrictive action concerning food containing analytically detectable traces of chemicals.

The effect of the Delaney clause [2] is to require the removal from interstate commerce of any food which contains analytically detectable amounts of a food additive shown to be capable of inducing cancer in experimental animals. This requirement would be excessively conservative if applied to foods containing unavoidable trace amounts of pesticides shown to be capable of inducing cancer in experimental animals when given in very high doses. If this clause were to be enforced for pesticide residues, it would outlaw most food of animal origin including all meat, all dairy products (milk, butter, ice cream, cheese, etc.), eggs, fowl, and fish. These foods presently contain and will continue to contain for years, traces of DDT despite any restrictions imposed on pesticides. Removal of these foods would present a far worse hazard to health than uncertain carcinogenic risk of these trace amounts.

Commonly consumed foodstuffs contain detectable amounts of unavoidable naturally occurring constituents which under certain experimental conditions are capable of inducing cancer in experimental animals. Yet, at the usual low level of intake of these constituents they are regarded as presenting an acceptable risk to human health.

Exquisitely sensitive modern analytical techniques which became available since enactment of the Delaney clause permit detection of extremely small traces of chemicals at levels which may be biologically insignificant. Positive response in carcinogenic testing has often been shown to be dose-related, in that the carcinogenic response increases with increasing dose levels of the carcinogen; when the dosage of a carcinogen is minimized, the risk for cancer is also minimized or eliminated.

The existence of such dose responses of carcinogens must be taken into account by evaluating the balance of benefits and risks as is commonly done in assessing any toxic chemical. Ignorance concerning the possible role of environmental chemicals in causation of human cancer reinforces the case for caution in making such judgments. In addition to the complexities of determining a "no effect" level of a weak carcinogen in a given experimental species, the extrapolation of the substance's effects to other species including man is of such intuitive nature that a wide margin of safety must be allowed. Nevertheless, compelling considerations of the increasing need for food may lead

[2] Federal Food, Drug, and Cosmetic Act, as amended, sec. 409(c)(3)(A).

to acceptance of an undetectable small risk in order to obtain the benefit of adequate food.

The recommendation for revision of the Delaney clause is made in order to permit determinations essential to the protection of human health, not to justify irresponsible increases in the exposure of the population to carcinogenic hazards.

Recommendation 9:

Establish a Department of Health, Education, and Welfare clearinghouse for pesticide information and develop pesticide protection teams.

The sources of information on pesticides are exertmely diverse and scattered, including Federal and State agencies, universities, private research centers, and industrial laboratories. The urgent problems of pesticide management require rapid access to scientific information. At the same time, a most serious information gap exists in the absence of reliable sources of data on local activities, progress, and problems throughout the Nation. Therefore, the establishment of a clearinghouse for pesticides is recommended and the organization of pesticide protection teams is strongly urged.

The clearinghouse should:

- Collect and organize information on pesticides and their relationships to human health and the quality of the environment in a modern system for storage, retrieval, and dissemination, and secure evaluations of such data.
- Provide bibliographies, reprints, and summaries upon request from the Secretary of the Department of Health, Education, and Welfare, appropriate Federal and State agencies, research centers and others with a valid need for knowledge.
- Receive continuously information from the pesticide protection teams and provide for its proper summary and distribution, with special attention to dangers or improvements related to methods of pest control.
- Maintain liaison with national and international bodies active in the field of pesticide safety.
- Receive, summarize, and distribute data from pesticide monitoring programs related to human health and welfare.

Pesticide protection teams should be developed from existing local personnel and coordinated with Federal and State personnel and facilities from agriculture, wildlife and public health. They would:

- Augment existing agricultural extension and fish and wildlife efforts relating to pesticides and thereby guide local usage and safeguards.

impact of changes in pesticide usage, and such information should be incorporated in this evaluation.

Cooperative Federal and State programs of research, training, and demonstration aimed at the solution of practical pest control problems should be expanded. The U.S. Department of Agriculture and the Department of the Interior should make greater use of cooperative agreements and grant support for these purposes. Such support would lead to:

a. Better evaluation of the benefits of pesticides used for various purposes in the context of alternative methods of pest control, including combinations of pest control methods;
b. Development of less hazardous pest control chemicals with high target specificity and minimal environmental persistence;
c. Comprehension of the nontarget effects of pesticides; and
d. Reduced damage to the environment.

Recommendation 11:

Provide incentives to industry to encourage the development of more specific pest control chemicals.

Incentives should be provided to industry to encourage the development of safer chemicals with high target specificity, minimal environmental persistence, and few, if any, side effects on nontarget species. Developmental costs will be disproportionately high in relation to profits from the lower volume of sales of more specific chemicals which will be used selectively. The high cost of development will discourage investments unless incentives are provided.

In order to encourage joint developmental efforts by Government and industry, consideration should be given to the applicability of the present patent laws and practices. The working life of a patent is in effect shortened by the extended period required to secure approval and registration. Moreover, the assignment of patents to public ownership rather than to licensees reduces the incentive for private enterprise to undertake the financial burden of approval and registration.

Recommendation 12:

Review and consider the adequacy of legislation and regulation designed to:

1. Improve the effectiveness of labeling and instructions to users.
    a. Advertising inconsistent with the label should be prohibited.
    b. An entirely new scheme of denoting relative toxicity should be devised. The average consumer does not understand the progression from caution, to warning, to poison. A need exists for a nonlanguage (graphic or numeri-

- Improve local surveillance of pesticide contamination, facilitate monitoring of human tissue residues of pesticides, and investigate usage patterns and episodes of human toxicity.
- Provide a rapid flow of local information based on the above activities, to and from the clearinghouse, especially concerning any emergency related to pesticides.
- Inform the public, users of pesticides, local government and enforcement agencies, and others in the proper and safe uses of pesticides, techniques for disposal, and other matters.
- Stimulate local awareness and constructive concern essential for optimal use of pesticides.

Recommendation 10:

Increase Federal support of research on all methods of pest control, the effects of pesticides on human health and on the ecosystems, and on improved techniques for prediction of human effects.

The scientific talent of the Nation should be mobilized more effectively to resolve the problems associated with the control of pests. This will require increased Federal support of intra- and extramural research and development by all Departments concerned with pesticides.

In order better to assess the toxic effects of pest control agents on nontarget organisms and on human health, research should be expanded relative to the metabolism and degradation of pesticides and their effects on the integrated systems by which organisms derive energy, build protoplasm, and reproduce. Additional studies on teratogenesis, mutagenesis, and carcinogenesis must be supported. Epidemiologic and pathologic relationships as may exist between pesticides and hematologic, metabolic, neurologic, cardiovascular, and pulmonary diseases, pregnancy losses, and cancer must be studied in appropriate communities and population groups.

The nature and extent of any interactions that may exist between pesticides and other factors in the environment require further elucidation. Improved scientific methods and protocols should be developed to assess dose-response and metabolic phenomena related to the biological effects of pest control chemicals in various species in order to increase the accuracy of extrapolative predictions concerning human effects.

The economic costs of pesticides should be evaluated. This should include the hidden costs to man resulting from the uses of pesticides. Accurate quantitative data on environmental contamination and damage to nontarget species by pesticides should be obtained in order to assess the impact of the total global burden of pesticide residues. The Department of Agriculture undertakes to determine the economic cal) representation of oral inhalation and dermal toxicity to enable consumers to select less hazardous materials.

c. Effective labeling practices and instructions to users require use of common (generic) names for all pesticides, and the conveying of clear directions for and information about proper use, dangers, and first aid. Printing should be readable and multilingual when that is appropriate.

d. When the chemical is known to be especially hazardous to some type of organism, as toxaphene is for fish, this should be stated.

e. Instructions should also offer clear directions for safe disposal of the empty container and of any unused material.

2. Extend the present concept of experimental permits as a mechanism to register pesticides initially on a restricted basis to enable close observation, documentation, and reassessment of direct and indirect effects under conditions of practical usage.
3. Improve packaging and transportation practices in order to minimize dangers of spillage and the contamination of vehicles and of other merchandise.
4. Provide for monitoring and control of effluents from plants manufacturing, formulating, and using pesticides.
5. Provide uniform indemnification to parties injured by mistakes in pesticide regulatory actions by Federal and State authorities.

Recommendation 13:

Develop, in consultation with the Council of State Governments, model regulations for the collection and disposal of unused pesticides, used containers, and other pesticide contaminated materials.

The current model pesticide law recommended by the Association of American Pesticide Control officials does not cover the important problems of disposal of surplus pesticides and of used pesticide containers. Regulations to control these important sources of contamination and of accidental poisonings properly belong in State or local codes.

An additional feature that should be included in a model law is registration, possibly by social security number, of all workers employed in manufacturing or applying especially hazardous pesticides. This would facilitate implementation of measures to protect them from hazardous exposures as well as to expedite epidemiological investigation of adverse effects of pesticides on human health.

Recommendation 14:

Increase participation in international cooperative efforts to promote safe and effective usage of pesticides.

DDT is widely distributed throughout the global environment. If present usage patterns continue, or if other persistent pesticides are used in large quantities, the contamination of the environment may increase with time. An international problem exists and it will require international cooperation to solve it.

Government and industry should increase their participation in international cooperative efforts to assist developing nations to promote safe and effective usage of pesticides for disease prevention and the production of essential foods and fibers. The risk-benefit considerations differ somewhat from country to country depending on the particular problems encountered. Both benefits and hazards of using a pesticide must be evaluated carefully in order to determine the appropriateness of use in a given area.

The U.S. Government should assume leadership in studying the inherent health hazards of pesticide usage and cooperate in the training of technical personnel from other countries.

SUMMARIES OF SUBCOMMITTEE REPORTS

USES AND BENEFITS OF PESTICIDES

Summary and Conclusions

The production and use of pesticides in the United States is expected to continue to grow at an annual rate of approximately 15 percent. Predictions are that insecticides will more than double in use by 1975 and herbicides will increase at an even more accelerated pace. The foreign use of pesticides will likewise continue to increase with the organochlorine and organophosphorus insecticides continuing to represent a significant part of the foreign market.

The use of DDT in domestic pest control programs is rapidly declining with the major need reported to be associated with cotton production in the Southeastern United States. Although the total production is declining, an increasing quantity is being purchased by AID and UNICEF for foreign malaria programs.

Most other persistent pesticides have continued to decline in use since 1957, a trend that will continue with the remaining uses being primarily nonagricultural. The shift to nonpersistent pesticides will continue at an accelerated rate, however, there will be a continued need for use of persistent materials for the control of selected pest problems.

Although imaginative and exciting research is in progress, non-insecticidal control techniques are not likely to have a significant impact on the use of insecticides in the foreseeable future. There is evidence of an increased appreciation for the use of integrated control in the management of pest populations with less persistent and more selective insecticides playing an important part.

There is a serious lack of information available on pesticide use patterns, particularly as they relate to nonagricultural uses. Likewise, available data are usually not obtainable for a proper evaluation of the economic implications of pesticide use. The United States activity in international pest control programs is complicated by the magnitude of involvement and the complexity of diplomatic and agency responsibilities. There are many factors that are influencing the changing use patterns of pesticides. In addition to new pest infestations, resistance to selected pesticides, alterations in the economics of crop production, and changing agricultural and social patterns, the impact of public opinion is having a growing influence on the use of pesticides. The increased concern for new legislation and regulation of the manufacture, sale, and use of pesticides must not be so structured as to destroy the incentive for development of new pesticides more compatible with other desirable environmental qualities.

CONTAMINATION

Summary and Conclusions

The subgroup on contamination has examined the present status of knowledge on the dissemination of pesticides into the environment, the mechanisms and rates at which they accumulate in various elements of the environment, and methods by which pesticides might be controlled so that their presence in the environment would pose a minimal hazard to society consistent with the benefits to be obtained from their use.

The subgroup has examined: a) The air route by which pesticides are applied and distributed in the biosphere; b) the water route; c) the food route; d) soil contamination; e) household uses of pesticides; f) occupational exposures resulting from the manufacture and application of pesticides, and accidents that may occur in their use; g) alternatives to the use of persistent pesticides; h) the monitoring of pesticides in the environment; i) systems analysis of pesticides in the environment.

Much contamination and damage results from the indiscriminate, uncontrolled, unmonitored and excessive use of pesticides, often in situations where properly supervised application of pesticides would confine them to target areas and organisms and at the concentrations necessary for their beneficial use without damage to the environment. Research investigations, demonstrations, and monitored operations reveal that the careful application of many of the pesticides and the use of techniques presently available and being developed can be expected to reduce contamination of the environment to a small fraction of the current level without reducing effective control of the target organisms.

The present piecemeal involvement of various Federal agencies in pesticide control requires more than the existing type of coordination.

As human health and welfare are the values of prime concern, the DHEW should provide a lead in the establishment of a mechanism for administering pesticide control programs.

Ad hoc studies of pesticides in the environment are not adequate to assess the inputs of pesticides to the biosphere, their degradation, translocation, movement and rates of accumulation. Monitoring is conducted by a large number of agencies, but in each instance the monitoring is related to a specific mission of the agency. Therefore, a single agency should take the initiative to insure the effective monitoring of the total environment, and the filling of gaps in data such as for oceans and ground water, as they are identified. A continuous systems analysis of pesticides in the environment needs to be conducted.

Aerial spraying should be confined to specific conditions of lapse and wind that will preclude drift. Regulations to limit aerial application to specific weather conditions would be helpful in providing guidance for regulatory programs. Increased engineering development effort is needed for the design of equipment for, and the adaptation of helicopters to the aerial spraying of pesticides.

The use of low volume concentrated sprays should be encouraged. Since this technique, if it is not properly controlled, can be more hazardous to workers, effective regulations must precede its increased use.

Increased information is needed on the degree of exposure of the general population to pesticides used for household, lawn, and gardening purposes. More effective means for regulation and control of pesticide use by the general public should be instituted, possibly by licensing of distribution outlets.

The use of lindane and similarly toxic materials which act by evaporation must be discontinued where humans or foods are subject to exposure, such as in homes, restaurants, and schools.

There is a vastly increased need for the education of the general public in the management of pesticides and in the training of professional applicators. Public communications media, schools and universities all have important roles to play.

Labeling regulations must also be improved. Print should be enlarged and language should be made intelligible for the lay public. A need exists for nonlanguage, internationally intelligible insignia or markings that will advise the user of the degree of toxicity and persistance of the product, its method of application, and the target organisms.

More vigorous effort is needed to replace the persistent, toxic, and broad-spectrum pesticides with chemicals that are less persistent and more specific. Certain of the less-persistent pesticides, however, may be more toxic to humans and therefore effective regulation of their application is required to insure against injury to personnel.

Integrated control techniques for the control of select pests promises to effect a reduced usage of pesticides. Such alternative techniques should be more widely applied.

Licensing of commercial pesticide applicators, as well as other large-scale applicators of hazardous materials should be required.

Analytical methods, although extremely good, require further development. Need exists for standardizing or referencing additional techniques, even on an international basis. There is need for both less sophisticated techniques for field use as well as for automated techniques for wide-scale monitoring.

Standards for selected pesticides should be included in the Public Health Service "Drinking Water Standards". Although guidelines and criteria for some pesticides have been delineated, they have never been officially established.

Prior to application of pesticides to waters for the control of weeds, snails, mosquitoes, and in other aquatic uses, a careful analysis should be made of the proposed pesticide characteristics with respect to the uses of the target area. Special concern is indicated where domestic water supply is involved, or where food-chain concentration may occur.

Steps should be taken to prevent the simultaneous shipment of pesticides and foodstuffs within the same vehicle. Comprehensive regulations for pesticide transportation are required.

Safe methods of disposal of pesticides, their wastes, and containers are needed to prevent the contaimination of the environment and to protect individuals from contamination and accidents.

Intensified research and development is needed in the following areas, among others:

a. Prediction of the micrometeorological conditions suitable for aerial spraying.

b. Application of systems analysis to the pesticide-enviroment problem.

c. Pesticide chemodynamics, with emphasis on reservoirs of storage.

d. More intensive development of less-persistent pesticides with narrow spectra of toxicity.

e. Continuing development of spray devices with narrow spectra of droplet sizes.

f. Continuing development of alternatives to chemical control of pests.

g. Creation of more suitable materials for pesticide packaging and containers to facilitate safe transfer, handling use, and disposal.

h. Treatment processes for the elimination of pesticides from domestic water supplies as well as from wastewaters.
i. Immediate studies of the effects of pesticide residues on algal photosynthetic activity.

EFFECTS ON NONTARGET ORGANISMS OTHER THAN MAN

Summary and conclusions

Man is an integral part of the living system, which includes about 200,000 species in the United States. Most of these are considered to be essential to the well-being of man. Pesticides are now affecting individuals, populations, and communities of natural organisms. Some, especially the persistent insecticidal chemicals such as DDT, have reduced the reproduction and survival of nontarget species.

Pesticides are dispersed via air, water, and the movements of organisms. The most significant concentrations are found in and near the areas of intensive use, but traces have been found in the Antarctic and other areas far from application. Pesticides have reduced the populations of several wild species. Both extensive field data and the results of excellent controlled experiments demonstrate that certain birds, fishes, and insects are especially vulnerable. There are suggestions that pesticides in the environment may adversely affect processes as fundamental to the biosphere as photosynthesis in the oceans.

However, the scarcity of information concerning the influences of pesticides on natural populations prevents adequate assessment of their total effects. Less than 1 percent of the species in the United States have been studied in this connection, and very few of these have been subjected to adequate observation. Present methods and programs for determining the influences of pesticides on nontarget organisms are inadequate. Little data exists on the distribution, location, and impact of various pest control chemicals in the natural living systems of the world.

The general nature of the effects of pesticides on nontarget species populations and communities can now be suggested. Although there is usually greater similarity of reaction between closely related species, each species reacts differently to specific pesticides. DDT, for example, causes egg shell thinning in ducks and falcons, but not in pheasants and quail. Pesticides from the air, water, and soil may be concentrated in the bodies of organisms. The concentrating effect is frequently enhanced as one species feeds on another and passes the pesticide from one link to another in the food chain. Hence, predators like some birds and fish may be exposed to levels several thousand times the concentration in the physical environment. Some nontarget organisms can, under highly selective pressure from pesticides, evolve resistance to them. The surviving resistant individuals may pass extremely high concentrations to their predators. In communities exposed to pesticides, the total number of species is usually reduced and the stability of populations within the community is upset. Often, beneficial species are unintentionally eliminated. Such a reduction in the number of species is frequently followed by outbreaks or population explosions in some of the surviving species, usually those in the lower parts of the food chain. When a vital link low in the food chain is eliminated, many predators and parasites higher in the food chain are often also destroyed.

The Committee has reached the following conclusions:

1. Adequate methods should be developed and utilized for evaluation of the hidden costs of the uses of pesticides.

Such evaluation is essential as part of the development of useful estimates of all of the benefits and costs to society. Some partial estimates of the direct benefits are available and useful. Adequate data are not available on such indirect costs as losses of useful fish and wildlife, damage to other species, and any esthetic effects. These are required to guide rational decisions on the proper uses and control of pesticides so that the net gains will be as great as possible while the net losses are minimal.

2. Persistent chlorinated hydrocarbons which have a broad spectrum of biological effects, including DDT, DDD, aldrin, chlordane, dieldrin, endrin, heptachlor, and toxaphene, should be progressively removed from general use over the next 2 years.

These pesticides are causing serious damage to certain birds, fish, and other nontarget species among world populations. Some of these species are useful to man for food or recreation, some are essential to the biological systems of which he is a part, and some merit special protection because they are already endangered.

These pesticides have value in specific circumstances, and we suggest that they be used only under license and with special permits. The system for assuring this careful use should be established as the unrestricted use of these materials is phased out over the 2-year period.

3. The release of biocidal materials into the environment should be drastically reduced.

In addition to restriction of the use of hazardous pesticides, many techniques can be applied which will minimize the release of pest control chemicals. In industry, improved chemical and engineering processes could reduce the quantity of contaminated wash water; more effective methods can be developed for disposal of unused stocks and residues of pesticides; and improved surveillance of effluents would be desirable. For home use, improved materials and methods of application can be created and employed with greater discretion on the part of the individuals involved. For large-scale applications, conversion to integrated methods of pest control, care in the selection and application of specific chemicals, and preference for short-lived pesticides would reduce release to the environment.

These efforts, combined with increased research and education, would slowly but effectively reduce the damage to nontarget species.

4. The U. S. Department of Health, Education, and Welfare or another Federal agency should negotiate a contract with a suitable national professional organization to develop a system, complete with standards of training, testing, and enforcement, for the effective restriction of use of selected pesticides known to be especially hazardous to man or to elements of the environment.

To achieve an adequate and prompt further reduction in the use of certain pesticides and still permit their use where no adequate substitute is acceptable, there must be a system of regulation based upon State or local authority but using uniform national standards. This system should provide for use of the selected pesticides only by or under the immediate supervision of a licensed operator meeting certain standards of training, competence, and ethics.

5. Educational efforts relating to the proper and improper usages of pesticides should be improved and expanded.

The most important element in the wise use of pesticides is the individual person who selects the chemical to be used and decides upon the methods of application. Suggestions have been provided elsewhere for the proper training of all large-scale applicators. It is equally important that homeowners, gardeners, students, legislators, civic officials, and others receive adequate and correct information and develop proper attitudes. Such education could contribute greatly to wise use of pesticides, and also to rational response to governmental efforts to protect public health and welfare while gaining as much advantage as possible from pest control methods.

6. All pertinent Federal and State agencies should review and improve policies and practices of pesticide use.

The beneficial uses of pesticides have been accompanied by a wide variety of policies and practices which have sometimes been wasteful, unnecessarily destructive, or ineffective. We offer the following suggestions to be included among the guidelines for wise use of pesticides:

a. Pesticides should be applied only when there is evidence that pest densities will reach a significant damage threshold.

b. Effective pest control does not usually require eradication of the pest species, and should be directed toward optimal management of pest densities.

c. Support for research and demonstrations should be provided to projects based on the systems approach to pest management and control.

d. Diversity of species is biologically desirable since it contributes to the stability and efficiency of life systems.

e. No species should be eradicated except as a carefully selected pest and when compensating human gains are ecologically sound and clearly established.

f. Special care must be taken to prevent any damage to the species and mechanisms which are of fundamental importance to biological systems. For example, oceanic phytoplankton produces most of the oxygen necessary for the earth's biological system.

g. Requirements for food quality should not be so high as to require excessive use of pesticides. Customer preference, and regulatory requirements, for unblemished fruit and vegetables and the complete absence of insect parts have encouraged heavy use of pesticides.

h. New pesticides should be given interim approval which permits contained use in limited but typical circumstances prior to general approval. The pattern of careful progressive risks would encourage new developments without endangering the public interest.

i. Effective incentives should be established to encourage the development of improved pest control techniques. The cost of entering a new product or testing a different control technique is high. Since effects on the national welfare are involved, proper governmental encouragement of private industrial efforts may be appropriate.

7. Registration requirements should be strengthened and redesigned to permit initial provisional approval, then general use approval, and to require periodic review and re-registration of materials.

Registration of pesticides offers the most important opportunity for estimating potential benefits and costs in advance of wide usage. In addition to present registration application information, useful estimates should be provided of the persistence of the pesticide, on the breadth of its biological impact, and on its fate. These will disclose the nature and possible magnitude of the nontarget effects. If approval is appropriate, we suggest that it be for a short-term period and for use under defined circumstances where risks are confined, and that general use be considered after such field experience. Since some of the significant effects in nontarget species are subtle, sublethal, and difficult to detect, we recommend that all pesticides be subject to periodic review and approval.

8. All commercial applications and other large-scale applications of pesticides should be performed under the supervision of competent trained persons.

The complex responsibilities of pesticide application involve both achievement of the greatest possible benefit and maximum prevention of damage. These require considerable knowledge of the management of crops, the biology of desirable and undesirable species, the effects of weather, and the effects of biocide in the ecosystems. They also require application of professional judgment and use of professional standards of conduct and responsibility. We suggest that all such applicators should be properly trained, required to demonstrate their competence, and awarded evidence of their ability. Incentives in the forms of salary and recognition will be needed to encourage such professional training.

Training programs for pest management specialists of all types, including applicators, should include the concepts of systems approaches to pest control and emphasize the relationships between pest management activities and the total biological community affected.

Since new information is emerging rapidly in pest management, refresher courses for county agricultural agents, applicators and others involved in the uses of pesticides and other control techniques would be of special value.

9. The production of additional information and comprehension should be encouraged and supported on many aspects of pesticide use and effects.

Experience with pesticides has revealed many serious gaps in available knowledge. Research is urgently needed on many general and specific problems. The following problems are all related to nontarget effects of pesticides, and many of them are also pertinent to other areas of pesticide use, to successful management of animals and plants, and to fundamental science.

a. What are the acute effects of the common pesticides when used on the many species of wildlife and other organisms which may be exposed to them?

b. What are the effects of indirect and chronic exposure?

c. What is the nature and magnitude of the effects of insecticides on beneficial insects and other species?

d. What are the normal patterns and variations in natural biotic communities, as baselines for understanding future pesticide pollution effects?

e. What mechanisms exert natural control on various pest populations?

f. How can we best estimate pest populations and predict their trends?

g. What are the full potentials and realistic limitations of the pest control methods which are suggested as alternatives to chemical pesticides, including predators, parasites, pathogens, cultural control, sterilization, attractants, repellants, genetic manipulation, and integrated approaches?

h. What improvements are possible for pesticide packaging and disposal (including degradable containers) to minimize threats to nontarget species?

10. A vigorous specific program should be created to bring the 100 most serious insect pest species of the United States under optimal control.

These require about 80 percent of the insecticides now in use. Dramatic focusing of attention on the "100 worst" could lead to rapid improvement in the species-specific insecticides, biological control methods, or integrated control programs.

11. The responsibilities of the several Federal agencies involved in pesticide regulation and control must be more clearly defined and certain specific activities should be improved or initiated by appropriate agencies.

Procedures and patterns for the regulation and control of pesticide use have emerged during the last 30 years in response to changes in law, evolving practices in agriculture, production of new chemical materials, changing public concern with health effects and nontarget damage, emerging scientific comprehension of benefits and costs, and other unstructured events. Both benefits and costs are now so large as to merit the national allocation of responsibilities. We suggest careful review and reassignment, by law if necessary, of the proper role of—

a. The Department of Interior, charged with protection and enhancement of nonagricultural resources and with water quality control.

b. The Department of Agriculture, charged with assisting in the maximum production of food, fibers, and other culturable crops in ways which are not detrimental to other interests.

c. The Department of Health, Education, and Welfare, charged with protection and improvement of human health and welfare.

d. The National Science Foundation, responsible for improved comprehension of fundamental processes and assisting in their application for human benefit.

e. The Environmental Quality Council, Federal Committee on Pest Control, and other coordinating agencies.

Other agencies are, of course, involved as users of pesticides and in other functions. Those listed above, however, appear to comprise the areas of primary attention. In addition to present programs and activities related to pesticides, we suggest the following services for new or additional emphasis:

a. A taxonomic and identification service should be established to provide increased knowledge and reference standards for biological investigations related to all fields of pest control.

b. Broader monitoring should be undertaken of the types and quantities of pesticide transmitted by various means and reaching nontarget species. Bioaccumulators like oysters and other molluscs can be unusually useful as indicators, and the levels of concentrations in predatory species are of special importance.

c. Early indications of undesirable effects must be detected effectively and followed by appropriate action. When the early warning system suggests a potential pollution hazard in the environment, the acquisition of additional pertinent information by the scientific community should be supported.

d. Multidisciplinary investigations of alternative control techniques should be carried out whenever present control methods are shown to contain potential hazards.

e. A single agency should assume the responsibility for assimilating information on the effects of pesticides on nontar-

371-074 O—69——4

get species and transmitting it to appropriate regulatory and educational centers.

f. Measurable predictors of potential hazards from pesticide use should be agreed upon and might be made the basis of a handicap tax to be applied to each pesticide in proportion to its pollution hazard.

EFFECTS OF PESTICIDES ON MAN

Summary and Conclusions

The scope of this report is intended to encompass the present state of knowledge concerning the nature, extent and consequences of human exposure to pesticides. Data relating to exposure of experimental animals have been reviewed only insofar as they contribute to our understanding of phenomena encountered in man or provide knowledge in areas where human data are meager or totally lacking.

No human activity is entirely without risk and this maxim holds for pesticide usage in the human environment just as it does for all other exposure to chemicals. There are formidable inherent difficulties in fully evaluating the risks to human health consequent upon the use of pesticides. In part, these difficulties stem from the complex nature of the problems involved, the fact that many facets of these problems have been recognized only recently, and the general backwardness in this area of research in *man*, as distinct from work in laboratory animals. Above all, one must not lose sight of the large number of human variables—such as age, sex, race, socio-economic status, diet, state of health—all of which can conceivably, or actually do, profoundly affect human response to pesticides. As yet, little is known about the effects of these variables in practice. Finally, one must realize that the components of the total environment of man interact in various subtle ways, so that the long-term effects of low-level exposure to one pesticide are greatly influenced by universal concomitant exposure to other pesticides as well as to chemicals such as those in air, water, food and drugs. While all scientists engaged in this field desire simple clear-cut answers to the questions posed by human exposure to pesticides, the complexity of the human environmental situation seldom allows such answers to be obtained. Attempts to extrapolate from the results of animal experiments to man are also beset with pitfalls. Hence, the greatest care needs to be exercised in drawing conclusions regarding cause-and-effect relationships in human pesticide exposure.

The available evidence concerning such human exposure to pesticides derives from three main sources: planned and controlled administration of pesticides to human subjects; case reports of episodes of accidental or other acute poisoning; and epidemiological studies, which in turn comprise surveys of occupationally-exposed groups (in accordance with a variety of retrospective and prospective approaches), and studies of the general population.

Indices of exposure of human beings to pesticides constitute a vital link in the chain of evidence that must be forged in order to reveal, interpret, and maintain effective surveillance of, pesticide exposures. Hitherto, the view that exposure of the general population was predominantly associated with the presence of pesticide residues in food has been reflected in the efficient monitoring of total diet samples and individual foods, but only sporadic attention to other sources of exposure. It is now evident that much can be learned by monitoring the end-product of human exposure in the form of pesticide levels in body fluids and tissues of people. The information thus obtained is quite distinct from, and at least as valuable as, the data on residues in food; the two types of data complement each other admirably. Provision of information on human levels, in adequately detailed coverage of various groups within the general population is seen as the single most immediate step towards a better understanding and surveillance of total exposure from all sources of pesticides.

Sophistication achieved through the use of modern techniques has made possible the study of absorption, disposition, metabolism and excretion of some pesticides in man. Experience derived from animal studies has provided guidance in directing the appropriate procedures to the investigation of the behavior of pesticides in the human body. To date, the most significant information of this sort relates mainly to two organochlorine pesticide groups, namely DDT and allied compounds as well as the aldrin-dieldrin group. Knowledge of the dynamic aspects of the behavior of these two pesticide groups in the human body is far from complete, but already some important facts have been established. In general, for any given level of pesticide intake, an equilibrium level of pesticide is attained in blood and body fat, despite continuing exposure. The precise concentration at which the plateau is established is directly related to the level of exposure but also to other determining factors. In the case of aldrin-dieldrin, the blood level appears to be a reliable measure of exposure. It appears further, that DDT in blood is directly related to recent exposure, while in contrast DDE in blood is a reflection of long term exposure.

A detailed survey of case reports of incidents involving accidental poisoning by organochlorine pesticides reveals that their general action is to increase the excitability of the nervous system. Some of these compounds also damage the liver. Their capacity to penetrate intact human skin varies from one compound to another; in the case of endrin, for example, percutaneous penetration plays an important part in clinical intoxication. Within the organochlorine group of compounds there is a wide range of potential for acute toxicity: DDT is relatively safe in terms of acute intoxication, while dieldrin and endrin have produced many cases of serious poisoning. Lindane presents a special problem, inasmuch as it has been implicated, largely on the basis of circumstantial evidence, in the causation of hematological disorders. A characteristic of organochlorine poisoning is the difficulty of establishing the correct diagnosis. This is especially true in cases of mild poisoning that result in nonspecific symptoms and signs, since except in the case of dieldrin there are no established criteria for diagnosis on the basis of blood levels. Specific therapeutic measures do not exist.

Inhibition of cholinesterase enzymes by the organophosphate pesticides appears to be the only important manifestation of acute or chronic toxicity produced by this class of compounds. Great variation in acute toxicity from one compound to another characterizes this group, which includes some of the most toxic materials used by man. Cholinesterase inhibition results in a well-defined clinical pattern of intoxication which can be readily diagnosed. Specific therapeutic measures are available and, provided they are pressed with sufficient speed and vigor, are highly effective. Skin penetration by organophosphates may be substantial. In view of the toxic potential of these compounds, protection of workers exposed to them assumes utmost importance. Protective measures should include education, training, proper equipment design, suitable personal protection devices, careful medical surveillance and well-organized facilities ready to treat cases of poisoning with a minimum of delay.

Carbamate pesticides are also cholinesterase inhibitors but, because of rapid in vitro reactivation of the enzyme, measurement of cholinesterase activity is not a reliable guide to exposure. As with organophosphates, the toxic potential of some members of the carbamate group is very great.

Controlled exposure of human volunteers to pesticides under close medical supervision constitutes the most reliable approach to the unequivocal evaluation of long-term effects of low levels of pesticide exposure. The difficulties involved in maintaining such studies have inevitably resulted in very small groups of subjects being exposed for any appreciable length of time. The longest studies on record have lasted less than four years and the results can only reflect the period of study. Consequently, the findings, especially when they are negative, are open to question when taken by themselves. It appears, however, that present levels of exposure to DDT among the general population have not produced any observable adverse effect in controlled studies on volunteers. The same is true of aldrin-dieldrin. These findings acquire greater force when combined with observations on other groups, such as occupationally-exposed persons.

With organophosphate pesticides, the problem of human residues does not arise because these compounds are not stored in body fat. Here the risk is one of acute poisoning. Much accidental poisoning is attributable to public ignorance of the toxicity of these chemicals and neglect of appropriate precautions in their use and storage. In developing countries serious accidents result from storage of pesticides in unlabeled bottles and of food in used pesticide containers. Epidemics of acute poisoning follow spillage of concentrated organophosphates into bulk food or water sources. The hazard to human life is shared by fish and wildlife. Regional pesticide protection teams are suggested as a means of investigating, recording and ultimately preventing accidents of this sort.

Industry has made much progress towards safe handling of pesticides. Nonetheless, a very real occupational hazard exists, and extension of preventive measures should include regular blood testing for evidence of organophosphate exposure. A limit for DDT and other organochlorine pesticides in blood should be established to prevent overexposure.

Pesticide exposure experienced by the population at large is in part the legacy of earlier excessive or injudicious use of persistent pesticides. Residues of these compounds have been, and are still being acquired from all articles of diet and a variety of other environmental sources. This is the major source of public concern. Although a number of persistent pesticides can be identified, attention is centered on DDT, and closely-related compounds, the most ubiquitous and predominant of all pesticide residues in man. The consequences of these prolonged exposures on human health cannot be fully elucidated at present. Evidence from workers who are subject to vastly greater exposure than the public is reassuring but far from complete. Animal experiments clarify certain issues but the results cannot be extrapolated directly to man. On the basis of present knowledge, the only unequivocal consequece of long-term exposure to persistent pesticides, at the levels encountered by the general population, is the acquisition of residues in tissues and body fluids. No reliable study has revealed a causal association between the presence of these residues and human disease.

Despite such reassurance, realization of the paucity of our knowledge in this area flows from increasingly sophisticated studies on human residues of DDT and related compounds. There appears to be marked geographical stratification of DDT residues in our population, the average levels in the cooler isotherms being one-half of those in the warmer climates. None of these observations apply to residues of dieldrin. Such findings cast serious doubt on accepted beliefs that food is the predominant source of DDT residues and that the entire general population has reached equilibrium as regards acquisition of such residues.

Reopening these questions emphasizes the inadequacy of present monitoring of exposure by relying mainly on analysis of food. This aspect was stressed above. It also renders more urgent the need to contain and eventually greatly reduce the extent of human and animal contamination by pesticide residues. Existing knowledge confirms the feasibility of inducing active withdrawal of pesticide residues from the human body but further research to achieve a practical means of attaining this goal is needed.

A survey of the reported effects of pesticides on laboratory animals has furnished information on factors and experimental conditions that could not easily be reproduced in human studies. For example, the influence of diet on pesticide toxicity, and particularly lack of dietary protein, has revealed substantial increases in acute toxicity of some pesticides. In this, as in some other sections of our report reference is made to the capacity of organochlorine pesticides to bring about a great increase in the activity of liver enzymes responsible for the metabolism of foreign compounds. This phenomenon of enzyme induction has been extensively studied in animals and is discussed in detail in the report of the Panel on Interactions. Comparable enzyme induction in the human liver is brought about by many drugs and also by DDT. It is a sad comment on the dearth of knowledge of human physiology to point out that the threshold dose of DDT for induction of metabolizing enzymes in human liver is unknown.

Special sections of the report deal with the possible effects of pesticides in bringing about heritable alterations in the genetic material (mutagenesis), effects on reproduction, including malformations in the fetus or newborn infant (teratogenesis) and increasing the incidence of various forms of cancer (carcinogenesis). The data available relate only to experimental animals or to lower forms of life. At the present time we do not know whether or not such results are applicable to man. While there is no evidence to indicate that pesticides presently in use actually cause carcinogenic or teratogenic effects in man, nevertheless, the fact that some pesticides cause these effects in experimental mammals indicates cause for concern and careful evaluation. It is prudent to minimize human exposure to substances producing these adverse effects in mammals while additional investigations are undertaken to assess the potential of such suspect pesticides for causing adverse effects in man. There is a need to develop standard protocols for safety evaluation that are sufficiently flexible to permit an individual approach to the particular and often unique problems presented by each pesticide. Assurance of safety to man demands special techniques, not only for extrapolation of animal data to man, but also for evaluation of controlled human expsure. Much effort will be required to attain these objectives. Research in these areas should be expanded and imbued with a greater sense of urgency than that manifested before.

The Panel on Interactions has provided a valuable analysis of the manner in which pesticides can interact with one another, and with drugs and other environmental agents, in exercising effects on man and animals. Once again one is struck by the complexity and importance of these interrelationships and by the extent of our ignorance of effects on man.

To sum up, the field of pesticide toxicology exemplifies the absurdity of a situation in which 200 million Americans are undergoing lifelong exposure, yet our knowledge of what is happening to them is at best fragmentary and for the most part indirect and inferential. While there is little ground for forebodings of disaster, there is even less for complacency. The proper study of mankind is man. It is to this study that we should address ourselves without delay.

VII

Pollution Associated with the Generation of Electricity

The Economy, Energy, and The Environment. A Background Study Prepared for the Use of the Joint Economic Committee, Congress of the United States, by the Environmental Policy Division, Legislative Reference Service, Library of Congress, September 1, 1970. Pages 92-117.

Much pollution is intimately connected with or caused by energy production and use. Strip mining of coal pollutes the land. The drilling of oil can produce a catastrophe like that of Santa Barbara in 1969 or that off the coast of Louisiana in 1970. The transportation of oil in ocean tankers can result in disastrous oil spills like that of the Torrey Canyon off the coast of southwest England in 1967, or the Arrow incident off the Nova Scotia coast in 1970. The internal combustion machine used in the automobile is a significant contributor to air pollution, and the generation of stack gasses as a by-product of electric production is a contributor. The list could be extended almost indefinitely. Nevertheless, a modern society depends on power production as an essential precondition for the good life. These very words are being typed on an electric typewriter! How can the pressing need for adequate power be balanced against the equally pressing need for a cleaner, more habitable environment?

In striking a proper balance between our need for a "power-full" environment, and our need for a clean environment the following must be considered:

- Our projected need for energy sources of all types
- Our projected supply of energy sources of all types
- The relative pollution characteristic of each source of energy
- The cost of alternative sources of energy.

These questions and many others are explored in the Background Study by the Environment Policy Division of the Legislative Reference Service (now the Congressional Research Service) of the Library of Congress. The study consists of 131 pages, not including the introductory matter, and comprehensively reviews the existing literature as of the date of its publication in late 1970. The excerpt reproduced below emphasises the polluting side effects of electrical energy production. The balance of the study, not reproduced here, emphasizes the possibility of shortages of key types of fuel in the years ahead.

Environmental Effects of Generating Electricity and Their Economic Implications

The generation of electric power inevitably must affect the environment. Wastes are generated and they must go to some place. Whether the effects of the wastes are beneficial, tolerable, undesirable or dangerous are value judgments determined by society.

In achieving a balance of interest between the users of electricity on the one hand and the users of the environment on the other, there is reason to avoid the two extreme positions that hold:

> The environment should not be available to wastes from generation electricity, or
>
> Powerplants may be constructed and operated without regard to their environmental effects.

In this section, the principal wastes of the electricity industry are identified, their effects briefly described, current regulation of such wastes summarized, and present technology for control is described. Cost information is also mentioned.

This section draws heavily upon the report of the Energy Study Group of the Office of Science and Technology on the siting of steam-electric powerplants and upon recent hearings of the Joint Committee on Atomic Energy and the Senate Committee on Government Operations.

CONVERSION OF HEAT ENERGY INTO ELECTRICITY

In a steam-electric powerplant, the heat energy released by burning fuel or fissioning nuclear materials heats water which turns to steam. The steam expands through a turbine, then flows to a condenser where it is condensed back into water which is returned to the boiler or the reactor to start the cycle again. The turbine turns a generator which produces the electricity. Ideally for every 3,413 British thermal units of heat energy released from the fuel, 1 kilowatt hour of electricity should be sent out from the plant.[1] However, present energy conversion technology is far from ideal. Depending upon its age, a steamplant may require from 19,000 to somewhat less than 9,000 B.t.u.'s of heat energy to generate 1 kilowatt-hour of electricity. The best heat rates [2] for 1968 ranged from 8,654 to 8,876 B.t.u.'s kilowatt-hour for the most efficient units in the country.[3] New thermal efficiency of powerplants, or efficiency, as we will call it, is the quotient of the plant electrical output, expressed in B.t.u., divided by the heat input in B.t.u. For technical reasons the best efficiency attainable with present steam powerplant technology is 40 percent. At this efficiency, 8,533 B.t.u.'s heat energy must be supplied for each kilowatt hour of electricity sent out. Present nuclear powerplants are less efficient. Those being built today are unlikely to have an efficiency better than 33 percent, meaning that 10,342 B.t.u.'s must be supplied for every

[1] 1 B.t.u. of heat energy will raise the temperature of 1 pound of water 1° Fahrenheit.
[2] The heat rate is the number of B.t.u. needed per kilowatt hour.
[3] Electrical World, Nov. 10, 1969, p. 26.

kilowatt-hour. The heat rates for the Vermont Yankee nuclear powerplant in New England, for example, is 10,560 B.t.u.'s per kilowatt-hour.

The amount of heat required by a steam-electric plant to generate a kilowatt-hour of electricity depends very much upon the temperature, pressure, and moisture content of the steam, which in turn depends upon the ability of materials to retain their strength in fireboxes and boilers when exposed to very high temperatures and to corrosive hot combustion gases. The higher the temperature and pressure of the steam and the less its moisture, the more heat energy is carried to the turbine by each pound of steam and the greater is the plant efficiency. The relative thermal inefficiency of nuclear powerplants derives from the lower temperature, pressure, and higher moisture of their steam. The technical reasons for these conditions are unlikely to be resolved with the type of nuclear powerplants now being sold to the utilities. More efficient nuclear plants are expected, but they are unlikely to come into operation until the mid or later 1980's.

The prospects of marked further improvements in powerplant thermal efficiency will have to await the outcome of current technical efforts to develop new energy conversion processes such as magnetohydrodynamics, electrogasdynamics and thermionic processes.

WASTES FROM STEAM-ELECTRIC GENERATING PLANTS

Electric powerplants that use the fossil fuels—coal, oil or gas, or nuclear fuels, all produce excess heat energy. Presently this heat is regarded as a waste to be disposed of to the environment. Fossil-fuel plants also produce solid and gaseous wastes that in certain quantities and concentrations are regarded as air pollutants. Nuclear plants do not omit wastes from combustion, but do produce some radioactive wastes that are routinely discharged into the air or water and may be regarded as pollutants if the releases exceed regulatory limits. The validity of limits set for emission of waste heat, some combustion wastes and radioactive wastes is being questioned by some scientists at this time.

WASTE HEAT

The heat energy released by burning fuel or fissioning atoms that does not leave a generating station as electricity must be discharged to the environment. It cannot be stored or kept within the powerplant. The air and water, in essence, are used as a sink for the waste heat. For a steam-electric powerplant operating at an efficiency of 40 percent, for each 100 B.t.u. released from the fuel, 60 B.t.u. must be thrown away; with an efficiency of 33 percent, for every 100 B.t.u. released, 67 must be disposed of. Thus from 60 to 67 percent of all the fuel consumed in a central powerplant ultimately serves only to heat up the air and water in the vicinity of a powerplant. It brings no income to the utility and may require capital investment and operating costs to disperse it in ways acceptable to the Government.

Removal of waste heat from a powerplant

For all steam-electric powerplants now built or contemplated, most or all of the excess heat which may either flow back into a river or other parent body of water or circulate through equipment that transfers the heat to the air is carried from the plant by water. For a conventional fossil-fueled plant, about 85 percent of the waste heat goes out in the cooling water with 15 percent going up the stack as hot flue gas. For a nuclear plant, virtually all of the waste heat, except for about 5 percent emitted to the air from hot surfaces, is in the cooling water. A modern fossil-fueled powerplant that requires 9,000 B.t.u. per kilowatt hour of electricity sent out would discharge 4,237 B.t.u. of waste heat for each kilowatt hour. A less efficient nuclear powerplant with a heat rate of 10,342 B.t.u. per kilowatt hour would discharge 6,400 B.t.u. of waste heat, almost half again as great, than for the fossil-fuel. This is the basis for the statement that a nuclear powerplant will discharge 50 percent more heat into the water than a conventional plant.

Many steam-electric plants use the once-through cooling system to dissipate the waste heat. In such a system the cooling water is taken from a river, lake, reservoir, or the sea, and passed through the powerplant whence it returns with an increased heat load and higher temperature. Once-through cooling is preferred at sites where there is an adequate supply of water and its use for cooling does not violate Federal or State water quality standards. This system has the advantage of low cost, minimum consumption of water and minimum intrusion upon the environment. If water is scarce or if compliance with water quality standards so requires, the waste heat from the cooling water can be transferred to the air by one or more processes.

The amount of cooling water required depends upon the heat rate of the plant and the permissible rise in water temperature. For an average fossil fueled plant with a heat rate of about 10,000 Btu per kilowatt-hour, and a temperature rise of 15 degrees Farenheit, the required flow is approximately 1.5 cubic feet per second for each megawatt of electrical capacity. At full load, this is equivalent to about 40 gallons per kilowatt-hour. A modern, more efficient plant would require a flow of about 1.5 cubic feet per second per megawatt for a 15-degree rise, equivalent at full load to about 30 gallons per kilowatt-hour. For a nuclear plant, the flow would be about 2.0 cubic feet per second per megawatt for a 15-degree rise. At full load this would come to about 55 gallons per kilowatt-hour.

In designing a power plant, the engineer seeks the most favorable economic balance between temperature rise in the cooling water, flow of the water, and size and cost of the equipment. With present technology he can choose a temperature rise between 12 to 27 degrees Farenheit. Temperature rises of less than 12 or greater than 27 degrees are considered impracticable from an engineering standpoint. If, for technical or regulatory reasons, it is desirable to keep the temperature of the cooling water below certain limits as it returns to its parent supply, unheated water can be mixed in to dilute the heat and lower the temperature.

HEAT AND WATER QUALITY

Discharging waste heat from steam electric plants into the waterways does not directly affect the public health. There is no danger of injury to persons. On the other hand, the waste heat can markedly change the quality of the water for further use and can drastically affect the marine life in the water.

What the specific effects from waste heat are remains a controversial matter. Some observers see only undesirable effects upon the quality of the water and the plant and animal life it sustains. Others see beneficial effects from waste heat. Depending upon the part of the country and the kinds of water life involved, effects of waste heat can range from fish-kills on one hand to speeding lobster growth on the other.

What follows is intended to briefly summarize the less than desirable effects of too much waste heat in a given body of water. It draws heavily upon the report of the Energy Policy Staff of the Office of Science and Technology.

Effects upon water life

The most pronounced effects of waste heat in the waters appear to be upon water life.

As a rule of thumb, the biochemical processes of aquatic life, including the critical rate of oxygen utilization, double for each 18 degree Fahrenheit rise in temperatures up to 86° to 95°. However, as water temperatures rise, the water can hold less oxygen in solution. Thus the potential supply of oxygen in the water diminishes with higher temperatures as the need for oxygen increases.

Up to a certain point, an increase in water temperature can cause more rapid development of eggs, faster growth of spat, fingerlings or juvenile fish and larger fish. Beyond that point, the hatch will be reduced and mortalities in the development stages will be higher. The temperatures at which maximum development occurs at each stage of the life cycle varies with the species. Over a period of several generations the composition of species in water bodies affected by waste heat can be expected to change if the temperature is changed, even though the change be small.

Even where a temperature change is not directly damaging to the development of desirable species, an increase is usually found to stimulate the more rapid development of less desirable or undesirable species.

While fish are generally available in the discharge areas for waste heat, sometimes in greater numbers than elsewhere, it is often found that an increase in temperature results in a loss of the more desirable species since they are unable to compete successfully for food, breeding areas or their lives. A warmer temperature is also considered to increase the occurence of disease in fish populations.

A particular problem exists for migratory fish since changes in water temperature are apparently important to some species as the stimulator of migratory activity. Changes in the normal times for migration triggered by heated water may put the fish at an environmental disadvantage later in their migratory cycle and adversely affect reproduction. Since the ability of each species to acclimate to changes in water temperature is different, each situation should be considered individually by fishery biologists.

On the other hand, techniques for forecasting ecological effects of heated waters are not as well advanced as the ability to forecast the patterns of heat dissipation in the receiving waters. We apparently know more about how and where the heated effluents from a powerplant will flow than we do about their specific effects in a particular situation.

Physical effects

Any increase in temperature of water because of waste heat will result in increased evaporation and a consequent reduction of available supply and an increase in the concentration of the minerals already present in the water, which do not leave with the evaporating vapors. While not ordinarily a problem, if a stream of water flows through a number of cooling cycles each with a loss from evaporation, a measurable increase in solids may result.

In northern climates, the discharge of heated water will tend to reduce ice cover, at least locally, and thus improve water quality by keeping the surface open to absorb oxygen from the air. The added heat may also result in local fogging on the water and adjacent land areas.

An increase in temperature may also make the waters more desirable for swimming and water sports if the normal temperatures are so cold as to limit use. If the water is already warm, however, further increase in temperature can reduce its recreational value.

The addition of waste heat to bodies of water may also reduce the value of the water for industrial cooling in those places where the local temperature has been increased substantially.

REGULATION OF WASTE HEAT IN WATER

Although water quality standards had previously been adopted by some States and interstate bodies, a major impetus to setting such standards was the Water Quality Act of 1965. That act encouraged the States to establish water quality standards for interstate streams and coastal waters by June 30, 1967. If the States failed to do so, the Secretary of the Interior was authorized to establish such standards. All 50 States have developed water quality standards and have submitted them to the Department of Interior for approval.

Provisions of the Water Quality Act

The act requires that the standards be such as to protect the public health or welfare and to enhance the quality of water. In establishing the standards, consideration is to be given to the use and value for public water supplies, propagation of fish and wildlife, recreation, agriculture, industry and other legitimate uses.

As interpreted by the Federal Water Pollution Control Administration of the Department of the Interior, the standards to be established include water use classifications, criteria to support these uses, and a plan to implement and enforce the criteria. The criteria include the quality characteristics of a physical, chemical or biological nature demanded by aquatic life, industrial process, or other intended uses. For streams expected to have more than one use, the criteria of the most sensitive use would govern in establishing standards. Thus, in most cases, the criteria applicable to fish and other aquatic life would be controlling.

State thermal criteria for waste heat

Pursuant to the act, all States have submitted water quality standards. The standards for all States and other jurisdictions have been approved by the Secretary of the Interior, although some approvals have been with reservations. Many of the reservations relate to temperature criteria.

The temperature criteria is water quality standards are established on the basis of the proposed uses of the water. Generally, maximum permissible temperature and maximum changes in temperature constitute the criteria. Some States have specified maximum rates of change in temperature. Several State standards provide for varied criteria depending on the time of year. Some waters are so designated as to allow no change from the natural conditions. In such cases, the limitations are usually determined by the requirements of fisheries.

Most States have established 68 degrees Farenheit as the maximum allowable temperature and from 0 to 5 degrees as the maximum allowable change in temperature for streams with cold water fisheries. For warm water fisheries, the maximum allowable temperatures are generally in the range of 83 to 93 degrees and the maximum allowable rise in the range of 4 to 5 degrees.

Turkey Point, a departure in Federal regulation

In Florida south of Miami, the Florida Power and Light Company has two large conventional steam-electric power plants and is building two large new nuclear power plants. The cooling waters from the new nuclear plants is to flow through a canal 6 miles long to mix with the waters of Card Sound, an adjunct to Biscayne Bay. The heated water from the canal presumably will meet a Dade County temperature limit of 95 degrees Ferenheit.

In February 1970, the Department of the Interior requested the Justice Department to take legal action to block construction of the canal. It asserted that the canal system with the proposed 150 percent dilution of the cooling water would not meet the temperature limits agreed upon by the State and Federal conferees at a meeting called at the request of Governor Kirk of Florida. Out of this meeting came a recommendation of 90 degrees as maximum temperature for water discharged from the canal.

The utility has argued that waste heat must be discharged to the Bay for other means of dissipating the heat are not feasible.

On March 13, 1970, the Justice Department filed suit in the U.S. District Court for the southern district of Florida to stop present and future thermal pollution of Biscayne Bay. Attorney General John N. Mitchell said the suit alleges that the heated water now being discharged from the present two powerplants is rapidly ruining marine life in the Bay, including an area encompassed by the Biscayne National Monument, and that the damage will be even greater when two planned nuclear powerplants are installed at the site.

The Government also filed a motion for a preliminary injunction. It asked that the powerplants be permitted to operate but to modify its operations which result in thermal pollution. It also asks for submission to the court within 45 days of a plan to eliminate the destruction of the natural environment by the powerplant operation; and a halt to construction to the canal.

BYPRODUCT USE OF WASTE HEAT

Ideally the excess heat energy from a steam electric powerplant should be put to productive use in industry, agriculture, dwellings or other places where large amounts of low-grade heat may be useful. To do so would reduce the waste heat discharged directly to the environment and save the additional fuel that otherwise would be consumed to supply such heat. Sale of byproduct waste heat might even become a souce of income for the utilities. The chemical and petroleum industries, for example, require large amounts of heat as does the desalting of water. Proposals have been made to use heated waters from powerplants in agriculture and aquaculture. However the benefits, of such applications, their technical and economic feasibility remain to be demonstrated.

The tendency towards building large powerplants outside of the cities and the impracticability of transporting low-grade heat for long distances will require new innovations in business and industry to make the use of waste heat as a byproduct an attractive alternative to discharging it to the environment.

DISSIPATION OF WASTE HEAT TO THE ENVIRONMENT

As noted earlier, the simplest and least expensive, and the traditional method for disposing of excess heat from a steam electric powerplant is to pump water from a river or some other body of water through the powerplant to pick up and carry away the waste heat. The heated waste water mixes with its parent and its burden of heat energy ultimately is transferred to the air by evaporation, conduction, and radiation. Depending upon the amount of waste heat and the characteristics of the receiving waters, the water temperatures in some places may exceed limits set in water quality standards. In such instances, some or all of the waste heat from a powerplant may have to be transferred directly to the air. For these, the cooling water from the powerplant is circulated through a man-made cooling pond or lake, or thorugh cooling towers.

Cooling ponds and lakes

The electricity industry makes wide use of cooling ponds in the Southwest and Southern States where available water supplies may not be wholly adequate to dissipate the waste heat. The extensive land areas necessary for the ponds and their drainage areas are available in these parts of the country at relatively low cost, and the low humidity in the Southwest promotes more effective transfer of waste heat from the pond to the air.

In many instances, cooling ponds and lakes may be quite large. Approximately 1 acre of pond plus 10 acres of drainage area to supply water for the pond is needed for each megawatt of generating capacity. Such ponds may be used for public benefits including water supply, flood control, recreation, and enhancement of fish and wildlife values.

Cooling ponds require a flow of water to replace that lost through evaporation. The loss is equivalent to about 1.5 percent of the flow of cooling water from the powerplant.

Where land is available at low costs, cooling ponds may be the least expensive alternative to direct discharge cooling. Capital cost estimates for cooling ponds and associated dams and structures range from $2.50 to $5 per kilowatt of generating capacity, and in some conditions $6 to $10 per kilowatt.

Cooling towers

Waste heat can be transferred to the air through two types of cooling towers. In the evaporative type, the water to be cooled falls over exposed surfaces within the tower and gives up its heat by evaporation. In the dry type, the water is pumped through the giant equivalent of an automobile radiator and gives up its heat by air convection.

Evaporative cooling towers

The performance of a cooling tower depends upon the movement of air through the structure to carry away the evaporated water. In some cooling towers, mechanical fans draw air through the structure, in others the flow of air depends upon natural air movements, or convection.

Environmental effects of evaporative cooling towers.—The mechanical cooling towers discharge large amounts of water vapor near the ground. Also droplets of water, or "windage" may be carried from the tower by air currents. Windage is troublesome because it may contain chemicals that are added to prevent biological fouling of the cooling system, chemicals resulting from corrosion or structural deterioration, and minerals that have become concentrated within the system. In some weather conditions, mist, fog or ice may result from these cooling towers.

As an alternative to the mechanical cooling tower, the natural draft tower discharges its moisture considerably higher off the ground. Such a tower for a large powerplant may rise as high as a 30-story building and measure more than a block in diameter. They are certainly a dominant feature of a power station and may be visible for miles. Some people consider them esthetically undesirable in certain locations.[1]

The windage effects of cooling towers would be aggravated were sea water to be used as the cooling water. With solids present in the amount of 35,000 parts per million, the salt in the windage could cause corrosion damage to adjacent structures and equipment and to nearby land.

Water requirements.—The water lost by evaporation within a cooling tower amounts to about 20 gallons per-kilowatt of generating capacity per day for an average steam electric plant, and about 13 gallons for one of high efficiency. A 1,000 megawatt nuclear powerplant, with its lower efficiency, would require about 20 million gallons of makeup water a day, in comparison with 14 million for a comparable fossil plant.[2]

Water pollution from cooling towers.—The evaporation of water in a cooling tower serves to build up a concentration of minerals present in the source of cooling water, and also to concentrate chemicals and solids from other sources. For technical reasons, the concentration cannot be permitted to increase without limit. Therefore part of the cooling water is routinely drained off and replaced. This is known as "blow-down." The concentration of minerals and chemicals in the blow-down water may exceed water quality standards. This waste water must either be processed to remove enough of the mineral and chemical contents to bring the effluent into compliance, or be diluted enough for this purpose.

---

[1] Cooling tower applications and technology are reviewed in detail in the Federal Power Commission report "Problems in Disposal of Waste Heat From Steam-Electric Plants," published in 1969.

[2] "Cut pollution at what price?" Electrical World, Jan. 19, 1970, p. 32.

Water brought into a cooling system to make up for evaporation may typically contain 50 parts per million of solids. The concentration within the cooling circuit may be held at 700 parts per million, meaning that blow-down waters contain this concentration.

Dry cooling towers

In principle the dry cooling tower should avoid the problems of fogging, mist, and icing of the evaporative types, and has no routine water loss. It discharges only dry heat to the air. Dry towers may either be mechanical, with a forced air draft, or natural draft. Dry towers are not able to cool the water as much as a wet or evaporative tower, which reduces the powerplant efficiency and requires more fuel for each kilowatt-hour of electricity generated.

Dry cooling tower technology has yet to be demonstrated in the United States for large steam-electric plants. The largest natural draft tower in operation today is one at a 120-megawatt powerplant in England. This tower was built in 1962 by the Central Electricity Generating Board, primarily to obtain comparative investment and performance data. It is reported that the performance of the tower has been satisfactory.

Environmental effects of dry cooling towers.—At present the environmental effects of discharging large quantities of dry heat from such cooling towers are not known.

Costs of cooling tower

The costs of various types of cooling systems depend upon the design of the system and the site conditions. The Federal Power Commission has estimated the ranges of costs based on data from electric utilities. Table 47 summarizes the estimated investment cost for evaporative cooling towers.

TABLE 47.—COMPARATIVE COSTS OF COOLING WATER SYSTEMS FOR STEAM-ELECTRIC PLANTS

Type of system	Investment cost, dollars per kilowatt	
	Fossil fueled plant [1]	Nuclear fueled plant [1]
Once through [2]	2.00–3.00	3.00–5.00
Cooling ponds [3]	4.00–6.00	6.00–9.00
Wet cooling towers:		
Mechanical draft	5.00–8.00	8.00–11.00
Natural draft	6.00–9.00	9.00–13.00

[1] Based on unit sizes of 600 mw and larger.
[2] Circulation from lake, stream, or sea and involving no investment in pond or reservoir.
[3] Artificial impoundments designed to dissipate entire heat load to environment. Cost data are for ponds capable of handling 1,200–2,000 mw of generating capacity.

Source: Federal Power Commission. "Problems in disposal of waste heat from steam-electric plants." A staff study supporting the Commission's 1970 National Power Survey. 1969, p. 15.

An operating cost common to all cooling systems is the cost of power used to pump water through the systems. For cooling towers, a greater pumping effort is required, with the additional power required being equivalent to one-half percent or more of the plant output. Power to drive the fans in a mechanical tower account for upward of 1 percent of the plant output. Annual operating and maintenance costs for cooling tower systems, exclusive of the costs of power, are equivalent to 1 or 2 percent or more of the capital investment in the cooling system.

According to Federal Power Commission estimates, the use of evaporative cooling towers rather than the once-through cooling could increase the cost of generating power as much as 5 percent. Also cooling towers ordinarily reduce turbine efficiency so that most estimates indicate a 1-percent capacity penalty chargeable against plants using wet cooling towers.

For a specific example, the cooling towers for the Monticello nuclear powerplant in Minnesota were recently reported to add $5 million to the capital costs and $1.9 million annually to the operating cost. At an 80-percent operating factor, the cooling towers thus would add about one-tenth of a mill per kilowatt-hour to the estimated generating cost of 7 mills per kilowatt-hour for the plant.[1]

Investment costs for dry-type cooling towers are largely conjectural because of limited experience with them. The FPC thinks a price range of $25 to $28 per kilowatt for mechanical draft and $27 to $30 per kilowatt for the natual draft appear to be reasonable. With these costs, dry-type cooling does not compare favorably with other types of cooling at places where adequate water supplies are available. Also, the plant electrical output may be from 6 to 8 percent less than it would be with on-through water cooling, which would increase the cost of power.

In one recent estimate, cooling ponds would be expected to increase generating costs by perhaps 15 percent, and dry towers perhaps 30 percent, with evaporate towers in between. In terms of billings to the public, installation of those heat dissipation methods could increase the retail[2] rate from 5 to 10 percent.

While such an increase probably would be accepted by the public, industries that use large amounts of electricity at low rates would be more seriously affected should the addition of such measures add 1 to 2 mills per kilowatt-hour to a price of 5 to 10 mills. Such an increase could have a significant effect on the prices of products that require large amounts of electricity to manufacture.

Cooling water requirements

A very practical question is how much water may be affected by waste heat from large steam-electric plants?

The Federal Power Commission expects that 59 new fossil-fueled plants or additions to existing plants of 500 megawatts or more, comprising 81 units and totaling 52,000 megawatts, will go into service during the years 1967 to 1973. An additional 41 nuclear plants or additions to existing plants comprising 57 units, totaling 42,000 megawatts, also are scheduled to go into service in that period.

The combined cooling water discharges from these 138 units with almost 100,000 megawatts of capacity will be a substantial addition to the waste heat discharged to the Nation's waters.

Looking ahead to 1990, the FPC estimates a total of 492 plantsites will be in use for large steam-electric installations. Of these, 255 will be new sites. Some 292 of the total number of sites would be for fossil-fueled plants and the remaining 200 for nuclear power. Most of the new plants are expected to be in the 1,000- to 4,000-megawatt size range, with the largest site approaching 10,000 megawatts, The total capacity at the 492 sites by 1990 would be about 1 million megawatts.

[1] Nucleonics Week, Mar. 5, 1970, pp. 4, 5.
[2] "Cut Pollution at What Price?" op. cit., p. 33.

The total estimated fresh water withdrawal in 1990 for these powerplants is estimated by the FPC to come to 300,000 cubic feet a second. Although this would be equivalent to one-sixth of the total annual rate of runoff of streams in the United States, much of the water can be used again at several sites along a particular river.

AIRBORNE WASTES FROM ELECTRIC POWERPLANTS

Future plans for generation of electricity in powerplants that burn fossil fuels are likely to be critically affected by the need to control emission to the air of wastes that have undesirable effects.

At the outset, one should note that powerplants are not the only source of air pollution. The combustion of fossil fuels for all purposes produce some 142 million tons of air pollutants, as shown in table 48. Automobiles and other forms of transportation discharge nearly 60 percent of the total emissions. However transportation is not a significant source of sulfur oxides because the fuels used are low in sulfur content. Stationary fuel combustion sources account for 75 percent of the sulfur oxides, while refineries, smelters, acid plants, and similar processes emit the remainder. Fossil-fueled powerplants which produce over 85 percent of the electricity generated in the United States in 1966 discharge almost 50 percent of the sulfur oxides, 25 percent of the particulates, and about 25 percent of the nitrogen oxide emissions.

TABLE 48.—SOURCES OF AIR POLLUTION

[In millions of tons annually (1965)]

	Carbon monoxide	Sulfur oxides	Nitrogen oxides	Hydrocarbons	Particulate matter	Totals
Motor vehicles	66	1	6	12	1	86
Industry	2	9	2	4	6	23
Powerplants	1	12	3	1	3	20
Space heating	2	3	1	1	1	8
Refuse disposal	1	1	1	1	1	5
Total	72	26	13	19	12	142

Source: "Considerations affecting steam powerplant site selection." Op. cit., p. 29.

Sulfur oxides

Recent projections estimate that by the year 1980 some 48 million tons of sulfur dioxide would be released to the air annually, assuming that control measures are not applied. Of this, 36 million tons would come from powerplants in comparison with 12 million tons in 1966.

Effects of sulfur dioxide

Gaseous sulfur dioxide from burning fossil fuels may later form droplets of sulfuric acid in moist air. These droplets are potentially injurious to the respiratory system. When combined with small particle pollution and stagnant air, the resulting air pollution may lead to the kinds of injury experienced in Donora, Pa., New York, and London when severe pollution episodes occurred. On the other hand, the precise quantitative biological effects of sulfur dioxide are not fully known, which complicates the setting of air quality standards. Regardless of specific biological effects, it appears generally agreed that sulfur dioxide air pollution can effect persons suffering from lung ailments of bronchitis, emphysema, or cancer.

The acid mists also may damage property and vegetation. In combination with other pollutants, for examples particulates, sulfur oxides have been shown to exhibit synergistic effects and produce results several times more severe than from comparable exposure to either pollutant alone.

Regardless of the completeness of present scientific information about the biological efforts of sulfur dioxide, the public regards it as a pollutant to be controlled.

Regulation of sulfur oxides

The regulation of sulfur dioxide and other air pollutants is primarily the responsibility of State, local, and regional agencies, backed up by the Department of Health, Education, and Welfare.

Federal legislation.—Pursuant to the Clean Air Act (Public Law 90–148), DHEW administers Federal aid grants to establish and maintain regional, State and local air pollution control programs. DHEW also is establishing air quality control regions, with a completion target date of September 1970. It has released air quality criteria for carbon monoxide, photochemical oxidants, and hydrocarbons, and also control techniques for stationary sources of emissions of carbon monoxide, nitrogen oxides, hydrocarbons and organic solvents. The Department has reported to Congress that national emission standards for stationary sources, which would include steam-electric powerplants, is not in the best interest of pollution abatement. Rather, DHEW favors national air quality standards with local, State, or regional agencies responsible for implementing them, and with national emission standards limited to application to new installations.

The act authorizes DHEW to recommend and establish standards if sufficient local standards are not adopted and in an emergency to enjoin the emission of contaminants.

State legislation.—State air pollution control laws empower State and local air pollution control agencies to promulgate standards for regulating sulfur compounds in the air. Typically, States enacting or amending air pollution control laws authorize the creation of a State air pollution control agency, which is instructed to issue rules and regulations pertaining to air quality and, in some instances, to issue sulfur emission standards and limits for sulfur content of fuels.

State regulation.—Rules and regulations of State air pollution control agencies have become increasingly specific for sulfur control. State regulations generally contain a sulfur dioxide emission limit for individual sources, using a figure of 2,000 parts per million by volume of sulfur dioxide as a limit for existing sources. This standard appears directed more toward regulation of sources such as sulfuric acid plants that may emit sulfur dioxide as a byproduct of manufacturing rather than from the combustion of fuel. Recent legislation in South Carolina, New York, Missouri, and other States, has set variable emission requirements for combustion sources. Consequently an electric power station may have quite a different sulfur dioxide emission limit in many jurisdictions than an industrial processing plant, and for electric powerplants there may be a wide variation in the emission limits prescribed.

Sulfur dioxide emission standards are being supplemented, and in some places preempted, by regulations limiting the sulfur content of fuels. This approach is more certain and less expensive to administer.

Current enactments set different fuel limits according to use. Fuel for steam and electric stations, heating and industrial may have different limits. Limits are usually expressed in terms of a maximum percentage of sulfur by weight, and there is little uniformity amongst them. Some authorities have set the maximum as low as 1 percent and by 1970 it may be as low as 0.37 percent. According to the National Coal Policy Conference, in every instance the sulfur limit set is significantly lower than the sulfur contained in the coal previously burned within the jurisdiction.

Action in California.—Perhaps the most severe limitation upon sulfur dioxide is to be found in California. There the State's environmental quality study council has recommended a moratorium on fossil-fueled powerplants. The Orange County Board of Supervisors subsequently voted against two 790-megawatt units at Huntington Beach. The Los Angeles Air Pollution Control Board has indicated it will not approve further applications for fossil-fueled powerplants.[1]

Technological alternatives to reduce sulfur dioxide emissions

Five technological approaches may be used, singly or in combination, to keep the sulfur dioxide emission from a steam electric powerplant within limits of air quality standards. These are to:

(1) Use fuels of low natural sulfur content.
(2) Remove or reduce the sulfur in fuels.
(3) Remove the sulfur dioxide from stack gases.
(4) Improve the combustion process.
(5) Disperse the stack gases sufficiently that the sulfur dioxide at ground levels stays within air quality limits.

Use of low sulfur fuels.—The ideal fuel of low sulfur content is natural gas, which explains why some air pollution control authorities specify the use of natural gas by steam electric plants. Some residual fuel oils also may have a naturally low sulfur content, depending upon their origin. The residual oils from Africa are the lowest in sulfur. Some coal deposits also are low in sulfur, but limited supply and strong competition for nonfuel uses greatly limits their use.

While the use of low sulfur fuels may provide some temporary relief from air pollution, in the long run ways must be found to reduce the sulfur content of fuels before they are burned and to remove enough sulfur dioxide from powerplant emissions to stay within air quality limits.

One noticeable result of the specification of sulfur emission standards has been to accelerate a trend away from coal into gas and residual fuel oils. According to the Office of Oil and Gas of the Department of the Interior, the use of gas in the utilities market of the east coast, for example, for the first 6 months of 1969 was 45 percent more than in 1968 and residual fuel oil was up 28 percent, while use of coal did not increase at all.

Low sulfur coal: Before addressing the availability of low sulfur coal, it should be noted that in some powerplants this kind of coal cannot be burned in existing furnaces without operational difficulties or incurring high capital costs for furnace modifications.

Sulfur, unfortunately, is universally present in coal not in elemental form but combined with the organic coal substances or in the form of

[1] Electrical World, Nov. 10, 1969, p. 25.

pyrite. In most U.S. coals, the total sulfur content varies from 0.5 to 6 percent. Much of the coal now burned by powerplants is high sulfur coal, that is, with a content of 1 percent or more.

In terms of national coal reserves of all classes, approximately 50 percent are located east of the Mississippi River and 50 percent on the western side. However, of the total reserves of low sulfur coal with less than 1 percent sulfur, almost 90 percent, including lignite, is located west of the Mississippi. The Office of Science and Technology asserts that the supply of low sulfur coal is costly and limited. The National Coal Policy Conference asserts the supply for power generation is wholly inadequate and is in extremely short supply.

Most of the low-sulfur coal in the East is of metallurgical grade coking quality and is largely dedicated to the steel industries, both domestic and foreign. These fine grade coals are produced in West Virginia and adjoining States and are in demand throughout the free world. They constitute a large source of export tonnage and income which makes an important contribution to the national balance of payments.

Even with a premium of $2 to $3 per ton, which would be required in the East, producers of low sulfur coal may not be able to supply the rapidly growing demand for this commodity. Even if supplies were available, the premium price would result in substantially higher costs of generation.

According to the Department of the Interior, about two-thirds of the coal produced east of the Mississippi River cannot meet present limits for sulfur content and virtually none of it will be able to meet the more restrictive standard of 0.37 percent that some States have scheduled by the end of the year 1971.

Low sulfur oil: Some residual oils from abroad are low enough in sulfur content to be used in steam electric powerplants. A decision announced by the Secretary of the Interior in July 1967, revised Government oil import controls to combat air pollution. The change allowed fuel users a greater supply of low-sulfur fuel oil by reclassifying No. 4 and other low-sulfur oil, previously subject to import quotas, to the category of "residual" fuel oil. This reclassification permitted the east coast to import low-sulfur oil with few import restrictions. And the Interior Department established a system to permit imports of low-sulfur fuel on the west coast and allowed U.S. refiners a special allocation for low-sulfur fuel they manufacture from imported oil.

The importing of residual fuel oil has recently become a matter of controversy before Congress, for the utilities are seeking to import larger quantities. For example, recently the Commonwealth Edison Co. asked for a special quota to import 6 million barrels of residual from Venezuela into the interior of the Nation via the Mississippi River. The company chose to do this, paying an estimated additional $5 million per year rather than attempt to remove sulfur dioxide from the furnace gases. The cost of heat energy from this imported residual is estimated at 44 to 52 cents per million B.t.u. in comparison with coal at 24 cents per million B.t.u.

Gas: Natural gas already appears to be in short supply and pipeline and distributing companies are experiencing difficulties, according to the National Coal Policy Conference, in meeting increased consumption of present customers. Some technical prospects exist for

46-366 O—70——8

producing gas from coal. The Bureau of Mines and the Department of the Interior are sponsoring pilot plant studies on gasification. According to the Bureau, if a decision were made to press ahead, a commercial coal gasification plant could be operating by 1977. However whether the demonstration and subsequent adoption of gasification technology could be carried out fast enough to help substantially with the anticipated gas shortage seems doubtful.

Removal of sulfur from fuels.—At present it seems unlikely that commercial processes to remove sulfur from coal will be available during the next few years, when many critical decisions will have to be made about fuel for large new powerplants. According to the Office of Science and Technology, research projects do show promise of removing as much as 70 percent of the sulfur, although the final product might still contain enough sulfur to be classified as a high sulfur fuel. What the technological and economic feasibility of such removal processes may be remains to be seen.

The sulfur content of fuel oil, on the other hand, can be brought within acceptable amounts either by removing some of the sulfur, or by diluting a high sulfur oil with low sulfur oil, or both.

Present technology indicates that the most economical means of removing sulfur from residual oil for use in electric powerplants may be at the refinery. The OST estimates that oil can be desulfurized for a cost of about 25 to 50 cents a barrel depending upon the original material, the amount of sulfur to be removed and processing methods. The capital investment to build a desulfurizing plant is estimated at about $260 per barrel of daily capacity.

The petroleum industry is investing heavily in ways to reduce sulfur content of fuel oil. Esso, for example, is installing such a plant in Venezuela to produce 100,000 barrels a day primarily for east coast powerplants. Prof. Thomas K. Sherwood of Massachusetts Institute of Technology estimates that the refining to reduce sulfur content from 2.6 to 0.5 percent will increase the price of residual fuel oil to the power station by 50 to 80 cents per barrel, an increase of 20 to 35 percent. For comparison, an increase of 50 cents per barrel would be expected to increase the cost of generation by about 0.7 mills per kilowatt-hour in a modern steamplant.[1]

While domestic crude oil is generally lower in sulfur than the imported oils, it is priced too high for fuel use in generating electricity. Only about one-third of the residual oil marketed in the United States is derived from domestic sources.

Removal of sulfur during combustion.—Five technical processes are in various stages of research, development, and demonstration for removal of sulfur dioxide from the furnace gases of a steam-electric powerplant. The remaining technological problems for this alternative appear much closer to solution than for reducing the sulfur content of coal. However, the search for an economic method of removing sulfur compounds from the gases has been going on for 30 years with no commercially available devices yet available for modern powerplants. The coal industry in particular would encourage Federal support of research into sulfur compound removal so as to insure the future use for coal in generating electricity.

[1] Thomas K. Sherwood, "Must We Breathe Sulfur Oxides?" Technology Review, January 1970, p. 27.

The Office of Science and Technology identifies the three post-combustion removal processes which show the most promise of eventual commercial success as the alkalized alumina process, the catalytic oxidation process, and the limestone/dolomite processes. Each is relatively expensive. The first requires large and complex equipment so that its application is limited to new, large powerplants. The third is less expensive, requires less equipment, and can be adapted to existing powerplants. All are in various stages of development.

It seems evident that regardless of the system chosen for removal of various offensive gases, additional space will be needed at a powerplant to erect the equipment and to provide storage for the extracted wastes. For instance, the waste produced by the limestone/dolomite process for a 1,250 megawatt powerplant is about 2,000 tons per day.

The Consolidated Coal Co. in February 1970 announced that it had developed a process for removing sulfur oxide from stack gas. According to the company, this process, which differs from others being developed, can be used in existing or new powerplants. It would produce elemental sulfur as a product, which can easily be stored, and should find a ready market. Whether it will be used remains to be seen.

In January 1970, an experimental installation of the limestone/dolomite process began an 18-month test at TVA's Shawnee steamplant near Paducah, Ky.

As for costs, these remain conjectural. One estimate for the limestone/dolomite process puts the initial capital cost at $10 per kilowatt. Figuring in operating and fixed charges, the costs come to the equivalent of 25 cents per barrel of oil burned for 1 percent sulfur oil and 30 cents per barrel for 3 percent fuel. If the price of 1 percent oil is $2 per barrel, this system would increase the equivalent fuel cost by 12½ percent, and the cost of generation by 0.4 mill per kilowatt-hour.

The National Coal Association estimates that the first generation of sulfur dioxide removal plants will operate at a cost range of 75 cents to $1 per ton of coal burned and that, as the technology improves, future costs should drop to about 20 to 25 cents per ton of coal burned.

Improving the combustion of coal.—Another strategy is to reduce sulfur emissions from coal-fired powerplants by radically changing the method of burning coal. Instead of burning pulverized coal, a so-called fluidized bed technique could be used. The Office of Coal Research is optimistic on this approach because it believes it can reduce air pollution, lower capital and operating costs of coal-fired plants. However, because of tight funds, the Office of Coal Research has terminated its support for this development. The Nationl Air Pollution Control Administration has indicated it believes the air pollution aspect of the fluidized bed process warrants further investigation and plans to provide some support. However, unless the development and demonstration of this technique is expedited, the chances that it can be used in large new powerplants ordered during the 1970's are slim.

Dispersing and diluting sulfur emissions in the air.—Since the effects of sulfur dioxide depend upon its concentration in the air, one way to reduce its effects is to dilute the emission from a large powerplant by discharging the furnace gases from very tall stacks. Such stacks may be effective in reducing the ground-level concentration of pollutants, but they do not reduce the amount of pollutants released into the air. Also, under some local weather they may not cause dispersion and high concentrations of sulfur dioxide may occur at the ground.

The 1,200 foot stack of a power plant in West Virginia is the highest power plant stack to date. The cost of tall stacks is considered to be about 10 to 20 percent of the estimated cost of some of the sulfur removal processes discussed above. The OST thinks it doubtful that the stacks will be able to afford the dilution necessary to meet stringent sulfur dioxide standards particularly for a large plant that burns high sulfur fuel.

Nitrogen oxides

The nitrogen compounds contained in fossil fuels are released to the air during combustion, usually in the form of oxides of nitrogen.

Among the fossil fuels, pulverized coal is the greatest producer of nitrogen oxides, with oil next and gas last. The Federal Power Commission estimates that the following amounts of nitrogen oxides can be expected from the combustion of coal, oil, and gas in generating 1000 kilowatt-hours of electric power: Coal, 8.6 pounds; oil, 7.6 pounds; and gas, 4.1 pounds.

Effects of nitrogen oxides

Until the 1950's, when these chemicals were implicated in the formation of eye-irritating smog in the Los Angeles area, nitrogen oxides were ignored as a pollutant from steam electric powerplants. Since then some research has been done on the formation of these oxides and general methods of reducing emission of nitrogen oxides have been suggested. However in comparison to the effort to control emission of sulfur oxides, the research on nitrogen oxides is practically nonexistant.

Since nitrogen oxides are produced by stationary and vehicular combustion sources, both of these sources contribute to smog. The exact role of each has not been clearly defined.

Regulation of nitrogen oxides

At present regulation of nitrogen oxide as a gaseous pollutant from powerplants has received only secondary attention. The National Air Pollution Control Administration will not issue criteria for their emissions until 1971.

Actual regulation now is carried out by State and local air pollution control agencies, as with sulfur oxides.

Control systems

No tested systems to control the emission of nitrogen oxides are commercially available for powerplants. In comparison to the massive effort now underway to control the oxides of sulfur, research on nitrogen oxide control is practically nonexistent.

Use of alternative fuels is not a real option because, as seen above, the combustion of fossil fuels all yield roughly comparable quantities.

Cost of control

At present any cost estimate for control of nitrogen oxide emissions would be purely speculative.

Solid wastes from powerplants

Fly ash and furnace ash are wastes from combustion of oil and coal in powerplants. Emissions are dependent upon fuel quality, type of equipment, size and method of firing, and maintenance and operation. Ash emission from burning of natural gas is insignificant in comparison with other fossil fuels.

Effects of fly ash

The principal environmental effect of fly ash which is discharged to the air is the dirt it deposits on surrounding homes and factories.

The 297 million tons of coal burned for electric power in 1968 produced approximately 29.6 million tons of this waste material. Until about 10 years ago, nearly all of this was stored in piles near the utility plant, resulting in destruction of the vegetation near the plant, creating an adverse esthetic effect, contributing to air pollution as the dried ash blew about, and damaging streams, crops and vegetation by the leaching of chemicals from the ash piles by rain water. It is estimated that 200 million tons have been stored on the surface in the past 10 years. If the storage piles averaged 40 feet in height, approximately 2000 acres would be covered with this material. One estimate of the ash to be generated by coal combustion from 1968 to the year 2000 is for 1.9 billion tons, which would occupy 20,000 acres if not otherwise disposed of.

Regulation of particulate emission

Regulation of the amount and characteristics of particulate emissions permitted from powerplants and other users of fossil fuels is the function of local air pollution control agencies.

Control of particulate emissions

Control of emissions from powerplants has, in the past, emphasized "smoke" and particulate control. Four fundamental types of control equipment have been developed: mechanical separators, electrostatic precipitators, bag houses, and scrubbers. There latter two are found most frequently in conventional manufacturing industries and are often included to recover otherwise valuable lost materials.

The technology to collect fly ash has shown a continuing improvement. The average efficiency of collectors being specified for modern powerplants ranges from 98 to 99 percent. The Office of Science and Technology expects this trend will continue.

Research in electrostatic precipitation is now focused mainly in the collection efficiency region above 99 percent. Despite an anticipated decrease in particulate emissions, some increases are anticipated in the emissions of very small particles. OST notes that these very small particulates may be found to be of particular significance in regard to health effects and possible long-term effects upon the climate.

The disposal of fly ash, as indicated above, presents some problems, particularly if the solid wastes from certain air cleaning processes are added. One approach to disposal of fly ash has been research to convert it into a useful byproduct. At a recent conference on fly ash disposal[1] it was forecast that in 1975 the electric utilities of the United States will be producing fly ash at a rate of approximately 29 million tons per year, together with approximately 13.5 million tons of ash and slag from the furnaces for a total ash production of 42.5 million tons. To dispose of this waste commercially will require improved technologies of use and marketing techniques.

Radioactive wastes from nuclear power

The fissioning of uranium or plutonium atoms in a nuclear power reactor produces large quantities of intensely radioactive materials. In fact, the weight of the radioactive waste products virtually equals

[1] "New Uses For Fly, Other Ash Told to 300 at Pittsburgh," Electrical World, Mar. 30, 1970, pp. 22–23.

the weight of the nuclear fuel atoms that fission. In addition, structural and other materials within a power reactor may become radioactive because of exposure to the neutrons emitted during fission.

Most of these wastes are enclosed within the fuel elements within the reactor, although some of them may escape from the fuel elements through small imperfections in their cladding. These escaped wastes remain within the reactor, which is a closed system.

For routine operations, radioactive wastes from a nuclear powerplant reach the environment in one of two ways. Radioactive gases are collected and routinely vented to the outside air, usually from a tall stack or from a blower atop the powerplant. These gases include radioactive krypton and xenon. Some vapors of iodine may also appear depending upon the amount of leakage from the fuel. Other radioactive wastes are routinely collected during powerplant operations. A small part of these may remain in plant waste waters after these have been filtered and in other ways treated to remove the greatest part of them. The waste water is mixed in with cooling water leaving the plant.

Effects of radioactive wastes

Radiation from radioactive wastes depending upon the amount and nature of the waste and the conditions of exposure to it, may produce noticeable biological effects. Large exposures to such radiation from wastes in the environment or that find their way into an organism, can cause injury or death. The exposures that produce these effects are well known and the nature of the effects are established. This kind of exposure is unlikely to result from the routine operation of a nuclear powerplant, except for an accident which might rupture the reactor and disperse its radioactive contents to the surroundings. The exposure which has prompted most recent concern is prolonged exposure to very small quantities of radioactive wastes which produce radiation less than much of the radiation which exists in nature from naturally radioactive minerals.

Although the radioactive wastes routinely discharged from a nuclear powerplant are within limits specified by the Atomic Energy Commission, some scientists have expressed concern that these small amounts if continuously emitted for long periods of time may find their way into the food chains and water supply. Some waterplants and animals tend selectively to remove and concentrate certain radioactive wastes. For example, radioactive species of cobalt, cesium, and manganese are concentrated in the edible tissues of shellfish, while in dairy country radioactive iodine vapors that condense on grass may appear in the milk of the cows that eat the grass.

The aspect of radiation which arouses the most concern and controversy is its postulated effects upon the genetic mechanism. It is well known that large exposures to radiation can cause mutations in animals such as fruit flies. What is not as well known is the effect of small amounts of radiation upon the inherited characteristics of human beings and other living things. The Federal Radiation Council in its first report had this to say about the genetic effects of radiation:

Although ionizing radiation can induce genetic and somatic effects (effects on the individual during his lifetime other than genetic effects), the evidence at the present time is insufficient to justify precise conclusions on the nature of the dose-effect relationship especially at low doses and dose rates. Moreover, the evidence is insufficient to prove either the hypothesis of a "damage threshold" (a point below which no damage occurs) or the hypothesis of "no threshold" in man at low doses.[1] Because of limitations of knowledge and the complexities of assessing the effects or radiation exposure, the FRC endorses the philosophy that all exposures should be kept as far below any arbitrarily selected levels as practicable. "There should not be any man-made radiation exposure without the expectation of benefits resulting from such exposure." [2]

Regulation of radioactive wastes

Until recently it was commonly assumed that Congress in the Atomic Energy Acts of 1946 and 1954 had preempted to the Atomic Energy Commission the authority to regulate emission of radioactive wastes from nuclear powerplants. The AEC's regulatory system takes a twofold approach. First, the nuclear powerplants each must obtain first a construction permit to build the plant and then an operating permit to put it into operation. The AEC review of the plant design and construction prior to issuing such permits looks into measures to control the discharge of radioactive wastes. Second, the AEC's regulations in part 20 to title 10 of the Code of Federal Regulations establishes specific limits for the emission of radioactive materials from nuclear powerplants. These latter regulations, however, do not extend to control of natural materials or those that are made artificially radioactive with machines other than nuclear reactors. Control of these substances remains with the States. The AEC on March 28, 1970 announced a proposed amendment to 10 CFR 20 which would require licensees of power reactors to make "* * * every reasonable effort to maintain radiation exposures and releases of radioactive materials in effluents to unrestricted areas as far below the limits specified * * * as practicable." Recently the Minnesota Pollution Control Agency in issuing a permit for the operation of a large nuclear powerplant, included a limitation upon discharge of radioactive wastes which is more restrictive than those of the AEC. The issue of whether this State agency can apply stricter controls than those of the AEC was still in Federal court for decision in April 1970.

The AEC regulations on emission of radioactive wastes are interpretations of guides laid down by the Federal Radiation Council. These guides, in turn, are largely derived from the judgement of scientists who are members of the semiofficial National Council on Radiation Protection and Measurements [which has a Federal charter but receives no Federal funds] and the unofficial but prestigious International Committee on Radiation Protection.

The scientific validity of present AEC regulations in 10 CFR 20 recently has been challenged before the Joint Committee on Atomic Energy and the Subcommittee on Air and Water Pollution of the Senate Committee on Public Works. The Secretary of Health, Education and Welfare also reportedly has called for a general review of the basis for the radiation standards.

---

[1] "Background Material For the Development of Radiation Protection Standards," Staff Report No. 1 of the Federal Radiation Council, May 13, 1960, p. 36.

[2] Ibid., p. 37.

The principal recent challenge has come from two scientists of the AEC's Lawrence Radiation Laboratory. Drs. John W. Gofman and Arthur R. Tamplin state that in their opinion the most crucial problem facing everyone concerned with atomic energy is to "* * * secure the earliest possible revision downward, by at least a factor of tenfold, of the allowable radiation dosage to the population from peaceful atomic energy activities." [1]

Consequences of a major nuclear accident

While the AEC asserts that the likelihood that a major accident with a nuclear reactor might release much of its contained radioactive wastes is very small, it did in 1957 publish a report on the theoretical possibilities and consequences of such an accident. The purpose of quoting the following excerpts is not to suggest that such an accident is probable, but to indicate what might be the range of results should the improbable accident occur.

According to this AEC report, and depending upon the type of accident and the amount of the radioactive wastes released, the effects might be as follows:

> * * * the theoretical estimates indicate that personal damage might range from a lower limit of none injured or killed to an upper limit, in the worst case, of about 3,400 killed and about 45,000 injured.
>
> Theoretical property damages ranged from a lower limit of about one-half million dollars to an upper limit in the worst case of about $7 billion. This latter figure is largely due to assumed contamination of land with fission products.
>
> Under adverse combinations of conditions considered, it was estimated that people could be killed at distances up to 15 miles and injured at distances of about 45 miles. Land contamination could extend for greater distances.
>
> In the large majority of theoretical reactor accidents considered, the total assumed losses would not exceed a few hundred million dollars.

The AEC has since declined to revise or update this study.

Disposal of high level radioactive wastes

The most likely places for large amounts of radioactive materials to escape to the environment during the routine generation of nuclear power appears to be not at a powerplant, but in the transportation of used fuel from a powerplant to a fuel reprocessing plant, during subsequent reprocessing, there, and in the long term disposal of the radioactive wastes.

After nuclear fuel has been in a power reactor for perhaps a year or more, or if it becomes too damaged for safe use, it is removed. After interim storage at the powerplant, to permit some of its radioactivity to diminish, the used fuel is carried by truck or rail in special containers to a fuel reprocessing plant. There the still usable uranium or plutonium is recovered from the used fuel for subsequent reuse in new fuel.

At present there is one operating commercial nuclear fuel reprocessing plant in the United States, near Buffalo, N.Y. Another is nearing completion near Chicago, Ill., and a third is supposed to start construction in South Carolina during 1970.

At the reprocessing plant, the used nuclear fuel is chopped up and dissolved. The radioactive gases released from the fuel generally would be emitted to the air in concentrations permissible under AEC regulations. Most of the intensely radioactive fission products remain in

[1] Testimony of Drs. Gofman and Tamplin before the Subcommittee on Air and Water Pollution, Senate Committee on Public Works, Nov. 18, 1969.

the waste liquors of the process. The weight of these radioactive wastes is virtually equal to the weight of the uranium that fissioned while the fuel was in the reactor. It is during the reprocessing that the intensely radioactive wastes are in forms which could most easily reach the environment in an accident.

What to do with the wastes is somewhat of an open question. The AEC expects they will be put into solid form and stored in worked-out salt mines. At the moment there is no commercial service for high level radioactive waste disposal. The word "disposal" itself is not accurate, for these wastes cannot be released to the environment. Thus they must be stored indefinitely.

In May 1966, a committee of the National Academy of Sciences in advising the Atomic Energy Commission on geologic aspects of radioactive waste disposal, reiterated the basic rule that "* * * concentrations of radionuclides in waste materials should not be allowed to appear in the earth's biosphere before they have decayed to innocuous levels." [1] This concept requires assurance that during any storage or disposal operations, hazardous amounts of radioactive wastes are isolated from the environment, and that upon completion of the reprocessing, the wastes will remain isolated as long as they might constitute a hazard. For some species of radioactive wastes, this means isolation for periods of six to ten centuries, periods so long, notes the committee, that neither perpetual care nor permanence of records can be relied upon. The committee did not object to radioactive materials reaching the environment in concentrations less than those specified in AEC regulations. Within those limits the committee said it had no concern. Rather it was the possibility of cumulative buildups of long-lived radioactive wastes that may exceed these limits after continued use of doubtful practices and the prospect of unforeseen concentrations in excessive amounts resulting from unexpected and uncontrollable alterations in the future environment that the committee wished to guard against. As for the economics of long term waste disposal, the committee observed that while these are of concern, "* * * they are relegated to second-rank consideration, safety being the matter of first concern always." [2]

The Atomic Energy Commission estimates that over the past 10 years, improvements in chemical processing have reduced the waste volumes from about 1,500 gallons per ton of used uranium processed to about 100 gallons per ton. Assuming an installed nuclear generating capacity of 123,000 megawatts by the year 1980, the AEC estimates the accumulated high level wastes in solution from nuclear power would be 3.5 million gallons, which could be reduced to solids with a volume of about 35,000 cubic feet, the equivalent of a cube of 32 feet to a side. Looking ahead to an installed nuclear capacity of 675,000 megawatts by the end of the century, the accumulated high level liquid wastes, if not previously solidified, would total 55 million gallons, and with a solid volume of 550,000 cubic feet. The AEC categorically states that disposal of high level wastes will pose no significant problem technically or economically.[3]

[1] "Report to the Division of Reactor Development and Technology, U.S. Atomic Energy Commission," National Academy of Sciences-National Research Council, Division of Earth Sciences, Committee on Geologic Aspects of Radioactive Waste Disposal. May 1966, p. 18.

[2] Ibid., p. 19.

[3] Testimony of Milton Shaw as excerpted in "Selected materials on environmental effects of producing electric power." Joint Committee on Atomic Energy, 91st Cong., 1st sess., 1969, p. 45.

The ultimate risk to the environment from reprocessing of nuclear fuels and storage of their wastes in the long run seems likely to depend upon how well the reprocessors comply with AEC regulations. The commercial fuel reprocessor, as is any other service industry, will be under financial pressure to reduce costs which might lead to an attitude of bare compliance, or even neglect of AEC regulations rather than a determined attitude to reduce emissions of radioactive wastes to the lowest level permitted by the fuel reprocessing technology.

Another open item is the question who will own and operate the salt mines or other places for the long term storage of the radioactive wastes.

ENVIRONMENTAL ASPECTS OF TRANSMITTING ELECTRICITY AND THEIR ECONOMIC ASPECTS

The trend toward very large steam-electric powerplants, the growing public insistence upon reliable supply of electricity, and a trend toward citing large powerplants outside of urban areas all combine to increase the demand for more transmission lines. Yet the scarcity of land in the areas of high population, which also are the large users of electricity, and increasing public resistance to transmission lines because of their environmental effects, are two factors that are likely to reduce the ability of the electricity industry to deliver electricity when and where needed during the coming decades.

Primary functions of a transmission system

The primary function of a transmission system is, of course, to carry electricity from generating stations to the areas where it is distributed to local customers. In addition, from the standpoint of bulk power supply—which is becoming more important because of the trend toward large plants—there are three more objectives for adequate transmission capacity. These are to—

(1) Provide additional support for any load areas as may be required in emergencies. The network must be able to handle the automatic flow of power within the system and through its associated interconnections.

(2) Transfer, without serious restrictions, capacity and energy within regions and when available between regions to meet power shortages.

(3) Exchange power and energy on a regional and interregional scale, and to achieve economies in capital and operating costs.

Some effects of transmission systems

In the early days of electric power systems, generating plants were located next to their customers and there was little long-distance movement of any large amounts of electricity. Then as distant customers began to use electricity and as transmission from remote hydroelectric plants became a reality, a trend set in toward higher voltage transmission systems.[1] As the practical transmission distances increased, it became feasible to consider placing new generating plants at places relatively remote from the load centers, which opened up an entirely new outlook upon the siting of powerplants. This was particularly true for the hydroelectric plants and there followed an

[1] As a general rule, doubling the voltage of a transmission system quadruples the electrical energy it can carry.

era of dam building and hydroelectric development. Later, the idea of placing a steam-electric powerplant at the mouth of a coal mine was made feasible by improvements in transmission technology.

The same increases in electric power transmission capability and reduction of unit costs for carrying electricity made it feasible to move large amounts of power between neighboring power systems under exchange or interchange arrangements. The recent trend toward joint-owned generating plants to permit use of larger installations than could be afforded or used by one system alone has been made possible by improved transmission. At the same time, however, joint ownership places greater emphasis on the transmission line costs and right-of-way problems which can be controlling in the selection of a site for such an installation.

The independence of nuclear powerplants from location of primary energy sources suggests the possibility of selecting sites in the vicinity of load centers which may somewhat reduce requirements for transmission. However, strong interconnections would still be needed to assure adequate reliability of interconnected systems.

Technological trends in transmission lines

From the introduction of 110 kilovolt alternating current transmission in the United States in 1908 to about 1950, there was a steady increase in the voltage of transmission lines. See table 49. Then during the 1950's the development of still higher voltage (in excess of 200 kilowatts), or extra-high voltage (EHV) transmission began. The first signification application of direct current EHV in the United States was expected to go into service late in 1969. It is an 800-mile line at 400 kilovolts between the Pacific Northwest and the Pacific Southwest which will be capable of transmitting about 1,330 megawatts.

TABLE 49.—MAXIMUM TRANSMISSION VOLTAGES IN THE UNITED STATES

Year	Kilovolts
1886	3
1892	10
1901	60
1908	110
1923	220
1934	287
1954	345
1964	500

Source: National Power Survey, pt. 1, p. 14.

Forecasts for transmission lines

The FPC report of 1967

In its 1967 report on the prevention of power failures, the Federal Power Commission projected a possible pattern of needed power transmission capability for 1975 and estimated the approximate cost.

About half of the added lines were already programed or then under consideration by utilities or pools for completion in the later 1960's or early 1970's. A major part were in the east-central, north-central, and far west regions of the United States. Additions in EHV lines beyond those scheduled for service in 1967 included 16,000 miles of 345 kilovolt line, 21,400 miles of 500 kilovolt line, 5,750 miles of 765 kilovolt lines and 1,665 miles of 400 kilovolt direct current transmission.

As for the comparative capacity of these lines for carriage of electricity, if 230 kilovolt transmission is taken as unity, a 345 kilovolt line can carry 2.75 times as much electricity; a 500 kilovolt line 6 times as much, and a 765 kilovolt line 16 times as much.

As estimate of the approximate cost of the transmission system additions from 1967 through 1975 was $8 billion.

The current forecast

The Federal Power Commission last year reported the projected general plans for transmission of its six regional advisory committees as shown in table 50. The FPC staff, which is independently examining projected requirements, considers these estimates as an appropriate guide for the general size of transmission needs.

Environmental effects of transmission lines

The most obvious environmental effect of electric transmission is the sight of the towers and their cables, and the accompanying withdrawal of land from other use. Lesser effects include interference with reception of radio and television signals under certain conditions and, in the case of direct current lines, the possibility of corrosion of underground metallic structures, such as sewer or water pipes, because of electrical currents within the earth.

The 300,000 miles of electric power transmission lines in service today occupy about 4 million acres of land, or the equivalent of more than 10,000 average sized farms. By 1990 the forecast 497,000 miles of transmission lines will require roughly 7,100,000 acres, or more than 11,000 square miles. In comparison, the area of the State of Connecticut is 5,000 square miles. The rights-of-way widths will probably average more than 142 feet for a single circuit line. The higher voltage transmission lines will require widths of 200 feet or more, and multiple line rights-of-way will be still wider.

TABLE 50.—PROJECTED TOTAL INSTALLATION OF MAJOR TRANSMISSION LINES IN CIRCUIT MILES

Voltage class (kilovolts)	1970	1980	1990
69 to 200	235,000	290,000	335,000
230	40,500	59,300	67,000
345	16,600	34,500	50,500
500	7,500	21,300	34,700
765	560	3,500	10,200
Total	300,160	408,600	497,400

Source: "Environmental effects of producing electric power." op. cit., p. 58.

The greater use of EHV transmission will minimize the total number of miles of overhead transmission, but the wider rights-of-way, the more massive and higher towers, and the larger conductors could, in view of the FPC, compound the problems in seeking to preserve environmental values.[1]

Through the 1990's it is expected that overhead transmission will dominate, for the technology for high voltage underground transmission is not expected to be available. A large, 2,400 megawatt powerplant typically would be the juncture of three rights-of-way, each 200 feet wide.

[1] Ibid., p. 59.

It is apparent, according to the FPC, that the more densely populated regions are generally expected to have the greatest increase in transmission requirements. Thus it should be anticipated that utilities serving these population centers will encounter increasing difficulties in acquiring new rights-of-way in these areas.[1]

Regulation of transmission lines

Governmental review of proposed transmission line construction is limited for both the Federal and State levels of government. FPC regulatory authority is largely limited to lines associated with Government licensed hydroelectric plants, or land of such projects crossed by transmission lines. As for the States, with few exceptions, State regulatory commissions are vested with little or no authority over the location of transmission lines. Less than a dozen States report they have significant jurisdiction over new transmission lines. The remainder either have no jurisdiction, or have jurisdiction in special cases only. Of the 51 State regulatory commissions, 25 have no jurisdiction of any kind over the routing of transmission lines. Of the 51 regulatory commissions, 16 indicated that esthetics and environmental matters were, or could be, among the factors taken into consideration. Others indicated their review was limited by law to matters such as safety, property of investment, and necessity for the line. In many States, transmission line construction is regulated piecemeal by local agencies.[2]

Corrective measures

Two different approaches to mitigating the effects of transmission lines are visible. One is to put the lines underground. The other is to encourage multiple use of the land required for their rights of way.

Underground transmission of electricity

Ideally there should be more underground transmission in urban areas, in locations of exceptional beauty, along scenic highways and rivers and through historic sites. However despite the mounting public desire for more underground transmission, the technology to do so is developing slowly. Even if EHV underground transmission technology is developed, it seems likely that the anticipated high costs for its use in the foreseeable future will preclude any significant shift from overhead transmission.

Multiple land use

One way to reduce the impacts of rights-of-way is to permit multiple use. For example, electricity, gas, oil, and rail traffic might move in the same corridors. Or, the rights-of-way might be used for recreation, or agriculture. However, as Vice Chairman Carl E. Bagge of the Federal Power Commission points out, the historical relations among utilities is one of independence and outright opposition to the idea of joint use. In his opinion, the Nation must evolve transportation and communication and energy corridors as an urgent matter of national policy. Yet there is still no effective communication between the rail, gas,[3] and electric interests to this end.

[1] This theme was explored in more detail by FPC Vice Chairman Carl E. Bagge before the Joint Committee on Atomic Energy in 1969. Cf. "Environmental Effects of Producing Electric Power," op. cit., pp. 449–451.

[2] Testimony of Chairman John N. Nassikas of the Federal Power Commission before the Joint Committee on Atomic Energy. Cf. "Environmental Effects of Producing Electric Power," op. cit., p. 66.

[3] Testimony of Vice Chairman Carl E. Bagge, Federal Power Commission, before the Joint Committee on Atomic Energy. Cf. "Environmental Effects of Producing Power," op. cit., pp. 472–473.

The Electric Power Council on Environment

One response of the electricity industry to the growing problems caused by the adverse effects of some powerplant operations was the formation of an Electric Power Council on Environment. Formed on September 25, 1969, the council's membership includes representatives of the privately, publicly and cooperatively owned power systems and the Federal operations of the Department of the Interior and the Tennessee Valley Authority. The council's objectives are:

Coordination of industry environmental programs;

Encouragement of cooperation between Government and industry; and

Stimulation of environmental research.

Four committees of the council will cover air pollution, water pollution, land use and esthetics. The representatives of the privately owned utilities are also members of the Edison Electric Institute's Committee on the Environment.